CHEMICAL ENGINEERING KINETICS
Second Edition

Second Edition

CHEMICAL ENGINEERING KINETICS

J. M. Smith

Professor of Chemical Engineering
University of California at Davis

McGRAW-HILL BOOK COMPANY   *New York   San Francisco   St. Louis*
Düsseldorf   London   Mexico   Panama   Sydney   Toronto

CHEMICAL ENGINEERING KINETICS, *Second Edition*

Printed in the United States of America.

Library of Congress catalog card number: 74–99204

11 12 13 14 15 16 KPKP 78321098

07–058693–4

PREFACE

The first edition of Chemical Engineering Kinetics appeared when the rational design of chemical reactors, as opposed to empirical scaleup, was an emerging field. Since then, progress in kinetics, catalysis, and particularly in engineering aspects of design, has been so great that this second edition is a completely rewritten version. In view of present-day knowledge, the treatment in the first edition is inadequate with respect to kinetics of multiple-reaction systems, mixing in nonideal reactors, thermal effects, and global rates of heterogeneous reactions. Special attention has been devoted to these subjects in the second edition. What hasn't changed is the book's objective: the clear presentation and illustration of design procedures which are based upon scientific principles.

Successful design of chemical reactors requires understanding of chemical kinetics as well as such physical processes as mass and energy transport. Hence, the intrinsic rate of chemical reactions is accorded a good measure of attention: in a general way in the second chapter and then with specific reference to catalysis in the eighth and ninth. A brief review of chemical thermodynamics is included in Chap. 1, but earlier study of the fundamentals of this subject would be beneficial. Introductory and theoretical material is given in Chap. 2, only in a manner that does not make prior study of kinetics mandatory.

The concepts of reactor design are presented in Chap. 3 from the viewpoint of the effect of reactor geometry and operating conditions on the form of mass and energy conservation equations. The assumptions associated with the extremes of plug-flow and stirred-tank behavior are emphasized. A brief introduction to deviations from these ideal forms is included in this chapter and is followed with a more detailed examination of the effects of mixing on conversion in Chap. 6. In Chaps. 4 and 5 design procedures are examined for ideal forms of homogeneous reactors, with emphasis upon multiple-reaction systems. The latter chapter is concerned with nonisothermal behavior.

Chapter 7 is an introduction to heterogeneous systems. The concept of a global rate of reaction is interjected so as to relate the design of heterogeneous reactors to the previously studied concepts of homogeneous reactor design. A secondary objective here is to examine, in a preliminary way, the method of combining of chemical and physical processes so as to obtain a global rate of reaction.

Chap. 8 begins with a discussion of catalysis, particularly on solid surfaces, and this leads directly into adsorption and the physical properties of porous solids. The latter is treated in reasonable detail because of the importance of solid-catalyzed reactions and because of its significance with respect to intrapellet transport theory (considered in Chap. 11). With this background, the formulation of intrinsic rate equations at a catalyst site is taken up in Chap. 9.

The objective of Chaps. 10 and 11 is to combine intrinsic rate equations with intrapellet and fluid-to-pellet transport rates in order to obtain global rate equations useful for design. It is at this point that models of porous catalyst pellets and effectiveness factors are introduced. Slurry reactors offer an excellent example of the interrelation between chemical and physical processes, and such systems are used to illustrate the formulation of global rates of reaction.

The book has been written from the viewpoint that the design of a chemical reactor requires, first, a laboratory study to establish the intrinsic rate of reaction, and subsequently a combination of the rate expression with a model of the commercial-scale reactor to predict performance. In Chap. 12 types of laboratory reactors are analyzed, with special attention given to how data can be reduced so as to obtain global and intrinsic rate equations. Then the modeling problem is examined. Here it is assumed that a global rate equation is available, and the objective is to use it, and a model, to predict the performance of a large-scale unit. Several reactors are considered, but major attention is devoted to the fixed-bed type. Finally, in the last chapter gas-solid, noncatalytic reactions are analyzed, both from a single pellet (global rate) viewpoint, and in terms of reactor design. These systems offer examples of interaction of chemical and physical processes under *transient* conditions.

No effort has been made to include all types of kinetics or of reactors. Rather, the attempt has been to present, as clearly and simply as possible, all the aspects of process design for a few common types of reactors. The material should be readily understandable by students in the fourth under-graduate year. The whole book can be comfortably covered in two semesters, and perhaps in two quarters.

The suggestions and criticisms of numerous students and colleagues have been valuable in this revision, and all are sincerely acknowledged. The several stimulating discussions with Professor J. J. Carberry about teaching chemical reaction engineering were most helpful. To Mrs. Barbara Dierks and Mrs. Loretta Charles for their conscientious and interested efforts in typing the manuscript, I express my thanks. Finally, the book is dedicated to my wife, Essie, and to my students whose enthusiasm and research accomplishments have been a continuing inspiration.

J. M. Smith

NOTATION

$[A]$, C_A	concentration of component A, moles/volume
$\mathbf{A}$	frequency factor in Arrhenius equation
A	area
a	activity or pore radius
a_m	external surface per unit mass
C	molal concentration, moles/volume
C_0	initial or feed concentration, moles/volume
C_b	concentration in bulk-gas stream, moles/volume
$\overline{C}$	concentration of component adsorbed on a catalyst surface, moles/mass
C_p	molal heat capacity at constant pressure, energy/(moles) (temperature)
C_s	concentration at catalyst surface, moles/volume
c_p	specific heat at constant pressure, energy/(mass) (temperature)
D	combined diffusivity, (length)^2/time
$\mathscr{D}_{AB}$	bulk diffusivity, (length)^2/time
$\mathscr{D}_K$	Knudsen diffusivity, (length)^2/time

x

D_e	effective diffusivity in a porous catalyst, $(length)^2/time$
d_p	diameter of pellet
E	activation energy/mole
F	feed rate, mass or moles/time
ΔF	free-energy change for a reaction, energy/mole
f	fugacity, atm
G	fluid mass velocity, mass/(area) (time)
H	enthalpy, energy/mass
ΔH	enthalpy change for a reaction, energy/mole
h	heat-transfer coefficient, energy/(time) (area) (temperature)
$J(\theta)$	residence-time distribution function
K	equilibrium constant for a reaction
K_a	adsorption equilibrium constant
k	specific reaction-rate constant
k'	reverse-reaction-rate constant
k_B	Boltzmann's constant, 1.3805×10^{-16} erg/°K
k_e	effective thermal conductivity, energy/(time) (length) (temperature)
k_m	mass-transfer coefficient (particle to fluid)
k_o	overall rate constant
L	length
M	molecular weight (W), mass/mole
m	mass
N	number of moles
N'	molal rate, moles/time
p	partial pressure, atmospheres
p_t	total pressure, atmospheres
Q	volumetric flow rate, volume/time
Q	heat-transfer rate, energy/time
q	heat flux, energy/(area) (time)
r	radius, radial coordinate
$\mathbf{r}$	reaction rate, moles/(volume) (time)
$\bar{\mathbf{r}}$	average rate of reaction, moles/(volume) (time)
$\mathbf{r}_P$	global reaction rate, moles/(mass catalyst) (time)
$\mathbf{r}_v$	global reaction rate, moles/(volume of reactor) (time)
R_g	gas constant, energy/(temperature) (mole) or (pressure) (volume)/(temperature) (mole)

S selectivity

S_o overall selectivity

S entropy, energy/(mole) (temperature)

ΔS entropy change for a reaction, energy/(mole) (temperature)

S_g · pore surface area (of catalyst) per unit mass

T temperature, absolute

t time

U overall heat-transfer coefficient, energy/(time) (length^2) (temperature)

u superficial velocity, length/time

V_g pore volume (of catalyst), volume/mass

V reactor volume

v specific volume, per unit mass

W molecular weight (M), mass/mole

w weight fraction

x conversion, yield, or distance

y mole fraction

z axial distance

γ ratio of global rate to rate evaluated at bulk-fluid conditions

ϵ void fraction

ϵ_S solids fraction

η effectiveness factor

θ residence time

$\bar{\theta}$ mean residence time

μ viscosity, mass/(length) (time)

ρ density, mass/volume

ρ_P density of catalyst pellet, mass/volume

Φ Thiele-type modulus for a porous catalyst

ρ_b density of bed of catalyst pellets, mass/volume

CONTENTS

CHEMICAL ENGINEERING KINETICS
Second Edition

1

INTRODUCTION

The design and operation of equipment for carrying out chemical reactions require analysis of both physical and chemical processes. The principles governing energy and mass transfer are often as important as those which govern chemical kinetics. This combination of physical and chemical operations is also a distinguishing feature of chemical engineering; the design of chemical reactors is an activity unique to the chemical engineer.

In designing a reactor the following factors must be considered: the type and size needed, provisions for exchange of energy (usually as heat) with the surroundings, and operating conditions (temperature, pressure, composition, flow rates). We identify this problem as the *process design* of the reactor. A cost analysis to determine the most profitable design introduces further questions about construction materials, corrosion, utility requirements, and maintenance. In order to maximize profits, the instrumentation and a control policy (ranging from manual to closed-loop computer control) for optimum operation must be determined. Optimum design will also depend indirectly on estimates of such market conditions as the price-volume relationships for reactants and products. While these factors are important in overall design and operation of reactors, they will not be discussed in this book. Our use of the term *design* will be restricted to *process design*.

The rate of transforming one chemical species into another cannot be predicted with accuracy. It is a specific quantity which must be obtained from experimental measurements. Measuring the *rate of chemical reactions* is the art and science of chemical kinetics (to be discussed in Sec. 1-3). Chemical kinetics is not concerned with physical processes, but only with the rate of transformation of atoms and molecules from one structural form to another. In contrast, the rates of physical processes such as heat and mass transfer can in many types of reactors be estimated reliably from the flow and geometrical arrangements. Hence it is desirable to consider the design of *types of reactors*, rather than approach each chemical system as a separate entity. Because of the specific nature of chemical processes, this generalization is not as successful as it would be, say, for designing heat exchangers. However, some conclusions are possible for reactors, because the kinetics of chemical reactions of a given type, such as homogeneous gas phase or gas-solid catalytic, have similarities.

Our chief objective is to determine how to answer process-design questions for various types of chemical reactions and reactors.

1-1 *Interpretation of Rate Data, Scale-up, and Design*

The chemical engineer depends on data from the chemist's laboratory, the pilot plant, or a large-scale reactor to help him in his design work. From this information he needs to extract, among other things, the rates of the chemical reactions involved, i.e., the chemical kinetics of the system. To do this he must separate the effects of physical processes from the observed data, leaving rate information for the chemical-transformation step alone. He can then reintroduce the influence of the physical steps for the particular reactor type and operating conditions chosen for the commercial plant. The interrelationship of the physical and chemical steps must be considered twice: once in obtaining rate-of-reaction expressions from the available laboratory or pilot-plant data, and again in using these rate-of-reaction equations to design the commercial-scale reactor. The first step, interpretation of the available data, is as important as the second and entails generally the same type of analysis. Hence the interpretation of laboratory data will sometimes be discussed in parallel with the reactor-design problem in the chapters that follow. Interpreting laboratory-reactor data is not necessarily as difficult and does not always entail the same steps (in reverse order) as reactor design. Because there are fewer restraints (economic ones, for example), there is more flexibility in choosing a laboratory reactor. It is common practice to design a laboratory reactor to minimize the significance of physical processes (see Chap. 12). This eliminates the need for accurate separation of physical and chemical effects and produces more accurate results for the rates of the chemical steps. For instance, a laboratory reactor may be operated at near-isothermal conditions, eliminating heat-transfer considerations, while such operation would be uneconomical in a commercial-scale system.

It is important to consider the relationship between reactor *scale-up* (projection of laboratory or pilot-plant data to the commercial reactor) and reactor design. In principle, if the rates of the chemical reactions are known, any type of reactor may be designed by introducing the appropriate physical processes associated with that type of equipment. Scale-up is an abbreviated version of this design process. The physical resistances are not separated from the measured laboratory data but are instead projected directly to a large unit which presumably has the same interrelationship of chemical and physical processes. If the dimensions and operating conditions for the large-scale reactor can be determined to ensure that the interrelationships of chemical and physical processes are the same as in the laboratory unit, then the laboratory results may be used directly to predict the behavior of the large-scale reactor. In a scale-up process no attempt is made to determine the rate of the chemical steps, that is, to evaluate the chemical kinetics of the process. However, when scale-up is applicable, it provides a rapid means of obtaining approximate reactor sizes, and also indicates the important parameters in the interrelationship between physical and chemical processes (see Sec. 12-5).

Scale-up will have a much better chance for success if the laboratory and commercial operations are carried out in the same type of system. Suppose that laboratory data for the thermal cracking of hydrocarbons are obtained in a continuous tube through which the reaction mixture flows. If a tubular-flow reactor of this type is also proposed for the commercial plant, it may be possible to scale-up the pilot-plant operation in such a way that the temperature and concentration gradients within the tube will be the same in both systems. Then performance of the large-scale reactor—for example, conversion of reactants to various products—can be predicted directly from the laboratory results. However, if the laboratory results were obtained from a batch reactor, a tank or vessel into which the reactants are initially charged (see Sec. 1-6), it is difficult to project them directly to a large-scale tubular reactor. In this case it would be necessary to analyze the laboratory data to obtain the rate equation for the chemical reactions and then use these results to design the commercial reactor. Our emphasis will be on this two-step process of determining the rates of reaction from laboratory data and then using these rates for design.

The foregoing comments do not imply that pilot-plant data from a small-scale replica of a proposed commercial unit are of no value. Such information provides an important evaluation of both the laboratory rate data and the procedures used to reintroduce the physical processes in the pilot reactor and, presumably, in the final equipment.

Our discussion thus far of the interrelationship of chemical and physical processes in a reactor has been general. Let us look further into the problem by considering a very simple reaction system, the conversion of ortho hydrogen to the para form.[1] Owing to thermodynamic restrictions (see Sec. 1-5), this reaction must be carried out at a low temperature in order to obtain a large conversion to para hydrogen. At low temperatures it is necessary to use a catalyst to obtain a rapid rate of reaction. The preferred type of reactor is a continuous steady-state system in which hydrogen flows through a tube packed with pellets of the solid catalyst. Consider the interpretation of rate measurements made with a laboratory version of this type of reactor. The observed data would consist of measurements of hydrogen compositions in the inlet and exit streams of the reactor. Probable variables would be the flow rate of hydrogen through the reactor, the mole fraction of para hydrogen in the feed to the reactor, and the temperature. The heat of reaction is negligible, so that the whole reactor system can easily be operated at isothermal conditions.

The first problem in designing a reactor for the production of para hydrogen is to obtain from the observed measurements a quantitative expression for the rate of reaction on the surface of the catalyst. Specifically, we must separate from the observed data the diffusional resistances between the point at which the composition is measured—the exit of the reactor—and the point at which the chemical transformation occurs—the gas-solid interface at the catalyst surface. There are three diffusional effects that may cause a difference between the conversion measured in the exit of the reactor and that predicted from the rate at the catalyst interface. The first arises from the mixing characteristics of the fluid as it flows around the particles in the fixed bed. There may be some bypassing or short circuiting, so that part of the flowstream does not come into contact with the catalyst; also, there may be diffusion or back-mixing of fluid as it flows through the bed. As a result, the observed amount of para hydrogen in the exit gas may be less than expected. The second factor is the tendency of the fluid to adhere to the catalyst pellet, so that the pellet is surrounded by a more or less stagnant layer of fluid which resists mass transfer. Thus a concentration gradient of para hydrogen must be established between the outer surface of the pellet and the bulk gas before the para hydrogen will

[1] This reactor-design problem has some practical significance because of the superior storage properties of liquid hydrogen when it is in the para form. Noriaki Wakao and J. M. Smith, *AIChE J.*, **8**, 478 (1962).

move on into the gas stream. This reduces the amount of para hydrogen available to the bulk-gas phase. A third factor is that most of the active surface of the catalyst is in the pores within the pellet. The reactant must reach this interior pore surface by diffusing into the pellet, and the product must diffuse out. This process is impeded by intraparticle resistance, causing another reduction in the para-hydrogen content of the gas stream. Thus to determine the rate of reaction on the surface of the catalyst (the chemical kinetics of the process) it is necessary to evaluate the concentration changes for each of these diffusional effects, ultimately arriving at the concentration of para hydrogen in the interior pore surface of the catalyst pellet. The interior concentration can then be used to establish a rate-of-reaction equation.

The second problem is use of the rate equation to design a commercial reactor. The individual diffusional resistances are now reintroduced so that we may determine the actual composition of para hydrogen in the exit stream from the reactor. Once the equation for the surface rate is known, it is possible, in principle, to predict the exit conversion for any type of re-actor, any size catalyst pellet, any conditions of gas flow around the pellet, and any condition of mixing of fluid around the particles in the fixed bed.

If the same problem were approached from a scale-up standpoint, the procedure would be to attempt to choose the operating conditions and reactor size for the large-scale reactor such that the diffusional resistances were the same as in the laboratory equipment.

1-2 Principles of Chemical-reactor Design

Now that we have discussed briefly the problems in chemical-reactor design, let us explore the concepts necessary to solve these problems.[1] First, we must know something about the formulation of equations for the rate of chemical transformation from one species to another. This is the subject of chemical kinetics. While we cannot predict the rate of a chemical reac-tion, chemical kineticists have developed a number of valuable generaliza-tions for formulating rate equations. It is necessary to understand these generalizations before proceeding with the design problem. Also, the rate equation must account for the thermodynamic limitations on the chemical reaction. For example, the ortho-para hydrogen conversion is a reaction in which the maximum conversion is less than 100%. An understanding of the equilibrium conversion as a function of operating conditions is necessary before a satisfactory rate equation can be formed. Thermodynamics also

---

[1] R. Aris [*Ind. Eng. Chem.*, **56**, 22 (1964)] and J. M. Smith [*Chem. Eng. Progr.*, **64**, 78 (1968)] have shown how physical and chemical sciences play a role in the design, analysis, and control of reactors.

comes into play in evaluating energy transfer in the reactor. Hence a primary requisite is a knowledge of chemical thermodynamics (Sec. 1-5).

Along with kinetics, the engineering concepts required to evaluate the conversion in the product from a reactor are of two types: the principles of conservation of mass and energy and rate equations for the physical processes. The conservation of mass and the rate of mass transfer determine the composition as a function of position in a continuous reactor and as a function of time in a batch reactor. Similarly, the conservation of energy and the rate of energy transfer determine the temperature as a function of position or time. The application of these principles is discussed in subsequent chapters, starting with Chap. 3 where mathematical formulations are considered for various types of reactors.

1-3 Chemical Kinetics

Chemical kinetics is the study of the rate and mechanism by which one chemical species is converted to another. The *rate* is the mass, in moles, of a product produced or reactant consumed per unit time. The *mechanism* is the sequence of individual chemical events whose overall result produce the observed reaction. Basolo and Pearson[1] have described the term "mechanism" as follows:

> By mechanism is meant all the individual collisional or elementary processes involving molecules (atoms, radicals and ions included) that take place simultaneously or consecutively in producing the observed overall rate. It is also understood that the mechanism of a reaction should give a detailed stereochemical picture of each step as it occurs. This implies a knowledge of the so-called activated complex or transition state, not only in terms of the constituent molecules but also in terms of the geometry, such as interatomic distances and angles. In most instances the postulated mechanism is a theory devised to explain the end results observed by experiments. Like other theories, mechanisms are subject to change over the years as new data is uncovered or as new concepts regarding chemical interreactions are developed.

It is not necessary to know the mechanism of a reaction in order to design a reactor. What is necessary is a satisfactory rate equation. A knowledge of the mechanism is of great value, however, in extending the rate data beyond the original experiments and in generalizing or systematizing the kinetics of reactions. Determining the mechanism of a reaction is a very difficult task and may require the work of many investigators over a period of many years. Reaction mechanisms are reliably known for only a

[1] F. Basolo and R. G. Pearson, "Mechanisms of Inorganic Reactions," John Wiley & Sons, Inc., New York, 1958.

few systems. However, postulated theories for mechanisms are available for a wide variety of reactions, ranging from simple, gas-phase homogeneous systems to complicated polymerization reactions involving initiation, propagation, and termination steps.

Since reactor design necessitates a reliable rate equation, this aspect of chemical kinetics is presented in detail in Chap. 2. Successful procedures for *predicting* rates of reactions will not be developed until reaction mechanisms are better understood. It is important for those involved in reactor design to be aware of developments in this area so that they may take advantage of new principles of chemical kinetics as they are developed. A brief discussion of theories of reaction and mechanisms is included in Chap. 2.

The rate of a chemical reaction can vary from a value approaching infinity to essentially zero. In ionic reactions, such as those that occur on photographic film, or in high-temperature combustion reactions, the rate is extremely fast. The rate of combination of hydrogen and oxygen in the absence of a catalyst at room temperature is immeasurably slow. Most industrially important reactions occur at rates between these extremes, and it is in these cases that the designer must apply data on kinetics to determine finite sizes of reaction equipment. It is particularly important to know how the rate changes with operating parameters, the most important of which are temperature, pressure, and composition of reaction mixture.

The science of kinetics is young. The first quantitative measurements of reaction rates were made in the middle of the nineteenth century by Wilhelmy,[1] Berthelot and St. Gilles,[2] and Harcourt and Esson.[3] The first attempt to develop a theory explaining the manner in which molecules of a substance react was that of Arrhenius[4] in 1889. He postulated that the reactants had both inert and active molecules and that only the active ones possessed sufficient energy to take part in the reaction. Since these early developments there have been a great many experimental studies of reaction rates for a wide variety of reactions, but few noteworthy advances in theory were made until the work of Eyring and Polanyi,[5] beginning in 1920. Using only such fundamental information as the configurations, dimensions, and interatomic forces of the reacting molecules, they postulated an activated-complex theory for predicting the rate of reaction. Lack of exact knowledge of the interatomic forces, and hence of energy-position relations, for any but the most simple molecules has prevented the activated-complex theory

[1] L. Wilhelmy, *Pogg. Ann.*, **81**, 413, 499 (1850).

[2] M. Berthelot and L. P. St. Gilles, *Ann. Chim. Phys.*, **63** (3), 385 (1862).

[3] A. V. Harcourt and W. Esson, *Proc. Roy. Soc. (London)*, **14**, 470 (1865).

[4] S. Arrhenius, *Z. Physik Chem.*, **4**, 226 (1889).

[5] H. Eyring and M. Polanyi, *Z. Physik Chem. B*, **12**, 279 (1931).

from being useful for predicting reaction-rate data accurately enough for engineering work. While these theoretical developments have been of great value in the search for an understanding of how and why a chemical reaction takes place, the quantitative evaluation of the rate remains an experimental problem.

The large amount of experimental data on rates of chemical reaction have established reliable empirical forms for the mathematical expression of the effects of such variables as temperature and composition on the rate. These results are interpreted for various types of reactions in Chap. 2.

1-4 Kinetics and Thermodynamics

From the principles of thermodynamics and certain thermodynamic data the maximum extent to which a chemical reaction can proceed may be calculated. For example, at 1 atm pressure and a temperature of 680°C, starting with 1 mole of sulfur dioxide and $\frac{1}{2}$ mole of oxygen, 50% of the sulfur dioxide can be converted to sulfur trioxide. Such thermodynamic calculations result in maximum values for the conversion of a chemical reaction, since they are correct only for equilibrium conditions, conditions such that there is no further tendency for change with respect to time. It follows that the net rate of a chemical reaction must be zero at this equilibrium point. Thus a plot of reaction rate [for example, in units of g moles product/(sec)(unit volume reaction mixture)] vs time would always approach zero as the time approached infinity. Such a situation is depicted in curve A of Fig. 1-1, where the rate approaches zero asymptotically. Of course, for some cases equilibrium may be reached more rapidly, so that the rate becomes almost zero at a finite time, as illustrated by curve B.

Similarly, the *conversion* (fraction of reactant transformed or converted) calculated from thermodynamic data would be the end point on a curve of conversion vs time such as that shown in Fig. 1-2. Again, curve A represents the case where the time required to reach equilibrium conditions is great, while in case B the equilibrium conversion is approached more rapidly and is attained essentially at a finite time. Curves A and B could apply to the same reaction; the difference between them reflects the fact that in case B the rate has been increased, for example, by use of a catalyst. The rate of the reaction is initially increased over that for the uncatalyzed reaction, but the equilibrium conversion as shown in Fig. 1-2 is the same for both cases.

The time available for carrying out a chemical reaction commercially is limited if the process is to be economically feasible. Hence the practical range of the curves in Figs. 1-1 and 1-2 is at the lower time values. However, the equilibrium conversion is important as a standard for evaluating the actual performance of the reaction equipment. Suppose a kinetics experi-

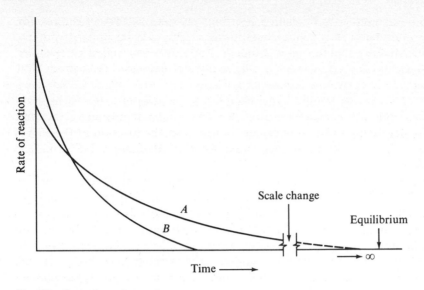

Fig. 1-1 *Rate of reaction vs time*

ment is carried out with a time corresponding to the dashed vertical line shown in Fig. 1-2. At this point the conversion for the noncatalytic reaction is about 25% (curve *A*). A comparison with the equilibrium value of 50% indicates that the noncatalytic rate is rather low and that a search for a catalyst is advisable. Curve *B*, giving a conversion of 45%, shows the benefit of using a catalyst and also indicates that additional effort to find a more

Fig. 1-2 *Conversion vs time*

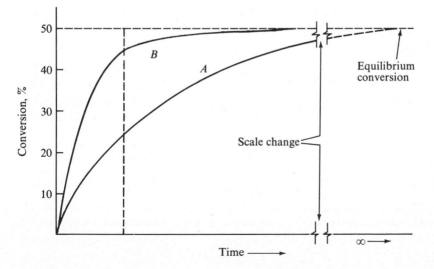

effective catalyst is unwarranted. Without prior knowledge of the equilibrium conversion, erroneous conclusions might well be drawn from the kinetic studies yielding curves A and B. For example, it might be reasoned that the catalyst giving curve B is only moderately effective, and considerable time might be spent in attempting to discover a catalyst which would give a conversion of 70 or 80%. Thermodynamic calculations are particularly valuable for such comparison of kinetic and equilibrium results. However, the actual design of reactors usually depends on the location of the curves shown in Figs. 1-1 and 1-2 and must therefore be determined by kinetic studies.

Prediction of the equilibrium conversion requires knowledge of the free-energy change for the reactions involved. Although the body of thermodynamic data is growing, it is still not possible to estimate the equilibrium conversion accurately for all reactions. The calculations and data available for gaseous systems are most reliable. The application of thermodynamics for such calculations is illustrated very briefly in the following section. More detailed treatment of the thermodynamics of chemical-reaction equilibrium is given in thermodynamics textbooks.[1]

The rate of energy transfer is important in determining the temperature distribution in reactors. Also, heats of reaction are significant in connection with equilibrium calculations. The following section deals with data and methods concerning heats of reaction, followed by a discussion of equilibrium conversion.

1-5 Thermodynamics of Chemical Reactions

Heat of Reaction   The *heat of reaction* is defined as the energy absorbed by the system when the products after reaction are restored to the same temperature as the reactants. The pressure must also be specified for a complete definition of the thermodynamic states of the products and reactants. If the same pressure is chosen for both, the heat of reaction is equal to the enthalpy change; this is the customary definition of the heat of reaction. The heat of any reaction can be calculated by combining heats of formation or heats of combustion of the products and reactants. Thus the basic information necessary for calculating heats of reaction are heats of formation and

[1]B. F. Dodge, "Chemical Engineering Thermodynamics," McGraw-Hill Book Company, New York, 1944; J. G. Kirkwood and Irwin Oppenheim, "Chemical Thermodynamics," McGraw-Hill Book Company, New York, 1961; G. N. Lewis and M. Randall, "Thermodynamics," 2d ed., rev. by K. S. Pitzer and Leo Brewer, McGraw-Hill Book Company, New York, 1961; J. M. Smith and H. C. Van Ness, "Introduction to Chemical Engineering Thermodynamics," McGraw-Hill Book Company, New York, 1959; H. C. Van Ness, "Classical Thermodynamics of Non-electrolyte Solutions," The Macmillan Company, Inc., New York, 1964.

Table 1-1 *Standard heats of formation and combustion for reaction products $H_2O(l)$ and $CO_2(g)$ at 25°C, in calories per gram mole*

Substance	Formula	State	ΔH°_{f298}	$-\Delta H^\circ_{c298}$
Normal paraffins				
Methane	CH_4	g	−17,889	212,800
Ethane	C_2H_6	g	−20,236	372,820
Propane	C_3H_8	g	−24,820	530,600
n-Butane	C_4H_{10}	g	−30,150	687,640
n-Pentane	C_5H_{12}	g	−35,000	845,160
n-Hexane	C_6H_{14}	g	−39,960	1,002,570
Increment per C atom above C_6	...	g	−4,925	157,440
Normal monoolefins (1-alkenes)				
Ethylene	C_2H_4	g	12,496	337,230
Propylene	C_3H_6	g	4,879	491,990
1-Butene	C_4H_8	g	−30	649,450
1-Pentene	C_5H_{10}	g	−5,000	806,850
1-Hexene	C_6H_{12}	g	−9,960	964,260
Increment per C atom above C_6	...	g	−4,925	157,440
Miscellaneous organic compounds				
Acetaldehyde	C_2H_4O	g	−39,760	
Acetic acid	$C_2H_4O_2$	l	−116,400	
Acetylene	C_2H_2	g	54,194	310,620
Benzene	C_6H_6	g	19,820	789,080
Benzene	C_6H_6	l	11,720	780,980
1,3-Butadiene	C_4H_6	g	26,330	607,490
Cyclohexane	C_6H_{12}	g	−29,430	944,790
Cyclohexane	C_6H_{12}	l	−37,340	936,880
Ethanol	C_2H_6O	g	−56,240	
Ethanol	C_2H_6O	l	−66,356	
Ethylbenzene	C_8H_{10}	g	7,120	1,101,120
Ethylene glycol	$C_2H_6O_2$	l	−108,580	
Ethylene oxide	C_2H_4O	g	−12,190	
Methanol	CH_4O	g	−48,100	
Methanol	CH_4O	l	−57,036	
Methylcyclohexane	C_7H_{14}	g	−36,990	1,099,590
Methylcyclohexane	C_7H_{14}	l	−45,450	1,091,130
Styrene	C_8H_8	g	35,220	1,060,900
Toluene	C_7H_8	g	11,950	943,580
Toluene	C_7H_8	l	2,870	934,500
Miscellaneous inorganic compounds				
Ammonia	NH_3	g	−11,040	
Calcium carbide	CaC_2	s	−15,000	
Calcium carbonate	$CaCO_3$	s	−288,450	
Calcium chloride	$CaCl_2$	s	−190,000	
Calcium chloride	$CaCl_2:6H_2O$	s	−623,150	
Calcium hydroxide	$Ca(OH)_2$	s	−235,800	
Calcium oxide	CaO	s	−151,900	

Table 1-1 (*Continued*)

Substance	Formula	State	ΔH°_{f298}	$-\Delta H^\circ_{c298}$
Carbon	C	Graphite	. . .	94,052
Carbon dioxide	CO_2	g	$-94,052$	
Carbon monoxide	CO	g	$-26,416$	67,636
Hydrochloric acid	HCl	g	$-22,063$	
Hydrogen	H_2	g	. . .	68,317
Hydrogen sulfide	H_2S	g	$-4,815$	
Iron oxide	FeO	s	$-64,300$	
Iron oxide	Fe_3O_4	s	$-267,000$	
Iron oxide	Fe_2O_3	s	$-196,500$	
Iron sulfide	FeS_2	s	$-42,520$	
Lithium chloride	LiCl	s	$-97,700$	
Lithium chloride	$LiCl \cdot H_2O$	s	$-170,310$	
Lithium chloride	$LiCl \cdot 2H_2O$	s	$-242,100$	
Lithium chloride	$LiCl \cdot 3H_2O$	s	$-313,500$	
Nitric acid	HNO_3	l	$-41,404$	
Nitrogen oxides	NO	g	21,600	
	NO_2	g	8,041	
	N_2O	g	19,490	
	N_2O_4	g	2,309	
Sodium carbonate	Na_2CO_3	s	$-270,300$	
Sodium carbonate	$Na_2CO_3 \cdot 10H_2O$	s	$-975,600$	
Sodium chloride	NaCl	s	$-98,232$	
Sodium hydroxide	NaOH	s	$-101,990$	
Sulfur dioxide	SO_2	g	$-70,960$	
Sulfur trioxide	SO_3	g	$-94,450$	
Sulfur trioxide	SO_3	l	$-104,800$	
Sulfuric acid	H_2SO_4	l	$-193,910$	
Water	H_2O	g	$-57,798$	
Water	H_2O	l	$-68,317$	

SOURCE: Most values have been selected from the publications of F. D. Rossini et al., Selected Values of Physical and Thermodynamic Properties of Hydrocarbons and Related Compounds, *Am. Petrol. Inst. Res. Proj.* 44, Carnegie Institute of Technology, Pittsburgh, 1953; F. D. Rossini et al., Selected Values of Chemical Thermodynamic Properties, *Natl. Bur. Stds. Circ.* 500, 1952.

combustion. Extensive tables of these data have been accumulated,[1] and a few values are summarized in Table 1-1.

[1] Selected Values of Chemical Thermodynamic Properties, *Natl. Bur. Stds. Circ.* 500, 1952; Selected Values of Physical and Thermodynamic Properties of Hydrocarbons and Related Compounds, *Am. Petrol. Inst. Res. Proj.* 44, Carnegie Institute of Technology, Pittsburgh, 1953.

In the absence of experimental data, there are procedures available for predicting heats of reaction.[1] These are all based on predictions of the effects of differences in the chemical structure of the reactants and products. One of the most useful procedures from an engineering standpoint is that proposed by Andersen et al. and described in detail by Hougen and Watson.[2] This method is applicable to compounds involving carbon, hydrogen, oxygen, nitrogen, and the halogens.

The variation of heat of reaction with temperature depends on the difference in molal heat capacities of the products and reactants. The following equation relates ΔH at any temperature T to the known value at the base temperature T_0:

$$\Delta H_T = \Delta H_{T_0} + \int_{T_0}^{T} \Delta C_p \, dT \tag{1-1}$$

Here ΔC_p is the difference in molal heat capacities,

$$\Delta C_p = \sum (N_i C_{p_i})_{\text{prod}} - \sum (N_i C_{p_i})_{\text{react}} \tag{1-2}$$

If mean heat capacities $\overline{C}_p$ are known for the reactants and products over the temperature range T_0 to T, it is not necessary to integrate Eq. (1-1). Under these conditions the relationship of ΔH_T and ΔH_{T_0} is

$$\Delta H_T = \Delta H_{T_0} + \sum (N_i \overline{C}_{p_i})_{\text{prod}} (T - T_0) - \sum (N_i \overline{C}_{p_i})_{\text{react}} (T - T_0) \tag{1-3}$$

When reactants and products enter and leave a reactor at different temperatures, it is usually simpler to bypass calculating ΔH_T and evaluate the desired energy quantity directly. This is illustrated in Example 1-1.

The effect of pressure on the heat of reaction for gaseous systems depends on the deviation of the components from ideal-gas behavior. If the reactants and products behave as ideal gases, there is no effect. Even for rather nonideal systems the effect of pressure is generally small. Details of the methods of calculating the effects of temperature and pressure are discussed in standard thermodynamics textbooks.

[1] J. W. Andersen, G. H. Beyer, and K. M. Watson, *Natl. Petrol. News, Tech. Sec.* 36 (R476), July 5, 1944; R. H. Ewell, *Ind. Eng. Chem.*, **32**, 778 (1940); F. D. Rossini, *Ind. Eng. Chem.*, **29**, 1424 (1937); Mott Souders, Jr., C. S. Mathews, and C. O. Hurd, *Ind. Eng. Chem.*, **41**, 1037, 1048 (1949); R. C. Reid and T. K. Sherwood "The Properties of Gases and Liquids," 2d ed., chap. 5, McGraw-Hill Book Company, New York, 1966.
[2] O. A. Hougen and K. M. Watson, "Chemical Process Principles," 2d ed., vol. II, John Wiley & Sons, Inc., New York, 1959.

The application of heat-of-reaction information for calculating energy-transfer rates in reactors is illustrated in the following example.

Example 1-1   Ethylene oxide is produced by direct oxidation with air using a bed of catalyst particles (silver on a suitable carrier). Suppose that the stream enters the flow reactor at 200°C and contains 5 mole % ethylene and 95% air. If the exit temperature does not exceed 260°C, it is possible to convert 50% of the ethylene to the oxide, although 40% is also completely burned to carbon dioxide. How much heat must be removed from the reaction, per mole of ethylene fed, in order not to exceed the limiting temperature? The average molal heat capacity of ethylene may be taken as 18 Btu/(lb mole) (°R) between 25 and 200°C and as 19 Btu/(lb mole)(°R) between 25 and 260°C. Similar values for ethylene oxide are 20 and 21 Btu/(lb mole)(°R). The pressure is essentially atmospheric.

Solution   Since heat effects at constant pressure are equal to enthalpy changes, the actual process may be replaced by one that utilizes the available heat-of-reaction data at 25°C (Table 1-1). The steps in this process are

1.   Cool the reactants and air from 200 to 25°C.
2.   Carry out the reactions at 25°C.
3.   Heat the products and the air from 25 to 260°C.

The sum of the enthalpy changes for each step will be the total heat absorbed by the reaction system.

STEP 1   With a basis of 1 mole of ethylene, there will be $^{95}/_5(1) = 19$ moles of air fed to the reactor. The mean heat capacity of air from 25°C to 200°C is 7.0. Hence

$$\Delta H_1 = 1(18)(+77 - 392) + 19(7.0)(+77 - 392) = -5,700 - 41,900$$
$$= -47,600 \text{ Btu/lb mole}$$

STEP 2   The only heat effect is due to the two reactions

$$C_2H_4 + \tfrac{1}{2}O_2 \rightarrow C_2H_4O(g)$$
$$C_2H_4 + 3O_2 \rightarrow 2CO_2 + 2H_2O(g)$$

Using the heat-of-formation data in Table 1-1, for the first reaction we obtain

$$\Delta H_{R_1} = -12,190 - 12,496 - 0$$
$$= -24,686 \text{ cal/g mole} \qquad \text{or} -44,500 \text{ Btu/lb mole}$$

and for the second

$$\Delta H_{R_2} = 2(-57,798) + 2(-94,052) - 12,496 - 0$$
$$= -316,196 \text{ cal/g mole} \qquad \text{or} -569,000 \text{ Btu/lb mole}$$

Since, per mole of ethylene, there will be 0.5 mole reacting to form ethylene oxide and 0.4 mole to be completely burned,

$$\Delta H_{T_0} = 0.5(-44,500) + 0.4(-569,000)$$
$$= -250,000 \text{ Btu/lb mole}$$

STEP 3   The products will consist of the following quantities:

Ethylene $= 1 - 0.5 - 0.4 = 0.1$ mole
Ethylene oxide $= 0.5$ mole
Water vapor $= 2(0.4) = 0.8$ mole ($\bar{C}_p = 8.25$)
Carbon dioxide $= 2(0.4) = 0.8$ mole ($\bar{C}_p = 9.4$)
Nitrogen $= 19(0.79) = 15.0$ moles ($\bar{C}_p = 7.0$)
Oxygen $= 19(0.21) - \frac{1}{2}(0.5) - 3(0.4) = 2.6$ moles ($\bar{C}_p = 7.25$)

The values shown for $\bar{C}_p$ are mean values between 25 and 260°C:

$$\Delta H_3 = [0.1(19) + 0.5(21) + 0.8(8.25) + 0.8(9.4) + 15(7.0)$$
$$+ 2.6(7.25)] (500 - 77)$$
$$= 150(500 - 77) = 63,500 \text{ Btu/lb mole}$$

Then the net heat absorbed will be

$$Q = -47,600 - 250,000 + 63,500$$
$$= -234,000 \text{ Btu/lb mole ethylene}$$

Hence the heat that must be removed is 234,000 Btu/lb mole of ethylene fed to the reactor.

Chemical Equilibrium   When a reaction occurs at equilibrium, the temperature and pressure in the system remain constant and the change in free energy is zero. These restraints can be used to develop the following relationship between the *standard* free energy change $\Delta F°$ and the equilibrium constant K:

$$\Delta F° = -R_g T \ln K \tag{1-4}$$

The standard free-energy change $\Delta F°$ is the difference between the free energies of the products and reactants when each is in a chosen standard state. These standard states are chosen so as to make evaluation of the free energy as simple as possible. For example, for gases the standard state is normally that corresponding to unit fugacity at the temperature of the reaction. If the gas is ideal, this standard state reduces to 1 atm pressure.

The equilibrium constant K is defined in terms of the equilibrium activities a_i of the reactants and products. For a general reaction

$$aA + bB = cC + dD \tag{1-5}$$

the equilibrium constant is

$$K = \frac{a_C{}^c a_D{}^d}{a_A{}^a a_B{}^b} \tag{1-6}$$

The activities refer to equilibrium conditions in the reaction mixture and are defined as the ratio of the fugacity in the equilibrium mixture to that in the standard state; that is,

$$a_i = \frac{f_i}{f_i^\circ} \tag{1-7}$$

For gaseous reactions with a standard state of unit fugacity the expression for the equilibrium constant becomes

$$K = \frac{f_C^c f_D^d}{f_A^a f_B^b} \tag{1-8}$$

If, in addition, the gases follow the ideal-gas law, the fugacity is equal to the pressure, and Eq. (1-8) reduces to

$$K = \frac{p_C^c p_D^d}{p_A^a p_B^b} \tag{1-9}$$

Here p (partial pressure) is the total pressure p_t times the mole fraction of the component in the mixture; for example,

$$p_A = p_t y_A \tag{1-10}$$

In many situations the assumption of ideal gases is not justified, and it is necessary to evaluate fugacities. This is the case for such reactions as the ammonia synthesis, where the operating pressure may be as high as 1,500 atm. The fugacity in Eq. (1-8) is that of the component in the equilibrium mixture. However, the fugacity of only the pure component is usually known. To relate the two we must know something about how the fugacity depends on the composition. Normally this information is not available, so that it is necessary to make assumptions about the behavior of the reaction mixture. The simplest and most common assumption is that the mixture behaves as an ideal solution. Then the fugacity at equilibrium, f, is related to the fugacity of the pure component, f', at the same pressure and temperature by

$$f_i = f' y_i \tag{1-11}$$

Substituting this expression in Eq. (1-8) leads to equations for the equilibrium constant in terms of pure-component fugacities and the composition of the equilibrium mixture,

$$K = \frac{(f_C')^c (f_D')^d}{(f_A')^a (f_B')^b} K_y \tag{1-12}$$

where

$$K_y = \frac{y_C^c y_D^d}{y_A^a y_B^b} \tag{1-13}$$

In gaseous reactions the quantity K_p is frequently used. This is defined as

$$K_p = \frac{(y_C p_t)^c (y_D p_t)^d}{(y_A p_t)^a (y_B p_t)^b} = K_y p_t^{(c+d)-(a+b)} \qquad (1-14)$$

From Eq. (1-9) it is clear that $K = K_p$ for an ideal-gas reaction mixture. For nonideal systems Eq. (1-14) may still be employed to calculate K_p from measured equilibrium compositions (K_y). However, then K_p is not equal to K determined from thermodynamic data, for example, from Eq. (1-4).

Equation (1-12) permits the evaluation of the composition ratio K_y in terms of the equilibrium constant. This is a necessary step toward evaluating the equilibrium conversion from free-energy data. The steps in the process are as follows:

1.   Evaluate ΔF°.
2.   Determine the equilibrium constant K, using Eq. (1-4).
3.   Obtain K_y from Eq. (1-12).
4.   Calculate the conversion from K_y.

The first and second steps require thermodynamic data. A brief tabulation of standard free-energy changes ΔF° at 25° is given in Table 1-2. More extensive data have been assembled.[1] Also, estimation procedures have been developed for use when the data are unavailable.[2]

Usually it is necessary to calculate the effect of temperature on ΔF° in order to obtain an equilibrium constant at reaction conditions. The *van't Hoff equation* expresses this relationship in differential form,

$$\frac{d(\ln K)}{dT} = \frac{\Delta H^\circ}{R_g T^2} \qquad (1-15)$$

where ΔH° is the standard-state enthalpy change for the reaction. Equation (1-15) has important implications in reactor design for reversible reactions. It shows that K will decrease with an increase in temperature for an exothermic reaction. Hence provisions must be made for removing the heat of reaction to avoid a thermodynamic limitation (decrease in K) to the potential conversion in exothermic systems. The oxidation of sulfur dioxide is a practical illustration. For endothermic reversible reactions energy must be added to maintain the temperature if a decrease in K is to be avoided. Dehydrogenation of hydrocarbons, such as butanes and butenes, is an example of a situation in which the addition of energy is important. If ΔH°

[1] Selected Values of Chemical Thermodynamic Properties, *Natl. Bur. Stds. Circ.* 500 1952; Selected Values of Physical and Thermodynamic Properties of Hydrocarbons and Related Compounds, *Am. Petrol. Inst. Res. Proj.* 44, Carnegie Institute of Technology, Pittsburgh, 1953.

[2] R. C. Reid and T. K. Sherwood, "The Properties of Gases and Liquids," 2d ed., chap. 5, McGraw-Hill Book Company, New York, 1966.

Table 1-2   Standard free energies of formation at 25°C

Substance	Formula	State	$\Delta F^{\circ}_{f\,298}$
Normal paraffins			
Methane	CH_4	g	$-12{,}140$
Ethane	C_2H_6	g	$-7{,}860$
Propane	C_3H_8	g	$-5{,}614$
n-Butane	C_4H_{10}	g	$-4{,}100$
n-Pentane	C_5H_{12}	g	$-2{,}000$
n-Hexane	C_6H_{14}	g	-70
n-Heptane	C_7H_{16}	g	$1{,}920$
n-Octane	C_8H_{18}	g	$3{,}920$
Increment per C atom above C_8		g	$2{,}010$
Normal monoolefins (1-alkenes)			
Ethylene	C_2H_4	g	$16{,}282$
Propylene	C_3H_6	g	$14{,}990$
1-Butene	C_4H_8	g	$17{,}090$
1-Pentene	C_5H_{10}	g	$18{,}960$
1-Hexene	C_6H_{12}	g	$20{,}940$
Increment per C atom above C_6		g	$2{,}010$
Miscellaneous organic compounds			
Acetaldehyde	C_2H_4O	g	$-31{,}960$
Acetic acid	$C_2H_4O_2$	l	$-93{,}800$
Acetylene	C_2H_2	g	$50{,}000$
Benzene	C_6H_6	g	$30{,}989$
Benzene	C_6H_6	l	$29{,}756$
1,3-Butadiene	C_4H_6	g	$36{,}010$
Cyclohexane	C_6H_{12}	g	$7{,}590$
Cyclohexane	C_6H_{12}	l	$6{,}370$
Ethanol	C_2H_6O	g	$-40{,}300$
Ethanol	C_2H_6O	l	$-41{,}770$
Ethylbenzene	C_8H_{10}	g	$31{,}208$
Ethylene glycol	$C_2H_6O_2$	l	$-77{,}120$
Ethylene oxide	C_2H_4O	g	$-2{,}790$
Methanol	CH_4O	g	$-38{,}700$
Methanol	CH_4O	l	$-39{,}750$
Methylcyclohexane	C_7H_{14}	g	$6{,}520$
Methylcyclohexane	C_7H_{14}	l	$4{,}860$
Styrene	C_8H_8	g	$51{,}100$
Toluene	C_7H_8	g	$29{,}228$
Toluene	C_7H_8	l	$27{,}282$
Miscellaneous inorganic compounds			
Ammonia	NH_3	g	$-3{,}976$
Ammonia	NH_3	aq	$-6{,}370$
Calcium carbide	CaC_2	s	$-16{,}200$
Calcium carbonate	$CaCO_3$	s	$-269{,}780$

Substance	Formula	State	$\Delta F^{\circ}_{f\,298}$
Miscellaneous inorganic compounds (*Continued*)			
Calcium chloride	$CaCl_2$	s	−179,300
Calcium chloride	$CaCl_2$	aq	−194,880
Calcium hydroxide	$Ca(OH)_2$	s	−214,330
Calcium hydroxide	$Ca(OH)_2$	aq	−207,370
Calcium oxide	CaO	s	−144,400
Carbon dioxide	CO_2	g	−94,260
Carbon monoxide	CO	g	−32,808
Hydrochloric acid	HCl	g	−22,769
Hydrogen sulfide	H_2S	g	−7,892
Iron oxide	Fe_3O_4	s	−242,400
Iron oxide	Fe_2O_3	s	−177,100
Iron sulfide	FeS_2	s	−39,840
Nitric acid	HNO_3	l	−19,100
Nitric acid	HNO_3	aq	−26,410
Nitrogen oxides	NO	g	20,719
	NO_2	g	12,390
	N_2O	g	24,760
	N_2O_4	g	23,491
Sodium carbonate	Na_2CO_3	s	−250,400
Sodium chloride	NaCl	s	−91,785
Sodium chloride	NaCl	aq	−93,939
Sodium hydroxide	NaOH	s	−90,600
Sodium hydroxide	NaOH	aq	−100,184
Sulfur dioxide	SO_2	g	−71,790
Sulfur trioxide	SO_3	g	−88,520
Sulfuric acid	H_2SO_4	aq	−177,340
Water	H_2O	g	−54,635
Water	H_2O	l	−56,690

NOTES: The standard free energy of formation $\Delta F^{\circ}_{f\,298}$ is the change in free energy when the listed compound is formed from its elements with each substance in its standard state at 298°K (25°C). Standard states are:

1. Gases (*g*), the pure gas at unit fugacity and 25°C
2. Liquids (*l*) and solids (*s*), the pure substance at atmospheric pressure and 25°C
3. Solutes in aqueous solution (*aq*), the hypothetical 1-molal solution of the solute in water at atmospheric pressure and 25°C

The units of ΔF° are calories per gram mole of the listed substance.

SOURCE: Selected mainly from F. D. Rossini et al., Selected Values of Properties of Hydrocarbons and Related Compounds, *Am. Petrol. Inst. Res. Proj.* 44, Carnegie Institute of Technology, Pittsburgh, 1953, and loose-leaf supplements (by permission); F. D. Rossini et al., in D. D. Wagman (ed.), Selected Values of Chemical Thermodynamic Properties, *Natl. Bur. Stds. Circ.* 500, 1952, and loose-leaf supplements.

is approximately independent of temperature, the integrated form of Eq. (1-15) is

$$\ln \frac{K_{T_2}}{K_{T_1}} = \frac{-\Delta H^\circ}{R_g} \left(\frac{1}{T_2} - \frac{1}{T_1} \right) \tag{1-16}$$

If ΔH° is not constant, but can be expressed by Eq. (1-1), the integrated form is

$$\ln K_T = -\frac{\Delta H_0}{R_g T} + \frac{\Delta a}{R_g} \ln T + \frac{\Delta b}{2R_g} T + \frac{\Delta c}{6R_g} T^2 + C \tag{1-17}$$

where ΔH_0, C, and Δa, Δb, and Δc are constants and Δa, Δb, and Δc arise from the expression

$$\Delta C_p = \Delta a + \Delta b\, T + \Delta c\, T^2 \tag{1-18}$$

K can be determined from Eqs. (1-17) and (1-18) for a gaseous reaction at any temperature, provided the constants C and ΔH_0 can be evaluated. Experimental data for K at two temperatures is sufficient for this evaluation. Alternately, ΔH_0 can be found from the known heat of reaction at one temperature with Eq. (1-1). In this case only one experimental value of the equilibrium constant is needed in order to determine the constant C. Of course, in both methods it is necessary to have heat-capacity data for reactants and products in order to evaluate the coefficients Δa, Δb, and Δc.

Application of the foregoing concepts in evaluating the equilibrium conversion from free-energy data is illustrated by the following examples.

Example 1-2 The following equilibrium data have been reported for the vapor-phase hydration of ethylene to ethanol:[1] at 145°C $K = 6.8 \times 10^{-2}$ and at 320°C $K = 1.9 \times 10^{-3}$. From these data develop general expressions for the equilibrium constant as a function of temperature.

Solution From the two values of K the constants ΔH_0 and C in Eq. (1-17) may be determined. First the values of Δa, Δb, and Δc must be obtained from heat-capacity data. For the reaction

$$C_2H_4(g) + H_2O(g) \rightarrow C_2H_5OH(g)$$

these are

$$\Delta = C_2H_5OH - C_2H_4 - H_2O$$

$$\Delta a = 6.990 - 2.830 - 7.256 = -3.096$$

$$\Delta b = 0.039741 - 0.028601 - 0.002298 = 0.008842$$

$$\Delta c = (-11.926 + 8.726 - 0.283) \times 10^{-6} = -3.483 \times 10^{-6}$$

[1] H. M. Stanley et al., *J. Soc. Chem. Ind.*, **53**, 205 (1934); R. H. Bliss and B. F. Dodge, *Ind. Eng. Chem.*, **29**, 19 (1937).

Substituting these values in Eq. (1-17), we have, at 145°C,

$$R_g \ln (6.8 \times 10^{-2}) = -\frac{\Delta H_0}{418} - 3.096 \ln 418 + \frac{0.00884}{2}(418)$$

$$-\frac{3.483 \times 10^{-6}}{6}(418)^2 + CR_g$$

or

$$\frac{\Delta H_0}{418} - CR_g = -R_g \ln (6.8 \times 10^{-2}) - 3.096 \ln 418$$

$$+ 0.00442(418) - (0.580 \times 10^{-6})(418)^2 = -11.59 \quad (A)$$

and at 320°C

$$\frac{\Delta H_0}{593} - CR_g = -R_g \ln (1.9 \times 10^{-3}) - 3.096 \ln 593$$

$$+ 0.00442(593) - (0.580 \times 10^{-6})(593)^2 = -4.91 \quad (B)$$

Equations (A) and (B) may be solved simultaneously for ΔH_0 and C. The results are

$$\Delta H_0 = -9460 \text{ cal}$$

$$C = -5.56$$

Then the general expression for K as a function of temperature is

$$\ln K = \frac{9,460}{R_g T} - \frac{3.096}{R_g} \ln T + \frac{0.00442}{R_g} T - \frac{0.580 \times 10^{-6}}{R_g} T^2 - 5.56$$

or

$$\ln K = \frac{4,760}{T} - 1.558 \ln T + 0.00222 T - 0.29 \times 10^{-6} T^2 - 5.56$$

Example 1-3   Estimate the maximum conversion of ethylene to alcohol by vapor-phase hydration at 250°C and 500 psia. Use the equilibrium data of Example 1-2 and assume an initial steam-ethylene ratio of 5.

Solution   The equilibrium constant at 250°C can be evaluated from the equation for K developed in Example 1-2:

$$\ln K = \frac{4,760}{523} - 1.558 \ln 523 + 0.00222(523) - 0.29 \times 10^{-6}(523^2) - 5.56$$

$$= -5.13$$

$$K = 5.9 \times 10^{-3}$$

It is necessary to assume that the gas mixture is an ideal solution. Then Eq. (1-12) is applicable, and

$$5.9 \times 10^{-3} = K_y \frac{f'_A}{f'_E f'_W} \tag{A}$$

The fugacities of the pure components can be determined from generalized correlations[1] and are evaluated at the temperature and pressure of the equilibrium mixture:

$$\frac{f'_A}{p_t} = 0.84 \qquad \text{for ethanol}$$

$$\frac{f'_E}{p_t} = 0.98 \qquad \text{for ethylene}$$

$$\frac{f'_W}{p_t} = 0.91 \qquad \text{for water}$$

Substituting these data in Eq. (A), we have

$$K_y = \frac{y_A}{y_E y_W} = 5.9 \times 10^{-3} \frac{0.98(0.91)}{0.84} \frac{500}{14.7} = 0.21 \qquad \text{(B)}$$

If the initial steam-ethylene ratio is 5 and a basis of 1 mole of ethylene is chosen, a material balance gives the following results for equilibrium conditions:

Ethanol $= z$
Ethylene $= 1 - z$

Water $= 5 - z$
Total moles $= 6 - z$

Then

$$y_A = \frac{z}{6 - z}$$

$$y_E = \frac{1 - z}{6 - z}$$

$$y_W = \frac{5 - z}{6 - z}$$

Substituting in Eq. (B), we have

$$0.21 = \frac{z(6 - z)}{(1 - z)(5 - z)}$$

$$z^2 - 6.0z + 0.868 = 0$$

$$z = 3.0 \pm 2.85 = 5.85 \qquad \text{or } 0.15$$

The first solution is greater than unity and is impossible. Therefore $z = 0.15$, indicating that 15% of the ethylene could be converted to ethanol, provided that equilibrium were achieved.

[1]For example from J. M. Smith and H. C. Van Ness, "Introduction to Chemical Engineering Thermodynamics," fig. 12-1, McGraw-Hill Book Company, New York, 1959.

In this reaction increasing the temperature decreases K and the conversion. Increasing the pressure increases the conversion. From an equilibrium standpoint the operating pressure should be as high as possible (limited by condensation) and the temperature as low as possible. A catalyst is required to obtain an appreciable rate, but all the catalysts that are presently available require a temperature of at least 150°C for a reasonably fast rate. Even at this temperature the catalysts which have been developed will give no more than a fraction of the equilibrium conversion. In this instance both equilibrium and reaction rate limit the commercial feasibility of the reaction process.

1-6 Classification of Reactors

Chemical reactors may have a great variety of sizes, shapes, and operating conditions. One of the most common is the small flask or beaker used in the chemical laboratory for liquid-phase reactions. At the other extreme in size are the large cylindrical vessels used in the petroleum industry (for example, in the cracking of hydrocarbons), which may be up to 40 ft in diameter. In the laboratory beaker a charge of reactants is added, brought to reaction temperature, held at this condition for a predetermined time, and then the product is removed. This *batch reactor* is characterized by the variation in extent of reaction and properties of the reaction mixture with time. The hydrocarbon-cracking reactor operates continuously with a steady flow of reactants in and products out. This is the *continuous-flow type*, in which the extent of reaction may vary with position in the reactor but not with time. Hence one classification of reactors is according to method of operation.

Another classification is according to shape. If the laboratory vessel is equipped with an efficient stirrer, the composition and temperature of the reaction mass will tend to be the same in all parts of the reactor. A vessel in which the properties are uniform is called a *stirred-tank* (or well-mixed) *reactor*. If there is no mixing in the direction of flow in the cylindrical vessel for hydrocarbon processing, another ideal type is realized: the *ideal tubular-flow*, or *plug-flow*, *reactor*. Here the reaction mass consists of elements of fluid that are independent of each other, each one having a different composition, temperature, etc. This classification is of basic significance in design, because simplified treatments of the physical processes of mass and energy transfer are applicable for each ideal reactor. We noted in Sec. 1-1 that evaluation of the importance of such processes is at the heart of the design problem.

The two classifications, batch or continuous and tank or tube, are independent. Thus the laboratory beaker can be made into a continuous-flow type, and often is, by adding tubes for continuous addition of reactants and withdrawal of products. Proper reactor shape and stirring arrangement will ensure ideal stirred-tank behavior.

A third classification is based on the number of phases in the reaction mixture. The significance of this classification is also due to the influence of physical processes. A common type of *heterogeneous* reaction is illustrated by the oxidation of sulfur dioxide with a vanadium pentoxide catalyst (solid phase). The overall production of sulfur trioxide can depend on the mass-transfer process in transporting sulfur dioxide from the gas phase to the surface of the solid catalyst. This physical process results from the fact that the reaction system is heterogeneous; it would not exist if the reaction were a single-phase, or *homogeneous*, system. Note also that the catalytic nature of the reaction is not responsible for the diffusional resistance. The same type of physical process is involved in smelting reactions, such as the gas-solid noncatalytic reaction between oxygen and zinc sulfide to form zinc oxide.

In summary, the three classifications of reactors of importance in design are (1) batch or continuous, (2) tank or tubular, and (3) homogeneous or heterogeneous. We shall consider them in detail beginning with Chap. 4. The classification by shape offers numerous possibilities. A few modifications of the tank and tube types are shown in Fig. 1-3. The tubular type can be made to approach the tank type as far as mixing is concerned by recirculating part of the product (Fig. 1-3a). The tubular-flow reactor can be arranged to give an increasing flow area by employing a radial-flow arrangement (Fig. 1-3c). This could be useful for a gaseous reaction accompanied by a large increase in number of moles (volume).

Equipment for heterogeneous reactions is particularly flexible, since each phase can be processed more or less independently. In the fluidized-bed reactor (Fig. 1-4) the reactants flow continuously through and out of the reactor, but the solid-catalyst phase is withdrawn, regenerated, and returned. In the lime kiln (an example of a gas-solid noncatalytic reactor) the two phases pass continuously and countercurrently through the reactor. In heterogeneous liquid-solid polymerization systems the slurry of catalyst and reaction mixture flow together through the reactors. Walas,[1] Brotz,[2] and particularly van Krevelen[3] have summarized the various types of

---

[1] S. M. Walas, "Reaction Kinetics for Chemical Engineers," McGraw-Hill Book Company, New York, 1959.

[2] W. Brotz, "Fundamentals of Chemical Reaction Engineering," translated by D. A. Diener and J. A. Weaver, Addison-Wesley Publishing Co., Inc., Reading, Mass., 1965.

[3] D. W. van Krevelen, *Chem. Eng. Sci.*, **8**, 5 (1958).

Fig. 1-3 *Typical reactors (a) tubular-flow recycle reactor, (b) multitube-flow reactor, (c) radial-flow catalytic reactor, (d) stirred-tank reactor with internal cooling, (e) loop reactor, (f) reactor with intercoolers (opposite)*

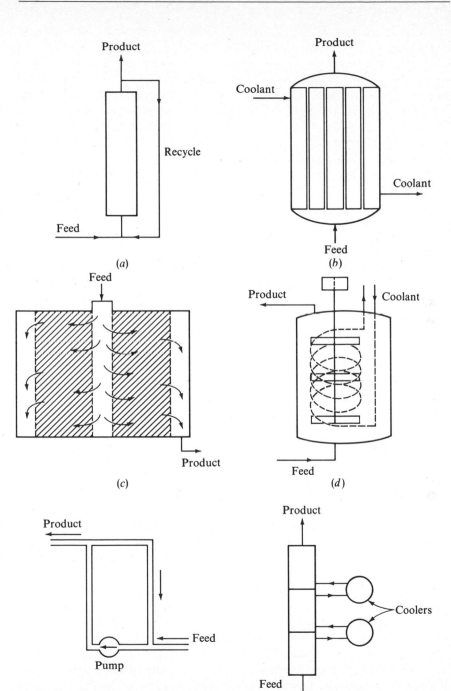

(a)

(b)

(c)

(d)

(e)

(f)

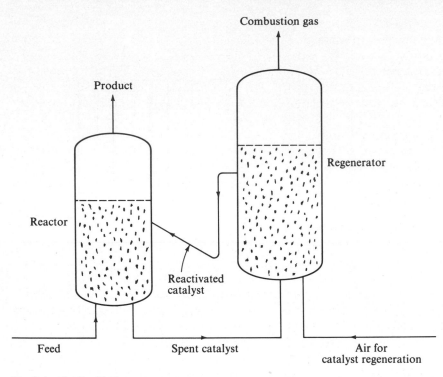

Fig. 1-4 *Fluidized-bed reactor-regenerator system*

heterogeneous reactors used industrially. Despite the multitude of possibilities, the performance of each reactor can be described adequately as a combination or adaptation of the three basic types.

There are some general relations between the physical nature of the reaction mixture and the type of reactor used in practice. Thus homogeneous gas-phase reactions are normally carried out in continuous-tubular-flow reactors rather than the tank type, either batch or flow. For liquid-phase and liquid-solid-phase heterogeneous reactions, both tank and tubular-flow reactors are employed. When a tank-type reactor is indicated but the operating pressure is so high that large-diameter vessels are too expensive, the same well-mixed condition can be obtained by circulating the reaction mixture in a loop (Fig. 1-3e). Batch-operated tank reactors are often used for small-scale production and when flexible operating conditions (temperature and pressure) are required. Such systems frequently involve costly reactants and products, as in the pharmaceutical industry.

Heat-transfer requirements can affect both the form of a reactor and its type. For example, the removal of large amounts of energy in a tank-type reactor can be achieved by introducing cooling coils to supply the required heat-transfer area. Similarly, in a tubular-flow reactor the heat-transfer rate can be increased by increasing the number and decreasing the diameter of the tubes used in parallel (Fig. 1-3b) or by using intercoolers (Fig. 1-3f). When it is necessary to approach isothermal conditions or when the heats of reaction are large, a fluidized-bed type is often used.

The chapters that follow deal with the interpretation of laboratory rate data for the design of large-scale reactors. We have seen that two aspects of chemistry, chemical thermodynamics (Sec. 1-5) and chemical kinetics, are necessary to provide the proper foundation for achieving this objective. Kinetics will be discussed in Chap. 2. We have also seen that an understanding of physical processes, particularly mass and energy transfer, is required. The interaction of chemical kinetics and physical processes depends on the type of reactor (recall the illustration for para-hydrogen conversion). Hence their study cannot be undertaken independently of the reactor-design problem. In this chapter the major classifications of reactors have been presented. Some concepts are common to all classifications, and these are discussed in Chap. 3. The remainder of the book is devoted to the design of different types of reactors. Design for homogeneous reactions is considered in Chaps. 4 to 6, and Heterogeneous systems are discussed in Chaps. 7 to 14. Since many heterogeneous reactions involve catalysts, the nature and kinetics of catalytic reactions are treated separately in Chaps. 8 and 9. In this chapter physical processes in the design of heterogeneous reactors has been emphasized; quantitative analysis of these processes is given in Chaps. 10 and 11.

Bibliography

There are several books which have as their objective the application of kinetics to chemical reactor design. The subject has been treated differently in each of these texts, and each is worthwhile for supplementary reading. A partial list is as follows:

1. Rutherford Aris, "Elementary Chemical Reactor Analysis," Prentice-Hall, Inc., Englewood Cliffs, N.J., 1969. A rather broad, analytical book in which theory of the general case is presented first and then followed with applications.
2. K. G. Denbigh, "Chemical Reactor Theory," Cambridge University Press, Cambridge, 1965. This small, very clearly written book provides an excellent treatment of the important problems in chemical reactor design.
3. O. A. Hougen and K. M. Watson, "Chemical Process Principles," part 3, Kinetics and Catalysis, John Wiley & Sons, Inc., New York, 1947. The principles of flow

reactors were first presented in this book. Design calculations are given for several industrially important reactions.

4. H. Kramers and K. R. Westerterp, "Chemical Reactor Design and Operation," Academic Press, Inc., New York, 1963. A concise treatment of a few selected subjects of reactor design, including optimization and residence-time distributions.

5. Octave Levenspiel, "Chemical Reaction Engineering," John Wiley & Sons, Inc., New York, 1962. A complete text for undergraduate students. The emphasis is rather more on homogeneous and noncatalytic reactors than on catalytic ones. Effects of residence-time distribution on reactor performance are treated in detail.

6. E. E. Petersen, "Chemical Reaction Analysis," Prentice-Hall, Inc., Englewood Cliffs, N.J., 1965. A more advanced text emphasizing fluid-solid catalytic reactions and reactor design.

Problems

1-1. Water vapor is 1.85 mole % dissociated at 2000°C and 1 atm total pressure, at equilibrium. Calculate the equilibrium dissociation at 25°C and 1 atm.

1-2. Consider a reactor in which two gas-phase reactions occur:

1. $A + B \rightarrow C + D$

2. $A + C \rightarrow 2E$

At the reaction temperature, $K_{p_1} = 2.667$ and $K_{p_2} = 3.200$. The total pressure is 10 atm, and the feed to the reactor has a composition of 2 moles of A and 1 mole of B. Calculate the composition of the reactor effluent if equilibrium is attained with respect to both reactions.

1-3. (a) Assuming that the only reaction involved is the dehydrogenation to vinyl alcohol, estimate the equilibrium constant and equilibrium yield of vinyl alcohol from ethanol at 400°F and 1 atm pressure. (b) Determine the composition of the gases at equilibrium obtained by dehydrogenating ethanol at 400°F and 1 atm pressure, this time considering the formation of both vinyl alcohol and acetaldehyde. The group-contribution method may be helpful in estimating thermodynamic properties. (See reference to O. A. Hougen and K. M. Watson p. 1004, Part II, Second Edition).

1-4. One mechanism that has been proposed for the conversion of ethyl alcohol to butadiene in the vapor phase consists of three steps: (a) dehydration of the ethyl alcohol; (b) dehydrogenation of the ethyl alcohol; (c) condensation of the ethylene and acetaldehyde in (a) and (b) to give butadiene, C_4H_6. At 400°C the following information is available for the three steps:

(a) $\Delta F° = -10,850$ cal/g mole

(b) $\Delta F° = -3610$ cal/g mole

(c) $\Delta F° = -1380$ cal/g mole

Determine the conversion of alcohol to butadiene at 400°C and 1 atm total pressure. For uniformity let α be the extent of reaction 1 and β the extent of reaction 2, and let γ be the moles of butadiene at equilibrium.

1-5. Another mechanism which has been proposed for the overall reaction in Prob. 1-4 is

1. $C_2H_5OH \rightarrow C_2H_4 + H_2O$

2. $C_2H_5OH \rightarrow CH_2{=}CHOH + H_2$

3. $CH_2{=}CHOH + C_2H_4 \rightarrow C_4H_6 + H_2O$

Is there any simple relationship between (a) the equilibrium constants for these reactions and the reactions in Prob. 1-4 and (b) the conversion to butadiene for these reactions and the reactions in Prob. 1-4.

1-6. Calculate the dissociation pressure of $Ag_2O(s)$ at 200°C. Use the following data:

$$\left.\begin{array}{l} \Delta H° = 6950 \text{ cal/g mole} \\ \Delta F° = 2230 \text{ cal/g mole} \end{array}\right\} \quad \text{at } 25°C$$

$$C_p = \begin{array}{ll} 5.60 + 1.5 \times 10^{-3}\,T & \text{for } Ag(s) \\ 6.50 + 1.0 \times 10^{-3}\,T & \text{for } O_2(g) \end{array}$$

where T is in degrees Kelvin.

1-7. Assuming that the value of K_p for the methanol-synthesis reaction is 9.28×10^{-3} at 1 atm pressure and 300°C, what are the numerical values of the following quantities at this temperature?

(a) K at $p_t = 1$ atm
(b) K_p at $p_t = 10$ atm
(c) K_p at $p_t = 50$ atm
(d) K at 10 and 50 atm total pressure
(e) K_y at 1, 10, and 50 atm total pressure

1-8. Newton and Dodge[1] and von Wettberg and Dodge[2] measured the composition of equilibrium mixtures of CO, H_2, and CH_3OH in the methanol synthesis. Compute the value of K and $\Delta F°$ at 309°C from the following data taken from their work:

$t = 309°C$

$p_t = 170$ atm

The equilibrium gas analysis, in mole %, is

Hydrogen = 60.9
Carbon monoxide = 13.5
Methanol = 21.3

[1] J. Am. Chem. Soc., **56**, 1287 (1934).
[2] Ind. Eng. Chem., **22**, 1040 (1930).

Inerts = 4.3
Total = 100.0

1-9.   The complete results referred to in Prob. 1-8 are as follows:

$1,000/T$, °K^{-1}	1.66	1.73	1.72	1.75	1.82	1.81	1.82	1.82
log K	-4.15	-3.75	-3.65	-3.30	-3.10	-3.20	-3.00	-2.90

	1.83	1.88	1.91	1.91	1.92	2.05	2.05	2.05
	-2.95	-2.60	-2.70	-3.00	-2.30	-2.30	-2.15	-2.35

From this information determine the best relationship between K and T in the form

$$\ln K = A\frac{1}{T} + B$$

1-10.   The determination of K in Prob. 1-8 was based on direct measurement of equilibrium compositions. Use the calorimetric data below and the third law to prepare a plot of log K vs $1/T$ for the methanol synthesis. Include a temperature range of 298 to 800°K. Compare the graph with the result obtained in Prob. 1-9.
        The entropy of CO gas at 298.16°K in the ideal-gas state at 1 atm is 47.30 cal/(g mole)(°K). A similar value for hydrogen is 31.21. The heat of vaporization for methanol at 298.16°K is 8943.7 cal/g mole, and the vapor pressure at 298.16°K is 0.1632 atm. Heat of formation of CH_3OH in the ideal-gas state at 1 atm is $-48,490$ cal/g mole. Low-temperature specific-heat and heat-of-transition data for methanol are as follows:

T, °K	18.80	21.55	24.43	27.25	30.72	34.33	37.64	40.87
C_p, cal/ (g mole)(°C)	1.109	1.512	1.959	2.292	2.829	3.437	3.962	4.427
	43.93	48.07	56.03	59.53	63.29	69.95	73.95	77.61
	4.840	5.404	6.425	6.845	7.252	8.001	8.392	8.735
	81.48	85.52	89.29	93.18	97.22	111.14	111.82	117.97
	9.001	9.295	9.693	9.939	10.23	11.23	11.48	11.64
	118.79	121.44	125.07	129.38	133.71	147.86	152.29	153.98
	11.64	11.74	12.18	12.28	12.64	12.97	13.69	14.12

164.14	166.23	167.75	181.09	185.10	189.06	196.77	210.34
11.29	11.63	11.68	16.60	16.67	16.77	16.78	16.97
235.84	256.34	273.58	285.15	292.01			
17.41	17.70	18.30	18.70	19.11			

Methanol crystals undergo a phase transition at 157.4°K for which $\Delta H = 154.3$ cal/g mole. The melting point is 175.22°K, and the heat of fusion is 757.0 cal/g mole. Specific-heat data at temperatures above 298.16°K are as follows:

T, °K	298.16	300	400	500	600	700	800
C_p, CH_3OH, cal/(g mole)(°C)	10.8	10.8	12.7	14.5	16.3	17.8	19.2

T, °C	25	100	200	300	400	500	600
(c_p) CO, cal/(g)(°C)	0.249	0.250	0.253	0.258	0.264	0.271	0.276
$(c_p)_{H_2}$, cal/(g)(°C)	3.42	3.45	3.47	3.47	3.48	3.50	3.53

2

KINETICS OF
HOMOGENEOUS REACTIONS

Kinetics is concerned with the rates of chemical reactions and the factors which influence these rates. The first kinetic measurements were made before 1820, but interpretation in terms of quantitative laws began with the studies on the inversion of sucrose by Wilhelmy,[1] the esterification of ethanol with acetic acid by Bethelot and St. Gilles,[2] and the reaction between oxalic acid and potassium permanganate by Harcourt and Esson.[3] These investigations established the relations between rate and concentration of reactants. The important contribution of Arrhenius[4] for the effect of temperature was also made in the nineteenth century.

In this chapter the definitions and concepts used in kinetics are presented, followed by a brief description of theories for reaction velocity.

[1]L. Wilhelmy, *Pogg. Ann.*, **81**, 413, 499 (1850).
[2]M. Berthelot and L. P. St. Gilles, *Ann. Phys.*, **63**, 385 (1862).
[3]A. V. Harcourt and W. Esson, *Proc. Roy. Soc. (London)*, **14**, 470 (1865); *Phil. Trans.*, **156**, 193 (1866); *Phil. Trans.*, **157**, 117 (1867).
[4]S. Arrhenius, *Z. Physik Chem.*, **4**, 226 (1889).

Then the use of rate equations for studying the kinetics of reactions is illustrated for simple systems and some complex ones. Only homogeneous reactions, devoid of any physical resistances, are considered here. The kinetics of heterogeneous reactions is taken up in Chaps. 8 and 9.

2-1 Rate of Reaction

The *rate of reaction* is formally defined as the change in moles of a component with respect to time, per unit volume of reaction mixture. This quantity is negative if the component is a reactant and positive if the component is a product. It is important that there be but one definition of rate, regardless of the type of reactor—flow or batch, tank or tube. Thus the rate must be a local, or point, value; that is, it must refer to a differential volume of reaction mixture. With this restriction the rate becomes a unique property for a given system. If the rate is to be the same throughout the volume of a tank reactor, the concentrations and temperature must be uniform. Otherwise the rate will vary from point to point in the reaction volume. In heterogeneous reactions, particularly those with solid phases, it may be convenient to base the rate on a unit mass or surface rather than volume, but it will always refer to a point in the reaction region.

For the *batch tank* reactor, with uniform concentrations and temperature, the independent variable is time, and the mathematical expression for the rate **r** is

$$\mathbf{r} = \frac{1}{V}\frac{dN}{dt} = \frac{\text{moles}}{\text{volume} \times \text{time}} \tag{2-1}$$

where V is the volume of the reaction mixture and N is the number of moles of a product species. If N refers to moles of reactant, so that dN/dt is negative, a minus sign is commonly used in front of the derivative so that the rate is always positive. For a *tubular* reactor with a steady flow of material in and out, the independent variable is position in the tube, or reactor volume, and the composition and rate change with this variable instead of time. To formulate the *point rate* a differential element of reactor volume, dV, must be chosen. If N' is the molal rate of flow of a component into the volume element, the rate will be

$$\mathbf{r} = \frac{dN'}{dV} \tag{2-2}$$

These two equations will be used throughout the book to describe the rate of reaction quantitatively.

The course of a reaction is normally measured by the change in *concentration* of a reactant or product. Particularly in gaseous reactions, volume changes may also occur (for example, because of a change in total moles). Then concentration changes arise from a change in volume as well as from reaction. The influence of volume change can be examined by writing $N = CV$. Thus Eq. (2-1) becomes

$$\mathbf{r} = \frac{1}{V} \frac{d(CV)}{dt} \tag{2-3}$$

If the volume, or density, of the reaction mixture is constant, Eq. (2-3) reduces to the common form

$$\mathbf{r} = \frac{dC}{dt} \tag{2-4}$$

Care must be exercised in deciding when Eq. (2-4) is applicable. In a flow reactor, used for a gaseous reaction with a change in moles, it is not correct (see Examples 4-3 and 4-4). However, it is correct for all gas-phase reactions in a tank-type reactor, since the gaseous reaction mixture fills the entire vessel, so that the volume is constant. For many liquid-phase systems density changes during the reaction are small, and Eq. (2-4) is valid for all types of reactors. The use of Eqs. (2-1) to (2-4) will become clear as we consider various kinds of reactions and reactors.

If the stoichiometric coefficients for two reactants are different, the rate expressed in terms of one reactant will not be the same as the rate expressed in terms of the other. Suppose the reaction

$$aA + bB \rightarrow cC + dD$$

occurs at constant volume, so that Eq. (2-4) is applicable. The concentration changes for the four reactants and products are related as follows:

$$-\frac{1}{a} \frac{d[A]}{dt} = -\frac{1}{b} \frac{d[B]}{dt} = \frac{1}{c} \frac{d[C]}{dt} = \frac{1}{d} \frac{d[D]}{dt} \tag{2-4'}$$

where $[A]$ is the concentration of reactant A, etc. For example, if the reaction is

$$2A + B \rightarrow C$$

$$-\frac{1}{2} \frac{d[A]}{dt} = -\frac{d[B]}{dt} = \frac{d[C]}{dt}$$

or

$$-\frac{d[A]}{dt} = -2 \frac{d[B]}{dt} = 2 \frac{d[C]}{dt}$$

For reactions with nonequal stoichiometric coefficients, Eq. (2-4′) shows that a rate which is the same for all components can be defined as[1]

$$\mathbf{r} = \frac{1}{a_i} \frac{dC_i}{dt} \tag{2-4″}$$

where the stoichiometric coefficient a_i is negative for reactant i and positive for product i, and C_i is the concentration.

2-2 Concepts of Kinetics

The early workers in kinetics found that simple relations existed between rates of reaction and concentrations of reactants. Thus Berthelot and St. Gilles[2] discovered that the rate of esterification was proportional to the first power of the concentration of ethanol and to the first power of the concentration of acetic acid. The rate is said to be *first order* with respect to each reactant. In general terms, suppose that the rate of the reaction

$$aA + bB \rightarrow cC + dD$$

may be written

$$\mathbf{r} = k[A]^\alpha [B]^\beta \tag{2-5}$$

Then α is the *order* of the reaction with respect to A, and β is the *order* with respect to B. In subsequent sections kinetics will be discussed with respect to a batch tank reactor, usually at constant density, so that Eq. (2-4) is applicable. Measured in terms of reactant A, this limitation means that Eq. (2-5) may be written

$$\mathbf{r} - -\frac{d[A]}{dt} = k[A]^\alpha [B]^\beta \tag{2-6}$$

The order of the reaction is determined by comparison of experimental data with Eq. (2-6) or integrated forms of it, as described in Secs. 2-7 to 2-10. As such, order is an empirical quantity, and α and β do not always have integer values.

There is no necessary connection between order and the stoichiometric coefficients in the reaction equation; that is, it is not required that $\alpha = a$ and $\beta = b$ in reaction (2-5). For example, the stoichiometry of the ammonia-synthesis reaction is

$$2N_2 + 3H_2 \rightarrow 2NH_3$$

[1] This concept of a unique rate has been used to develop a measure of the amount of reaction which is the same regardless of the component used to follow the course of the reaction. This so-called *extent* of reaction is described in Michel Boudart, "Kinetics of Chemical Processes," p. 10, Prentice-Hall, Inc., Englewood Cliffs, N.J., 1968.

[2] M. Berthelot and L. P. St. Gilles, *Ann. Phys.*, **63**, 385 (1862).

but for many catalysts the rate equation that best fits experimental data is first order in N_2 and zero order in hydrogen.

As kinetic studies of many kinds of reactions have accumulated, it has become increasingly clear that formation of the final products from the original reactants usually occurs in a series of relatively simple steps. The number of overall reactions that take place in a single step is small. Herein lies the explanation for the difference between order and stoichiometric coefficients. The rates of the individual steps will normally be different, and the rate of the overall reaction will be determined primarily by the slowest of these steps. The *mechanism* of a reaction is the sequence of steps that describe how the final products are formed from the original reactants. If the mechanism is known, it is usually possible to evaluate a rate equation such as (2-6) and, hence, the order of the reaction. In contrast, it is generally not possible to infer a mechanism from the rate equation alone.

These points are well illustrated by the gaseous photochlorination of propane,

$$C_3H_8 + Cl_2 \rightarrow C_3H_7Cl + HCl$$

At certain concentrations stoichiometry and order do not agree, for the rate is found[1] to be second-order in chlorine and independent of propane concentration (zero order) according to the rate equation

$$-\frac{d[Cl_2]}{dt} = k[Cl_2]^2 \tag{2-7}$$

However, the actual mechanism of the reaction probably entails many steps comprising an initiation reaction in which light is absorbed, propagation steps involving the free radicals C_3H_7 and Cl, and termination steps in which the radicals are eliminated.

A fortuitous example in which order and stoichiometry are identical is the decomposition of N_2O_5,

$$N_2O_5 \rightarrow 2NO_2 + \tfrac{1}{2}O_2$$

Ogg[2] has found that while the rate is first order in N_2O_5, the mechanism probably consists of three steps:

1. $N_2O_5 \rightleftharpoons NO_2 + NO_3$

2. $NO_2 + NO_3 \rightarrow NO + O_2 + NO_2$

3. $NO + NO_3 \rightarrow 2NO_2$

If the second step is second order and very slow with respect to the others,

[1] A. E. Cassano and J. M. Smith, *AIChE J.*, **12**, 1124 (1966).
[2] R. A. Ogg, Jr., *J. Chem. Phys.*, **15**, 337, 613 (1947).

the rate of the overall process will be proportional to the product of NO_2 and NO_3 concentrations. Furthermore, equilibrium will be quickly attained in the first step, so that $[NO_2][NO_3]$ will be equal to $K_1[N_2O_5]$, where K_1 is the equilibrium constant for the first step. These postulates and the three-step mechanism thus explain the first-order rate equation. Note that it would be incorrect to assume a single step for the overall reaction on a basis of the observed first-order dependency on the rate.

The individual steps, which together describe the overall reaction, are called *elementary processes*. Theories about kinetics (discussed in Secs. 2-4 to 2-6) refer to these elementary processes. Order and stoichiometric numbers are usually identical for elementary processes, but not always. The *molecularity* of an elementary step is the number of reactant molecules that take part in the reaction. This is usually equal to the total order, but exceptions exist for unimolecular reactions. For example, unimolecular reactions (molecularity $= 1$) are not necessarily first order; in fact, gaseous reactions which involve one molecule always become second order at low pressures (see Sec. 2-6).

Up to this point we have considered the influence only of concentration on the rate. The *specific rate constant k* in Eq. (2-6) includes the effects of all other variables. The most important of these is temperature, but others may be significant. For example, a reaction may be primarily homogeneous but have appreciable wall or other surface effects. In such cases k will vary with the nature and extent of the surface. A reaction may be homogeneous but also require a catalyst. An example is the reaction for the inversion of sugar, where the acid acts as a catalyst. In these instances k may depend on the concentration and nature of the catalytic substance. However, when the concentration effect of the catalyst is known, it is better to include the catalyst concentration in Eq. (2-5), so that k is independent of all concentrations.

The dependency of k on temperature for an elementary process follows the Arrhenius equation (see Sec. 2-3),

$$k = \mathbf{A}e^{-E/R_gT} \qquad (2\text{-}8)$$

where $\mathbf{A}$ is the frequency (or preexponential) factor and E is the activation energy. Combining Eqs. (2-8) and (2-6) yields

$$-\frac{d[A]}{dt} = \mathbf{A}e^{-E/R_gT}[A]^\alpha[B]^\beta \qquad (2\text{-}9)$$

This provides a description of the rate in terms of the measurable variables, concentration and temperature. It is rigorously limited to an elementary process because the Arrhenius equation is so restricted. However, the exponential effect of temperature often accurately represents experimental rate data for an overall reaction, even though the activation energy is not clearly defined and may be a combination of E values for several of the elementary steps.

2-3 The Arrhenius Law

Arrhenius developed his concepts about the variation of rate with temperature through thermodynamic arguments.[1] For a reaction whose rates are rapid enough to achieve a dynamic equilibrium the *van't Hoff equation* states that

$$\frac{d(\ln K)}{dT} = \frac{\Delta H^{\circ}}{R_g T^2} \tag{2-10}$$

If the reaction may be written

$$A + B \rightleftarrows C$$

the equilibrium constant[2] is

$$K = \frac{[C]}{[B][A]} \tag{2-11}$$

Since this is an elementary process, the rates of the forward and reverse reactions may be formulated [by Eq. (2-6)] with order and stoichiometric numbers identical:

$$\text{Forward rate} = k_2[A][B]$$

$$\text{Reverse rate} = k'_1[C]$$

At equilibrium the two rates are equal. This fact, plus Eq. (2-11), yields

$$k_2[A][B] = k'_1[C] \tag{2-12}$$

or[3]

$$\frac{[C]}{[A][B]} = \frac{k_2}{k'_1} = K \tag{2-13}$$

Using this result in Eq. (2-10) gives

$$\frac{d(\ln k_2)}{dT} - \frac{d(\ln k'_1)}{dT} = \frac{\Delta H}{R_g T^2} \tag{2-14}$$

[1] S. Arrhenius, *Z. Physik Chem.*, **4**, 226 (1889).
[2] To simplify the reasoning the complexities introduced by the differences between activities, upon which K in Eq. (2-10) is based, and concentrations, which express K in Eq. (2-11), are ignored.
[3] The form of Eq. (2-13) for an overall reaction is considered in Sec. 2-12.

The right-hand side of Eq. (2-14) may be divided into two enthalpy changes, ΔH_1 and ΔH_2, such that

$$\Delta H = \Delta H_2 - \Delta H_1 \tag{2-15}$$

Then Eq. (2-14) may be split into two equations, one for the forward reaction and another for the reverse reaction, which will have a difference in agreement with (2-15):

$$\frac{d(\ln k_2)}{dT} = \frac{\Delta H_2}{R_g T^2} \tag{2-16}$$

$$\frac{d(\ln k_1')}{dT} = \frac{\Delta H_1}{R_g T^2} \tag{2-17}$$

Integrating either equation and setting the integration constant equal to ln **A** gives a result of the form of the Arrhenius equation, Eq. (2-8):

$$k = \mathbf{A}e^{-\Delta H / R_g T} \tag{2-18}$$

An alternate derivation is based on the concept of an intermediate state, often called a *transition* or *activated state*, which is a postulate of the transition-state theory (Sec. 2-5). Suppose that product C of the reaction

$$A + B \leftrightharpoons C$$

is formed only by decomposition of an activated form of reactants A and B, which will be designated $(AB)^*$. Then the reaction occurs by two elementary steps,

1. $A + B \rightleftharpoons (AB)^*$

2. $(AB)^* \rightarrow C$

If the first step is comparatively rapid in both forward and reverse directions, $(AB)^*$ will be in equilibrium with A and B so that its concentration is given by

$$[(AB)^*] = K^*[A][B] \tag{2-19}$$

where K^* is the equilibrium constant for the formation of $(AB)^*$. The rate of reaction (rate of formation of C) is then given by the rate of the first-order decomposition step. With Eq. (2-19), this may be expressed as

$$\mathbf{r} = k^*[(AB)^*] = k^* K^*[A][B] \tag{2-20}$$

If we integrate the van't Hoff equilibrium equation for K^*,

$$\frac{d(\ln K^*)}{dT} = \frac{\Delta H^*}{R_g T^2} \tag{2-21}$$

we obtain

$$K^* = I e^{-\Delta H^*/R_g T} \qquad (2\text{-}22)$$

where I is the constant of integration. Combining Eqs. (2-20) and (2-22) gives

$$\mathbf{r} = k^* I e^{-\Delta H^*/R_g T} [A][B] \qquad (2\text{-}23)$$

Comparison with Eq. (2-6) shows that

$$k = \mathbf{A} e^{-\Delta H^*/R_g T} \qquad (2\text{-}24)$$

where $\mathbf{A} = k^* I$. Equation (2-24) is also of the form of the Arrhenius equation.

Since ΔH^* is the energy required to form the activated state $(AB)^*$ from A and B, $e^{-\Delta H^*/R_g T}$ is the *Boltzmann expression* for the fraction of molecules having an energy ΔH^* in excess of the average energy. This gives some meaning to the activation energy E in the Arrhenius equation. The diagram in Fig. 2-1 shows that this value is the energy barrier that must be overcome to form $(AB)^*$, and ultimately product C.

The value of Eq. (2-8) rests substantially on the accuracy with which it represents experimental rate-temperature data (see Example 2-1). When measured rates do not agree with the theory it is usually found that the reaction is not an elementary step or that physical resistances are affecting the measurements. In other words, Eq. (2-8) correlates remarkably well the rate measurements for single reactions free of diffusion and thermal resistances. The Arrhenius equation provides no basis for discerning the value of E. However, Fig. 2-1 indicates that the activation energy must be greater than the heat of the overall reaction, ΔH, for an endothermic case.

In view of the success of the Arrhenius equation, there have been many attempts to develop theoretical interpretations for the frequency

Fig. 2-1   *Energy levels of initial, activated, and final states*

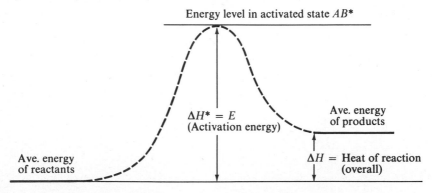

Energy level in activated state AB^*

$\Delta H^* = E$
(Activation energy)

Ave. energy
of products

Ave. energy
of reactants

ΔH = Heat of reaction
(overall)

factor A and the activation energy E. The collision theory (see Sec. 2-4), in which the frequency factor is treated as a collision rate, was inspired by the kinetic theory of gases. Subsequently the more sophisticated activated-complex theory was developed to take advantage of the more detailed description of collisions made possible with statistical thermodynamics and quantum chemistry.

Example 2-1   Wynkoop and Wilhelm[1] studied the rate of hydrogenation of ethylene, using a copper–magnesium oxide catalyst, over restricted pressure and composition ranges. Their data may be interpreted with a first-order rate expression of the form

$$\mathbf{r} = (k_1)_p \, p_{H_2} \tag{A}$$

where $\mathbf{r}$ is the rate of reaction, in g moles/(cm^3)(sec), and p_{H_2} is the partial pressure of hydrogen, in atmospheres. With this rate equation $(k_1)_p$ will be reported in g moles/(cm^3)(sec)(atm). The results for $(k_1)_p$ at various temperatures are given in Table 2-1. (a) What is the activation energy from rate equation (A)? (b) What would it be if the rate equation were expressed in terms of the concentration of hydrogen rather than the partial pressure?

Solution   (a) In the last column of Table 2-1 the reciprocal of the absolute temperature is shown for each run. Figure 2-2 is a plot of $(k_1)_p$ vs $1/T$ on semilogarithmic coordinates. It is apparent that the data describe a straight line, except for runs 8, 20, 21, and 22. It has been suggested that water vapor may have caused the low rates in these cases.[1] The line shown in the figure was located by fitting the data points by the least-mean-squares technique. This requires writing Eq. (2-8) in logarithmic form,

$$\ln (k_1)_p = \ln A - \frac{E_p}{R_g} \frac{1}{T} \tag{B}$$

This is a linear relation between $\ln (k_1)_p$ and $1/T$ with a slope of $-E_p/R_g$. If (T_i, k_i) represents one of n data points, the values of A and E_p/R_g which describe the least-mean-square fit are

$$-\frac{E_p}{R_g} = \frac{n \sum_{i=1}^{n} (\ln k_i)(1/T_i) - \left(\sum_{i=1}^{n} 1/T_i\right) \sum_{i=1}^{n} \ln k_i}{n \sum_{i=1}^{n} (1/T_i)^2 - \left(\sum_{i=1}^{n} 1/T_i\right)^2} \tag{C}$$

$$\ln A = \frac{\sum \ln k_i \sum (1/T_i)^2 - \sum 1/T_i \sum (1/T_i \ln k_i)}{n \sum (1/T_i)^2 - (\sum 1/T_i)^2} \tag{D}$$

Carrying out the summations indicated for all the data points, with the data in Table 2-1 for k_i and T_i, we find

$$-\frac{E_p}{R_g} = -6,460$$

$$E_p = 6,460 R_g = 12,800 \text{ cal/g mole}$$

[1] Raymond Wynkoop and R. H. Wilhelm, *Chem. Eng. Prog.*, **46**, 300 (1950).

Table 2-1   Data for hydrogenation of ethylene

Run	$(k_1)_p \times 10^5$, g moles/(sec)(atm)(cm³)	T, °C	$1/T \times 10^3$, °K^{-1}
1	2.70	77	2.86
2	2.87	77	2.86
3	1.48	63.5	2.97
4	0.71	53.3	3.06
5	0.66	53.3	3.06
6	2.44	77.6	2.85
7	2.40	77.6	2.85
8	1.26	77.6	2.85
9	0.72	52.9	3.07
10	0.70	52.9	3.07
11	2.40	77.6	2.85
12	1.42	62.7	2.98
13	0.69	53.7	3.06
14	0.68	53.7	3.06
15	3.03	79.5	2.83
16	3.06	79.5	2.83
17	1.31	64.0	2.97
18	1.37	64.0	2.97
19	0.70	54.5	3.05
20	0.146	39.2	3.20
21	0.159	38.3	3.21
22	0.260	49.4	3.10
23	0.322	40.2	3.19
24	0.323	40.2	3.19
25	0.283	40.2	3.19
26	0.284	40.2	3.19
27	0.277	39.7	3.20
28	0.318	40.2	3.19
29	0.323	40.2	3.19
30	0.326	40.2	3.19
31	0.312	39.9	3.19
32	0.314	39.9	3.19
33	0.307	39.8	3.19

SOURCE: Raymond Wynkoop and R. H. Wilhelm, *Chem. Eng. Progr.*, **46**, 300 (1950).

A quicker but less accurate method is to draw, visually, a straight line through the data plotted as ln $(k_1)_p$ vs $1/T$, measure its slope, and multiply by R_g to obtain the activation energy.

(*b*) For gaseous reactions the rate equation can be expressed in terms of concentrations or pressures. Equation (A) is the pressure form for this example. In terms of concentrations, the rate is

$$\mathbf{r} = (k_1)_C \, [H_2]$$

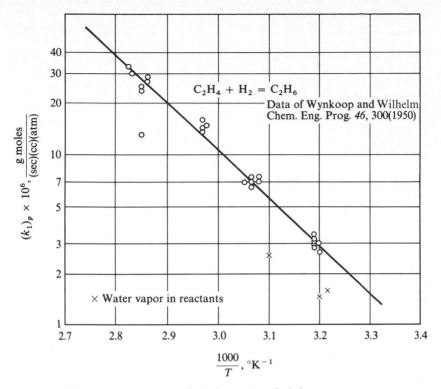

Fig. 2-2   *Plot of Arrhenius equation for hydrogenation of ethylene*

Expressing $(k_1)_C$ in the Arrhenius form and then differentiating gives

$$(k_1)_C = A_C\, e^{-E_C/R_g T}$$

and

$$\frac{d[\ln (k_1)_C]}{d(1/T)} = -\frac{E_C}{R_g} \tag{E}$$

This may be related to Eq. (A) by noting that for an ideal-gas mixture the concentration of H_2 is

$$[H_2] = \frac{p_{H_2}}{R_g T}$$

Then

$$r = \frac{(k_1)_C}{R_g T}\, p_{H_2} \tag{F}$$

Comparison of Eqs. (A) and (F) gives the relationship between the two rate constants,

$$(k_1)_C = (k_1)_p \, R_g T \tag{G}$$

Differentiating the logarithmic form of Eq. (G) and using Eqs. (B) and (E), we obtain

$$\frac{d[\ln (k_1)_C]}{d(1/T)} = \frac{d[\ln (k_1)_p]}{d(1/T)} + \frac{d(\ln T)}{d(1/T)}$$

$$-\frac{E_C}{R_g} = -\frac{E_p}{R_g} - T$$

or

$$E_C = E_p + R_g T \tag{H}$$

Thus the activation energy, in principle, depends on whether the rate equation is expressed in terms of concentrations or partial pressures. Also, the difference between E_C and E_p depends on the temperature. In practice this difference is not significant. In this example, at a temperature of 77°C,

$$E_C - E_p = 2(350°K) = 700 \text{ cal/g mole}$$

This difference of 6% is too small to be discerned from rate measurements of the usual precision. Hence it generally makes little difference whether E is evaluated from a rate equation expressed in terms of pressures or in terms of concentrations.

RATE THEORIES

2-4 Collision Theory

The Arrhenius concept as pictured in Fig. 2-1 requires that the molecules of reactants have an energy E above their normal, or average, energy. There is a possibility that some molecules will possess this excess because of the wide range over which the energy is distributed and the large number of molecules that make up the system. According to classical kinetic theory, some gaseous molecules will possess much larger amounts of translational energy than others because of variations in their molecular velocities. It is logical to suppose that collisions between these reactant molecules would provide the activation energy necessary for the reaction to occur. By assuming that the molecules behave as hard spheres, it is possible to develop simple expressions for the rate. This approach, originally advanced by Lewis[1] and Polanyi,[2] has become known as the *collision theory*.

The theory has a number of weaknesses and has been extended and supplemented by later developments. However, it offers a simple picture

[1] W. C. McC. Lewis, *J. Chem. Soc. (London)*, **113**, 471 (1918).
[2] M. Polanyi, *Z. Elektrochem.*, **26**, 48 (1920).

of the mechanism of reactions. According to the collision theory, the number of molecules of product formed per unit time per unit volume—i.e., the rate—is equal to the number of collisions multiplied by a factor f. This factor takes into account the fact that only a fraction of the collisions involve molecules that possess the excess energy (activation energy) necessary for reaction. For a gaseous reaction, such as $A + B \rightarrow C + D$, this may be stated mathematically as

$$\mathbf{r} = zf \tag{2-25}$$

where z is the number of collisions between molecules A and B in 1 cm^3 of reaction mixture per second. From kinetic theory (with the assumption that molecules are hard spheres), the number of collisions is given by

$$z = [A][B]\sigma_{AB}^2 \left(8\pi R_g T \frac{M_A + M_B}{M_A M_B} \right)^{1/2} \tag{2-26}$$

where the concentration is in molecules (*not* moles) per cubic centimeter, and

σ_{AB} = effective diameter of A plus B upon collision
M = molecular weight
R_g = gas constant = $k_B N_0$, the product of Boltzmann's constant and Avogadro's number, ergs/($^\circ$K)(g mole)

Then the rate equation (2-25) may be written

$$\mathbf{r} = f[A][B]\sigma_{AB}^2 \left(8\pi R_g T \frac{M_A + M_B}{M_A M_B} \right)^{1/2} \tag{2-27}$$

The rate may also be expressed in terms of the specific rate constant,

$$\mathbf{r} = k[A][B] \tag{2-28}$$

Using the Arrhenius equation (2-8) for k, we may write this as

$$\mathbf{r} = A e^{-E/R_g T} [A][B] \tag{2-29}$$

Combining Eqs. (2-27) and (2-29) gives the following result for the frequency factor $\mathbf{A}$:

$$A e^{-E/R_g T} = f\sigma_{AB}^2 \left(8\pi R_g T \frac{M_A + M_B}{M_A M_B} \right)^{1/2}$$

The fraction of the molecules that possess the required excess energy for reaction should not depend on the number of collisions but on the magnitude of the energy itself. If a Maxwellian distribution is assumed, the fraction of the total molecules having an energy at least equal to E

can be shown to be $e^{-E/R_g T}$. Hence f may be taken as $e^{-E/R_g T}$, and then the frequency factor is given by

$$A = \sigma_{AB}^2 \left(8\pi R_g T \frac{M_A + M_B}{M_A M_B} \right)^{1/2} \tag{2-30}$$

Finally, substitution of this value of frequency factor A in Eq. (2-8) gives the collision-theory expression for the specific reaction rate,

$$k = \sigma_{AB}^2 \left(8\pi R_g T \frac{M_A + M_B}{M_A M_B} \right)^{1/2} e^{-E/R_g T} \qquad cm^3/(molecule)(sec)$$

$$\tag{2-31}$$

The first part of the equation represents the number of collisions per unit time per unit volume (when $[A] = [B] = 1$), and $e^{-E/R_g T}$ represents the fraction of the collisions that involve molecules with the necessary activation energy.

Example 2-2   Use the collision theory to estimate the specific reaction rate for the decomposition of hydrogen iodide, $2HI \rightarrow I_2 + H_2$. Assume that the collision diameter σ is $3.5A$ (3.5×10^{-8} cm), and employ the activation energy of 44,000 cal/g mole determined experimentally by Bodenstein.[1] Also evaluate the frequency factor.

Solution   According to the collision theory, the specific reaction rate is given by Eq. (2-31) in units of $cm^3/(molecule)$ (sec). For the reaction $2HI \rightarrow H_2 + I_2$

$$M_A = M_B = M_{HI} = 128$$

The other numerical quantities required are

$$R_g = k_B n = (1.38 \times 10^{-16})(6.02 \times 10^{23})$$

$$= 8.30 \times 10^7 \; ergs/(°K)(g \; mole) \qquad or \; 1.98 \; cal/(g \; mole) \; (°K)$$

$$\sigma_{AB} = 3.5 \times 10^{-8} \; cm$$

$$E = 44,000 \; cal/g \; mole$$

$$T = 273 + 321.4 = 594.6°K$$

Substituting these values in Eq. (2-31) yields

$$k = (3.5 \times 10^{-8})^2 \left[8\pi(8.30 \times 10^7) \; 594.6 \left(\frac{2}{128} \right) \right]^{1/2} e^{-44,000/R_g T}$$

$$= 1.70 \times 10^{-10} \; e^{-37.4} \qquad cm^3/(molecule)(sec)$$

To convert this result to the usual units of liters/(g mole)(sec) it should be multi-

---

[1] M. Bodenstein, *Z. Physik Chem.*, **100**, 68 (1922).

plied by Avogadro's number, 6.02×10^{23} molecules/mole, and divided by 1,000 cm^3/liter:

$$k = \frac{6.02 \times 10^{23}}{1,000} 1.70 \times 10^{-10} e^{-37.4}$$

$$= 1.02 \times 10^{11} e^{-37.4} = 5.7 \times 10^{-6} \text{ liter/(g mole)(sec)} \tag{A}$$

As we shall see in Example 2-6, the rate constant from Kistiakowsky's data is found to be 2.0×10^{-6} liter/(g mole)(sec). For reactions involving more complex molecules the experimental rates are usually much less than the theory predicts.

Comparison of the form of Eq. (A) and the Arrhenius expression shows that the frequency factor is

$$\mathbf{A} = 1.0 \times 10^{11} \text{ liters/(g mole)(sec)}$$

2-5 Activated-complex (Transition-state) Theory

The collision theory has been found to give results in good agreement with experimental data for a number of bimolecular gas reactions. The decomposition of hydrogen iodide considered in Example 2-2 is an illustration. The theory has also been satisfactory for several reactions in solution involving simple ions. However, for many other reactions the predicted rates are much too large. Predicted frequency factors lie in the rather narrow range of 10^9 to 10^{11}, while measured values may be several orders of magnitude less. The deviation appears to increase with the complexity of the reactant molecules. (Moreover, unimolecular decompositions are difficult to rationalize by the collision theory.) As a means of correcting for this disagreement it has been customary to introduce a probability, or *steric*, factor (having a value less than unity) in Eq. (2-31). To retain the hard-sphere concept we must then explain why all the collisions supplying the necessary energy do not result in reaction.[1]

Beginning in about 1930 the principles of quantum mechanics were applied to this problem by Eyring, Polanyi, and their coworkers, and the result is known as the *activated-complex theory*.[2] In this theory reaction is still presumed to occur as a result of collisions between reacting molecules, but what happens after collision is examined in more detail. This examination is based on the concept that molecules possess vibrational and rotational, as well as translational, energy levels.

---

[1] For a more detailed and complete description of the collision theory and its limitations see E. A. Moelwyn-Hughes, "Kinetics of Reactions in Solution," Oxford University Press, New York, 1946.

[2] Samuel Glasstone, K. J. Laidler, and Henry Eyring, "The Theory of Rate Processes," McGraw-Hill Book Company, New York, 1941.

The essential postulate is that an activated complex (or transition state) is formed from the reactant, and that this subsequently decomposes to the products. The activated complex is assumed to be in thermodynamic equilibrium with the reactants. Then the rate-controlling step is the decomposition of the activated complex. The concept of an equilibrium activation step followed by slow decomposition is equivalent to assuming a time lag between activation and decomposition into the reaction products. It is the answer proposed by the theory to the question of why all collisions are not effective in producing a reaction.

These ideas may be illustrated by a simple reaction between A and B to form a product C. If the activated complex is designated by $(AB)^*$, the overall process can be written as

$$A + B \rightleftharpoons (AB)^* \rightarrow C$$

Since equilibrium is assumed for the first step, the concentration of $(AB)^*$ is that determined by the equilibrium constant. Then the rate of the overall reaction is equal to the product of the frequency of decomposition of the complex and its equilibrium concentration, or

$$\mathbf{r} = v[(AB)^*] \qquad \text{molecules/(sec)(cm}^3) \tag{2-32}$$

where v is in units per second and the concentration $[(AB)^*]$ is in molecules per cubic centimeter. If the equilibrium constant for the formation of AB is K^*, then in terms of activity a,

$$K^* = \frac{a_{(AB)^*}}{a_A a_B} = \frac{\gamma_{AB}[(AB)^*]}{(\gamma_A[A])(\gamma_B[B])} \tag{2-33}$$

where γ is the activity coefficient. The concentration of the activated complex can be substituted in Eq. (2-32) to give

$$\mathbf{r} = v\frac{\gamma_A \gamma_B}{\gamma_{AB}} K^*[A][B] \tag{2-34}$$

The equilibrium constant is related to the standard free-energy change for the formation of the activated complex. From Eq. (1-4), this relationship is

$$K^* = e^{-\Delta F^*/R_g T} \tag{2-35}$$

It can be shown[1] that the decomposition frequency is

$$v = \frac{k_B T}{h} \tag{2-36}$$

[1] Samuel Glasstone, K. J. Laidler, and Henry Eyring, "The Theory of Rate Processes," McGraw-Hill Book Company, New York, 1941.

where k_B = Boltzmann's constant, 1.380×10^{-16} erg/°K, and h = Planck's constant, 6.624×10^{-27} ergs(sec).

Substituting Eqs. (2-35) and (2-36) in Eq. (2-34) gives

$$\mathbf{r} = \frac{k_B T}{h} \left(\frac{\gamma_A \gamma_B}{\gamma_{AB}} \right) e^{-\Delta F^*/R_g T} [A][B] \qquad (2\text{-}37)$$

Hence the specific reaction rate is

$$k = \frac{k_B T}{h} \left(\frac{\gamma_A \gamma_B}{\gamma_{AB}} \right) e^{-\Delta F^*/R_g T} = \frac{k_B T}{h} \left(\frac{\gamma_A \gamma_B}{\gamma_{AB}} \right) e^{\Delta S^*/R_g - \Delta H^*/R_g T} \qquad (2\text{-}38)$$

The latter form follows from the thermodynamic relation $\Delta F = \Delta H - T \Delta S$.

Comparison of Eq. (2-38) with the Arrhenius equation shows that

$$\mathbf{A} = \frac{k_B T}{h} \left(\frac{\gamma_A \gamma_B}{\gamma_{AB}} \right) e^{\Delta S^*/R_g} \qquad (2\text{-}39)$$

$$E = \Delta H^* \qquad (2\text{-}40)$$

These two relations are the predictions of the activated-complex theory for the frequency factor and the energy of activation.

The collision theory [Eq. (2-31)] does not offer a method for calculating the activation energy. The activated-complex theory suggests that E is the enthalpy change for formation of the activated complex from the reactants [Eq. (2-40)]. To predict this enthalpy we must know exactly what the activated complex is; i.e., we must know its structure. Even then the prediction of enthalpy from molecular-structure data by statistical mechanics is an uncertain operation for any but the simplest molecule. Eckert and Boudart[1] have illustrated the calculation procedures with the hydrogen iodide decomposition reaction. If an activation energy is available from experimental measurements, the theory need be used only for estimating the frequency factor from Eq. (2-39). Again the structure of the activated complex is necessary, this time to calculate the entropy of activation, ΔS^*.

Uncertainties about the structure of the activated complex and the assumptions involved in computing its thermodynamic properties seriously limit the practical value of the theory. However, it does provide qualitative interpretation of how molecules react and a reassuring foundation for the empirical rate expressions inferred from experimental data. The effect of temperature on the frequency factor is extremely difficult to evaluate from rate measurements. This is because the strong exponential function in the Arrhenius equation effectively masks the temperature dependency of $\mathbf{A}$.

[1] C. A. Eckert and M. Boudart, *Chem. Eng. Sci.*, **18**, 144 (1963).

Equation (2-39) suggests that **A** is proportional to T, the collision theory indicates a $T^{1/2}$ dependency [Eq. (2-30)], and the Arrhenius relationship [Eq. (2-8)] implies that **A** is unaffected by temperature. It is normally impossible to measure rates of reaction with sufficient sensitivity to evaluate these differences.

In Eq. (2-32) it is supposed that the rate is proportional to the *concentration* of the activated complex. Similarly, in the collision theory, Eq. (2-25), it is tacitly assumed that the concentration determines the collision frequency and the rate. However, if the results of thermodynamics were followed, the rate might be assumed proportional to activity. If the activity replaced concentration in Eq. (2-32), the activity coefficient of the activated complex would not be needed in Eq. (2-34). The final expression for the rate constant would then be

$$k = \frac{k_B T}{h} \gamma_A \gamma_B \, e^{\Delta S^*/R_g - \Delta H^*/R_g T} \tag{2-41}$$

instead of Eq. (2-38). Since the activity coefficient is a function of pressure, k values predicted from the two equations would vary differently with pressure.

Eckert and Boudart analyzed rate data for the decomposition of HI in this way. Their results were more compatible with Eq. (2-38), suggesting that the rate is proportional to the *concentration* of the activated complex. In this text the rate equation will be written in terms of concentrations.

2-6 Unimolecular Reactions

The activated-complex theory provides a plausible explanation of the first-order rate of unimolecular gaseous reactions. In such a reaction the reacting molecules gain the energy of activation by collision with other molecules. This might be thought of as a second-order process, since the number of collisions is proportional to the square of the concentration. However, Lindemann[1] showed in 1922 that activation by collision could result in first-order rates. If A^* is an activated molecule of reactant, the equilibrium between A and A^* and reaction to products B can be represented as

$$A + A \underset{}{\overset{K_1}{\rightleftharpoons}} A^* + A \tag{2-42}$$

$$A^* \overset{k_2}{\rightarrow} B \tag{2-43}$$

If the pressure (e.g., concentration) in the reaction mixture is high enough,

[1] F. A. Lindemann, *Trans. Faraday Soc.*, **17**, 598 (1922).

the number of collisions between A and A^* will be so large that equilibrium exists between A and A^*, according to Eq. (2-42). If the equilibrium constant is K_1, the concentration of A^* is given by

$$K_1 = \frac{[A^*][A]}{[A]^2}$$

$$[A^*] = K_1[A]$$

(2-44)

The rate of formation of products is determined by the second reaction, the decomposition of activated molecules. No collisions are needed for reaction, so the rate is first order in A^*, that is,

$$\mathbf{r} = k_2[A^*]$$

(2-45)

Substituting $[A^*]$ from Eq. (2-44), the rate is seen to be first order in A,

$$\mathbf{r} = k_2 K_1[A]$$

(2-46)

At low pressures the collision rate is low and equilibrium is not attained between A and A^*, so that Eq. (2-42) should be written

$$A + A \overset{k_1}{\rightarrow} A^* + A$$

(2-47)

If the A^* has a short lifetime, its concentration will be low with respect to that of A or products. Then $[A^*]$ will rapidly reach a constant, low value after which $d[A^*]/dt = 0$.[1] Summing the rates of formation and destruction of A^* by reactions (2-47) and (2-43) yields

$$\frac{d[A^*]}{dt} = 0 = k_1[A]^2 - k_2[A^*]$$

or

$$[A^*] = \frac{k_1}{k_2}[A]^2$$

(2-48)

Substituting this result into Eq. (2-45) gives a second-order rate

$$\mathbf{r} = k_1[A]^2$$

(2-49)

This changeover from first-order to second-order kinetics at low pressures has been observed experimentally for a number of unimolecular reactions.

---

[1] This statement is termed the stationary-state hypothesis—a concept widely used for reactions involving species of a transitory nature, such as free radicals and atoms. See Sec. 2-11 for a discussion of its use in photochemical kinetics. It is equally applicable here to the high-pressure case and also leads to Eq. (2-46), but by a more complicated route.

ANALYSIS OF RATE EQUATIONS

The most common experimental procedure for establishing rate equations is to measure the composition of the reaction mixture at various stages during the course of reaction.[1] In a batch system this means analysis at various times after the reaction begins.[2] Then the data are compared with various types of rate equations to find the one giving the best agreement. The comparison can be made in two ways:

1. The *integration* method, comparison of predicted and observed compositions. For this approach it is necessary to integrate the rate expression [for example, Eq. (2-6)] to give concentration as a function of time.
2. The *differential* method, comparison of predicted and observed rates. The latter are obtained by differentiating the experimental data.

In this chapter we are concerned only with the rate equation for the chemical step (no physical resistances). Also, it will be supposed that the temperature is constant, both during the course of the reaction and in all parts of the reactor volume. These ideal conditions are often met in the stirred-tank reactor (see Sec. 1-6). Data are invariably obtained with this objective, because it is extremely hazardous to try to establish a rate equation from nonisothermal data or data obtained in inadequately mixed systems. Under these restrictions the integration and differential methods can be used with Eqs. (2-1) and (2-5) or, if the density is constant, with Eq. (2-6). Even with these restrictions, evaluating a rate equation from data may be an involved problem. Reactions may be simple or complex, or reversible or irreversible, or the density may change even at constant temperature (for example, if there is a change in number of moles in a gaseous reaction). These several types of reactions are analyzed in Secs. 2-7 to 2-11 under the categories of simple and complex systems.

2-7 Integrated Rate Equations for Irreversible Reactions

If all the reactants and products can be explained by a single reaction, the reaction system is *simple*. Within this category the reaction may be of any order and may be reversible or irreversible.

[1] Many other methods are used for studying kinetics, particularly for fast reactions. For a description of various methods see F. J. W. Roughton and B. Chance, in S. L. Friess and A. Weissberger (eds.), "Rates and Mechanisms of Reactions," chap. 10, Interscience Publishers, New York, 1953.

[2] In continuous reactors the course of the reaction is measured at different flow rates, or reactor volumes. The analysis is different from that for batch reactors and involves Eq. (2-2). Analysis of such systems is discussed in Chaps. 4 and 8.

Zero Order   *Zero order*, meaning that the rate is independent of the concentration, may occur in two situations: when the rate is intrinsically independent of concentration and when the species is in such abundant supply that its concentration is nearly constant during reaction. In the latter case the dependency of the rate on concentration cannot be detected, and apparent zero order prevails. Thus in the oxidation of NO to NO_2 in the presence of a large excess of O_2, the rate is zero order in O_2.

For a zero-order reaction at constant density Eq. (2-6) becomes

$$-\frac{d[A]}{dt} = k_0 \tag{2-50}$$

Integrating from an initial condition of $[A] = [A]_0$ yields

$$[A] = [A]_0 - k_0 t \tag{2-51}$$

This result shows that the distinguishing feature of a zero-order reaction is that the concentration of reactant decreases linearly with time. It is difficult to cite a homogeneous reaction that is intrinsically zero order, although many reactions have apparent zero-order characteristics when the concentration of the species is large. However, in some heterogeneous reactions where the solid phase acts as a catalyst the rate is zero order. An example is the decomposition of NH_3 on platinum and tungsten surfaces.[1]

Equation (2-51) can be used with measurements of concentration vs time to determine if a reaction is zero order and to evaluate k. If two reactants, A and B, are involved, experiments can be carried out with A in large excess, so that the rate equation is independent of $[A]$. Then the concentration of B can be varied and its order determined. In this way the concentration of one reactant can be rendered ineffective in order to study the effect of another.

It may be simpler to measure the time when a certain fraction of reactant has disappeared than to obtain concentration-vs-time data. Common practice is to obtain the time required for one-half of the reactant to disappear. Defining this half-life as $t_{1/2}$, we have from Eq. (2-51)

$$\tfrac{1}{2}[A]_0 = [A]_0 - k_0 t_{1/2}$$

or

$$t_{1/2} = \frac{[A]_0}{2k_0} \tag{2-52}$$

Half-life data can be used with Eq. (2-52) to evaluate k_0 as an alternate to Eq. (2-51).

[1]C. N. Hinshelwood and R. E. Burk, *J. Chem. Soc.*, **127**, 1051, 1114 (1925).

First Order   Equation (2-6) for a first-order rate is

$$-\frac{d[A]}{dt} = k_1[A] \tag{2-53}$$

If the initial condition is $[A] = [A]_0$, integration yields

$$\ln\frac{[A]}{[A]_0} = -k_1 t \tag{2-54}$$

A plot of $\ln [A]/[A]_0$ vs t should be a straight line of slope equal to $-k_1$. The half-life is given by

$$-k_1 t_{1/2} = \ln \tfrac{1}{2} \tag{2-55}$$

or

$$t_{1/2} = \frac{1}{k_1} \ln 2 \tag{2-56}$$

Equations (2-54) and (2-56) show that the half-life and fraction of reactant remaining are independent of initial concentration for a first-order reaction.

Among the numerous examples of homogeneous first-order reactions are the rearrangement of cyclopropane to propylene,[1] certain cis-trans-isomerizations, and the inversion of sucrose.

Second Order   Two types of second-order reactions are of interest:

TYPE I $A + A \rightarrow P$

$$-\frac{d[A]}{dt} = k_2[A]^2 \tag{2-57}$$

TYPE II $A + B \rightarrow P$

$$-\frac{d[A]}{dt} = k_2[A][B] \tag{2-58}$$

For type I reactions integration of Eq. (2-57) yields

$$\frac{1}{[A]} - \frac{1}{[A]_0} = k_2 t \tag{2-59}$$

In terms of the half-life, this becomes

$$t_{1/2} = \frac{1}{k_2[A]_0} \tag{2-60}$$

Note that for a zero-order reaction $t_{1/2}$ is directly proportional to $[A]_0$; for

[1] T. S. Chambers and G. B. Kistiokowsky, *J. Am. Chem. Soc.*, **56**, 399 (1934).

a first-order reaction it is independent of $[A]_0$, and for a second-order reaction inversely proportional to $[A]_0$. Two well-known examples of type I reactions are the decomposition of HI in the gas phase and dimerization of cyclopentadiene in either gas or liquid phase.

When order and stoichiometry of type II reactions do not agree, the analysis is somewhat different. Suppose the stoichiometry may be represented as

$$aA + bB \rightarrow P$$

and the rate expression is Eq. (2-58). Initially $[A] = [A]_0$ and $[B] = [B]_0$; let $[X]$ be the amount of A that has reacted, expressed as a concentration. Then at any time

$$[A] = [A]_0 - [X] \tag{2-61}$$

$$[B] = [B]_0 - \frac{b}{a}[X] \qquad \frac{d[A]}{dt} = k_c[A][B] \tag{2-62}$$

$$-\frac{d[A]}{dt} = \frac{d[X]}{dt} \tag{2-63}$$

Substituting Eqs. (2-61) to (2-63) in (2-58) gives

$$-\frac{d[A]}{dt} = \frac{d[X]}{dt} = k_2([A]_0 - [X])\left([B]_0 - \frac{b}{a}[X]\right) \tag{2-64}$$

Integration from $t = 0$ gives the concentration of X, or A and B, at any time:

$$\ln \frac{[A]_0 - [X]}{[B]_0 - (b/a)[X]} = \ln \frac{[A]}{[B]} = \frac{b[A_0] - a[B_0]}{a} k_2 t + \ln \frac{[A]_0}{[B]_0}$$

$$\tag{2-65}$$

In terms of A, $t_{1/2}$ corresponds to $[A]/[A]_0 = \frac{1}{2}$, and Eq. (2-65) can be solved to yield

$$t_{1/2} = \frac{a}{k_2(b[A]_0 - a[B]_0)} \ln \frac{a[B]_0}{2a[B]_0 - b[A]_0} \tag{2-66}$$

If initial concentrations are stoichiometric, then

$$[B]_0 = \frac{b}{a}[A]_0$$

and Eq. (2-64) becomes

$$-\frac{d[A]}{dt} = k_2 \frac{b}{a}([A]_0 - [X])^2 = k_2 \frac{b}{a}[A]^2 \tag{2-67}$$

This is the same as Eq. (2-57), with k_2 multiplied by b/a. Numerous reactions of type II have been observed: for example, HI formation from gaseous H_2 and I_2,[1] dimerization of cyclopentadiene in benzene solution, and certain esterification reactions in aqueous solution.

Concentration-time data can easily be analyzed to test second-order kinetics. For type I (or type II with stoichiometric proportions of A and B initially), Eq. (2-59) indicates that the data should give a straight line if $1/[A]$ is plotted against t. For type II, Eq. (2-65) shows that a plot of log $[A]/[B]$ vs t should be linear. In this case the slope $(b[A]_0 - a[B]_0)k_2/a$ will be positive or negative, depending on the stoichiometric coefficients a and b and the initial concentrations.

In the following example first- and second-order rate equations are used to interpret data for an isothermal, constant-density, liquid system.

Example 2-3   The liquid-phase reaction between trimethylamine and n-propyl bromide was studied by Winkler and Hinshelwood[2] by immersing sealed glass tubes containing the reactants in a constant-temperature bath. The results at 139.4°C are shown in Table 2-2. Initial solutions of trimethylamine and n-propyl bromide in benzene, 0.2-molal, are mixed, sealed in glass tubes, and placed in the constant-temperature bath. After various time intervals the tubes are removed and cooled to stop the reaction, and the contents are analyzed. The analysis depends on the fact that the product, a quaternary ammonium salt, is completely ionized. Hence the concentration of bromide ions can be estimated by titration.

Table 2-2

Run	t, min	Conversion, %
1	13	11.2
2	34	25.7
3	59	36.7
4	120	55.2

From this information determine the first-order and second-order specific rates, k_1 and k_2, assuming that the reaction is irreversible over the conversion range covered by the data. Use both the integration and the differential method, and compare the results. Which rate equation best fits the experimental data?

Solution   The reaction may be written

$$N(CH_3)_3 + CH_3CH_2CH_2Br \rightarrow (CH_3)_3(CH_2CH_2CH_3)N^+ + Br^-$$

[1] M. Bodenstein, *Z. Physik Chem.*, **13**, 56 (1894); *Z. Physik Chem.*, **22**, 1 (1897); *Z. Physik Chem.*, **29**, 295 (1899).
[2] C. A. Winkler and C. N. Hinshelwood, *J. Chem. Soc.*, 1147 (1935).

Since the concentrations of reactants and products are small and the temperature is constant, the density may be assumed constant without serious error. Then the rate equations (2-53) and (2-58) are applicable for the first- and second-order possibilities. If T denotes trimethylamine and P n-propyl bromide, the rate expressions are

$$r = -\frac{d[T]}{dt} = k_1[T] \tag{A}$$

$$r = -\frac{d[T]}{dt} = k_2[T][P] \tag{B}$$

INTEGRATION METHOD For the first-order case, the integrated form of Eq. (A) is Eq. (2-54); that is,

$$\ln \frac{[T]_0}{[T]} = k_1 t \tag{C}$$

In the second-order case it is noted that $a = b$ and $[T]_0 = [P]_0 = 0.1$ molal. Hence $[T] = [P]$, and Eq. (B) reduces to a type I second-order equation,

$$-\frac{d[T]}{dt} = k_2[T]^2 \tag{D}$$

The solution is Eq. (2-59), which may be solved for k_2 to give

$$k_2 = \frac{1}{t}\left(\frac{1}{[T]} - \frac{1}{[T]_0}\right) \tag{E}$$

The conversion x is the fraction of the reactant that has been consumed. In this problem

$$x = \frac{[T]_0 - [T]}{[T]_0}$$

or

$$[T] = [T]_0 (1 - x) \tag{F}$$

The calculation of k_1 and k_2 will be illustrated for the first run. From Eq. (F),

$$[T] = [T]_0 (1 - 0.112) = 0.1(0.888)$$

Substituting in Eq. (C), we find

$$k_1 = \frac{1}{t} \ln \frac{[T]_0}{[T]} = \frac{1}{13(60)} \ln \frac{0.1}{0.0888} = 1.54 \times 10^{-4} \text{ sec}^{-1}$$

Then, using Eq. (E) for the second-order possibility, we obtain

$$k_2 = \frac{1}{t[T]_0}\left(\frac{1}{1-x} - 1\right) = \frac{0.112}{60(13)(0.1)(1 - 0.112)}$$

$$= 1.63 \times 10^{-3} \text{ liters/(g mole)(sec)}$$

Table 2-3 shows the results obtained in a similar way for the four runs. The k_1 values show a definite trend with time, and therefore the first-order mechanism does not

Table 2-3 *Specific reaction rates for trimethylamine and n-propyl bromide reaction*

Run	t, sec	$k_1 \times 10^4$, sec^{-1}	$k_2 \times 10^3$, liters/(g mole)(sec)	$[B]$, g mole/liter
1	780	1.54	1.63	0.0112
2	2,040	1.46	1.70	0.0257
3	3,540	1.30	1.64	0.0367
4	7,200	1.12	1.71	0.0552
			(1.67 av)	

satisfactorily explain the kinetic data. The k_2 values not only are more nearly identical, but the variations show no definite trend.

DIFFERENTIAL METHOD   The moles of bromide ions B produced are equal to the trimethylamine reacted. Hence

$$[B] = [T]_0 - [T]$$

and from Eq. (F)

$$[B] = [T]_0 - [T]_0(1 - x) = x[T]_0$$

Thus $[B]$ can be calculated from the conversion data. A plot of $[B]$ vs time of reaction is shown in Fig. 2-3. The slope of this curve at any point is equal to the rate of reaction, since

$$\mathbf{r} = -\frac{d[T]}{dt} = \frac{d[B]}{dt}$$

Slopes determined from the curve are given in Table 2-4.

Fig. 2-3   *Concentration vs time for reaction between* $(CH_3)_3N$ *and* $CH_3CH_2\,CH_2Br$

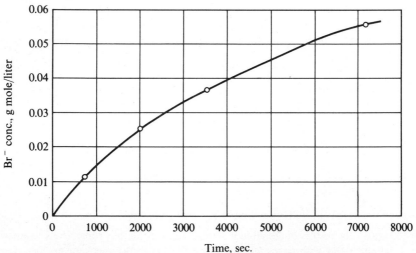

Table 2-4

Concentration, g mole/liter		$r = -d[T]/dt$,
[B]	[T]	g mole/(liter)(sec)
0.0	0.10	1.58×10^{-5}
0.01	0.09	1.38×10^{-5}
0.02	0.08	1.14×10^{-5}
0.03	0.07	0.79×10^{-5}
0.04	0.06	0.64×10^{-5}
0.05	0.05	0.45×10^{-5}

If the reaction is first order, the rate is given by Eq. (A), which may be written

$$\log r = \log k_1 + \log [T] \tag{G}$$

Similarly, if the reaction is second order, from Eq. (D) we have

$$\log r = \log k_2 + \log [T]^2 = \log k_2 + 2 \log [T] \tag{H}$$

For the first-order case $\log r$ plotted against $\log [T]$ should yield a straight line with a slope of 1.0. For the second-order case the result should be a straight line of slope of 2.0, in accordance with Eq. (H). A plot of the data in Table 2-4 is shown in Fig. 2-4. While there is some scattering, the points do suggest a straight line of a slope approximately equal to 2.0. For comparison purposes lines with slopes of both 2.0 and 1.0 have been included on the plot. The equation of the solid line (slope 2.0) is

$$\log r = -2.76 + 2.0 \log [T]$$

Fig. 2-4   Rate vs concentration of trimethylamine

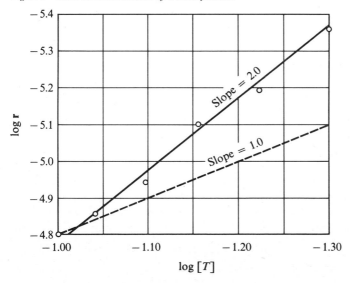

By comparison with Eq. (H),

$$\log k_2 = -2.76$$

$$k_2 = 1.73 \times 10^{-3} \text{ liter/(g mole)(sec)}$$

This value agrees well with the average result 1.67×10^{-3} obtained by the integration method.

Both methods show that the second-order mechanism is preferable. However, the failure of the first-order assumption is perhaps more clearly shown by the differential method than the integration approach. The data in Fig. 2-4 do not approach a slope of 1.0 at all closely, but the k_1 values in Table 2-3 are of the same magnitude, differing from an average value by not more than 17%. This is because the integration process tends to mask small variations.

Third-order reactions are uncommon. Fractional orders exist when the reaction represents a sequence of several elementary steps. Procedures for establishing the order and rate constants for these cases are similar to those given above. Experimental data that suggest fractional-order rate equations should be examined carefully for effects of physical resistances. Sometimes these effects, rather than a sequence of elementary processes, can be responsible for the fractional order. An example is the study of the hydrochlorination of lauryl alcohol with zinc chloride as a homogeneous catalyst:[1]

$$CH_3(CH_2)_{10} CH_2OH(l) + HCl(g) \rightarrow CH_3(CH_2)_{10} CH_2Cl(l)$$

$$+ H_2O(l)$$

The reaction was carried out by dissolving gaseous HCl in a stirred vessel containing the alcohol. The resulting concentration-time data could be correlated with a rate equation half-order in alcohol concentration. However, the rate constant was found to vary with the gas (HCl) flow rate into the reactor, suggesting that the observed rate was influenced by the resistance to diffusion of dissolved HCl in the liquid phase. A method of analysis which took into account the diffusion resistance indicated that the chemical step was probably first order in dissolved HCl and zero order with respect to lauryl alcohol.

2-8 Reversible Reactions

For an *elementary* process the ratio[2] of the forward- and reverse-rate constants is equal to the equilibrium constant, Eq. (2-13). Hence the net rate of reaction can be expressed in terms of one k and the equilibrium constant. Then the integrated form of this rate expression can be used with

[1] H. A. Kingsley and H. Bliss, *Ind. Eng. Chem.*, **44**, 2479 (1952).
[2] See Sec. 2-12 for a discussion of this ratio for nonelementary reactions.

concentration-time data to evaluate k, just as for irreversible reactions. However, the evaluation procedure is more complex.

First Order   If k_1 and k_1' are the forward- and reverse-rate constants for the elementary process

$$A \rightleftarrows B$$

then

$$-\frac{d[A]}{dt} = k_1[A] - k_1'[B] \tag{2-68}$$

The concentration of B can be expressed in terms of $[A]$ by a simple mass balance. At constant density, and since the number of moles is constant, the concentration of B is its initial concentration $[B]_0$ plus the concentration of A that has reacted; that is,

$$[B] = [B]_0 + ([A]_0 - [A]) \tag{2-69}$$

Combining this result with Eq. (2-68) gives

$$-\frac{d[A]}{dt} = (k_1 + k_1')[A] - k_1'([A]_0 + [B]_0) \tag{2-70}$$

At equilibrium the forward and reverse reaction rates are equal, and Eq. (2-68) becomes

$$k_1[A]_{eq} = k_1'[B]_{eq}$$

or

$$\frac{[B]_{eq}}{[A]_{eq}} = K = \frac{k_1}{k_1'} \tag{2-71}$$

where K is the equilibrium constant. Eliminating k_1' from Eq. (2-70) by using Eq. (2-71) yields

$$-\frac{d[A]}{dt} = k_1 \left\{ \frac{K+1}{K}[A] - \frac{1}{K}([A]_0 + [B]_0) \right\} \tag{2-72}$$

Now, applying Eq. (2-69) at equilibrium conditions to find $[B]_{eq}$ and substituting this result in Eq. (2-71), we have

$$K = \frac{[B]_{eq}}{[A]_{eq}} = \frac{[B]_0 + [A]_0 - [A]_{eq}}{[A]_{eq}}$$

or

$$[A]_{eq}(K+1) = [B]_0 + [A]_0 \tag{2-73}$$

With this value of $[B]_0 + [A]_0$ we can express Eq. (2-72) in terms of $[A]$ $- [A]_{eq}$,

$$-\frac{d[A]}{dt} = \frac{k_1(K + 1)}{K}([A] - [A]_{eq})$$

or

$$-\frac{d[A]}{dt} = k_R([A] - [A]_{eq}) \tag{2-74}$$

where

$$k_R = \frac{k_1(K + 1)}{K} \;\; \simeq\; k_1\left(1 + \tfrac{1}{k}\right) \tag{2-75}$$
$$\simeq\; k_1\left(1 + \tfrac{k_1'}{k_1}\right)$$

Written in this form, the rate expression is similar to that for an irreversible first-order reaction. Thus if $[A'] = [A] - [A]_{eq}$, Eq. (2-74) becomes

$$-\frac{d[A']}{dt} = k_R[A'] \tag{2-76}$$

Then, according to Eq. (2-54), the integrated result is

$$\ln\frac{[A] - [A]_{eq}}{[A]_0 - [A]_{eq}} = -k_R t \tag{2-77}$$

If experimental concentration-time data are plotted as $\ln ([A] - [A]_{eq}/([A]_0 - [A]_{eq})$ vs t, the result is a straight line with a slope of $-k_R$. If the equilibrium constant is known, k_1 can be evaluated from Eq. (2-75). Note that $[A]_{eq}$ is determined solely by K and the initial concentrations by means of Eq. (2-73).

Illustrations of first-order reversible reactions are gas-phase cis-trans isomerizations, isomerizations in various types of hydrocarbon systems, and the racemization of α and β glucoses. An example of a catalytic reaction is the ortho-para hydrogen conversion on a nickel catalyst. This reaction is used to illustrate other forms of Eq. (2-74) in the following example.

Example 2-4   The ortho-para hydrogen reaction has been studied at $-196°C$ and constant pressure in a flow reactor, with a nickel-on-Al_2O_3 catalyst.[1] The rate data can be explained with an expression of the form

$$\mathbf{r} = k(y_{eq} - y)_p \tag{A}$$

where y_p is the mole fraction of para hydrogen. Show that this expression follows from the first-order reversible rate equation (2-74).

[1] N. Wakao, P. W. Selwood, and J. M. Smith, *AIChE J.*, **8**, 478 (1962).

Solution   The reaction is

$$o\text{—}H_2 \rightleftharpoons p\text{—}H_2$$

so that $[A]$ in Eq. (2-74) refers to the concentration of ortho hydrogen. Assuming hydrogen is an ideal gas at reaction conditions and that P is the total pressure, we have

$$[A] = \frac{P}{R_g T} y_o \tag{B}$$

Since the system is a binary one, $y_p = 1 - y_o$, and Eq. (B) becomes

$$[A] = \frac{P}{R_g T} (1 - y)_p$$

and

$$[A]_{eq} = \frac{P}{R_g T} (1 - y_{eq})_p$$

Substituting these results in Eq. (2-74) yields

$$-\frac{d[A]}{dt} = \mathbf{r} = k_R \frac{P}{R_g T} (y_{eq} - y)_p = \frac{k_1(K + 1)}{K} \frac{P}{R_g T} (y_{eq} - y)_p \tag{C}$$

Equation (C) is equivalent to Eq. (A), with

$$k = \frac{k_1(K + 1)}{K} \frac{P}{R_g T} \tag{D}$$

Second Order   For a second-order reversible reaction, where the stoichiometry is

$$A + B \rightleftharpoons C + D$$

the rate equation will be

$$-\frac{d[A]}{dt} = k_2 [A][B] - k_2' [C][D] \tag{2-78}$$

If $[X]$ is the amount of A which has reacted, expressed as a concentration, the concentration of A at any time will be

$$[A] = [A]_0 - [X]$$

Rewriting Eq. (2-78) in terms of $[X]$ and replacing k_2' with k_2/K gives

$$\frac{d[X]}{dt} = k_2 \left\{ ([A]_0 - [X])([B]_0 - [X]) - \frac{1}{K} ([C]_0 \right.$$

$$\left. + [X])([D]_0 + [X]) \right\}$$

$$= \alpha + \beta[X] + \gamma[X]^2 \tag{2-79}$$

where

$$\alpha = k_2 \left\{ [A]_0 [B]_0 - \frac{1}{K} [C]_0 [D]_0 \right\} \tag{2-80}$$

$$-\beta = k_2 \left\{ ([A]_0 + [B]_0) + \frac{1}{K} ([C]_0 + [D]_0) \right\} \tag{2-81}$$

$$\gamma = k_2 \left(1 - \frac{1}{K} \right) \tag{2-82}$$

Integration of Eq. (2-79) with the initial condition

$$[A] = [A]_0 \qquad [B] = [B]_0 \qquad [C] = [C]_0 \qquad [D] = [D]_0$$

$$\text{at } t = 0$$

gives

$$\ln \left| \frac{\{2\gamma [X]/(\beta - q^{1/2})\} + 1}{\{2\gamma [X]/(\beta + q^{1/2})\} + 1} \right| = q^{1/2} t \tag{2-83}$$

where

$$q = \beta^2 - 4\alpha\gamma \tag{2-84}$$

In practice it is difficult to use this result with concentration-time data to determine k_2. The solution is a trial-and-error one, requiring successive choices of k_2 and comparison of the data with Eq. (2-83) for each choice to establish the best value. For special initial conditions the equations are simplified so that direct evaluation of k_2 is possible. This case is illustrated in Example 2-5.

Rate and integrated equations for the various forms of irreversible and reversible reactions are summarized in Table 2-5. The complex analysis for reversible reactions, indicated by the complicated equations, can be avoided if measurements are made before much reaction has occurred. Under these conditions the concentrations of the products will be small, making the reverse rate negligible. Then the data can be analyzed as though the system were irreversible to determine the forward-rate constant. With this result and the equilibrium constant, the rate constant for the reverse reaction can be obtained. This *initial-rate* approach is frequently used to simplify kinetic studies. Besides the fact that the reverse reaction is eliminated, the composition of the reaction system is usually known more precisely at the initial state than at subsequent times. This is because compositions at later times are generally evaluated from a limited experimental analysis plus assumptions that certain reactions have occurred. Particularly in complex systems, knowledge of the reactions taking place may not be exact. When the rate equation determined from the concentration depen-

Table 2-5   Rate equations for simple reactions

Reaction	Order	Rate equation	Integrated forms
IRREVERSIBLE REACTIONS			
$A \to B$	Zero	$-\dfrac{d[A]}{dt} = k_0$	$[A] = [A]_0 - k_0 t$
			$t_{1/2} = \dfrac{[A]_0}{2k_0}$
$A \to B$	First	$-\dfrac{d[A]}{dt} = k_1[A]$	$\ln\dfrac{[A]}{[A]_0} = -k_1 t$
			$t_{1/2} = \dfrac{1}{k_1}\ln 2$
$A + A \to P$	Second, type I	$-\dfrac{d[A]}{dt} = k_2[A]^2$	$\dfrac{1}{[A]} - \dfrac{1}{[A]_0} = k_2 t$
			$t_{1/2} = \dfrac{1}{k_2[A]_0}$
$aA + bB \to P$	Second, type II	$-\dfrac{d[A]}{dt} = k_2[A][B]$	$\ln\dfrac{[A]_0 - [X]}{[B]_0 - b/a[X]} = \ln\dfrac{[A]}{[B]}$
			$= \dfrac{b[A]_0 - a[B]_0}{a}k_2 t + \ln\dfrac{[A]_0}{[B]_0}$
			$t_{1/2} = \dfrac{a}{k_2(b[A]_0 - a[B]_0)}$
			$\times \ln\dfrac{a[B]_0}{2a[B]_0 - b[A]_0}$
REVERSIBLE REACTIONS			
$A \rightleftharpoons B$	First $\rightleftharpoons$ first	$-\dfrac{d[A]}{dt} = k_1[A] - k_1'[B]$	$\dfrac{[A] - [A]_{eq}}{[A]_0 - [A]_{eq}} = e^{-k_R t}$
			$k_R = \dfrac{k_1(K + 1)}{K}$
			$[A]_{eq} = \dfrac{[B]_0 + [A]_0}{K + 1}$
$A + B$ $\rightleftharpoons C + D$	Second $\rightleftharpoons$ second	$-\dfrac{d[A]}{dt} = k_2[A][B]$ $\qquad - k_2'[C][D]$	$q^{1/2}t = \ln\dfrac{\{2\gamma[X]/(\beta - q^{1/2})\} + 1}{\{2\gamma[X]/(\beta + q^{1/2})\} + 1}$ $[X] = [A]_0 - [A]$
			$\alpha, \beta, \gamma,$ and q defined by Eqs. (2-80) to (2-82) and (2-84)

dency at initial conditions is different from that established by data taken at different times, some change in the reaction scheme arising from the presence of products is indicated. When the order of a reaction determined from data plotted against time is less than the order at initial conditions, the rate is decreasing less rapidly with time than expected. This suggests that the products of the reaction somehow speed up the rate. Such systems are called *autocatalytic*. Conversely, when the order with respect to time is greater than that established at initial conditions, the products of reaction must inhibit the reaction.[1]

Example 2-5[2]   The reaction between methyl iodide and dimethyl-*p*-toluidine in nitrobenzene solution forms an ionized quaternary ammonium salt. It can be studied kinetically in the same manner as the trimethylamine reaction considered in Example 2-3. The data shown in Table 2-6 were obtained from an initial solution containing methyl iodide and dimethyl-*p*-toluidine in concentrations of 0.05 g mole/liter.

Table 2-6

Run	t, min	Fraction of toluidine reacted
1	10.2	0.175
2	26.5	0.343
3	36.0	0.402
4	78.0	0.523

In view of the results for Example 2-3, and assuming that the equilibrium constant for this reaction is 1.43, what rate equation best fits the experimental data?

Solution   The reaction may be written as

$$CH_3I + N—R \rightleftharpoons CH_3—N^+—R + I^-$$

If it is supposed to be second order and reversible, Eq. (2-78) is applicable. Expressing the rate in terms of the iodide concentration, we have

$$\frac{d[I]}{dt} = k_2[M][T] - k_2'[I][N] \tag{A}$$

The initial concentrations of reactants are equal and those of the products are zero. If $[X]$ in Eq. (2-79) is taken as the iodide concentration, the result for these simplified conditions is

$$\frac{d[I]}{dt} = k_2([M]_0 - [I])^2 - k_2'[I]^2 = k_2\left\{([M]_0 - [I])^2 - \frac{1}{K}[I]^2\right\} \tag{B}$$

where $[M]_0$ is the initial concentration of either reactant.

---

[1] M. Letort, *Bull. Soc. Chim. France*, **9**, 1 (1942).
[2] From K. J. Laidler, "Chemical Kinetics," McGraw-Hill Book Company, New York, 1950.

INTEGRATION METHOD   The solution of Eq. (B) is given by Eqs. (2-83) and (2-84), where $[X] = [I]$ and

$$\alpha = k_2[M]_0^2$$

$$\beta = -2k_2[M]_0$$

$$\gamma = k_2 \frac{K-1}{K}$$

$$q = 4k_2^2[M]_0^2 - 4k_2^2[M]_0^2 \frac{K-1}{K} = \frac{4k_2^2[M]_0^2}{K}$$

Substituting these values in Eq. (2-83) and simplifying gives

$$\frac{2k_2[M]_0}{K^{1/2}} t = \ln \frac{[M]_0 + [I](K^{-1/2} - 1)}{[M]_0 - [I](K^{-1/2} + 1)} \tag{C}$$

The data are given as the fraction x of the toluidine (or, in this example, methyl-iodide) reacted. Hence $[I] = x[M]_0$. Then Eq. (C) becomes

$$k_2 = \frac{K^{1/2}}{2[M]_0 t} \ln \frac{1 + x(K^{-1/2} - 1)}{1 - x(K^{-1/2} + 1)} \tag{D}$$

Substituting numerical values for K and $[M]_0$ gives the following expression relating x and t:

$$k_2 = \frac{1.43^{1/2}}{2(0.05)t} \ln \frac{1 - 0.165x}{1 - 1.835x}$$

This expression and the data in Table 2-6 can be used to calculate a value of k_2 for each run. The results are shown in Table 2-7.

As a matter of interest, the values of k_2 evaluated with the assumption that the reaction is irreversible are shown in the last column of the table. They were computed from Eq. (2-59); for example,

$$k_2 t = \frac{[I]}{[M]_0([M]_0 - [I])}$$

The steady trend in the values of k_2 indicates that the irreversible assumption is a poor one.

Table 2-7

Run	θ, sec	k_2, liters/(g mole)(sec)	k_2 (neglecting reverse reaction)
1	612	7.05×10^{-3}	6.93×10^{-3}
2	1,590	7.06×10^{-3}	6.57×10^{-3}
3	2,160	7.06×10^{-3}	6.23×10^{-3}
4	4,680	7.97×10^{-3}	4.68×10^{-3}

In studies of rate data trends in k are of more significance than random variations. The former suggest that the assumed order is open to question, while the latter suggest errors in the experimental data. Of course, if the precision of the data is poor, random variations can mask trends in computed k values.

DIFFERENTIAL METHOD   Replacing $1/K$ and $[M]_0$ with their numerical values and taking logarithms of the rate equation (B) yields

$$\log \mathbf{r} = \log \frac{d[\mathrm{I}]}{dt} = \log k_2 + \log\left\{ (0.05 - [\mathrm{I}])^2 - 0.70[\mathrm{I}]^2 \right\} \tag{E}$$

According to Eq. (E), a plot on logarithmic coordinates of the rate vs $(0.05 - [\mathrm{I}])^2 - 0.70[\mathrm{I}]^2$ should yield a straight line of slope equal to unity.

Figure 2-5 shows a plot of $[\mathrm{I}]$ vs t. Slopes of this curve give the rate values shown in Table 2-8. Also shown is a plot of Eq. (E). Observe that the first four points establish a line with a slope close to unity, as required by the second-order mechanism. The last point deviates from the line, just as the k value for this point obtained by the integration method was not in agreement with the other values.

Example 2-6   The interpretation of kinetic data for gaseous reactions is similar to that for liquid systems. The analysis for a reversible case is well illustrated by the vapor-phase decomposition of hydrogen iodide,

$$2\mathrm{HI} \rightarrow \mathrm{H}_2 + \mathrm{I}_2$$

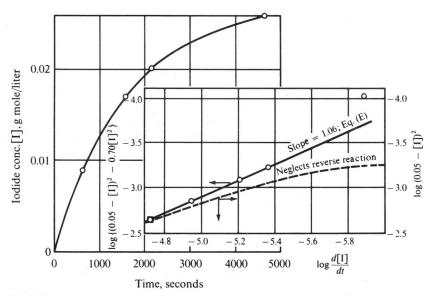

Fig. 2-5

Table 2-8

t, sec	$[I]$ g mole/liter	$\dfrac{d[I]}{dt}$	$\log \dfrac{d[I]}{dt}$	$(0.05 - [I])^2$ $- 0.70[I]^2$	$\log\{(0.05 - [I])^2$ $- 0.70[I]^2\}$
0	0	1.93×10^{-5}	-4.71	25.0×10^{-4}	-2.60
612	0.00875	1.12×10^{-5}	-4.95	16.5×10^{-4}	-2.78
1,590	0.0171	0.62×10^{-5}	-5.20	8.75×10^{-4}	-3.05
2,160	0.0201	0.42×10^{-5}	-5.37	6.13×10^{-4}	-3.21
4,680	0.0261	0.13×10^{-5}	-5.89	0.96×10^{-4}	-4.01

This reaction has been carefully studied by a number of investigators,[1] and it is generally considered one of the most certain examples of a second-order reaction, at least at low pressures. The equilibrium values of the fraction x of HI decomposed can be accurately represented by Bodenstein's equation,

$$x_{eq} = 0.1376 + 7.22 \times 10^{-5}t + 2.576 \times 10^{-7}t^2 \qquad t = {}^\circ C$$

Kistiakowsky used a static experimental method to study the reaction. Pure hydrogen iodide was sealed in glass bulbs, immersed in a constant-temperature bath for various time intervals, and then removed and cooled, and the contents were analyzed for all three chemical species. The initial pressure (and hence the initial concentration) of HI and the size of the reaction bulb were varied over a wide range. The data obtained at an average temperature of 321.4°C are given in Table 2-9.

(a) From this information estimate the specific-reaction-rate constants [liters/ (g mole)(sec)] for the forward and reverse reactions, both of which may be taken as second order. (b) What would be the values of the specific-reaction-rate constants in units of g moles/(liter)(sec)(atm)^2? (c) Could the course of this reaction be followed by measuring the total pressure in the reaction vessel at various times?

Solution This reaction system is somewhat simpler than the second-order reversible system whose solution is given by Eq. (2-83), since the concentrations of products are very low and there is only one reactant. The extent of reaction, noted by the percentage of HI decomposed, is always low. The series of runs constitute *initial rate* data, with each run corresponding to a different concentration of HI.

If the rate is followed by the concentration of iodine, Eq. (2-79) becomes

$$\frac{d[I_2]}{dt} = k_2([HI]_0 - 2[I_2])^2 - k_2'[I]^2 = k_2\left\{([HI]_0 - 2[I_2])^2 - \frac{1}{K}[I_2]^2\right\}$$

where $[HI]_0$ is the initial concentration of HI and the second equality indicates the assumption that the reaction is an elementary one.

[1] M. Bodenstein, *Z. Physik Chem.*, **13**, 56 (1894); *Z. Physik Chem.*, **22**, 1 (1897); *Z. Physik Chem.*, **29**, 295 (1898); H. A. Taylor, *J. Phys. Chem.*, **28**, 984 (1924); G. B. Kistiakowsky, *J. Am. Chem. Soc.*, **50**, 2315 (1928).

Table 2-9

Run	t, sec	% of HI decomposed	Volume of reaction bulb, cm^3	$[HI]_0$, g moles/liter
1	82,800	0.826	51.38	0.02339
2	172,800	2.567	59.80	0.03838
3	180,000	3.286	51.38	0.04333
4	173,100	3.208	51.38	0.04474
5	81,000	2.942	7.899	0.1027
6	57,560	2.670	7.899	0.1126
7	61,320	4.499	7.899	0.1912
8	19,200	2.308	7.899	0.3115
9	18,000	2.202	7.899	0.3199
10	16,800	2.071	7.899	0.3279
11	17,400	2.342	7.899	0.3464
12	17,700	2.636	7.899	0.4075
13	18,000	2.587	7.899	0.4228
14	23,400	4.343	7.899	0.4736
15	6,000	2.224	3.28	0.9344
16	5,400	1.903	0.778	0.9381
17	8,160	3.326	0.781	1.138
19	5,400	2.741	0.713	1.231

(*a*) If the rate equation is integrated, with the initial condition $[I_2] = 0$ at $t = 0$, the result is very similar to Eq. (C) of the previous example:

$$\frac{2k_2[HI]_0}{K^{\frac{1}{2}}} t = \ln \frac{[HI]_0 + [I_2](K^{-\frac{1}{2}} - 2)}{[HI]_0 - [I_2](K^{-\frac{1}{2}} + 2)} \tag{A}$$

The equilibrium constant is related to the concentrations at equilibrium by

$$K = \left(\frac{[I_2][H_2]}{[HI]^2} \right)_{eq}$$

If the fraction of HI decomposed is x, then

$$[I_2] = [H_2] = \frac{1}{2}[HI]_0 x$$
$$[HI] = [HI]_0(1 - x)$$

Hence

$$K = \frac{1}{4} \frac{x_{eq}^2}{(1 - x_{eq})^2}$$

From the expression given for x_{eq} at 321.4°C,

$$x_{eq} = 0.1376 + 7.221 \times 10^{-5}(321.4) + 2.576 \times 10^{-7}(321.4)^2 = 0.1873$$

Hence

$$K = \frac{1}{4} \frac{(0.1873)^2}{(1 - 0.1873)^2} = 0.0133$$

Substituting this value of K in Eq. (A) and introducing the fraction decomposed, we have

$$k_2 = \frac{1}{2(8.67)[HI]_0 t} \ln \frac{[HI]_0 + \frac{1}{2}[HI]_0(8.67 - 2)x}{[HI]_0 - \frac{1}{2}[HI]_0(8.67 + 2)x}$$

or

$$k_2 = \frac{1}{2(8.67)[HI]_0 t} \ln \frac{1 + 3.335x}{1 - 5.335x} \qquad (B)$$

The experimental data for x can be used directly in Eq. (B) to compute values of the specific reaction rate k_2. However, another form of the expression is more useful when the x values are very low, as in this case (the maximum value of x is 0.04499 for run 7). Equation (B) may be written in the form

$$k_2 = \frac{1}{2(8.67)[HI]_0 t} \ln \left(1 + \frac{8.67x}{1 - 5.335x}\right)$$

Expanding the logarithmic term in a power series and retaining the first two terms gives

$$k_2 = \frac{1}{2(8.67)[HI]_0 t} \left[\frac{8.67x}{1 - 5.335x} - \frac{1}{2}\left(\frac{8.67x}{1 - 5.335x}\right)^2\right] \qquad (C)$$

The use of Eq. (C) may be illustrated with run 1:

$$k_2 = \frac{1}{2(8.67)(0.02339)(82,800)}$$

$$\times \left\{\frac{8.67(0.00826)}{1 - 5.335(0.00826)} - \frac{1}{2}\frac{[(8.67)(0.00826)]^2}{[(1 - 5.335)(0.00826)]^2}\right\}$$

$$= 2.97 \times 10^{-5}(0.0749 - 0.0028)$$

$$= 2.14 \times 10^{-6} \text{ liter/(g mole)(sec)}$$

The results for the other runs are summarized in Table 2-10. The average value of k_2 is 1.99×10^{-6}. For the reverse reaction.

$$k_2' = \frac{k_2}{K_1} = \frac{1.99 \times 10^{-6}}{0.0133} = 1.50 \times 10^{-4} \text{ liter/(g mole)(sec)}$$

(b) The rate equations could also have been written in terms of partial pressures. These are proportional to the concentrations in a gas-phase reaction, provided the gas mixture follows perfect-gas behavior. In terms of pressures the net rate may be written

$$\frac{d[I]}{dt} = k_p p_{HI}^2 - k_p' p_{I_2} p_{H_2}$$

Table 2-10

Run	Conversion	$k_2 \times 10^6$, liters/(sec)(g mole)
1	0.00826	2.14
2	0.02567	2.01
3	0.03286	2.20
4	0.03208	2.17
5	0.02942	1.92
6	0.02670	2.08
7	0.04499	2.04
8	0.02308	1.99
9	0.0202	1.80
10	0.02071	1.77
11	0.02342	2.00
12	0.02636	1.90
13	0.02587	1.75
14	0.04343	2.08
15	0.02224	2.05
16	0.01903	1.93
17	0.03326	1.87
19	0.02741	2.15

According to the perfect-gas law, the partial pressure of I_2, for example, is given by the expression

$$p_{I_2} = \frac{N_i R_g T}{V} = [I_2] R_g T$$

Substituting partial-pressure relationships in the rate equation, we have

$$\frac{d[I]}{dt} = k_p (R_g T)^2 [HI]^2 - k'_p (R_g T)^2 [I_2][H_2]$$

Comparison with the original rate equation shows that k_p is related to k_2 as follows:

$$k_p = \frac{k_2}{(R_g T)^2} = \frac{1.99 \times 10^{-6}}{0.082^2 (273.1 + 321.4)^2}$$

$$= 0.84 \times 10^{-9} \text{ g mole/(liter)(atm)}^2 (\text{sec})$$

and

$$k'_p = \frac{k'_2}{(R_g T)^2} = 6.40 \times 10^{-8} \text{ g mole/(liter)(atm)}^2 (\text{sec})$$

Although no pressures are given in the tabulated data, approximate values can be computed from the perfect-gas law. Thus the total pressure will be given by

$$p_t = \frac{N_t R_g T}{V}$$

Initially the only substance present is HI, so that N_t/V represents the initial concentration of HI. For run 1

$$p_t = [\text{HI}]_0 R_g T = 0.02339(0.082)(594.5) = 1.14 \text{ atm}$$

The highest pressure will be for run 19, where the initial concentration is the greatest. In this case

$$p_t = 1.2310(0.082)(594.5) = 60 \text{ atm}$$

(c) In this reaction the number of moles does not change with reaction. Hence the total pressure does not change with the extent of the reaction.

2-9 The Total-pressure Method of Studying Kinetics

The treatment in this chapter has been limited to nonflow systems at constant temperature and constant volume so that the rate can be represented by the rate of change of concentration of a reactant or product. With these restrictions there will be little change in the total pressure for a gaseous reaction unless the total number of moles changes.[1] However, if there is an increase in number of moles as the reaction proceeds, there will be an increase in pressure. This increase is uniquely related to the extent of reaction. Hence measuring the total pressure as a function of time is a suitable method for studying the kinetics of such a system and has been widely used.

As an illustration of the method, consider the reaction

$$2\text{NO}_2 \rightarrow \text{N}_2\text{O}_4$$

In this instance the pressure decreases, and this change will depend on the extent of the reaction. If the initial number of moles of NO_2 is N_0 (initial concentration $[\text{NO}_2]_0$) and no N_2O_4 is present, the total moles at any time is

$$N_t = N_{\text{N}_2\text{O}_4} + N_{\text{NO}_2} = N_{\text{N}_2\text{O}_4} + N_0 - 2N_{\text{N}_2\text{O}_4} = N_0 - N_{\text{N}_2\text{O}_4}$$

Then the total pressure will be

$$p_t = \frac{N_t R_g T}{V} = \frac{N_0 - N_{\text{N}_2\text{O}_4}}{V} R_g T \tag{2-85}$$

If the initial total pressure is p_0,

$$p_0 = \frac{N_0 R_g T}{V}$$

---

[1] Any change in pressure for a reaction with no change in number of moles would be due to deviations from ideal-gas behavior. If the deviation of the reaction mixture changes with composition, a change in total pressure with extent of reaction would occur, but it would be relatively small.

Hence the expression for p_t may be written in terms of p_0 and the concentration of N_2O_4 as

$$p_t = p_0 - \frac{N_{N_2O_4}}{V} R_g T = p_0 - [N_2O_4] R_g T \tag{2-86}$$

To compute the rate constant from total-pressure measurements we first write the rate equation in the usual way,

$$\frac{d[N_2O_4]}{dt} = k_2[NO_2]^2 = k_2([NO_2]_0 - 2[N_2O_4])^2 \tag{2-87}$$

Then Eq. (2-86) is used to replace the concentration of N_2O_4 with the total pressure. For example, differentiation of Eq. (2-86) yields

$$\frac{d[N_2O_4]}{dt} = -\frac{1}{R_g T} \frac{dp_t}{dt}$$

Also,

$$[NO_2]_0 = \frac{p_0}{R_g T}$$

$$[N_2O_4] = \frac{p_0 - p_t}{R_g T}$$

Substituting these equalities in Eq. (2-87), we have

$$\mathbf{r} = \frac{d[N_2O_4]}{dt} = -\frac{1}{R_g T} \frac{dp_t}{dt} = k_2 \left(\frac{p_0}{RT} - 2\frac{p_0 - p_t}{RT} \right)^2$$

or

$$-\frac{dp_t}{dt} = \frac{k_2}{R_g T} (2p_t - p_0)^2$$

This differential equation can be integrated, using the initial condition $p_t = p_0$ at $t = 0$, to give

$$\frac{1}{2} \left(\frac{1}{2p_t - p_0} - \frac{1}{p_0} \right) = \frac{k_2}{R_g T} (t - 0)$$

or

$$k_2 = \frac{R_g T}{2t} \left(\frac{1}{2p_t - p_0} - \frac{1}{p_0} \right) \tag{2-88}$$

The total pressure-vs-time data can be used directly in Eq. (2-88) to evaluate k_2.

ANALYSIS OF COMPLEX RATE EQUATIONS

A *complex* system is one in which more than one reaction occurs. This can lead to multiple products, some of which are more desirable than others from a practical standpoint. For example, in the air oxidation of ethylene the desired product is ethylene oxide, but complete oxidation to carbon dioxide and water also occurs. The important performance factor is the production rate of ethylene oxide and its purity in the reaction products, rather than the total amount of ethylene reacted. To characterize this performance two parameters are used: yield and selectivity. The *yield* of a specific product is defined as the fraction of reactant converted to that product. The *point selectivity* is the ratio of the rate of production of one product to the rate for another product. With multiple products there is a separate selectivity based on each pair of products. The *overall*, or integrated, *selectivity* is the ratio of the amount of one product produced to the amount of another. Selectivity and yield are related to each other through the total conversion, i.e., the total fraction of reactant converted to all products.

As an illustration of these terms consider the simultaneous reaction system

$$A \xrightarrow{k_1} B \tag{2-89}$$
$$ \searrow^{k_2}$$
$$ C$$

Suppose the total conversion of A is x_t, consisting of the fraction x_B of reactant A converted to B and the fraction x_C converted to C. The *yield* of B is simply x_B and that of C is x_C. The amount of a product produced is proportional to the yield. Hence the *overall selectivity* of B is the ratio of the yields of B and C,

$$S_o = \frac{x_B}{x_C} \tag{2-90}$$

If both reactions are first order and irreversible, the *point selectivity* is

$$S_p = \frac{d[B]/dt}{d[C]/dt} = \frac{k_1[A]}{k_2[A]} = \frac{k_1}{k_2} \tag{2-91}$$

Under the restriction of constant density, the amount of a product produced is proportional to its concentration. Thus the overall selectivity can also be written as

$$S_o = \frac{[B]}{[C]} \tag{2-92}$$

The simple form of Eq. (2-91) shows that selectivity and yield calculations can advantageously be carried out by dividing the rate for one reaction by that for another, eliminating time in the process. Since yield and selectivity are usually more important than total conversion for complex-reaction systems, this procedure will be emphasized in the following section. The possible combinations of simultaneous, parallel, and consecutive reactions are very large. A few irreversible first-order cases will be analyzed in Sec. 2-10 to illustrate the method of approach. Then in Sec. 2-11 a different type of complex system, chain reactions, will be discussed.

2-10 First-order Complex Reactions

Consider first the *simultaneous* reactions described by Eq. (2-89). The rates of formation of the components are given by the following three equations:

$$-\frac{d[A]}{dt} = (k_1 + k_2)[A] \tag{2-93}$$

$$\frac{d[B]}{dt} = k_1[A] \tag{2-94}$$

$$\frac{d[C]}{dt} = k_2[A] \tag{2-95}$$

Our goal is to integrate these equations to establish $[B]$ and $[C]$ for any $[A]$. Then all yields and selectivities can be obtained from equations like (2-90). In this simple example Eq. (2-93) can be immediately integrated to give $[A] = f(t)$. This result, used in Eqs. (2-94) and (2-95), allows these two equations to be integrated, so that $[B]$ and $[C]$ are also known as a function of time. However, in more complicated cases (see Example 2-8) integrating with respect to time is not as easy. Therefore the solution will be obtained by dividing Eqs. (2-94) and (2-95) by Eq. (2-93), to eliminate time:

$$\frac{d[B]}{d[A]} = -\frac{k_1}{k_1 + k_2} \tag{2-96}$$

$$\frac{d[C]}{d[A]} = -\frac{k_2}{k_1 + k_2} \tag{2-97}$$

If Eqs. (2-96) and (2-97) are integrated with the conditions that at $t = 0$, $[A] = [A]_0$, and $[B] = [C] = 0$, then the yields of B and C are

$$x_B = \frac{[B]}{[A]_0} = \frac{k_1}{k_1 + k_2}\left(1 - \frac{[A]}{[A]_0}\right) = \frac{k_1}{k_1 + k_2}x_t \tag{2-98}$$

$$x_C = \frac{[C]}{[A]_0} = \frac{k_2}{k_1 + k_2} \left(1 - \frac{[A]}{[A]_0}\right) = \frac{k_2}{k_1 + k_2} x_t \qquad (2\text{-}99)$$

where x_t is the total conversion of A to B plus C.

From Eqs. (2-98) and (2-99), the overall selectivity is

$$S_o = \frac{x_B}{x_C} = \frac{k_1}{k_2} \qquad (2\text{-}100)$$

The point selectivity, given by Eq. (2-91) for first-order simultaneous reactions, is also equal to k_1/k_2. Although point and overall selectivities are identical for this type of first-order system, the two selectivities differ for most complex reactions.

Next let us consider a *consecutive* set of reactions,

$$A \xrightarrow{k_1} B \xrightarrow{k_3} D$$

and take $[B] = [D] = 0$ and $[A] = [A]_0$ at $t = 0$. The rates are

$$\frac{d[A]}{dt} = -k_1[A] \qquad (2\text{-}101)$$

$$\frac{d[B]}{dt} = k_1[A] - k_3[B] \qquad (2\text{-}102)$$

$$\frac{d[D]}{dt} = k_3[B] \qquad (2\text{-}103)$$

Dividing Eqs. (2-102) and (2-103) by Eq. (2-101) yields

$$\frac{d[B]}{d[A]} = -1 + \frac{k_3[B]}{k_1[A]} \qquad (2\text{-}104)$$

$$\frac{d[D]}{d[A]} = -\frac{k_3[B]}{k_1[A]} \qquad (2\text{-}105)$$

Since Eq. (2-104) is a linear first-order differential equation, it has an analytic solution. With the stated initial condition the result can be expressed in terms of the yield of B,

$$x_B = \frac{[B]}{[A]_0} = \frac{k_1}{k_1 - k_3} \left\{ \left(\frac{[A]}{[A]_0}\right)^{k_3/k_1} - \frac{[A]}{[A]_0} \right\} \qquad (2\text{-}106)$$

This expression for $[B]$ can be substituted in Eq. (2-105). Then integration of Eq. (2-105) gives

$$x_D = \frac{[D]}{[A]_0} = \frac{k_1}{k_1 - k_3}\left\{1 - \left(\frac{[A]}{[A]_0}\right)^{k_3/k_1}\right\} - \frac{k_3}{k_1 - k_3}\left(1 - \frac{[A]}{[A]_0}\right)$$

(2-107)

From these two expressions the overall selectivity x_B/x_D is seen to depend on the fraction unconverted, $[A]/[A]_0$, as well as on the rate constants. This means that the yield of B and the overall selectivity will vary with time. This is in contrast to the result for simultaneous reactions, Eq. (2-100).

Example 2-7 If $[B] = [D] = 0$ initially for the consecutive reaction system described by Eqs. (2-101) to (2-103), what is the time at which the yield of B is a maximum? What is the maximum yield?

Solution   Equation (2-106) gives the yield of B in terms of $[A]/[A]_0$. This ratio can be expressed as a function of time by integrating Eq. (2-101). The result is

$$\frac{[A]}{[A]_0} = e^{-k_1 t}$$

(A)

Then Eq. (2-106) becomes

$$x_B = \frac{k_1}{k_1 - k_3}(e^{-k_3 t} - e^{-k_1 t})$$

(B)

To obtain the maximum value of x_B we differentiate Eq. (B) with respect to time and set the derivative equal to zero:

$$\frac{dx_B}{dt} = 0 = \frac{k_1}{k_1 - k_3}(-k_3 e^{-k_3 t} + k_1 e^{-k_1 t})$$

$$t_{\max B} = \frac{\ln(k_1/k_3)}{k_1 - k_3}$$

(C)

Substituting this value for the time in Eq. (B) gives the maximum yield,

$$(x_B)_{\max} = \left(\frac{k_1}{k_3}\right)^{k_3/(k_3 - k_1)}$$

(D)

The form of the concentration-vs-time curves are shown in Fig. 2-6.

Solutions of assemblies of ordinary differential equations with time as the independent variable are ideally suited for solution by analog computation. Hence complex kinetics equations of the type considered in this section may conveniently be solved with an analog computer. This is illustrated in the problems at the end of this chapter.

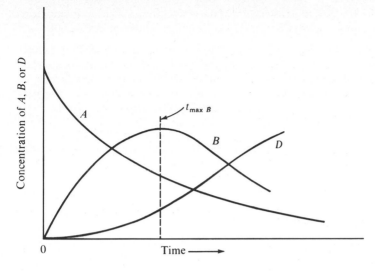

Fig. 2-6   *Concentration vs time for consecutive-reaction system* $A \rightarrow B \rightarrow$ *D for* $B = D = 0$ *at* $t = 0$

Example 2-8   Benzene is chlorinated in the liquid phase in a kettle-type reactor operated on a semibatch basis; i.e., the reactor is initially charged with liquid benzene, and then chlorine gas is bubbled into the well-agitated solution. The reactor is equipped with a reflux condenser which will condense the benzene and chlorinated products but will not interfere with the removal of hydrogen chloride. Assume that the chlorine is added sufficiently slowly that (1) the chlorine and hydrogen chloride concentrations in the liquid phase are small and (2) all the chlorine reacts.

At the constant operating temperature of 55°C the significant reactions are the three substitution ones leading to mono-, di-, and trichlorobenzene.

1. $C_6H_6 + Cl_2 \overset{k_1}{\rightarrow} C_6H_5Cl + HCl$

2. $C_6H_5Cl + Cl_2 \overset{k_2}{\rightarrow} C_6H_4Cl_2 + HCl$

3. $C_6H_4Cl_2 + Cl_2 \overset{k_3}{\rightarrow} C_6H_3Cl_3 + HCl$

MacMullin[1] found the ratios of the rate constants to have the following values at 55°C:

$$\frac{k_1}{k_2} = 8.0 \qquad \frac{k_2}{k_3} = 30$$

[1] R. B. MacMullin, *Chem. Eng. Progr.*, **44**, 183 (1948).

Find the yields of each product as a function of the moles of chlorine added per mole of benzene charged to the reactor.[1] The holdup in the reflux condenser is negligible.

Solution[2]   The process described is neither flow nor batch, but semibatch in nature. However, with assumptions which are reasonably valid, the problem can be reduced to that for a constant-density batch reactor. If the density of the solution remains constant and the hydrogen chloride vaporizes and leaves the solution, the volume of the liquid-phase reaction will be constant. Then the relationship between the composition of the substances in the liquid phase is governed by rate expressions of the type used in this chapter. Assume that the reactions are second order. Then the rate of disappearance of benzene, determined entirely by the first reaction, is

$$-\frac{d[\mathrm{B}]}{dt} = k_1[\mathrm{B}][\mathrm{Cl}_2] \tag{A}$$

Similarly the net rates of formation of the mono-, di-, and trichlorobenzenes are

$$\frac{d[M]}{dt} = k_1[\mathrm{B}][\mathrm{Cl}_2] - k_2[M][\mathrm{Cl}_2] \tag{B}$$

$$\frac{d[D]}{dt} = k_2[M][\mathrm{Cl}_2] - k_3[D][\mathrm{Cl}_2] \tag{C}$$

$$\frac{d[T]}{dt} = k_3[D][\mathrm{Cl}_2] \tag{D}$$

These four rate equations, along with the mass balance, can be solved for the desired yields of the products (in terms of the amount of benzene reacted) by eliminating time as a variable. The expressions cannot be solved directly to give compositions as a function of time because the magnitudes of the individual rate constants are not known (only their ratios are known). Although the rate equations are second order, the chlorine concentration appears in all the expressions and cancels out. Hence the reaction system is equivalent to the consecutive first-order system already considered, except that three reactions are involved.

If Eq. (B) is divided by Eq. (A), we obtain

$$\frac{d[M]}{d[\mathrm{B}]} = -1 + \frac{k_2}{k_1}\frac{[M]}{[\mathrm{B}]} \tag{E}$$

This is similar to Eq. (2-104), and so its solution can be written immediately by reference to Eq. (2-106). Thus the yield of monochlorobenzene is

$$x_M = \frac{[M]}{[\mathrm{B}]_0} = \frac{k_1}{k_1 - k_2}\left\{\left(\frac{[\mathrm{B}]}{[\mathrm{B}]_0}\right)^{k_2/k_1} - \frac{[\mathrm{B}]}{[\mathrm{B}]_0}\right\}$$

[1] From the standpoint of determining rate equations from experimental data it would be more appropriate to reverse this example, i.e., to require the evaluation of the ratios of the rate constants from given composition curves. Actually, the calculations are essentially the same in both cases.

[2] This problem was originally solved by MacMullin in a somewhat different manner.

or

$$x_M = [M] = \frac{[B]}{1 - \alpha} ([B]^{\alpha - 1} - 1) \tag{F}$$

where $\alpha = k_2/k_1$.

Similarly, Eq. (C) can be divided by Eq. (A) to give

$$\frac{d[D]}{d[B]} = -\alpha \frac{[M]}{[B]} + \beta \frac{[D]}{[B]} \tag{G}$$

where $\beta = k_3/k_1$. Equation (F) can be used in Eq. (G) to replace $[M]$ with a function of $[B]$. Then another linear first-order differential equation in $[D]$ results. Integrating this, and noting that $[D]_0 = 0$ when $[B]_0 = 1$, gives the yield of D,

$$x_D = \frac{[D]}{[B]_0} = [D] = \frac{\alpha}{1 - \alpha} \left(\frac{[B]}{1 - \beta} - \frac{[B]^\alpha}{\alpha - \beta} \right) + \frac{\alpha[B]^\beta}{(\alpha - \beta)(1 - \beta)} \tag{H}$$

The concentration of trichlorobenzene can be determined by difference, with a mass balance of aromatic components. Since the initial concentration of benzene is 1.0, and 1 mole of each chlorobenzene is produced per mole of benzene, the total molal concentration is constant and equal to unity. Hence

$$1.0 = [B] + [M] + [D] + [T] \tag{I}$$

Equations (F), (H), and (I) give the concentrations of $[M]$, $[D]$, and $[T]$ in terms of $[B]$. The corresponding amount of chlorine added can be determined from a mass balance on chlorine. If $[Cl_2]$ represents the total moles of chlorine added (or reacted) per mole of benzene, then

$$[Cl_2] = [M] + 2[D] + 3[T] \tag{J}$$

As an illustration of the numerical calculations choose the point at which one-half the benzene has reacted. Then $[B] = 0.5$. It is given that

$$\alpha = \frac{1}{8} = 0.125$$

$$\beta = \frac{k_3}{k_1} = \frac{k_3}{k_2} \frac{k_2}{k_1} = \frac{1}{30} \left(\frac{1}{8} \right) = 0.00417$$

Then in Eq. (F)

$$x_M = [M] = \frac{1}{0 - 0.125} (0.5^{0.125} - 0.5) = 0.477$$

Equations (H) and (I) give

$$x_D = [D] = \frac{0.125}{1 - 0.125}\left(\frac{0.5}{1 - 0.00417} - \frac{0.50^{0.125}}{0.125 - 0.00417}\right)$$

$$+ \frac{0.125(0.50)^{0.00417}}{(0.125 - 0.00417)(1 - 0.00417)} = 0.022$$

$$[T] = 1 - [B] - [M] - [D] = 1 - 0.50 - 0.477 - 0.022 = 0.001$$

Finally, from Eq. (J),

$$[Cl_2] = 0.477 + 2(0.022) + 3(0.001) = 0.524$$

Hence, with 0.524 mole of chlorine reacted per mole of benzene, most of the product is monochlorobenzene, with a little dichlorobenzene, and a negligible quantity of trichlorobenzene.

To obtain the composition at a much later time we choose $[B] = 0.001$. Then, proceeding in the same manner, we find most of the product to be mono- and dichlorobenzene, with a little trichlorobenzene substituted product. The results for a range of $[Cl_2]$ up to 2.14 are summarized in Table 2-11. Note that the maximum yield of monochlorinated product is obtained when approximately 1 mole of Cl_2 has been reacted, and the maximum yield of dichlorinated product results when about 2 moles of Cl_2 have been reacted. Selectivities for any two products can easily be found by taking ratios of the yields, Eq. (2-90). Since the problem has been solved on a basis of 1 mole of initial benzene, $[M]$, $[D]$, $[T]$ also are equivalent to mole fractions.

Table 2-11   Composition of chlorinated benzenes

Compound	Mole fractions, or yields per mole of benzene							
	1.0	0.50	0.10	0.01	0.001	10^{-4}	10^{-10}	10^{-20}
Monochlorobenzene	0	0.477	0.745	0.632	0.482	0.362	0.064	0.004
Dichlorobenzene	0	0.022	0.152	0.353	0.509	0.625	0.877	0.852
Trichlorobenzene	0	0.001	0.003	0.005	0.008	0.013	0.059	0.144
Total	1.000	1.000	1.000	1.000	1.000	1.000	1.000	1.000
Moles of chlorine used per mole of original benzene	0	0.524	1.06	1.35	1.52	1.65	1.99	2.14

2-11 Chain Reactions

Chain mechanisms are a significant class of complex reactions since they explain many photochemical, combustion, and polymerization processes.[1]

[1] For more complete discussions of chain reactions see A. A. Frost and R. G. Pearson "Kinetics and Mechanisms," John Wiley & Sons, Inc., New York, 1961; S. A. Benson, "Foundations of Chemical Kinetics," McGraw-Hill Book Company, New York, 1960; S. W. Benson, *Ind. Eng. Chem.*, **56**, 18 (1964).

Chain reactions consist of three kinds of elementary steps: initiation (or activation), propagation, and termination. As an illustration, consider the chlorination of propane (PrH). The experimental evidence suggests the following sequence:

INITIATION

$$Cl_2 \overset{k_1}{\to} 2Cl$$

PROPAGATION

$$Cl + PrH \overset{k_2}{\to} Pr + HCl$$

$$Pr + Cl_2 \overset{k_3}{\to} PrCl + Cl$$

TERMINATION

$$\left.\begin{array}{l} Cl + Cl \overset{k_4}{\to} Cl_2 \\[1em] Cl + Pr \overset{k_5}{\to} PrCl \end{array}\right\} \text{homogeneous}$$

$$\left.\begin{array}{l} Cl + W \overset{k_6}{\to} \quad \text{end product} \\[1em] Pr + W \overset{k_7}{\to} \quad \text{end product} \end{array}\right\} \text{heterogeneous}$$

In the initiation step the activated chlorine atoms can be obtained from intermolecular collisions of molecules heated to a high temperature (thermal activation). Alternately, the chlorine molecule can absorb radiant energy of the proper wavelength and dissociate (photochemical activation). The propagation steps are generally reactions which produce unstable intermediates (the Pr radical in our illustration) and also regenerate the activated reactant (Cl). These two reactions produce stable products (PrCl, HCl) and continuously regenerate activated reactant. Pr and Cl are *chain carriers*. Only a small amount of initiation is required to start the chain and lead to large rates of product formation. Hence chain reactions may be very fast. The overall rate is reduced by the termination processes, which convert the unstable intermediate and activated reactant into stable, end products. Termination may be homogeneous (occurring by intermolecular collisions) or heterogeneous (occurring by collision with the wall W, of the reactor).

If the propagation steps are rapid with respect to the termination process, the overall rate of product formation can be very large. When the propagation steps produces *two* chain carriers for every one consumed, the extra radical may cause further propagation or it may be destroyed by termination processes. If it is not destroyed, the growth in propagation (and product formation) approaches infinity, resulting in an explosion. The oxidation of hydrogen is an example. After initiation produces H, this may react with O_2 producing *two* chain carriers, OH and O. Both chain carriers then propagate further to form product and regenerate themselves. Thus the sequence of reactions is

1. $H + O_2 \rightarrow OH + O$

2. $OH + H_2 \rightarrow H_2O + H$

3. $O + H_2 \rightarrow OH + H$

Reactions such as reaction 1 are called *chain-branching steps*.

Polymerization does not occur by a normal chain-reaction sequence, since the activated reactant is not regenerated. There is no chain carrier. Nevertheless, the overall process can be effectively analyzed in many instances as a combination of initiation, propagation, and termination steps. For example, suppose P_r represents a reactive polymer containing r molecules of monomer and M_{r+n} represents an inactive polymer of $r + n$ molecules of monomer. The polymerization process might then be described by the following reactions, in which M is the monomer feed and P_1 an activated form of the monomer:

INITIATION

$$M \rightarrow P_1$$

PROPAGATION

$$M + P_1 \rightarrow P_2$$

$$M + P_2 \rightarrow P_3$$

$$\cdots\cdots\cdots\cdots$$

$$M + P_{r-1} \rightarrow P_r$$

TERMINATION

$$P_r + P_n \rightarrow M_{n+r}$$

$$P_{r-1} + P_n \rightarrow M_{r-1+n}$$

$$\cdots\cdots\cdots\cdots\cdots$$

The kinetics problems of interest in chain reactions, as in all complex systems, are to predict the conversion and product distribution as a function of time from the rate equations for the individual reactions, or to decide on the reactions involved and evaluate their rate constants from experimental data on the conversion and product distribution. The methods are illustrated with a photochemical reaction in Example (2-9). Polymerization will be discussed in more detail in Chap. 4, after we have considered the nature of reactor performance.

Example 2-9 Experimental measurements for the photochlorination of propane at 25°C and 1 atm pressure showed that the rate of consumption of Cl_2 was independent of propane, second order in chlorine, and first order in light intensity.[1] If the controlling termination step is the heterogeneous termination of propyl radicals, show that the description of elementary steps given earlier for the chlorination of propane satisfactorily explains the experimental data.

Solution If h is Planck's constant and v is the frequency of radiation, the initiation reaction for a photochemical activation may be written as

$$Cl_2 + hv \rightarrow 2Cl$$

The rate of production of Cl atoms depends on the volumetric rate of adsorption of radiant energy, Ia, according to the expression

$$r_i = 2\varphi_i\, Ia \tag{A}$$

where φ_i is the quantum yield of the initiation step.[2] The rate of energy absorption is equal to the intensity of radiation, I, multiplied by the absorptivity α and the chlorine concentration, so that Eq. (A) becomes

$$r_i = 2\varphi_i\, \alpha[Cl_2]\, I$$

Suppose that the concentrations of intermediates rapidly attain constant low values; that is, the stationary-state hypothesis is valid.[3] Then a steady state will quickly be reached, after which the production and destruction rates of Pr and Cl will balance each other, or

$$\frac{d[Pr]}{dt} = \frac{d[Cl]}{dt} = 0 \tag{B}$$

We can write two independent equations from Eq. (B), one for Pr and the other for Cl, using the elementary steps described in the text.[4]

$$\frac{d[Pr]}{dt} = 0 = k_2[Cl][PrH] - k_3[Pr][Cl_2] - k_7[Pr]$$

$$\frac{d[Cl]}{dt} = 0 = 2\varphi_i\, \alpha[Cl_2]I - k_2[Cl][PrH] + k_3[Pr][Cl_2]$$

[1] A. E. Cassano and J. M. Smith, *AIChE J.*, **12**, 1124 (1966).
[2] The quantum yield is the molecules of product produced per quantum of energy absorbed. For a primary step φ_i approaches unity. Note that the coefficient 2 in Eq. (A) takes account of the production of two atoms of chlorine for each molecule of Cl_2.
[3] Note that the requirement for the stationary-state hypothesis to be valid is that the concentrations of intermediates be small with respect to those of the reactants and products.
[4] In writing these expressions only the one termination step involving k_7 is needed, since it is assumed to control the termination process.

Adding these two expressions gives $[Pr]$ in terms of the concentration of stable species and the rate constants,

$$[Pr] = \frac{2\varphi_i \, \alpha[Cl_2]I}{k_7} \tag{C}$$

The rate of consumption of Cl_2 or production of propyl chloride is obtained from the second of the propagation reactions. Thus

$$-\frac{d[Cl_2]}{dt} = k_3[Pr][Cl_2] \tag{D}$$

The concentration of $[Pr]$ is small but unknown. However, the stationary-state hypothesis has allowed us to develop Eq. (C) for $[Pr]$ in terms of known quantities. Using Eq. (C) in Eq. (D) gives

$$-\frac{d[Cl_2]}{dt} = \frac{2k_3\varphi_i}{k_7} \alpha[Cl_2]^2 \, I \tag{E}$$

This is the desired result, giving the rate as second order in chlorine and first order in light intensity. By choosing different termination steps to be controlling, we can obtain different expressions for $-d[Cl_2]/dt$, some involving $[Pr]$, $[Cl_2]$, and $[I]$ to other powers. Comparison of the various results with experimentally determined rates allows us to choose the best form of the rate equation and to evaluate the ratio of rate constants; for example, $k_3\varphi_i/k_7$ in Eq. (E). From measurements only on stable species, individual values of k cannot, in general, be established.

No mention has been made of the effects of wavelength, measurement of light intensity, and other complications in photochemical studies. These questions and others are discussed in the literature.[1]

2-12 Rate Constants and Equilibrium

In Sec. 2-8 it was shown that for an elementary process the ratio of rate constants in the forward and reverse directions is equal to the equilibrium constant. When order and stoichiometry do not agree this is not necessarily true. To explain with a simple example, suppose the reversible, overall reaction

$$A_2 + 2B \underset{k'}{\overset{k}{\rightleftharpoons}} 2C$$

is found to be first order in A_2 and B in the forward direction, and first order in C in the reverse direction. The net rate of consumption of A_2 is

$$\mathbf{r}_{A_2} = k[A_2][B] - k'[C] \tag{2-108}$$

[1] J. G. Calvert and J. N. Pitts, Jr., "Photochemistry," John Wiley & Sons, Inc., New York, 1966; W. A. Noyes, Jr., and P. A. Leighton, "The Photochemistry of Gases," Reinhold Publishing Corporation, New York, 1941.

At equilibrium the net rate will be zero, so that

$$\frac{k}{k'} = \frac{[C]}{[A_2][B]}$$
(2-109)

The stoichiometry of the reaction requires that the equilibrium constant be defined as

$$K = \frac{[C]^2}{[A_2][B]^2}$$
(2-110)

Comparison of Eqs. (2-109) and (2-110) shows that k/k' is not equal to K for this case.

It has been suggested[1] that if the mechanism of the overall reaction is known, and if one of the elementary steps controls the rate, the relationship between rate constants and the overall equilibrium constant is

$$\frac{k}{k'} = K^{1/n}$$
(2-111)

The coefficient n is defined by Horiuti as the stoichiometric number of the particular elementary process which controls the rate. This is the number of times this step must occur to accomplish the overall reaction. Suppose that the mechanism of the reaction we have chosen consists of the elementary steps

1. $A_2 \rightleftarrows 2A$

2. $A + B \rightleftarrows C$

For the overall reaction $A_2 + 2B \rightleftarrows 2C$ to result, the first step must occur once and the second step twice. Accordingly, the stoichiometric numbers are 1 and 2 for these steps.

2-13 Precision of Kinetic Measurements

Errors in experimental data may arise from random causes or inherent difficulties in the system. The latter type can be corrected once the performance of the system is fully understood. For example, in kinetics studies erroneous rates may be caused by some appreciable unknown reaction not accounted for in the treatment of the data. Random errors, such as temperature fluctuations in a thermostat, can be reduced by improvements in apparatus and technique but usually cannot be eliminated. These residual random errors may be evaluated from the known precision of the experi-

[1] M. Manes, L. Hofer, and S. Weller, *J. Chem. Phys.*, **18**, 1355 (1950); J. Horiuti, in *Advan. Catalysis*, **9**, 339 (1957).

mental observations. It is important in kinetics to be able to calculate the precision of rates of reaction, rate constants, and activation energies from the errors in the measurements.

Let us consider as an illustration the precision of rate constants. The fractional error in dependent variable Ω, which is a function of independent variables α_i, is given by

$$\left(\frac{\Delta\Omega}{\Omega}\right)^2 = \sum_{i=1}^{m} \left[\frac{\partial(\ln \Omega)}{\partial(\ln \alpha_i)}\right]^2 \left(\frac{\Delta\alpha_i}{\alpha_i}\right)^2 \tag{2-112}$$

The starting point in using this expression is the relation between the quantity we want to know the precision of and the experimental observations used. For our example this is the relation between the rate constant and the rate and concentrations. Suppose the reaction is second order and of the form

$$\mathbf{r} = k_2[A][B]$$

Then the required relationship is

$$k_2 = \frac{\mathbf{r}}{[A][B]} \tag{2-113}$$

where $\mathbf{r}$, $[A]$, and $[B]$ are the three independent variables (α_i) and k is the dependent variable (Ω). Evaluating the partial derivatives from Eq. (2-113) and substituting them in Eq. (2-112) yields

$$\left(\frac{\Delta k}{k}\right)^2 = \left[\frac{\partial(\ln k)}{\partial(\ln \mathbf{r})}\right]^2 \left(\frac{\Delta \mathbf{r}}{\mathbf{r}}\right)^2 + \left\{\frac{\partial(\ln [k])}{\partial(\ln [A])}\right\}^2 \left(\frac{\Delta[A]}{[A]}\right)^2$$

$$+ \left\{\frac{\partial(\ln k)}{\partial(\ln [B])}\right\}^2 \left(\frac{\Delta[B]}{[B]}\right)^2$$

or

$$\left(\frac{\Delta k}{k}\right)^2 = \left(\frac{\Delta \mathbf{r}}{\mathbf{r}}\right)^2 + \left(\frac{\Delta[A]}{[A]}\right)^2 + \left(\frac{\Delta[B]}{[B]}\right)^2 \tag{2-114}$$

This result shows that the squares of the fractional errors in the individual measurements are additive. If the precision of the rate measurements is 8% and that of each concentration is 4%, the error in k will be

$$\left(\frac{\Delta k}{k}\right)^2 = 0.08^2 + 0.04^2 + 0.04^2 = 0.0096$$

$$\frac{\Delta k}{k} = 0.098 \qquad \text{or } 9.8\%$$

The rate is not a direct measurement but is itself calculated from observations of such variables as time and concentration. Its precision, arbitrarily chosen in the example as 8%, should be based on an evaluation similar to that illustrated for obtaining the error in k.

The evaluation of k from data for rates and concentrations should be carried out by statistically sound methods, provided sufficient data points are available. For example, Eq. (2-113) shows that a linear relationship exists between the product $[A][B]$ and $\mathbf{r}$. Hence k_2 should be determined from the slope of the line of $\mathbf{r}$ vs $[A][B]$ determined by a least-mean-square fit of the data points. This technique was illustrated in Example 2-1, where the best fit of the Arrhenius equation to k-vs-T data was used to evaluate the activation energy.

Errors in activation energy can be evaluated from the Arrhenius equation by the same procedure described for finding the errors in k. The precision of E will depend on the uncertainty in k and T. Since E is based on differences in k and T values, the same errors in k and T will result in smaller errors in E as temperature range covered by the data is increased.

Bibliography

1. S. W. Benson, "The Foundations of Chemical Kinetics," McGraw-Hill Book Company, New York, 1960. The stationary-state hypothesis mentioned in Secs. 2-6 and 2-11 is defined and illustrated on pp. 50–53. In chap. XII the collision theory is considered in detail, complications related to the energy distribution of molecules and the steric factor are discussed, and the results are compared in depth with those from the transition-state theory.
2. S. W. Benson, "Thermochemical Kinetics," John Wiley & Sons, Inc., New York, 1968. Proposes means for estimating rate constants for a variety of homogeneous reactions.
3. Michel Boudart, "Kinetics of Chemical Processes," Prentice-Hall, Inc., Englewood Cliffs, N.J., 1968. A concise presentation of the fundamental concepts of kinetics for homogeneous and heterogeneous reactions, including a chapter on the application and validity of the stationary-state hypothesis.
4. A. A. Frost and R. G. Pearson, "Kinetics and Mechanism," John Wiley & Sons, Inc., New York, 1961. Provides an excellent discussion of reaction mechanisms with applications.
5. N. N. Semenov, "Problems in Chemical Kinetics and Reactivity," vols. I and II translated by M. Boudart, Princeton University Press, Princeton, N.J., 1958. Provides large reservoirs of information on theories of kinetics and experimental data.

Problems

2-1. A common rule of thumb is that the rate of reaction doubles for each 10°C rise in temperature. What activation energy would this suggest at a temperature of 25°C?

2-2.   The rate of an overall reaction $A + 2B \rightarrow C$ has been found to be first order with respect to both A and B. What mechanism do these results suggest?

2-3.   The overall reaction for the thermal decomposition of acetaldehyde is

$$CH_3CHO \rightarrow CH_4 + CO$$

A chain-reaction sequence of elementary steps proposed to explain the decomposition is as follows:

INITIATION

$$CH_3CHO \xrightarrow{k_1} CH_3 \cdot + CHO \cdot$$

PROPAGATION

$$CH_3 \cdot + CH_3CHO \xrightarrow{k_2} CH_3CO \cdot + CH_4$$

$$CH_3CO \cdot \xrightarrow{k_3} CH_3 \cdot + CO$$

TERMINATION

$$CH_3 \cdot + CH_3 \cdot \xrightarrow{k_4} C_2H_6$$

Use the stationary-state hypothesis to derive an expression for the overall rate of decomposition. Do order and stoichiometry agree in this case?

2-4.   Using the collision theory, calculate the rate constant at 300°K for the decomposition of hydrogen iodide, assuming a collision diameter of $3.5A$ and an activation energy of 44 kg cal (based on a rate constant in concentration units). To what entropy of activation does the result correspond?

2-5.   The frequency factor for the gas-phase dissociation of the dimer of cyclopentadiene is 1.3×10^{13} sec^{-1} and the activation energy is 35.0 kg cal (based on k_c). Calculate (a) the entropy of activation, (b) the rate constant at 100°C, and (c) the rate at 100°C and 1 atm pressure.

2-6.   The homogeneous dimerization of butadiene has been studied by a number of investigators[1] and found to have an experimental activation energy of 23,960 cal/g mole, as indicated by the specific-reaction rate,

$$k = 9.2 \times 10^9 e^{-23\,960/R_gT} \qquad cm^3/(g\ mole)(sec)$$

(based on the disappearance of butadiene). (a) Use the transition-state theory, to predict a value of **A** at 600°K for comparison with the experimental result of 9.2×10^9. Assume that the structure of the activated complex is

$$\overset{|}{CH_2}-CH=CH-CH_2-CH_2-\overset{|}{CH}-CH=CH_2$$

and use the group-contribution method (see Sec. 1-5) to estimate the thermodynamic properties required. (b) Use the collision theory to predict a value of

[1] W. E. Vaughan, *J. Am. Chem. Soc.*, **54**, 3863 (1932); G. B. Kistiakowsky and F. R. Lacher, *J. Am. Chem. Soc.*, **58**, 123 (1936); J. B. Harkness, G. B. Kistiakowsky, and W. H. Mears, *J. Chem. Phys.*, **5**, 682 (1937).

A at 600°K and compare it with the experimental result. Assume that the effective collision diameter is 5×10^{-8} cm.

2-7. From the transition-state theory and the following thermodynamic information, calculate the rate constant for the given unimolecular reactions. Assume ideal-gas behavior.

Reaction	T, °K	ΔH^*, cal/g mole	ΔS^*, cal/(g mole)(°K)
Decomposition of methyl azide, CH_3N_3	500	42,500	8.2
Decomposition of dimethyl ether, CH_3OCH_3	780	56,900	2.5

SOURCE: O. A. Hougen and K. M. Watson, "Chemical Process Principles," vol. III, "Kinetics and Catalysis," John Wiley & Sons, Inc., New York, 1947.

2-8. A gaseous second-order reversible reaction of the form

$$A + B \rightleftarrows C + D$$

has forward-rate constants as follows:

$$k_2 = \begin{cases} 10.4 \text{ liter/(g mole)(sec)} & \text{at } 230°C \\ 45.4 \text{ liter/(g mole)(sec)} & \text{at } 260°C \end{cases}$$

The standard-state entropy and enthalpy changes for the overall reaction are approximately independent of temperature and are given by $\Delta H° = 8,400$ cal/g mole and $\Delta S° = -2.31$ cal/(g mole)(°K). Derive expressions for the forward- and reverse-rate constants as functions of temperature.

2-9. The reaction

$$CH_3COCH_3 + HCN \rightleftarrows (CH_3)_2C{\overset{\displaystyle CN}{\underset{\displaystyle OH}{\Big<}}}$$

was studied in aqueous solution by Svirbely and Roth.[1] In one run with initial concentrations of 0.0758 normal for HCN and 0.1164 normal for acetone, the following data were obtained:

t, min	4.37	73.2	172.5	265.4	346.7	434.4
[HCN], normal	0.0748	0.0710	0.0655	0.0610	0.0584	0.0557

Determine a reasonable rate equation from these data. ($K_c = 13.87$ liter/mole)

2-10. With HCl as a homogeneous catalyst the rate of esterification of acetic acid and alcohol is increased. At 100°C the rate of the forward reaction is

$$\mathbf{r}_2 = k_2[H][OH] \qquad \text{g moles/(liter)(min)}$$

[1] *J. Am. Chem. Soc.*, **75**, 3109 (1953).

$$k_2 = 4.76 \times 10^{-4} \text{ liter/(g mole)(min)}$$

and the rate of the reverse reaction is

$$\mathbf{r}'_2 = k'_2[E][W] \qquad \text{g moles/(liter)(min)}$$

$$k'_2 = 1.63 \times 10^{-4} \text{ liter/(g mole)(min)}$$

where $[H]$ = concentration of acetic acid
$[OH]$ = concentration of alcohol
$[E]$ = concentration of ester
$[W]$ = concentration of water

An initial mixture consists of equal masses of 90 wt % aqueous solution of acid and 95 wt % solution of ethanol. For constant-volume conditions calculate the conversion of acid to ester for various times of reaction. Assuming complete miscibility, estimate the equilibrium conversion.

2-11.   Smith[1] has studied the gas-phase dissociation of sulfuryl chloride, SO_2Cl_2, into chlorine and sulfur dioxide at 279.2°C. The total-pressure method was employed to follow the course of the reaction. Under constant-volume conditions the results were as follows:

t, min	3.4	15.7	28.1	41.1	54.5	68.3	82.4	96.3
p_t, mm Hg	325	335	345	355	365	375	385	395

What reaction order do these data suggest? The conversion is 100% at infinite time.

2-12.   For two consecutive reversible reactions (liquid-phase)

1. $A \leftrightarrows B$

2. $B \rightleftarrows C$

the forward-rate constants k and equilibrium constants K are

$$k_1 = 1 \times 10^{-3} \text{ min}^{-1} \qquad K_1 = 0.8$$

$$k_2 = 1 \times 10^{-2} \text{ min}^{-1} \qquad K_2 = 0.6$$

If the initial concentration of A is 1.0 molal, plot the concentration of A vs time from 0 to 1,000 min. Both reactions are first order.

2-13.   The thermal decomposition of nitrous oxide (N_2O) in the gas phase at 1030°K is studied in a constant-volume vessel at various initial pressures of N_2O. The half-life data so obtained are as follows:

p_0, mm Hg	52.5	139	290	360
$t_{1/2}$, sec	860.0	470	255	212

Determine a rate equation that fits these data.

[1] D. F. Smith, *J. Am. Chem. Soc.*, **47**, 1862 (1925).

2-14.   It has been postulated that the thermal decomposition of diethyl ether occurs by the following chain mechanism:

INITIATION

$$(C_2H_5)_2O \xrightarrow{k_1} CH_3\cdot + \cdot CH_2O\, C_2H_5$$

PROPAGATION

$$CH_3\cdot + (C_2H_5)_2O \xrightarrow{k_2} C_2H_6 + \cdot CH_2OC_2H_5$$

$$\cdot CH_2OC_2H_5 \xrightarrow{k_3} CH_3\cdot + CH_3CHO$$

TERMINATION

$$CH_3\cdot + \cdot CH_2OC_2H_5 \xrightarrow{k_4} \text{end products}$$

Show that the stationary-state hypothesis indicates that the rate of decomposition is first order in ether concentration.

2-15.   Consider the reaction sequence

$$A \xrightarrow{k_1} B \xrightarrow{k_2} C \xrightarrow{k_3} D$$

Determine the profiles of concentration vs time for A, B, C, and D on an analog computer. Solve the problem in the following steps:

STEP 1   To limit the variables range from 0 to 1.0, first put the differential equations expressing the rate in dimensionless form by using the new variables

$$A^* = \frac{[A]}{[A]_0} \qquad B^* = \frac{[B]}{[A]_0} \qquad C^* = \frac{[C]}{[A]_0} \qquad D^* = \frac{[D]}{[A]_0} \qquad t^* = k_1 t$$

where $[A]_0$ is the initial concentration of A. Note that the initial conditions are $t^* = 0$, $A^* = 1$, and $B^* = C^* = D^* = 0$.

STEP 2   Next prepare a block diagram showing the hookup of integrators, inverters, summers, and potentiometers needed to solve the differential equations for A^*, B^*, C^*, and D^*. Suppose $k_2/k_1 = k_3/k_1 = 1.0$.

STEP 3   Hook up an analog computer according to the block diagram and display the outputs A^*, B^*, C^*, and D^* vs t^* on an oscilloscope or xy plotter.

2-16.   Solve Prob. 2-15 by analytical integration of the differential equations (dimensionless form).

2-17.   Aqueous solutions of diazobenzene decompose irreversibly according to the reaction

$$C_6H_5N_2Cl(aq) \rightarrow C_6H_5Cl(aq) + N_2(g)$$

The kinetics are first order. In one experiment,[1] at 50°C the initial concentration

[1]O. A. Hougen and K. M. Watson, "Chemical Process Principles," vol. III, "Kinetics and Catalysis," John Wiley & Sons, Inc., New York, 1947.

of $C_6H_5N_2Cl$ was 10 g/liter and the following amounts of N_2 were liberated:

Reaction time, min	6	9	12	14	18	20	22	24	26	30
N_2 liberated, cm^3 at 50°C, 1 atm				19.3	26.0	32.6	36.0	41.3	43.3	45.0	46.5	48.4	50.3

Complete decomposition of the diazo salt liberated 58.3 cm^3 of N_2. Calculate an accurate value for the rate constant.

2-18. The thermal decomposition of dimethyl ether in the gas phase was studied by Hinshelwood and Askey[1] by measuring the increase in pressure in a constant-volume reaction vessel. At 504°C and an initial pressure of 312 mm Hg the following data were obtained:

t, sec	390	777	1,195	3,155	∞
p_t, mm Hg	408	488	562	779	931

Assuming that only ether was present initially and that the reaction is

$$(CH_3)_2O \rightarrow CH_4 + H_2 + CO$$

determine a rate equation for the decomposition. What is the numerical value of the specific-reaction rate at 504°C?

2-19.[2] Suppose that a gaseous reaction between A and B is studied kinetically by making isothermal measurements of the half-life period for several initial compositions of reactants. The results for each of four different initial conditions are as follows:

$(p_A)_0$, mm Hg	500	125	250	250
$(p_B)_0$, mm Hg	10	15	10	20
$(t_{1/2})_B$, min	80	213	160	80

If the rate is first order with respect to component A and second order with respect to B, what is the numerical value of the specific reaction rate?

2-20. The reaction mechanism for the decomposition of nitrogen pentoxide is complex, as described in Sec. 2-2. However, a satisfactory rate equation can be developed by considering the two reactions

1. $2N_2O_5 \rightarrow 2N_2O_4 + O_2$

2. $N_2O_4 \rightarrow 2NO_2$

Reaction 2 is rapid with respect to reaction 1, so that nitrogen dioxide and

[1] C. N. Hinshelwood and P. J. Askey, *Proc. Roy. Soc. (London)*, **A115**, 215 (1927).

[2] From A. A. Frost and R. G. Pearson, "Kinetics and Mechanism," John Wiley & Sons, Inc., New York, 1953 (by permission).

nitrogen tetroxide may be assumed to be in equilibrium. Hence only reaction 1 need be considered from a kinetic standpoint. Calculate the specific reaction rate for this reaction (which is essentially irreversible) from the following total-pressure data[1] obtained at 25°C:

t, min	0	20	40	60	80	100	120	140	160	∞
p_t, mm Hg	268.7	293.0	302.2	311.0	318.9	325.9	332.3	338.8	344.4	473.0

It may be assumed that only nitrogen pentoxide is present initially. The equilibrium constant K_p for the dissociation of nitrogen tetroxide into nitrogen dioxide at 25°C is 97.5 mm Hg.

2-21.[2] The decomposition of nitrogen dioxide follows a second-order rate equation. Data at different temperatures are as follows:

T, °K	592	603	627	651.5	656
k_2, cm³/(g mole)(sec)	522	755	1,700	4,020	5,030

Compute the energy of activation E_c from this information. If the reaction is written

$$2NO_2 \rightarrow 2NO + O_2$$

also evaluate the activation energy E_p.

2-22. Reactants A and B are placed in a reaction vessel at zero time, where $[A]_0 = [B]_0$. The following reactions occur at constant density:

1. $A + B \xrightarrow{k_1} C$

2. $A + C \xrightarrow{k_2} D$

where C is the desired product. If both reactions are second order, derive an expression for the selectivity of C with respect to D in terms of the total conversion of A. Also determine the total conversion at which the selectivity will be a maximum if $k_2/k_1 = 1.0$. Will the maximum conversion of A to C occur at the same total conversion as that for which the selectivity of C with respect to D is a maximum?

2-23. The parallel, first-order irreversible reaction system

1. $A \xrightarrow{k_1} B$

2. $A \xrightarrow{k_2} C$

[1] F. Daniels and E. H. Johnston, *J. Am. Chem. Soc.*, **43**, 53 (1921).
[2] From A. A. Frost and R. G. Pearson, "Kinetics and Mechanism," John Wiley & Sons, Inc., New York, 1953 (by permission).

consists of three components, so that the reaction path can be conveniently represented on a triangular diagram.[1] Suppose that initially only A is present at a concentration $[A]_0$. Assume constant density, so that the sum of the concentrations of all the components will be constant and equal to $[A]_0$. Let one apex of the equilateral triangle represent a reaction mixture containing 100% A, the second 100% B, and the third 100% C. Use Eq. (2-100) to show how the composition of the reaction mixture can be represented on the diagram. Specifically, draw the reaction path from zero to complete conversion of A when $k_2/k_1 = 2$. What path would be followed if $k_2/k_1 = 0$?

2-24.   Suppose the consecutive first-order reactions described on page 77 occur at constant density in a batch reactor, with an initial mixture containing only A at a concentration $[A]_0$. Show on a triangular diagram the reaction paths for three cases: $k_3/k_1 = 0.5$, 1.0, and 2.0.

2-25.   The second-order reactions

1. $A + B \xrightarrow{k_1} C$

2. $A + A \xrightarrow{k_2} A_2$

occur at constant density, with an initial mixture containing only A and B, each at the same concentration. For a batch reactor, show the reaction path on a triangular diagram for $k_2/k_1 = 1.0$.

2-26.   Consider that the photochlorination of propane occurs according to the reactions in Sec. 2-11. If the controlling termination step is the heterogeneous termination of chlorine radicals,

Cl + wall → end product

derive a rate equation for the overall reaction. What would be the form of the rate equation if the second-order homogeneous termination of Cl were controlling (i.e., according to the reaction Cl + Cl → Cl_2)?

2-27.   The kinetics of a second-order irreversible liquid-phase reaction of the form $A + B \to C$ are studied in a constant-volume apparatus. Starting with equal concentrations of 1.0 mole/liter for A and B, the reaction is stopped after 30 min, at which time about 20% of the reactants have disappeared. Random errors will amount to about 5 sec in the time readings and 0.002 moles/liter in the concentration measurements. Estimate the fractional error in the rate constants computed from such data.

[1] The representation of reaction paths on triangular diagrams is described in detail in Kramers and Westerterp, "Chemical Reactor Design and Operation," p. 47, Academic Press, Inc., New York, 1963.

3

DESIGN FUNDAMENTALS

In this chapter three main subjects are discussed: (1) general aspects of reactor design, in preparation for design problems for specific reactors in later chapters, (2) mass and energy balances for the two ideal types of reactors, and (3) an introduction to deviations from these ideal types.

3-1 Reactor Design and Laboratory Rate Data

Designing a reactor entails determining the *size* of vessel needed to obtain the specified amounts of products and determining the temperature, pressure, and composition of the reaction mixture (i.e., the *operating conditions*) in various parts of the equipment. Required information includes the initial, or entrance, conditions of temperature, pressure, and reactants composition and the method of operating the reactor—batch or flow, isothermal, adiabatic, etc. Such information, along with the flow rates of feed and products, provides the *design conditions*. Usually there are many combinations of reactor size and operating conditions which satisfy the design conditions. The optimum design—that which will produce the greatest profit—depends on raw materials, initial and operating costs, and the market value of the end products.

The ideal approach to design is apparent: the method of designing the reactor should be established, and then an optimization technique should be used to find the most profitable design. Such a procedure should lead to an optimum solution for a set of constant values of the design conditions.

Generally some of the conditions, such as feed composition, will change with time. Such changes may be abrupt, as in switching to feed from another storage tank that contains material of a different composition, or they may be mild cyclical changes. The problem in *reactor control* is determining how to change operating conditions so that the reactor returns to optimum performance as quickly as possible. The first decision is what operating conditions to use as control points; then a policy or strategy is developed for responding to fluctuations in such a way as to maximize profit. The resulting control procedure may be manual; or it may be a semicomputerized procedure, in which a computer is used to determine rapidly the value of the profit function but the indicated changes in operating conditions are made manually; or it may be a direct-digital or closed-loop procedure, in which adjustments to operating conditions are made automatically on signal from the computer. All control procedures, to be successful, require a knowledge of how to design the reactor for a set of constant design conditions.

The optimum design requires iterative numerical work; hence machine computation greatly simplifies the optimization task. In this book we shall limit the problem to answering design questions for a single set of design conditions. Even for such a limited scope we shall find that the numerical integrations often require repetitive calculations well suited for machine solution.

As mentioned in Sec. 1-1, the first step in a logical design procedure is to obtain a suitable expression for the rate of the chemical reaction process, and this requires experimental data. The data can be obtained in several ways:

1. From a bench-scale laboratory reactor designed to operate at nearly constant temperature and composition. Usually operating conditions are chosen to facilitate separating the effects of diffusion and heat transfer (the physical processes) from the observed measurements, so that the rate of the chemical step can be accurately evaluated. This is the most successful of the three methods.

2. From a small-scale reactor (pilot plant) in which the composition, temperature, and pressure may change. Here calculations similar to,

but the reverse of, the design steps are required to evaluate the rate of the chemical reaction. Accurate separation of the diffusion and heat-transfer effects from the chemical step may be difficult.

3. From a commercial-scale reactor which happens to be already available. The problems in obtaining an expression for the chemical rate are similar to those in method 2 but are usually even more severe, because there is less instrumentation, and hence fewer data.

As an illustration of the first two methods of arriving at the data necessary for design calculations, consider the oxidation of sulfur dioxide. Suppose that an air-SO_2 mixture flows over particles of solid catalyst in a tubular-flow reactor. In the bench-scale study the reactants would be passed over a very small amount of catalyst, and the rate of production of sulfur trioxide would be determined by measuring the rates of flow and composition of the inlet and exit streams.[1] This production rate, divided by the mass of catalyst, would represent the rate of reaction,[2] for example, in grams of sulfur trioxide per hour per gram of catalyst. It would approach a point rate rather than an average value, because the amount of catalyst is small enough that the temperature, pressure, and composition changes in passing over the catalyst bed are similarly small.[3] In the second approach the amount of catalyst in the pilot-plant reactor is sufficient to cause considerable conversion, and the temperature and composition may change appreciably as the mixture flows through the reactor. Since the rate of reaction is a function of these variables, it varies from location to location, and the measured production of sulfur trioxide represents an integrated average of all the point rates. To reduce the measured result (called *integral-conversion,* or *integral-reactor, data*) to the rate of the chemical step requires a procedure that is reverse of the design calculations. A promising rate equation is assumed. Then the point rates are integrated through the catalyst bed, with the effects of diffusion and heat transfer in causing composition and temperature changes taken into account. Finally, the predicted conversions are compared with the experimental results. Repetition of this procedure will result in an equation for the rate of the chemical step. Because of the

[1] In flow systems such small-scale reactors are commonly called *differential reactors,* since the changes in temperature, pressure, and composition in the reactor are small.

[2] For a reaction requiring a solid catalyst the rate is usually based on a unit mass of catalyst rather than a unit volume, as defined in Chap. 2. The two rates are directly related through the bulk density of the bed of catalyst.

[3] Note that the change in composition between the inlet and exit streams must be large enough for precise measurement; otherwise the rate of conversion in the reactor cannot be accurately established. This restriction imposes a limitation on the applicability of the method. If precise analytical methods of determining small composition changes are not available for the particular reaction, a close approach to a point value of the rate cannot be ascertained.

inaccuracy of the calculations, particularly if there are significant temperature changes, it is generally not possible to arrive at a rate equation that is wholly satisfactory.

Data obtained in both bench and pilot-plant equipment are valuable, and it is common practice to carry out investigations with both before building the commercial-scale reactor. The first yields a better rate equation and more knowledge about the kinetics of the reaction; i.e., it tells the engineer more accurately just what variables affect the rate of the chemical step and how they influence the course of the reaction. This information is particularly valuable in case it is necessary to predict how the commercial-scale plant will be affected by a change of operating conditions not specifically considered in the pilot-plant work.

The bench-scale study alone leaves the engineer largely dependent on prediction methods for the effects of the physical processes, whereas data obtained on the scale of a pilot plant can provide a check on the suitability of the prediction methods.[1] As an illustration, consider the extreme case where the kinetic studies are carried out in a batch-tank reactor with diffusion resistances, and the commercial unit is to be a tubular-flow reactor. The diffusion rates will not be the same under batch and flow conditions. Hence the observed rate in the batch reactor will not be directly applicable for design calculations in the commercial unit. While the importance of such effects can be estimated for different types of reactors, as explained in Chap. 10, the uncertainties in the estimates are sometimes so large that experimental verification from pilot-plant data is desirable. There are similar problems with temperature differences arising from heat-transfer considerations. Even where the kinetic studies are carried out in a flow system similar to that to be employed in the large plant, pilot-plant investigations provide invaluable information on such important factors as temperature distribution in the reactor and the effect of specialized designs on improving the process.[2]

3-2 Mass and Energy Balances

Mass and energy balances on the reactor provide the basis for relating the production rate and composition of the products to the chemical-reaction rate. If the operation is not steady, changes with time are also involved. In a single-reaction system one reactant is usually critical because of cost

---

[1] From this standpoint the objective of the pilot plant is to obtain a model of the reactor, that is, an understanding of how the physical processes affect the performance of a reactor. In contrast, the objective of the bench-scale reactor is to obtain a model for the chemical kinetics, that is, a rate equation.

[2] A more detailed discussion of the functions of bench-scale and pilot-plant reactors is given by J. M. Smith, *Chem. Eng. Progr.*, **64**, 78 (1968).

or limited availability, and the mass balance is applied to this limiting reactant. For example, in the air oxidation of SO_2 the sulfur dioxide is the limiting reactant. Of course, the balance may be written for each component, and for the total mass, but this is not necessary because the composition of the reaction mixture can be expressed in terms of one variable, the conversion, using the composition of the original reactants and the stoichiometry of the reaction (see Example 4-1). For multiple-reaction systems mass balances may need to be written for additional components.

The conversion x is the fraction of a reactant that has been converted into products. When there is only one reaction, there is no uncertainty in this definition. When a reactant can undergo simultaneous or successive reactions to multiple products, both the total conversion of reactant and the conversions to specific products are important. The conversion to a given product was defined in Chap. 2 as the yield of that product. For example, in the air oxidation of ethylene both ethylene oxide and carbon dioxide are products. It is customary to speak of the *yield* of ethylene oxide or *conversion to* ethylene oxide.

Example 3-1 Liquid benzene is chlorinated by bubbling gaseous chlorine into a well-stirred tank reactor containing the benzene. Three reactions can occur:

1. $C_6H_6 + Cl_2 \rightarrow C_6H_5Cl + HCl$

2. $C_6H_5Cl + Cl_2 \rightarrow C_6H_4Cl_2 + HCl$

3. $C_6H_4Cl_2 + Cl_2 \rightarrow C_6H_3Cl_3 + HCl$

Initially the reactor contains $(N_B)_0$ moles of benzene. Then a total of N_{Cl_2} $[N_{Cl_2} < 3(N_B)_0]$ moles of chlorine per mole of benzene is added to the reactor slowly, so that no unreacted chlorine leaves the reactor. Also, the concentrations of dissolved unreacted chlorine and dissolved HCl are small.[1] If the density of the reaction mixture is constant, express the concentrations of mono-, di-, and trichlorobenzene in terms of their corresponding conversions x_M, x_D, and x_T.

Solution Chlorine is the limiting reactant, so the conversions will be based on this component. Let N_B, N_M, N_D, and N_T represent the moles of benzene and mono-, di-, and trichlorobenzene per mole of original benzene. Since 1 mole of chlorine is required for 1 mole of monochlorobenzene,

$$x_M = \frac{N_M}{N_{Cl_2}} \tag{A}$$

However, 2 moles of chlorine are needed to produce 1 mole of dichlorobenzene. Therefore

$$x_D = \frac{2N_D}{N_{Cl_2}} \tag{B}$$

---

[1] The HCl produced leaves the reactor as a gas.

and similarly,

$$x_T = \frac{3N_T}{N_{Cl_2}} \tag{C}$$

The number of moles of unreacted benzene will be

$$N_B = 1 - (N_M + N_D + N_T) = 1 - (x_M + \tfrac{1}{2}x_D + \tfrac{1}{3}x_T)N_{Cl_2} \tag{D}$$

The initial concentration of benzene in the reactor is $[B]_0 = (N_B)_0/V$. Equations (A) to (D) give the number of moles of each component per mole of initial benzene. Therefore the concentrations at any conversion will be

$$[M] = \frac{(N_B)_0 N_M}{V} = [B]_0 N_{Cl_2} x_M$$

$$[D] = \tfrac{1}{2}[B]_0 N_{Cl_2} x_D$$

$$[T] = \tfrac{1}{3}[B]_0 N_{Cl_2} x_T$$

$$[B] = \frac{(N_B)_0 N_B}{V} = [1 - (x_M + \tfrac{1}{2}x_D + \tfrac{1}{3}x_T)N_{Cl_2}][B]_0$$

These four equations give the desired relationships for the concentrations in terms of the conversions for a constant-density system. All three conversions are not independent, because it is supposed that there is no unreacted chlorine; that is, their sum must be unity. This is made clear by writing a mass balance of chlorine,

Total chlorine fed = total chlorine in products

$$N_{Cl_2} = N_M + 2N_D + 3N_T \tag{E}$$

Introducing the conversions with Eqs. (A) to (C), we have

$$N_{Cl_2} = x_M N_{Cl_2} + x_D N_{Cl_2} + x_T N_{Cl_2}$$

or

$$1 = x_M + x_D + x_T$$

Example 3-2   When ethylene is oxidized with air by means of a silver catalyst at low temperature (200 to 250°C), two reactions occur:

1. $C_2H_4(g) + \tfrac{1}{2}O_2(g) \rightarrow C_2H_4O(g)$

2. $C_2H_4(g) + 3O_2(g) \rightarrow 2CO_2(g) + 2H_2O(g)$

If the conversion of ethylene is x_1 by reaction 1 and x_2 by reaction 2, express the molal composition of the reaction mixture in terms of the conversions and the ratio a, moles of air per mole of ethylene in the feed. What is the yield and selectivity of ethylene oxide?

Solution   If a basis of 1 mole of C_2H_4 is chosen, and $N_{C_2H_4O}$ and N_{CO_2} represent the moles of these components at any conversions, x_1 and x_2, then

$$x_1 = \frac{N_{C_2H_4O}}{1} \qquad \text{or } N_{C_2H_4O} = x_1$$

and

$$x_2 = \frac{\frac{1}{2}N_{CO_2}}{1} \qquad \text{or } N_{CO_2} = 2x_2$$

The moles of the other components will be

$H_2O = 2x_2$

$C_2H_4 = 1 - x_1 - x_2$

$N_2 = 0.79a$

$O_2 = 0.21a - (\frac{1}{2}x_1 + 3x_2)$

Total moles $= 1 + a - \frac{1}{2}x_1$

Then the mole fractions of each component are as follows:

$$C_2H_4O = \frac{x_1}{1 + a - \frac{1}{2}x_1}$$

$$CO_2 = \frac{2x_2}{1 + a - \frac{1}{2}x_1}$$

$$H_2O = \frac{2x_2}{1 + a - \frac{1}{2}x_1}$$

$$C_2H_4 = \frac{1 - x_1 - x_2}{1 + a - \frac{1}{2}x_1}$$

$$N_2 = \frac{0.79a}{1 + a - \frac{1}{2}x_1}$$

$$O_2 = \frac{0.21a - (\frac{1}{2}x_1 + 3x_2)}{1 + a - \frac{1}{2}x_1}$$

The total conversion of ethylene is $x_1 + x_2$. The yield of ethylene oxide is x_1, and its overall selectivity is x_1/x_2.

Mass Balance   The mass balance for a reactant can be written in a general form applicable to any type of reactor. For a time element Δt and a volume element ΔV this general form is

$$\begin{Bmatrix} \text{Mass of reactant} \\ \text{fed to volume} \\ \text{element} \end{Bmatrix} - \begin{Bmatrix} \text{mass of reactant} \\ \text{leaving volume} \\ \text{element} \end{Bmatrix} - \begin{Bmatrix} \text{mass of reactant} \\ \text{converted in the} \\ \text{volume element} \end{Bmatrix}$$

$$= \begin{Bmatrix} \text{accumulation of reactant} \\ \text{in the volume element} \end{Bmatrix} \quad (3\text{-}1)$$

The first two terms represent the mass of reactant entering and leaving the reactor in the time Δt. The third term depends on the rate of reaction applicable to the volume element ΔV. The fourth term expresses the resultant change in mass of reactant in time Δt caused by the other three terms. The third term will have the form $\mathbf{r} \, \Delta V \, \Delta t$, where $\mathbf{r}$ is the rate of disappearance of the reactant per unit volume. It is important to note that $\mathbf{r}$ is equal to the rate of the chemical step (used exclusively in Chap. 2) only if there are no physical resistances, i.e., no temperature or concentration gradients within ΔV.

Energy Balance   Energy balances are needed solely because the rate of the chemical reaction may be a strong function of temperature (Arrhenius equation, Chap. 2). The purpose of the energy balance is to describe the temperature at each point in the reactor (or at each time for a batch reactor), so that the proper rate may be assigned to that point.

For a volume element ΔV and time period Δt conservation of energy requires

$$\left\{ \begin{array}{l} \text{Energy in streams} \\ \text{fed to volume element} \end{array} \right\} - \left\{ \begin{array}{l} \text{energy in streams} \\ \text{leaving volume element} \end{array} \right\} + \left\{ \begin{array}{l} \text{energy transferred} \\ \text{from surroundings} \\ \text{to volume element} \end{array} \right\}$$

$$= \left\{ \begin{array}{l} \text{accumulation of} \\ \text{energy in the volume} \\ \text{element} \end{array} \right\} \quad (3\text{-}2)$$

To be complete, all forms of energy—internal, potential, and kinetic—should be included in the terms referring to the energy of a fluid.[1] However, in chemical reactors only the internal and, rarely, the mechanical work-forms are quantitatively important. We can evaluate only *changes* in the internal energy of a fluid, so that the first, second, and fourth terms must be referred to the same reference state of zero energy. The difference between the first and second terms reflects temperature differences and differences in energy due to differences in composition of the entering and leaving streams (i.e., the heat of reaction). The third term accounts for possible exchange of energy with the surroundings by heat transfer through the reactor walls. The solution of Eq. (3-2) gives the temperature as a function of position and/or time in the reactor.

The form of Eqs. (3-1) and (3-2) depends on the type of reactor. In most cases one or more of the terms do not exist. More important, the possibility of solving the equations depends on the assumptions made about

---

[1] For more complete and detailed formulation of macroscopic mass and energy balances see R. B. Bird, W. E. Stewart, and E. N. Lightfoot, "Transport Phenomena," John Wiley & Sons, Inc., New York, 1960.

the conditions of mixing, or diffusion, in the reactor. This explains the signif-icance of the classification of reactors into stirred tank or tubular flow, because it is for these ideal types that extreme assumptions about the nature of mixing are valid. In the following sections Eqs. (3-1) and (3-2) are applied to these types of reactors.

3-3 The Ideal Stirred-tank Reactor

The *stirred-tank reactor* may be operated as a steady-state flow type (Fig. 3-1a), a batch type (Fig. 3-1b), or as a non-steady-state, or semibatch, reactor (Fig. 3-1c). The key feature of this reactor is that the mixing is com-plete, so that the properties of the reaction mixture are uniform in all parts of the vessel and are the same as those in the exit (or product) stream. This means that the volume element chosen for the balances can be taken as the volume V of the entire reactor. Also, the composition and temperature at which reaction takes place are the same as the composition and tempera-ture of any exit stream.

Fig. 3-1   *Ideal stirred-tank reactors classified according to method of operation: (a) flow (steady-state), (b) batch, (c) semibatch (non-steady-flow)*

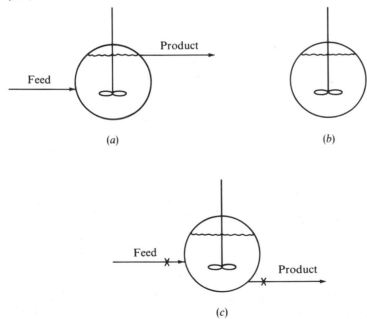

Steady-state Flow   Consider the simple case of a single feed stream and a single product stream, as shown in Fig. 3-1a. The properties of these streams do not change with time. Hence the first and second terms in Eq. (3-1) are constants, equal to the mass-flow rate of limiting reactant multiplied by Δt. Suppose there is only one reaction occurring. If the mass-flow rate[1] of *reactant* corresponding to zero conversion is F, and its conversion in the feed stream is x_F, then $F(1 - x_F)\,\Delta t$ is the first term and, similarly, $F(1 - x_e)\,\Delta t$ is the second. Since the reaction mixture in the vessel is at uniform temperature and composition, the rate of reaction is constant and should be evaluated at the temperature and composition of the *product* stream. If the rate of conversion of reactant is $\mathbf{r}_e$, with the subscript e indicating exit or product conditions, the third term in Eq. (3-1) is $\mathbf{r}_e V\,\Delta t$. There can be no accumulation of mass of reactant in the reactor at steady-state conditions, so the fourth term is zero. Then Eq. (3-1) can be written

$$F(1 - x_F)\,\Delta t - F(1 - x_e)\,\Delta t - \mathbf{r}_e V\,\Delta t = 0$$

or

$$\frac{V}{F} = \frac{x_e - x_F}{\mathbf{r}_e} \tag{3-3}$$

This expression can be used to evaluate the volume of reactor needed to produce a given conversion x_e for known design conditions F and x_F, provided $\mathbf{r}_e$ is available. Notice that the exit conversion can be determined from Eq. (3-3) without integration through the volume of the reactor. This is because the rate is constant throughout the vessel—a direct consequence of the assumption of complete mixing.

There are many variations of the simple form of stirred-tank reactor just considered: multiple units in parallel, series, split streams, more than one reaction, etc. Some of these will be discussed in Chap. 4. Deviations from the assumption of complete mixing are mentioned in Sec. 3-5 and discussed in more detail in Chap. 6.

To evaluate $\mathbf{r}_e$ it is necessary to know the temperature of the product stream leaving the reactor. This requires an energy balance, which again may be written over the total volume element V. The fourth term in Eq.

---

[1] Mass and energy balances may be written either in terms of mass or moles as long as all quantities are on the same basis. For example, if mass units are used, both the rate and the feed rate must be in mass units. For isothermal reactors molal units are more convenient, and the examples in Chap. 4 are solved using these units. Mass units may be more suitable for energy balances so that some examples in Chap. 5 are treated on a mass basis. In this chapter all the equations are written with the supposition that mass units are employed. Also note that the conversion x refers to the fraction of the reactant feed rate F that is converted, not to the fraction of the total feed rate F_t.

(3-2) is zero. Next we choose a base state (temperature, composition, and pressure) from which to evaluate the energy. Suppose the enthalpy[1] (per unit mass) above the base state of the feed stream is H_F and that of the product H_e. If F_t is the total mass feed rate, the first and second terms in Eq. (3-2) are $F_t H_F \, \Delta t$ and $F_t H_e \, \Delta t$, and the balance becomes

$$F_t(H_F - H_e)\, \Delta t + UA_h(T_s - T_e)\, \Delta t = 0$$

or

$$F_t(H_F - H_e) + UA_h(T_s - T_e) = 0 \qquad (3\text{-}4)$$

Heat exchange with the surroundings is expressed here in terms of the surroundings temperature T_s, the overall heat-transfer coefficient U, and the effective area for heat transfer A_h. The heat of reaction ΔH and the rate of reaction do not appear directly in Eq. (3-4). However, their effects are reflected in the difference in the enthalpies of the feed and product streams, since these streams will have different compositions. That is, to calculate the enthalpy H_e of the product stream it is necessary to know ΔH and the rate $\mathbf{r}$. Therefore it is more convenient to replace $H_F - H_e$ with the appropriate T, $\mathbf{r}$, and ΔH. This is accomplished by considering the enthalpy change when feed at T_F is first heated to T_e and then changed in composition to that of the product.[2] Expressed mathematically, the first contribution is $c_p(T_e - T_F)$ and the second is $(x_e - x_F)\,\Delta H/M$. Hence

$$H_e - H_F = c_p(T_e - T_F) + (x_e - x_F)\frac{\Delta H}{M}\frac{F}{F_t} \qquad (3\text{-}5)$$

where M is the molecular weight of the limiting reactant and ΔH is heat of reaction per mole of limiting reactant. Combining Eqs. (3-5) and (3-4) yields

$$F_t(T_F - T_e)c_p - F(x_e - x_F)\frac{\Delta H}{M} + UA_h(T_s - T_e) = 0 \qquad (3\text{-}6)$$

This result can also be expressed in terms of the rate of reaction, rather than the conversion, by using the mass balance. Replacing $x_e - x_F$ with its value from Eq. (3-3) gives

$$F_t(T_F - T_e)c_p - \mathbf{r}_e V\frac{\Delta H}{M} + UA_h(T_s - T_e) = 0 \qquad (3\text{-}7)$$

[1] The internal energy plus the pV term, or enthalpy, is the correct form of the energy in the feed and exit streams because the reactor is operated on a steady-flow basis.

[2] Since enthalpy is a state property, $H_e - H_F$ can be evaluated by any conveniently chosen path. Rigorously, for the path described, ΔH applies at the product temperature T_e, and the specific heat c_p is for a mixture of the composition of the feed. Except for simple gaseous systems, the thermodynamic data available are insufficient to take into account variations in ΔH with temperature and c_p with composition. Often these variations are small.

Equations (3-4), (3-6), and (3-7) are equivalent forms of the energy balance. They are useful in homogeneous-reactor design (Chap. 5).

Batch Operation (Non–steady State)   In batch operation there are no streams entering and leaving the reactor (see Fig. 3-1*b*). Suppose again that the limiting reactant undergoes only one reaction. If m is the mass of reactant corresponding to zero conversion and Δx is the conversion in the time Δt, the accumulation of reactant in the vessel in time Δt is $-m\,\Delta x$. Then the mass balance may be written[1]

$$-\mathbf{r}V\Delta t = -m\,\Delta x \tag{3-8}$$

The properties of the reaction mixture will vary with time, so that the mass balance becomes a differential equation. If we divide by Δt and take the limit as $\Delta t \to 0$, Eq. (3-8) becomes

$$\frac{dx}{dt} = \mathbf{r}\frac{V}{m} \tag{3-9}$$

For isothermal operation $\mathbf{r}$ is dependent only on the composition (or conversion, for a single reaction), so that a solution for conversion with respect to time is obtainable from the mass balance alone. We can show the result by solving Eq. (3-9) for dt and integrating formally:

$$t = m\int_{x_1}^{x_2}\frac{dx}{\mathbf{r}V} = \frac{m}{m_t}\int_{x_1}^{x_2}\rho\;\frac{dx}{\mathbf{r}} \tag{3-10}$$

where x_1 is the initial conversion and x_2 the conversion at any time t. The second equality follows by expressing the volume of the reaction mixture in terms of its total mass m_t and density ρ. The importance of arranging the expression in this way is that time, the independent design variable, is separated from the dependent variables characteristic of the chemical reaction. If the rate $\mathbf{r}$ and the density of the reaction mixture are known as a function of conversion, the value of the integral for any desired conversion can be evaluated without reference to reaction equipment. Then the various combinations of time and mass of charge that will give the required production rate of product can be examined separately. Expressed differ-

---

[1] Note that this does not require that the conversion be zero initially. If the initial conversion is greater than zero, m will be the actual mass of reactant in the charge plus an amount equivalent to the initial conversion. For example, suppose the initial conversion is 20% and the mass of reactant in the initial charge to the reactor is 1,000 lb. Evidently the 1,000 lb is equivalent to 80% of the reactant that would have been present with no conversion. Hence $m = 1{,}000 + (0.20/0.80)(1{,}000) = 1{,}250$ lb.

ently, Eq. (3-10) relates the time to an integral dependent on the series of intensive states experienced by the reaction mixture.

If the reactor does not operate isothermally, an energy balance is required. In this case the first and second terms of Eq. (3-2) do not exist, but the other two can be finite. The accumulation term should express the energy change with time due to the change in composition and the change in temperature of the mixture. The energy accumulated from the change in composition is due to the heat of reaction and may be written $(\Delta H/M)\mathbf{r}V\,\Delta t$. If the change in temperature of the reaction mixture in time Δt is ΔT, the accumulated energy due to the temperature change is $m_t c_v \Delta T$.[1] Then Eq. (3-2) may be written

$$UA_h(T_s - T)\,\Delta t = \frac{\Delta H}{M}\mathbf{r}V\,\Delta t + m_t c_v \Delta T$$

where T is the temperature and c_v the specific heat of the reaction mixture. Dividing by Δt and taking the limit as $\Delta t \to 0$ gives

$$m_t c_v \frac{dT}{dt} = \frac{-\Delta H}{M}\mathbf{r}V + UA_h(T_s - T) \qquad (3\text{-}11)$$

Equation (3-11) relates the temperature to the reaction variables and, with Eq. (3-9), establishes the relationships of conversion, temperature, and time for nonisothermal operation.

It is possible to express Eq. (3-10) for isothermal operation in simpler forms when assumptions such as constant density are permissible. These will be considered in Chap. 4. The constant-density form of Eq. (3-10) was used in Chap. 2 to calculate rate constants from measured conversions or concentrations as a function of time (see, for example, Sec. 2-7). It is important to recall that we could determine the rate equation for the chemical step from a form of Eq. (3-10) because the reactor is assumed to be an ideal stirred-tank type, with no physical resistances involved.

Semibatch Operation   In semibatch operation the rates of mass flow into and out of the system are unequal (see Fig. 3-1c). For example, benzene may be chlorinated in a stirred-tank reactor by first adding the charge of liquid benzene and catalyst and then continuously adding chlorine gas until the required ratio of chlorine to benzene has been obtained. Operation of this kind is batch from the standpoint that the composition of the reaction mixture changes with time. However, from a process standpoint the chlorine is added continuously. The system is still an ideal stirred-tank reactor if the

---

[1] The volume of the reaction mixture should be constant to make the use of c_v rigorously correct.

mixing in the vessel is complete. An advantage of semibatch operation is that a small concentration of one reactant (the chlorine in this example) can be maintained at all times. This may be important when it is desired to obtain a predominant amount of one product from several possible products. Thus a large concentration of chlorine would favor the formation of di- and trichlorobenzenes, while the monochlorination of the benzene ring is favored when a small amount of chlorine is present.

Semibatch operation is also advantageous when a high heat of reaction would cause excessive temperature changes in normal batch operation. For example, hexamethylenetetramine is manufactured by reacting ammonia and formaldehyde. If the reaction is carried out in the liquid phase by using ammonium hydroxide and formalin solutions, ordinary batch operation (in which the two reactants are mixed initially) may result in a large increase in temperature because of the highly exothermic heat of reaction. The temperature rise can be reduced and controlled by adding the ammonium hydroxide continuously, at a controlled rate, to a reactor containing formalin. The rate of this chemical step is extremely rapid; the ammonia reacts just as soon as it is mixed with the formalin. Hence the rate of production of product is controlled by the rate at which the ammonia is added rather than by the rate of the chemical reaction.

These two examples are special cases in which there was no exit stream and no change in flow rate of the feed stream. In its general form semibatch operation includes variations with respect to time in both the rate and state of the inlet and exit streams. The combustion zone in liquid rockets is an example of such a case; as the fuel and oxidant are consumed, their state (temperature, pressure, and composition) and flow rates from the fuel tanks may change.

All the terms in Eqs. (3-1) and (3-2) may be significant in semibatch operation. However, the exact formulation of terms depends on the specific case. A few of these are considered in detail in Chaps. 4 and 5.

3-4 The Ideal Tubular-flow (Plug-flow) Reactor

In a *tubular-flow reactor* the feed enters one end of a cylindrical tube and the product stream leaves at the other end (Fig. 3-2). Such equipment is normally operated at steady state (except at startup or shutdown), so that the properties are constant with respect to time. The long tube and lack of provision for stirring prevents complete mixing of the fluid in the tube. Hence the properties of the flowing stream will vary from point to point. In general, variations in properties will occur in both the longitudinal and radial directions. Under these conditions the volume element ΔV chosen in writing Eqs. (3-1) and (3-2) must be small enough that the properties

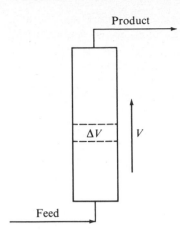

Fig. 3-2   *Ideal tubular-flow reactor*

are constant within the element. It is clear that ΔV will have to approach zero for this condition to be satisfied. Then Eqs. (3-1) and (3-2) become partial-differential equations expressing gradients in the longitudinal and radial directions.

The extent of mixing will affect how the properties vary with position; we are particularly interested in the concentrations and temperature, since they determine the rate of reaction. In the *ideal* tubular-flow reactor specific assumptions are made about the extent of mixing: no mixing in the axial direction (the direction of flow), complete mixing in the radial direction, and a uniform velocity across the radius. The absence of longitudinal mixing is the special characteristic of this type of reactor, also called a *plug-flow reactor*. It is an assumption at the opposite extreme from the complete-mixing supposition of the ideal stirred-tank reactor.

The validity of the assumptions will depend on the geometry of the reactor and the flow conditions. Deviations are frequent but not always important. Methods of describing the deviations and their origins are discussed briefly in Sec. 3-5 and considered in more detail in Chap. 6.

Mass Balance   Equations (3-1) and (3-2) can be written for a volume element ΔV extending over the entire cross section of the tube, as shown in Fig. 3-2. This is because there is no variation in properties or velocity in the radial direction. Suppose the mass feed rate of *reactant* to the reactor is F and the conversion of this reactant at the entrance to the volume element is x. In the absence of axial mixing, reactant can enter the element only by bulk flow of the stream. Hence the first term in Eq. (3-1) is $F(1 - x)\,\Delta t$. If the conversion leaving the element is $x + \Delta x$, the second term is $F(1 - x - \Delta x)$ Δt. Since the operation is at steady state, the fourth term is zero. The third term is $\mathbf{r}\,\Delta V\,\Delta t$, where $\mathbf{r}$ is the rate of reaction. If there are no physical resistances within ΔV, this rate will be determined solely by the chemical step. If physical resistances are possible, as in a fluid-solid heterogeneous reaction, then $\mathbf{r}$ will be determined by the combined resistances of the chemical and physical processes.

Equation (3-1) takes the form

$$F(1 - x)\,\Delta t - F(1 - x - \Delta x)\,\Delta t - \mathbf{r}\,\Delta V\,\Delta t = 0$$

or

$$F\,\Delta x - \mathbf{r}\,\Delta V = 0$$

Dividing by ΔV and taking the limit as $\Delta V \to 0$ gives

$$\frac{dx}{dV} = \frac{\mathbf{r}}{F} \tag{3-12}$$

This is the relationship between conversion and size of reactor for the ideal tubular-flow type. The rate is a variable, as in Eq. (3-9), but now $\mathbf{r}$ varies with longitudinal position (volume) in the reactor, rather than with time.

Since F is a constant, the integrated form may be formally written

$$\frac{V}{F} = \int_{x_1}^{x_2} \frac{dx}{\mathbf{r}} \tag{3-13}$$

This method of writing the mass balance, like Eq. (3-10) for batch operation of a tank reactor, separates the extensive variables V and F and relates them to an integral dependent on the intensive conditions in the reaction mixture. It is worthwhile to note the similarity between Eq. (3-13) and the more familiar design equation for heat-transfer equipment based on an energy balance. This may be written

$$\frac{A_h}{F_t} = \int_{T_1}^{T_2} \frac{c_p\,dT}{U\,\Delta T} \tag{3-14}$$

where A_h is the required heat-transfer area [analogous to V in Eq. (3-13)] and F_t is the mass-flow rate of one fluid through the exchanger (analogous to F).

The rate of reaction plays the same role in reactor design that the product of the overall heat-transfer coefficient U and the temperature difference ΔT plays in sizing heat-transfer equipment. Thus the numerical value of the integrals in Eqs. (3-10), (3-13), and (3-14) represents the degree of difficulty of the job to be done—whether it be chemical conversion or heat transfer. Similarly, the design equation for the absorption of a gas in a packed column through which an absorbent is passed can be written

$$\frac{V}{L} = \int_{C_1}^{C_2} \frac{dC}{K_l a_v (C_{eq} - C)} \qquad (3\text{-}15)$$

where V = required volume of the absorption tower
  C = concentration
  L = volumetric rate of flow of liquid absorbent
  $K_l a_v$ = rate of absorption of liquid per unit concentration difference
  $C_{eq} - C$

The similarity to Eq. (3-13) is apparent here as well. Again the integral measures the difficulty of the job.

The design equations for ideal tubular-flow reactors involve no new concepts but simply substitute a rate of reaction for a heat-transfer rate or mass-transfer-rate function. The increased complexity of reactor design in comparison with the design of equipment for the purely physical processes lies in the difficulty in evaluating the rate of reaction. This rate is dependent on more, and less clearly defined, variables than a heat- or mass-transfer coefficient. Accordingly, it has been more difficult to develop correlations of experimental rates, as well as theoretical means of predicting them.

Energy Balance   If the enthalpy of the reaction mixture per unit mass above the base state is H at the entrance to the element and $H + \Delta H$ at the exit, the energy balance becomes

$$F_t H \Delta t - F_t (H + \Delta H) \Delta t + U(\Delta A_h)(T_s - T)\Delta t = 0$$

or

$$-F_t \Delta H + U(\Delta A_h)(T_s - T) = 0 \qquad (3\text{-}16)$$

where F_t is the total mass-flow rate. The change in enthalpy ΔH in the volume element will depend on the heat of reaction and the rate of reaction. As in the development of Eq. (3-5), we add the change in enthalpy due to the temperature change, $c_p \Delta T$, and the change due to the reaction $(\Delta H/M)\Delta x$. Hence

$$\Delta H = c_p \Delta T + \frac{\Delta H}{M} \Delta x \frac{F}{F_t}$$

Using this expression for ΔH in Eq. (3-16) yields

$$-F_t c_p \Delta T - F \frac{\Delta H}{M} \Delta x + U(\Delta A_h)(T_s - T) = 0 \qquad (3\text{-}17)$$

If we divide each term by ΔV and take the limit as $\Delta V \to 0$, we obtain[1]

$$F_t c_p \frac{dT}{dV} = U(T_s - T)\frac{dA_h}{dV} + F\left(\frac{-\Delta H}{M}\right)\frac{dx}{dV} \tag{3-18}$$

This result can be written in terms of the rate instead of the conversion if we substitute r/F for dx/dV, according to Eq. (3-12). Note that dA_h/dV is the rate of change of heat-transfer area with reactor volume; for a cylindrical tube this would be

$$\frac{dA_h}{dV} = \frac{d(\pi d_t z)}{d(\pi d_t{}^2 z/4)} = \frac{4}{d_t} \tag{3-19}$$

where z is the axial distance along the tube.

Equation (3-18) establishes the temperature at any element of volume in the reactor. It is useful in conjunction with the mass balance, Eq. (3-12), in solving nonisothermal problems.

Example 3-3   The thermal (noncatalytic) decomposition of acetaldehyde,

$$CH_3CHO(g) \to CH_4(g) + CO(g)$$

is studied in an ideal tubular-flow reactor at a constant total pressure of p_t atm. Suppose pure acetaldehyde vapor enters the reactor at $T_0°K$ and a steady rate of F g/sec. The heat of reaction and specific heat of the reaction mixture can be assumed constant and equal to ΔH cal/g mole and c_p cal/(g mole)(°C). The rate of reaction is given by the second-order equation

$$\mathbf{r} = \mathbf{A}(e^{-E/R_gT})\, p_A{}^2 \qquad \text{g moles/(sec)(cm}^3) \tag{A}$$

where E is the activation energy, in calories per gram mole, and p_A is the partial pressure of acetaldehyde, in atmospheres. If the reactor operates adiabatically, express the rate of reaction in terms of the conversion as the sole variable.

Solution   The energy balance relates the temperature to conversion, providing a method of eliminating T from the rate equation. Variables other than x and T are not involved in the energy balance because the operation is adiabatic. In Eq. (3-18) the second term is zero, so that the volume of the reactor does not enter into the problem. Thus

$$Fc_p \frac{dT}{dV} = F\left(\frac{-\Delta H}{M}\right)\frac{dx}{dV}$$

---

[1] The term $F_t c_p$ is the heat capacity of the reaction mixture per unit time. While the total mass-flow rate F_t is constant throughout the reactor, c_p may vary with temperature and composition. Where the specific heats of the various components differ, it may be convenient to evaluate $F_t c_p$ as $\Sigma_i F_i c_{pi}$, where F_i is the mass-flow rate of component i. This quantity varies with location in the reactor. The evaluation of the term is illustrated in Example 5-2.

or

$$dT = \frac{-\Delta H}{c_p M}\, dx \tag{B}$$

Since the coefficient of dx is constant, Eq. (B) may be integrated to give the absolute temperature T at any conversion x,

$$T - T_0 = \frac{-\Delta H}{c_p M}\,(x - 0) = \frac{-\Delta H(x)}{c_p M} \tag{C}$$

where $x = 0$ at $T = T_0$ (entrance to reactor).

Through Eq. (C) the part of the rate equation involving the temperature can be expressed in terms of x. The partial pressure of CH_3CHO is related to the conversion by the stoichiometry of the reaction. If we choose 1 mole of acetaldehyde as a basis, at any conversion x the number of moles of the constituents will be

Acetaldehyde $= 1 - x$
Methane $= x$
Carbon monoxide $= x$
Total moles $= 1 + x$

Hence the mole fraction and partial pressure of acetaldehyde are

$$y_A = \frac{1 - x}{1 + x}$$
$$p_A = \frac{1 - x}{1 + x}\, p_t \tag{D}$$

Substituting Eqs. (C) and (D) into Eq. (A) gives the required expression for the rate in terms of x,

$$\mathbf{r} = \Lambda \exp\left[-\frac{E}{R_g}\left(\frac{1}{T_0 - [\Delta H(x)/c_p M]} \right) \right]\left(\frac{1 - x}{1 + x} \right)^2 p_t{}^2 \tag{E}$$

This expression could be used in the mass-balance equation (3-12) and integrated (p_t is constant) to obtain the exit conversion in terms of the volume of the reactor. It is possible here to obtain a single equation relating x and V only because the reactor operated adiabatically. Otherwise the volume of the reactor would have been included in the energy balance, and both Eq. (3-18) and Eq. (3-12) would have to be solved simultaneously. Such solutions are illustrated in Chap. 5.

3-5 Deviations from Ideal Reactors: Mixing Concepts

Imagine a tank reactor of poor design such that pockets of stagnant fluid exist, as shown by regions marked S in Fig. 3-3a. The conversion will become very high in the stagnant fluid, but this fluid does not leave the reactor. The remainder of the fluid will spend less time in the reactor than it would if the pockets did not exist, and hence it will have less time to react.

The result will be an average conversion in the exit stream which is less than that for the ideal type. Figure 3-3*b* shows another type of deviation, caused by bypassing or short-circuiting of the fluid. Here a part of the fluid entering the tank takes a shortcut to the exit and maintains its identity (does not mix) while doing so. Again, the conversion in the exit stream is reduced below that of the ideal stirred-tank reactor. While these are extreme cases attributable to poor design, it is clear that actual reactors may deviate to some degree from ideal behavior, and the behavior of a specific reactor will depend on the extent of mixing.

Deviations from ideal tubular-flow behavior are also possible. Two kinds of deviations are: (1) some mixing in the longitudinal direction and (2) incomplete mixing in the radial direction. Figure 3-4*a* and *b* illustrates these two effects. In Fig. 3-4*a* the inlet and exit nozzles are such that vortices and eddies produce mixing in the longitudinal direction. Figure 3-4*b* represents the situation where the fluid is in laminar flow, forming a parabolic velocity profile across the tube. Since the molecular-diffusion process is relatively slow, the annular elements of fluid flow through the reactor only slightly mixed in the radial direction. Also, the fluid near the wall will have a longer residence time in the reactor than for ideal tubular-flow performance, while the fluid near the center will have a shorter residence time. The result again is a decrease in conversion. Bypassing or short-circuiting can also occur in a fixed-bed, fluid-solid catalytic reactor, as shown in Fig. 3-4*c*. The nonuniform packing arrangement results in a higher flow rate about one pellet diameter from the wall. The composition of the fluid will be different at this radius, so that the requirement of complete radial mixing is not satisfied.

All deviations from ideal performance fall into two classifications. The first is a flow arrangement in which elements of fluid do not mix, but

Fig. 3-3 *Deviations from ideal stirred-tank performance: (a) stagnant regions, (b) bypassing*

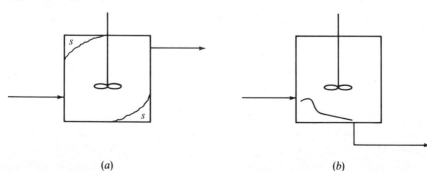

(*a*)   (*b*)

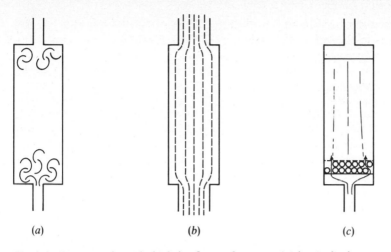

(a)   (b)   (c)

Fig. 3-4 *Deviations from ideal tubular-flow performance: (a) longitudinal mixing due to vortices and turbulence, (b) laminar-flow (poor radial mixing), (c) bypassing in fixed-bed catalytic reactor*

follow separate paths through the reactor (segregated flow). These elements are retained in the reactor for different times; that is, they have different residence times. The second is a flow arrangement whereby adjacent elements of fluid partially mix (micromixing). The effects of these deviations on the conversion can be evaluated, provided we know the distribution of residence times in the fluid leaving the reactor *and* the extent of micromixing. Such complete information is seldom available. However, for well-defined cases of micromixing the effect of a distribution of residence times on the conversion can be evaluated. Procedures for treating these factors quantitatively are considered in Chap. 6. Usually the effects are relatively small, although for special reactors, such as a stirred-tank type with internal coils and a viscous reaction mixture, they may be large. Because of the great influence of temperature on the rate of reaction, unaccounted-for temperature profiles in a reactor are often more significant.[1]

3-6 Space Velocity and Space Time in Tubular-flow Reactors

The term V/F evaluated from Eqs. (3-3) and (3-13) determines the size of the flow reactor necessary to process a given feed rate F. The inconsistency of comparing volume with mass can be overcome by converting the feed rate to a volumetric basis. If ρ_F is the density of the feed, then F_t/ρ_F represents

[1] See K. G. Denbigh, "Chemical Reactor Theory," pp. 44, 63, Cambridge University Press, New York, 1965.

the total feed rate as volume per unit time. The ratio of the feed rate in volumetric units to the volume of reactor is the space velocity,

$$\vartheta_F = \frac{F_t}{V\rho_F} \tag{3-20}$$

Space velocity, a term commonly used to describe the extensive operating characteristics of a tubular-flow reactor, is the maximum feed rate per unit volume of reactor that will give a stated conversion. Also space velocity is a measure of the ease of the reaction job. A high space velocity means that the reaction can be accomplished with a small reactor or that a given conversion can be obtained with a high feed rate.

The reciprocal of the space velocity is the space time θ_F

$$\theta_F = \frac{V\rho_F}{F_t} = \frac{1}{\vartheta_F} \tag{3-21}$$

If the feed consists only of reactant, then $F_t = F$ and the mass balance, Eq. (3-13), may be written

$$\theta_F = \frac{1}{\vartheta_F} = \rho_F \int_{x_1}^{x_2} \frac{dx}{\mathbf{r}} \tag{3-22}$$

The space time is usually not equal to the actual time an element of fluid resides in the reactor. Variations in the temperature, pressure, and moles of reaction mixture can all cause the local density to change through the reactor and be unequal to ρ_F. Also, feed rates frequently are measured under conditions grossly different from those in the reactor. For example, the feed may be measured as a liquid and then vaporized before it enters the reactor. Also, as observed in Sec. 3-5, there may well be a distribution of residence times in the fluid leaving, so that we must use the concept of a *mean residence time*. The mean residence time is equal to θ_F only when the following conditions are met:

1. The temperature and pressure are constant throughout the reactor.
2. The feed rate is measured at the temperature and pressure in the reactor.
3. The density of the reaction mixture is constant (for a gaseous reaction this requires, in addition to 1, that there be no change in number of moles).

The effect of temperature and pressure variations on the residence time for an *ideal* tubular-flow (plug-flow) reactor can be evaluated by comparing an actual residence time θ with θ_F. The actual time required for an element of fluid to pass through the volume of reactor dV is

$$d\theta = \frac{\text{distance}}{\text{velocity}} = \frac{\text{volume}}{\text{volumetric flow rate}} = \frac{dV}{F_t/\rho} \qquad (3\text{-}23)$$

where ρ is the local mass density at the point dV in the reactor. For a gaseous reaction it is more appropriate to express $d\theta$ in terms of the total molal flow rate N_t' and the volume per mole v. In this form Eq. (3-23) may be written

$$d\theta = \frac{dV}{N_t'v} \qquad (3\text{-}24)$$

In Eq. (3-24) both N_t' and v can change with position. When ρ, or N_t' and v, are constant, Eqs. (3-23) and (3-21) show that $\theta = \theta_F$. When this is not true, the actual time must be obtained by integration. Since N_t' and v are more easily related to the conversion than to V, it is convenient to substitute for dV the value from Eq. (3-12). Then Eq. (3-24) becomes

$$d\theta = \frac{F\,dx}{N_t'v\mathbf{r}}$$

Integrating formally, we obtain

$$\theta = F \int_{x_1}^{x_2} \frac{dx}{N_t'v\mathbf{r}} \qquad (3\text{-}25)$$

Example 3-4 Acetaldehyde vapor is decomposed in an *ideal* tubular-flow reactor according to the reaction

$$CH_3CHO \rightarrow CH_4 + CO$$

The reactor is 3.3 cm ID and 80 cm long and is maintained at a constant temperature of 518°C. The acetaldehyde vapor is measured at room temperature and slightly above atmospheric pressure. For consistency, the measured flow rate is corrected to standard conditions (0°C and 1 atm) before the space velocity is reported. In one run, at a reported space velocity of 8.0 hr^{-1}, 35% of the acetaldehyde is decomposed in the reactor. The second-order specific-rate constant is 0.33 liter/(scc)(g mole) at 518°C, and the reaction is irreversible. The pressure is essentially atmospheric. Calculate the actual residence time and compare it with the space time determined from Eq. (3-21).

Solution The rate equation may be written

$$\mathbf{r} = k[CH_3CHO]^2 = \text{g moles/(liter)(sec)}$$

At a point in the reactor where the conversion is x, the molal flow rates are

$$CH_3CHO = F(1 - x)$$

$$CH_4 = Fx$$

$$CO = Fx$$
$$N_t' = F(1 + x)$$

The molal concentration of CH_3CHO is the ratio of $F(1 - x)$ to the total volumetric flow rate Q_t,

$$[CH_3CHO] = \frac{F(1 - x)}{Q_t} = \frac{F(1 - x)p_t}{N_t R_g T} = \frac{1 - x}{1 + x}\frac{p_t}{R_g T}$$

Then the rate in terms of x is

$$\mathbf{r} = k\left(\frac{1 - x}{1 + x}\right)^2 \left(\frac{p_t}{R_g T}\right)^2$$

Under perfect-gas conditions the volume per mole of reaction mixture will be

$$v = \frac{1 R_g T}{p_t}$$

and

$$N_t' = F(1 + x)$$

Substituting these values for $\mathbf{r}$, v, and N_t' in Eq. (3-25) gives

$$\theta = F \int_0^{x_2} \frac{p_t(1 + x)^2 \, dx}{F(1 + x)R_g Tk(1 - x)^2 (p_t/R_g T)^2}$$

$$= \frac{R_g T}{p_t k} \int_0^{x_2} \frac{(1 + x) \, dx}{(1 - x)^2} = \frac{R_g T}{p_t k}\left[\frac{2}{1 - x} + \ln(1 - x)\right]_0^{x_2 = 0.35}$$

$$= \frac{0.082(518 + 273)}{1(0.33)}\left[\frac{2}{1 - 0.35} + \ln(1 - 0.35) - 2\right] = 127 \text{ sec}$$

The space time, from Eq. (3-21), is

$$\theta_F = \frac{1}{\vartheta_F} = \frac{1}{8.0} = 0.125 \text{ hr} \qquad \text{or } 450 \text{ sec}$$

The major difference between θ and θ_F arises because the space velocity was based on a flow rate at a standard temperature of 273°C. If the density of feed ρ_F were based on the actual reaction temperature of 518°C (791°K), the residence time would be much lower. Thus if the space velocity were corrected to the reactor temperature, it would be

$$\vartheta_F \text{ (at 791°K)} = 8.0\frac{\rho_F(\text{at } 273°\text{K})}{\rho_F(\text{at } 791°\text{K})} = 8.0\left(\frac{791}{273}\right) = 23.2 \text{ hr}^{-1}$$

and the space time would be

$$\theta_F = \frac{1}{23.2}(3,600) = 155 \text{ sec}$$

The difference between this value and the actual θ (127 sec) is due to the increase in the number of moles occurring as a result of reaction.

3-7 *Temperature Effects*

In a few cases the heat of reaction is so low that heat exchange with the surroundings is sufficient to eliminate temperature changes. The design problem for such isothermal reactors is greatly simplified because the variation in rate of reaction with temperature need not be considered. The isomerization of *n*-butane (ΔH at $25°C = -1600$ cal/g mole) is an example of this class of reactions. Even when the heat of reaction is moderate, it may be possible to approach isothermal operation by adding or removing heat from the reactor. In the sulfuric acid alkylation process for producing iso-octane from isobutane and butenes the heat of reaction is about $-17,000$ cal/g mole at $25°C$. However, by cooling the liquid mixture in the reactor with external cooling jackets it is possible to reduce the temperature variation to 20 to $40°F$.

When the heat of reaction is large, sizable temperature variations may be present even though heat transfer between the reactor and surroundings is facilitated. In such cases it is necessary to consider the effect of temperature on the rate of reaction. Reactors operating in this fashion are termed *nonisothermal* or *nonadiabatic*.

Reactors (both flow and batch) may also be insulated from the surroundings so that their operation approaches adiabatic conditions. If the heat of reaction is significant, there will be a change in temperature with time (batch reactor) or position (flow reactor). In the flow reactor this temperature variation will be limited to the direction of flow; i.e., there will be no radial variation in a tubular-flow reactor. We shall see in Chap. 13 that the design procedures are considerably simpler for adiabatic operation.

In addition to heat exchange with the surroundings, there are other methods of approaching isothermal operating conditions. For example, in the dehydrogenation of butylenes to butadiene the temperature must be maintained at a rather high level (1200 to $1400°F$) for a favorable equilibrium conversion. However, the endothermic nature of the reaction means that the reaction mixture will cool as it flows through the reactor bed. It is both difficult and expensive to transfer heat to the reaction mixture at this high temperature level by external heating. Instead, high-temperature steam is added directly to the butenes entering the reactor. The large quantity of steam serves to maintain the reaction mixture at a high temperature level.[1] Another device frequently employed for tank reactors is internal cooling

[1] Steam has other advantages; in particular, it reduces polymerization.

or heating coils. A modification of the same principle is illustrated by the Downs[1] reactor for the oxidation of naphthalene to phthalic anhydride. The flow reactor is divided into a large number of small tubes rather than a single large-diameter tube. Each small tube is surrounded with cooling fluid which absorbs the heat of reaction. In this particular operation boiling mercury is employed as the cooling medium. Under actual operating conditions some naphthalene is oxidized completely to carbon dioxide and water vapor, so that the heat of reaction per gram mole of naphthalene is as high as $-570,000$ cal.

In summary, the operation of commercial reactors falls into three categories: isothermal, adiabatic, and the broad division of nonadiabatic, where attempts are made to approach isothermal conditions, but the magnitude of the heat of reaction or the temperature level prevents attaining this objective. Quantitative calculations for isothermal and nonisothermal homogeneous reactors are given in Chaps. 4 and 5.

3-8 Mechanical Features

Batch Reactors   The batch reactor is a kettle or tank, or it may be a closed loop of tubing provided with a circulating pump. It should have a number of accessories to be operated satisfactorily. It generally must be closed, except for a vent, to prevent loss of material and danger to the operating personnel. For reactions carried out under pressure the vent is replaced by a safety valve.

High-pressure conditions frequently introduce complications in the design and greatly increase the initial cost. For tanks the top closure must be able to withstand the same maximum pressure as the rest of the auto- clave. At medium pressures a satisfactory closure can be assembled by using bolts or studs and suitable flanges and gaskets. The seal is obtained by tightening the six or more bolts holding the flange to the head. Such a closure is illustrated in the batch reactor shown in Fig. 3-5. For higher pressure (above approximately 5,000 psia) this type of construction is not desirable because of the very high stresses that the bolts must withstand. The preferred design is one in which the pressure itself seals the vessel, and increases in pressure do not cause a corresponding increase in stress in the bolts. An example of this self-sealing closure is shown in Fig. 3-6. The pressure acting on the head is transmitted to the gasket, which is confined by the reactor wall, the head, and a retaining ring. The internal pressure pushes the head against the gasket, augmenting the force exerted by the bolts through the lifting collar. The problems encountered in designing batch

[1] C. R. Downs, *Ind. Eng. Chem.*, **32**, 1294 (1940); U.S. Patent 1.604,739, 1926.

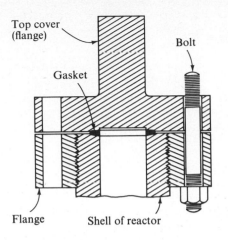

Top cover (flange)

Bolt

Gasket

Flange   Shell of reactor

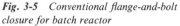

Fig. 3-5   Conventional flange-and-bolt closure for batch reactor

reactors for medium- and high-pressure operation have been studied in some detail.[1] Loop reactors are well suited for high pressure, because the tubing has a smaller diameter than a tank.

It is necessary to agitate the reaction mixture in tank reactors. This can be done mechanically with stirrers operated by a shaft extending through the reactor wall. In high-pressure reactors rather complicated packing glands are needed around the shaft to prevent leakage. A typical design is illustrated in Fig. 3-7, where the mechanical details of a reactor, which is also jacketed, are shown.

[1] D. B. Gooch, *Ind. Eng. Chem.*, **35**, 927 (1943); E. L. Clark, P. L. Golber, A. M. Whitehouse, and H. H. Storch, *Ind. Eng. Chem.*, **39**, 1955 (1947); F. D. Moss, *Ind. Eng. Chem.*, **45**, 2135 (1945).

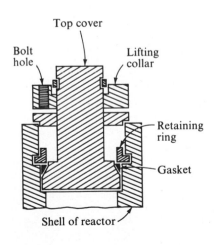

Top cover

Bolt hole

Lifting collar

Retaining ring

Gasket

Shell of reactor

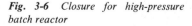

Fig. 3-6   Closure for high-pressure batch reactor

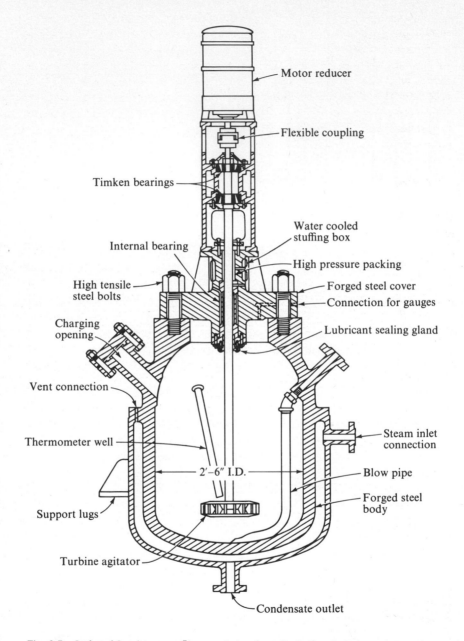

Fig. 3-7   *Jacketed batch reactor* [*by permission from D. B. Gooch*, Ind. Eng. Chem., *35, 927 (1943)*]

Batch reactors vary in construction from ordinary steel tanks to glass-lined equipment, depending on the properties of the reaction mixture for which they are to be used. In pilot-plant operations either stainless-steel or glass-lined reactors are ordinarily used because of their corrosion resistance, and hence their general applicability to a variety of systems. In commercial-scale equipment it may be more economical to use ordinary steel even though corrosion is significant. In the food and pharmaceutical industries it is frequently necessary to use glass-lined or stainless-steel equipment to ensure purity of the product.[1]

Flow Reactors   Flow reactors may be constructed in a number of ways. The conventional thermal cracking units in the petroleum industry are examples of a noncatalytic type. The petroleum fraction is passed through a number of lengths of alloy-steel tubes placed in series on the walls and roof of a furnace. Heat is transferred by convection and radiation to the tube surface in order to raise the temperature of the fluid to the reaction level (600 to 1300°F) and to supply the endothermic heat of reaction. Flow reactors may consist of a tank or kettle, much like a batch reactor, with provision for continuously adding reactants and withdrawing product. The tank type is not suitable for reactions such as thermal cracking, where large quantities of thermal energy must be supplied, because of the low-heat-transfer surface per unit volume of reactor. Tank-type flow reactors are advantageous for conversions that require a long reaction time. In such reactors it is possible to obtain essentially complete mixing by mechanical agitation. Under these conditions the composition, temperature, and pressure are uniform through the vessel. In the tubular type, where the length is generally large with respect to the tube diameter, the forced velocity in the direction of flow is sufficient to retard mixing in the axial direction; that is, it is possible to approach plug-flow performance.

A large number of commercially important reactions are of the fluid-solid catalytic class. Examples are the catalytic cracking of petroleum, oxidation of sulfur dioxide, ammonia synthesis, dehydrogenation of butenes to butadiene, and oxidation of naphthalene to phthalic anhydride. In this group of reactions the solid catalyst may be held in a fixed position while the fluid moves through it (fixed-bed reactors), or much smaller catalyst particles may be suspended in the fluid phase by the motion of the fluid (fluidized-bed reactor), or the solid particles may be in point-to-point contact and fall slowly by gravity through the fluid (moving-bed reactor).

[1] See J. H. Perry, "Chemical Engineers' Handbook," McGraw-Hill Book Company, New York, 1950.

A fixed-bed reactor built as a single large-diameter tube is less costly than a multitubular design. However, the latter arrangement may be required when large quantities of heat must be transferred to the surroundings, as in the case of a highly exothermic reaction. The smaller the tube diameter, the larger the ratio of heat-transfer surface to mass of reaction mixture in the tube, and the easier it is to limit temperature changes between inlet and exit. Of course, the low capacity of small tubes means that a larger total number of tubes must be built into the reactor in parallel to achieve a given production rate. Other means of preventing large temperature variations in fixed-bed reactors may be used. In addition to the devices suggested earlier, the catalyst bed may be divided into sections, with heating or cooling coils placed between each section.

All these devices to reduce temperature gradients in the fixed-bed reactor are corrective rather than preventive. In solid catalytic reactors the potentially large temperature variations in the direction of flow are due to the fact that the solid catalyst is unable to mix and reach a more uniform temperature. Near the entrance to the bed the rate of reaction is high and large quantities of heat are evolved (for an exothermic reaction), while near the exit, where the rate is low, there is a relatively small evolution of heat. Because the heat-transfer rate from pellet to pellet and between pellet and gas is small, each layer of catalyst in the bed is, in effect, partially insulated from adjacent layers. This effectively prevents the flow of heat from the entrance to the exit of the catalyst bed, which in turn causes significant temperature gradients. The fluidized-bed reactor does away with the source of this difficulty, the stationary condition of the bed. The rapid movement of the small catalyst particles goes a long way toward eliminating temperature variations within the solid phase. Any one particle may be near the entrance of the reactor at one instant and near the exit at the next. This rapid mixing tends to equalize both fluid- and solid-phase temperatures, so that the entire reactor system approaches a uniform temperature. The small size (generally 5 to 100 microns) of the fluidized particles provides a large heat-transfer area per unit mass and in this way increases the heat-transfer rates between solid and gas phases. Although the resultant motion of the gas is upward through the reactor, usually neither plug-flow nor complete mixing describes the flow pattern (see Chap. 13).

An important advantage of the fluidized-bed reactor over the fixed-bed type is that the catalyst can be regenerated without disturbing the operation of the reactor. In fluidized-bed catalytic cracking units a portion of the solid particles is continuously removed from the reactor and regenerated in a separate unit. The regeneration is accomplished by burning off the carbon with air, and the reactivated catalyst is continuously returned to the reactor proper. In the fixed-bed reactor the closest approach to continuous operation obtainable with a catalyst of limited life is to construct two or more identical reactors and switch streams from one to the other when the catalyst needs to be regenerated.

A disadvantage of fluidized-bed reactors is that the equipment is large. So that the solid particles will not be blown out the top of the reactor, the gas velocity must be low. This in turn necessitates large-diameter vessels and increases the initial cost. There are also losses of catalyst fines from the reactor, necessitating expensive dust-collection equipment in the exit streams.

Moving-bed systems do not permit the uniformity of temperature achieved in fluidized-bed reactors, but they do allow continuous handling of the solid phase. This is advantageous in some operations, such as catalyst or adsorbent regeneration. For example, the regeneration of the charcoal adsorbent for the hypersorption process[1] of separating hydrocarbons has been accomplished in a moving-bed reactor. The deactivated charcoal is added to the top of the regenerator, and steam is passed upward through the slowly moving bed of solid to strip out and react with the adsorbed hydrocarbons. Moving-bed systems are also employed for fluid-solid noncatalytic reactions; e.g., blast furnaces, lime kilns, and smelters (see Chap. 14).

Problems

3-1. A pellet of uranium dioxide actually consists of a mixture of UO_2 and UO_3 with a composite formula $UO_{2.14}$. The pellet is reacted with HF gas, and the extent of reaction is followed by weighing the pellet at successive time increments. The reactions are

1. $UO_2(s) + 4HF(g) \rightarrow UF_4(s) + 2H_2O(g)$

2. $UO_3(s) + 2HF(g) \rightarrow UO_2F_2(s) + H_2O(g)$

Derive a relationship between the initial weight of the pellet, m_0, its weight at any time, m, and a composite conversion expressed as the fraction of $UO_{2.14}$ reacted.

3-2.† Desired product C is produced in an ideal tubular-flow reactor by the second-order reaction

$$A + B \rightarrow C \qquad \mathbf{r}_C = k_1[A][B]$$

Unwanted product A_2 is also produced by a second-order reaction

$$A + A \rightarrow A_2 \qquad \mathbf{r}_{A_2} = k_2[A]^2$$

For the maximum selectivity of C to A_2 it is necessary to keep the concentration

[1] Clyde Berg, *Trans. AIChE*, **42**, 685 (1946).
†This problem was supplied by Professor J. B. Butt.

of A as low as possible. To approach this ideal, only B is added to the inlet to the reactor, and A is added uniformly along the length, as indicated in the figure. The molal flow rate of B is $Q_B[B]_0$, where Q_B is the volumetric flow rate entering the reactor and $[B]_0$ is the concentration of B in the feed. Reactant A is added

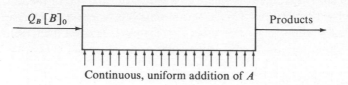

Continuous, uniform addition of A

through the reactor wall in a stream whose concentration is $[A]_0$. The uniform rate of addition of this stream is dQ_A/dV, as volumetric flow rate per unit volume of reactor. Assuming that the density is constant, write mass balances for reactants A and B for this reactor operating at steady state.

3-3. (a) An endothermic gaseous reaction is carried out in an ideal tubular-flow reactor operated adiabatically. If the heat capacity of the reaction mixture is constant, develop an integrated relationship [starting with Eq. (3-18)] between the conversion and temperature at any axial position in the reactor. The feed temperature is T_0 and the conversion in the feed is zero. (b) To reduce the temperature drop along the reactor it is proposed to add inert nitrogen at temperature T_0 to the feed. If sufficient nitrogen is added to double the heat capacity of the reaction mixture, how much will the temperature drop be reduced?

3-4. In an otherwise ideal stirred-tank reactor it is estimated that 10% of the reactor volume is occupied by stagnant fluid (Fig. 3-3a). If the reactor had no stagnant regions, the conversion in the effluent stream would be 60%. What conversion is expected in the actual reactor?

3-5. In studying the kinetics of the homogeneous gas-phase reaction between sulfur vapor and methane, Fisher[1] reported conversions for various space velocities. These space velocities were defined as the volumetric flow rate in milliters per hour divided by the total volume of empty reactor in cubic centimeters. The flow rate is based on all the sulfur being considered S_2 and is referred to 0°C and 1 atm pressure.

From the fact that the operating pressure was 1.0 atm and the temperature 600°C, compute the values of V/F corresponding to the space velocities given in the reference for runs 55, 58, 57, 78, and 79. V/F is the ratio of the volume of reactor to the molal feed rate, in gram moles per hour. Also determine the true contact time for a slug of reaction mixture for each of the runs; i.e., the time it takes for a slug of gas to pass through the reactor.[2]

3-6. The production of toluene from benzene and xylenes was studied by Johanson and Watson[3] in a standard 1-in.-pipe reactor, with a silica-alumina catalyst. At the reactor temperature of 932°F the reaction mixture is in the vapor phase.

[1] R. A. Fisher, *Ind. Eng. Chem.*, **42**, 704 (1950).
[2] See Fisher's work (or Table 4-5) for the conversion values for the runs.
[3] L. N. Johanson and K. M. Watson, *Natl. Petrol. News*, Aug. 7, 1946.

However, the benzene and xylenes were measured and pumped separately into the system as liquids by means of a proportioning pump. Hence the space velocity was reported on a liquid-hourly basis; that is, as the ratio of the feed rate, in cubic centimeters of liquid per hour, to the total volume of the reactor, in cubic centimeters. The feed consisted of an equimolal mixture of reactants, and the liquid rates were corrected to 60°F before reporting the following information:

Liquid-hourly space velocity, hr^{-1}	0.5	0.25	1.0	2.0	2.0	4.0
Reactor pressure, psia	20	20	65	65	115	115

The reactor contained 85 g of catalyst packed in a volume of 135 cm³, and the densities of benzene and xylenes at 60°F may be taken as 0.879 and 0.870 g/cm³, respectively.

From the data determine corresponding ratios of mass of catalyst to feed rate, expressed in units of g catalyst/[(g mole)/(hr)].

3-7. Convert the liquid-hourly space velocities in Prob. 3-6 to a gas basis, that is, to space velocities defined as the ratio of the gas-flow rate at reaction conditions to the total reactor volume. Then calculate the actual contact time for each run. The gases may be assumed to obey the perfect-gas law. The reaction does not result in a change in number of total moles,

$$C_6H_6 + C_6H_4(CH_3)_2 = 2C_6H_5CH_3$$

4

HOMOGENEOUS REACTOR DESIGN: ISOTHERMAL CONDITIONS

The conservation principles applied to mass and energy were used in Chap. 3 to develop equations for *ideal* forms of batch and flow reactors. Now that we have some knowledge of kinetics of homogeneous reactions (Chap. 2), we can design ideal batch, tubular-flow, and tank reactors. Isothermal behavior is considered here; nonisothermal operation is in Chap. 5. Nonideal flow complications are taken up in Chap. 6 for homogeneous reactions. Physical resistances are usually most significant for heterogeneous reactions (for example, fluid-solid catalytic reactions) carried out in tubular-flow reactors. Hence a more general approach to tubular-flow reactors, which includes the effect of mass- and energy-transfer resistances, is given in Chap. 13.

There are *homogeneous* reactions of commercial importance of both the catalytic and the noncatalytic type. For example, one process for the production of ethylene dichloride consists of the reaction between ethylene and chlorine in the presence of bromine, with all three materials in the vapor phase. The bromine reacts with chlorine to form the unstable bromine chloride, which behaves as a catalyst in accordance with the following reactions:

1. $C_2H_4 + BrCl \rightarrow C_2H_4BrCl$

2. $C_2H_4BrCl + Cl_2 \rightarrow C_2H_4Cl_2 + BrCl$

Illustrations of noncatalytic homogeneous reactions are numerous and include the thermal cracking of hydrocarbons, the combustion of gaseous fuels such as natural gas, and various inorganic reactions in aqueous solutions.

IDEAL BATCH REACTORS

Batch reactors are seldom employed on a commercial scale for gas-phase reactions because the quantity of product that can be produced in a reasonably sized reactor is small. The chief use of batch systems for gaseous reactions is in kinetic studies. Batch reactors are often used, however, for liquid-phase reactions, particularly when the required production is small. Batch reactors are generally more expensive to operate than continuous units for the same production rate. However, the initial cost of a continuous system may be higher owing to the instrumentation required. Therefore, for relatively high-priced products (such as pharmaceuticals) where operating expense is not a predominant factor in the total cost, batch reactors are commonly employed.

4-1 Simplified Design Equation for Constant Volume

In Chap. 3 Eq. (3-10) was developed to relate conversion and time in a batch reactor. If the volume of the reaction mixture (i.e., the density) is constant, this expression can be written in a simpler way in terms of concentration. At constant volume the conversion and concentration of *reactant* are related by

$$x = \frac{C_0 - C}{C_0} \tag{4-1}$$

where C_0 is the initial concentration and is equal to m/V. Equation (3-10) then becomes

$$t = C_0 \int_0^x \frac{dx}{\mathbf{r}} = - \int_{C_0}^c \frac{dC}{\mathbf{r}} \tag{4-2}$$

Note that the differential form of Eq. (4-2),

$$r = -\frac{dC}{dt}$$

is identical to Eq. (2-4), which was used throughout Chap. 2 to interpret laboratory kinetic data.

4-2 Design Procedure

The difficulty in integrating Eqs. (3-10) and (4-2) depends on the number of variables influencing the rate of reaction. For example, if the rate of formation of the desired product depends on only one irreversible reaction, the expression for **r** will be simpler than if reversible or multiple reactions are involved. The integration of Eq. (4-2) for various reaction networks under constant-temperature conditions was considered in Chap. 2. At that point the objective was determination of the rate constant k. In reactor design the situation is reversed: k, and hence **r**, is known, and it is the time necessary to obtain a given conversion that is required.

The following example concerning the rate of esterification of butanol and acetic acid in the liquid phase illustrates the design problem of predicting the time-conversion relationship for an isothermal, single-reaction, batch reactor.

Example 4-1   Leyes and Othmer[1] studied the formation of butyl acetate in a batch-operated reactor at 100°C, with sulfuric acid as catalyst. The original feed contained 4.97 moles butanol/mole acetic acid, and the catalyst concentration was 0.032% by weight H_2SO_4. The following rate equation was found to correlate the data when an excess of butanol was used:

$$r = k[A]^2$$

where $[A]$ is concentration of acetic acid, in gram moles per milliliter, and **r** is the rate of reaction, in gram moles of acid disappearing per milliliter per minute. For a ratio of butanol to acid of 4.97 and a sulfuric acid concentration of 0.032% by weight, the reaction-velocity constant was

$$k = 17.4 \text{ ml/(g mole)(min)}$$

Densities of mixtures of acetic acid, butanol, and butyl acetate are not known. Reported densities for each of the three compounds at 100°C were

Acetic acid = 0.958 g/ml
Butanol = 0.742
Butyl acetate = 0.796

Although the density of the reaction mixture varies with conversion, the excess of

[1] C. E. Leyes and D. F. Othmer, *Ind. Eng. Chem.*, **36**, 968 (1945).

butanol reduces the magnitude of the change. Therefore, as an approximation, the density of the mixture was assumed constant and equal to 0.75 g/ml.

(a) Calculate the time required to obtain a conversion of 50%. (b) Determine the size of reactor and the original mass of reactants that must be charged to the reactor in order to produce ester at the average rate of 100 lb/hr. Only one reactor will be used, and this unit will be shut down 30 min between batches for removal of product, cleaning, and startup. Assume that the reaction mixture is well mixed.

Solution   The molecular weights are

> Ester = 116
> Butanol = 74
> Acetic acid = 60

(a) The concentration of acetic acid, $[A]$, is related to the conversion by

$$[A] = [A]_0(1 - x)$$

where $[A]_0$ is the initial acid concentration. Substituting this expression in the rate equation gives

$$\mathbf{r} = k[A]_0^2(1 - x)^2$$

The design expression [Eq. (4-2)], which is applicable at constant volume or density, can be written in terms of the variables t and x as

$$t = [A]_0 \int_0^{x_1} \frac{dx}{k[A]_0^2(1 - x)^2} = \frac{1}{k[A]_0} \int_0^{x_1} \frac{dx}{(1 - x)^2} \tag{A}$$

Integrating and substituting a final value of $x_1 = 0.50$ lead to the result

$$t = \frac{1}{k[A]_0} \left(\frac{1}{1 - x_1} - \frac{1}{1 - 0} \right) = \frac{1}{k[A]_0}(2 - 1) = \frac{1}{k[A]_0} \tag{B}$$

The initial concentration of acetic acid is[1]

$$[A]_0 = \frac{1(0.75)}{4.97(74) + 1(60)} = 0.0018 \text{ g mole/ml}$$

From Eq. (B), the time required for a conversion of 50% is

$$t = \frac{1}{17.4(0.0018)} = 32 \text{ min} \qquad \text{or } 0.53 \text{ hr}$$

(b) The production rate of the reactor, in pounds of ester per hour, in terms of the pounds of acid charged, m_A, will be

$$100 = \frac{(m_A/60)(116)(0.5)}{0.53 + 0.50}$$

[1] Note that in this example $\mathbf{r}$ and x and the concentration are expressed in molar units rather than mass units. Design equations can be used with either molar or mass units as long as $\mathbf{r}$ and x are consistent.

This expression allows 30 min for shutdown time per charge and takes into account that the conversion is 50%. Thus

$m_A = 106$ lb acetic acid/charge

Total charge $= 106 + 4.97(\frac{74}{60})(106) = 756$ lb

The volume occupied by the charge will be

$$V = \frac{756}{0.75(62.4)(0.1337)} = 121 \text{ gal}$$

The reactor must be large enough to handle 121 gal of reaction mixture. The charge would consist of 106 lb of acid and 650 lb of butanol.

When the rate of the reverse reaction is significant (i.e., when equilibrium is approached in the reactor) or when more than one reaction is involved, the mechanics of solving the design equation may become more complex, but the principles are the same. Equation (4-2) is applicable, but the more complicated nature of the rate function may make the mathematical integration difficult.

The calculations for a reversible reaction are illustrated in Example 4-2 for the esterification of ethyl alcohol.

Example 4-2 In the presence of water and hydrochloric acid (as a catalyst) the rate of esterification, in g moles/(liter)(min), of acetic acid and ethyl alcohol at 100°C is given by

$$\mathbf{r}_2 = k[H][OH] \qquad k = 4.76 \times 10^{-4} \text{ liter/(min)(g mole)}$$

The rate of the reverse reaction, the hydrolysis of the ester in the same concentration of catalyst, is

$$\mathbf{r}_2' = k'[E][W] \qquad k' = 1.63 \times 10^{-4} \text{ liter/(min)(g mole)}$$

$$CH_3COOH + C_2H_5OH \underset{k'}{\overset{k}{\rightleftharpoons}} CH_3COOC_2H_5 + H_2O$$

(a) A reactor is charged with 100 gal of an aqueous solution containing 200 lb of acetic acid, 400 lb of ethyl alcohol, and the same concentration of HCl as used to obtain the reaction-velocity constants. What will be the conversion of acetic acid to ester after 120 min of reaction time? The density may be assumed constant and equal to 8.7 lb/gal; neglect the water vaporized in the reactor. (b) What is the equilibrium conversion?

Solution (a) The net rate of formation of acid is obtained by combining the rate expressions for the forward and reverse reactions,

$$\mathbf{r} = k[H][OH] - k'[E][W]$$

The initial concentrations of acid H, alcohol OH, and water W are as follows:

$$[H]_0 = \frac{200}{100(60)} \frac{454(1,000)}{0.1337(30.5)^3} = 4.00 \text{ g moles/liter}$$

$$[OH]_0 = \frac{400}{100(46)} \frac{454(1,000)}{0.1337(30.5)^3} = 10.8 \text{ g moles/liter}$$

$$[W]_0 = \frac{8.7(100) - (200 + 400)}{100(18)} \frac{454(1,000)}{0.1337(30.5)^3} = 18.0 \text{ g moles/liter}$$

With the conversion based on the acid, as required by the statement of the problem, the concentrations at any time are

$$[H] = 4.0(1 - x)$$

$$[OH] = 10.8 - 4x$$

$$[E] = 4.0x$$

$$[W] = 18 + 4x$$

These relationships between concentrations and conversion rest on the assumption of a constant density during the reaction. The problem could still be solved without this assumption, provided data on the variation in density with conversion were available. Substituting the expressions for concentration in the rate equation, we have

$$\mathbf{r} = k[4(1 - x)(10.8 - 4x)] - k'[4x(18 + 4x)]$$

The numerical values for k and k' result in the following equation for $\mathbf{r}$, in g moles/(liter) (min):

$$\mathbf{r} = (0.257 - 0.499x + 0.062x^2)(8 \times 10^{-2})$$

Substituting this in the design equation for constant volume [Eq. (4-2)] yields

$$t = \frac{[H]_0}{8 \times 10^{-2}} \int_0^{x_1} \frac{dx}{0.257 - 0.499x + 0.062x^2}$$

$$= 50 \int_0^{x_1} \frac{dx}{0.257 - 0.499x + 0.062x^2}$$

This expression can be integrated to give

$$t = \frac{50}{0.430} \left[\ln \frac{0.125x - 0.499 - 0.430}{0.125x - 0.499 + 0.430} \right]_0^{x_1}$$

$$= \frac{50}{0.430} \ln \frac{(0.125x_1 - 0.929)(0.069)}{(0.125x_1 - 0.069)(0.929)}$$

The conversion is desired for a time of 120 min. Hence

$$\frac{120(0.430)}{50} = 1.03 = \ln \frac{0.125x_1 - 0.929}{0.125x_1 - 0.069} \frac{0.069}{0.929}$$

Thus $x_1 = 0.365$, or 36.5% of the acid is converted to ester. It is interesting to compare this result with that based on neglecting the reverse reaction. Under this condition the rate equation is

$$\mathbf{r} = k[H][OH] = k[H]_0(1 - x)([OH]_0 - [H]_0 x)$$

The design equation becomes

$$t = [H]_0 \int_0^{x_1} \frac{dx}{k[H]_0(1 - x)([OH]_0 - [H]_0 x)}$$

$$= \frac{1}{k[OH]_0 - [H]_0} \int_0^{x_1} \left(\frac{1}{1 - x} - \frac{[H]_0}{[OH]_0 - [H]_0 x} \right) dx$$

$$= \frac{1}{k([OH]_0 - [H]_0)} \left[-\ln (1 - x_1) + \ln \frac{[OH]_0 - [H]_0 x_1}{[OH]_0} \right]$$

Simplifying, we have

$$\frac{[OH]_0 - [H]_0 x_1}{[OH]_0(1 - x_1)} = e^{tk([OH]_0 - [H]_0)} = e^{120(4.76 \times 10^{-4})(10.8 - 4.0)} = 1.474$$

Solving for x_1, we find

$$x_1 = \frac{0.474[OH]_0}{1.474[OH]_0 - [H]_0} = \frac{0.474(10.8)}{1.474(10.8) - 4.0} = 0.43$$

If the reverse reaction is neglected, the conversion is in error by

$$[(43 - 36.5)/36.5]100 \qquad \text{or } 18\%$$

This deviation would increase as equilibrium is approached.

(b) The conventional method of evaluating the equilibrium conversion is first to calculate the equilibrium constant from the forward- and reverse-reaction rates ($K = k/k'$), and then to use this value of K for the equilibrium ratio of the concentrations. However, it is easier in this case to utilize the expression already available for the net rate of reaction,

$$\mathbf{r} = (0.257 - 0.499x + 0.0626x^2)(8 \times 10^{-2})$$

At equilibrium the net rate must be zero. Hence the equilibrium conversion is determined by the expression

$$0.257 - 0.499x_e + 0.0626x_e^2 = 0$$

Solving, we obtain

$$x_e = 0.55 \qquad \text{or } 55\%$$

In these examples it has been assumed that a negligible amount of reaction occurs during heating and cooling periods. If only one reaction is involved, the time computed on the assumption that reaction takes place only at the operating temperature will be a conservative value. However, in reactions of industrial interest it is more likely that byproducts will be created. Then the yield of desired product will be reduced during heating and cooling cycles in a batch operation, provided an undesired reaction is favored at temperatures below the operating temperature. The magnitude of this effect will depend on the time spent for heating and cooling in relation to reaction time. In serious cases the problem may be solved by preheating the reactants separately prior to mixing, or by feeding one reactant continuously to a batch charge of the coreactant (semibatch operation). Rapid quenching after reaction may be realized by rapid discharge of the mass (blowdown) or by rapid coolant injection (directly or indirectly contacted with the product mass).

IDEAL TUBULAR-FLOW (PLUG-FLOW) REACTORS

Tubular-flow reactors are used both as laboratory units, where the purpose is to obtain a rate equation, and for commercial-scale production. In fact, it is a desirable and common practice to develop a rate equation for data obtained in a laboratory-sized reactor of the same form as the proposed large-scale unit.

4-3 Interpretation of Data from Laboratory Tubular-flow Reactors

Interpretation of data to obtain a rate equation requires calculations which are the reverse of those for design. The two procedures differ because in the laboratory it may be feasible to operate at nearly constant temperature, and possibly nearly constant composition; in the commercial unit it may be possible to approach isothermal conditions, but constant composition in a tubular-flow reactor is impossible. These differences simplify the analysis of laboratory results, as illustrated in Example 4-3.

   The starting point for evaluating the rate of reaction in a *flow reactor* is Eq. (2-2),

$$\mathbf{r} = \frac{dN'}{dV}$$

where dN' is the molal rate of production of product in a reactor volume dV. If the reactor is small enough, the change in composition of the fluid as it flows through the volume will be slight. In addition, suppose that heat-transfer conditions are such that the temperature change is also slight. Since composition and temperature determine the rate, $\mathbf{r}$ will also be nearly constant throughout the reactor. In other words, a *point rate* is measured, corresponding to the average composition and temperature in the reactor. An apparatus of this type is called a *differential reactor*. Since **r** is a constant, integration of Eq. (2-2) is simple and gives

$$\mathbf{r} = \frac{N'}{V} \tag{4-3}$$

where V is the volume of the differential reactor. The following example illustrates how the rate is calculated and the error that is involved in assuming that it is constant.

Example 4-3   The homogeneous reaction between sulfur vapor and methane has been studied in a small silica-tube reactor of 35.2 ml volume.[1] In a particular run at 600°C and 1 atm pressure the measured quantity of carbon disulfide produced in a 10-min run was 0.10 g. Assume that all the sulfur present is the molecular species S_2. The sulfur-vapor (considered as S_2) flow rate was 0.238 g mole/hr in this steady-state run.

(*a*) What is the rate of reaction, expressed in g moles of carbon disulfide produced/(hr)(ml of reactor volume)? (*b*) The rate at 600°C may be expressed by the second-order equation

$$\mathbf{r} = k p_{CH_4} p_{S_2}$$

where p is partial pressure, in atmospheres. Use the rate determined in (*a*) and this form of the rate equation to calculate the specific reaction rate in units of g moles/(ml) (atm^2)(hr). The methane flow rate was 0.119 g mole/hr, and the H_2S and CS_2 concentrations in the reactants were zero. (*c*) Also compute the value of k without making the assumption that the rate is constant and that average values of the partial pressures may be used; that is, consider the equipment to operate as an integral, rather than a differential, reactor. Compare the results and comment on the suitability of the apparatus as a differential reactor.

Solution   Consider the reaction

$$CH_4 + 2S_2 \rightarrow CS_2 + 2H_2S$$

(*a*) The carbon disulfide formation, in gram moles per hour, is

$$N' = \frac{0.10}{76} \left(\frac{60}{10}\right) = 0.0079$$

Then the rate of reaction, according to Eq. (4-3), will be

$$\mathbf{r} = 0.0079 \left(\frac{1}{35.2}\right) = 2.2 \times 10^{-4} \text{ g mole/(hr)(ml)}$$

(*b*) With the assumption that at 600°C and 1 atm pressure the components

[1] R. A. Fisher and J. M. Smith, *Ind. Eng. Chem.*, **42**, 704 (1950).

behave as perfect gases, the partial pressure is related to the mole fraction by the expression

$$p_{CH_4} = p_t y_{CH_4} = 1 y_{CH_4}$$

where y_{CH_4} represents the mole fraction of methane. The average composition in the reactor will be that corresponding to a carbon disulfide rate of $(0 + 0.0079)/2 = 0.0040$ g mole/hr. At this point the molal rates of the other components will be

$CS_2 = 0.0040$ g mole/hr

$S_2 = 0.238 - 2(0.0040) = 0.230$

$CH_4 = 0.119 - 0.0040 = 0.1150$

$H_2S = 2(0.0040) = 0.0079$

Total $= 0.357$ g mole/hr

The partial pressures will be

$$p_{CH_4} = \frac{0.1150}{0.357} = 0.322 \text{ atm}$$

$$p_{S_2} = \frac{0.230}{0.357} = 0.645 \text{ atm}$$

From the rate equation, using the value of $\mathbf{r}$ obtained in (a), we have

$$k = \frac{\mathbf{r}}{p_{CH_4} p_{S_2}} = \frac{2.2 \times 10^{-4}}{0.322(0.645)}$$

$$= 1.08 \times 10^{-3} \text{ g mole/(ml)(atm}^2)(\text{hr})$$

(c) If the variations in rate through the reactor are taken into account, the integral design expression developed in Chap. 3 for a tubular-flow reactor should be used; this is Eq. (3-13),

$$\frac{V}{F} = \int_0^{x_1} \frac{dx}{\mathbf{r}} \qquad\qquad \text{(A) or (3-13)}$$

where the conversion x is based upon methane and F is the methane feed rate. At a point in the reactor where the conversion of methane is x, the molal flow rate of each component will be

$CS_2 = 0.119x$

$S_2 = 0.238(1 - x)$

$CH_4 = 0.119(1 - x)$

$H_2S = 0.238x$

Total $= 0.357$ g mole/hr

With this information, the rate equation may be written in terms of x as

$$\mathbf{r} = k y_{CH_4} y_{S_2} = k \frac{0.119(0.238)(1 - x)^2}{0.357^2} \tag{B}$$

Equation (A) may be integrated from the entrance to the exit of the reactor by means of Eq. (B). Thus

$$\frac{V}{F} = \frac{4.5}{k} \int_0^{x_1} \frac{dx}{(1 - x)^2} = \frac{4.5}{k} \frac{x_1}{1 - x_1}$$

Solving for k, we have

$$k = \frac{4.5}{V/F} \frac{x_1}{1 - x_1}$$

The conversion of methane at the exit of the reactor, in moles of methane reacted per mole of methane in the feed, is

$$x_1 = \frac{0.0079}{0.119} = 0.0664$$

Hence the specific reaction rate is

$$k = \frac{4.5}{35.2/0.119} \left(\frac{0.0664}{1 - 0.0664} \right) = 1.08 \times 10^{-3} \text{ g mole/(ml)(atm}^2)(hr)$$

in good agreement with the result in (b).

In the foregoing example the change in composition in the reactor is sufficiently small that a rate corresponding to the average composition may be used to evaluate the rate. In other words, the concept of a differential reactor is satisfactory. If the conversion had been considerably larger than $x = 0.066$, this would not have been true. By the type of calculation illustrated in part (c) the error in the use of the differential-reactor assumption can be evaluated for any conversion level and for any chosen reaction order. It is surprising how large the conversion may be and still introduce only a small error by the assumption of differential-reactor operation.

While the differential-reactor system was satisfactory in Example 4-3, in many cases there are difficulties with this approach. If the small differences in concentration between entering and exit streams cannot be determined accurately, the method is not satisfactory. Also, when the heat of reaction or the rate, or both, is particularly high, it may not be possible to operate even a small reactor at conditions approaching constant temperature and constant composition. It is interesting that the limitation to *ideal* flow is not significant in a differential reactor. This is because for low conversion (small composition changes) the effects of deviations from ideal flow are negligible, regardless of the cause (axial dispersion, radial velocity, or composition gradients). When the conversion is large, this is not true; these effects are evaluated in Chap. 6.

When the conditions of approximately constant rate cannot be met, the measured conversion data will represent the integrated value of the rates in all parts of the reactor. A reactor operated in this way is termed an *integral reactor*. The problem of obtaining a rate equation is essentially one of either *differentiating* the measured conversion data to give point values of the rate, or *integrating* an assumed form for the rate equation and comparing the results with the observed conversion data. To give either process a reasonable chance of success, as many variables as possible should remain constant. The ideal situation is one in which the only variable is composition, i.e., where the temperature and pressure are the same in all parts of the reactor. This type of integral operation is just one step removed from that of a differential reactor. If both the temperature and the composition change significantly, the process of differentiating the experimental data to obtain a rate equation is of doubtful accuracy and, indeed, is seldom successful.

These conclusions concerning the ease of interpretation of laboratory data in terms of a rate equation are summarized in Table 4-1.

Table 4-1 Experimental methods of obtaining rate data for tubular-flow systems

Type of reactor	Characteristics	Interpretation of data
Differential	Constant temperature, composition, and pressure (i.e., constant rate)	Rate data obtained directly, interpretation simple.
Integral (A)	Constant temperature and pressure (rate depends upon composition only)	Interpretation of integral data usually satisfactory by graphical differentiation or fitting of integral-conversion curves.
Integral (B)	Constant pressure only	Interpretation complicated by temperature variation; if effect of temperature on rate is known from independent measurements, interpretation possible in principle.

A complete set of integral-reactor data consists of measurements of the conversion for different flow rates through the reactor, with each run made at constant reactants ratio, pressure, and temperature (if possible). Then an additional set of conversion-vs-flow runs is made at a different reactants ratio but the same pressure and temperature. This procedure is continued until data are obtained over the entire range of reactants ratio, temperature, and pressure that may be employed in the commercial reactor. A useful means of summarizing the results is to prepare graphs of V/F vs conversion at constant values of reactants ratio and p. Such graphs for representing the experimental data are suggested by the form of the design equation for flow reactors [Eq. (3-13)]. The shape of these curves is determined by the nature of the rate, and they provide some qualitative information about the rate equation.

The two procedures for treating integral-reactor data are equivalent to the integration and differential methods first described in Chap. 2 for batch systems. In the integration approach, a rate equation is assumed and then the design expression is integrated. This integration gives a relationship between V/F and x which may be compared with the experimental data. The final step is the choice of the rate equation that gives the best agreement with the experimental V/F-vs-x curves for all conditions of reactants ratio, temperature, and pressure. The differential method entails differentiating the V/F-vs-x curves graphically to obtain the rate of reaction as a function of composition. Various assumed rate equations can then be tested for agreement with the rate-vs-composition data.

Perhaps the clearest way of explaining these methods of interpreting laboratory data is to carry out specific examples in some detail, pointing out the features of the method which are general and applicable to any reaction. In Example 4-4 the integration and differential methods are applied to a single-reaction system. Example 4-5, which is concerned with the homogeneous reaction of sulfur vapor and methane to produce carbon disulfide, involves multiple reactions.

Example 4-4 A kinetic study is made of the decomposition of acetaldehyde at 518°C and 1 atm pressure in a flow apparatus. The reaction is

$$CH_3CHO \rightarrow CH_4 + CO$$

Acetaldehyde is boiled in a flask and passed through a reaction tube maintained by a surrounding furnace at 518°C. The reaction tube is 3.3 cm ID and 80 cm long. The flow rate through the tube is varied by changing the boiling rate. Analysis of the products from the end of the tube gives the results in Table 4-2.

Table 4-2

Rate of flow, g/hr	130	50	21	10.8
Fraction of acetaldehyde decomposed	0.05	0.13	0.24	0.35

What is a satisfactory rate equation for these data?

Solution A second-order rate equation, $\mathbf{r} = k_2[A]^2$, will be tested by both the integral and the differential methods. To utilize either, it is necessary to express the rate in terms of the conversion of acetaldehyde. This may be accomplished by applying material balances and the ideal-gas law.

The molal flow rate of acetaldehyde entering the reaction tube is F. At a point where the conversion is x it will be

$$N'_A = F - Fx$$

The molal rates of the other components will be

$$N'_{CH_4} = xF$$

$$N'_{CO} = xF$$

Total flow rate $N'_t = F(1 + x)$

From the ideal-gas law,

$$[A] = \frac{N'_A}{Q} = \frac{N'_A}{N'_t R_g T / p_t} = \frac{N'_A}{N'_t} \frac{p_t}{R_g T} = \frac{1 - x}{1 + x} \frac{p_t}{R_g T}$$

The second-order rate expression in terms of conversion is

$$\mathbf{r} = k_2[A]^2 = k_2 \left(\frac{p_t}{R_g T}\right)^2 \left(\frac{1 - x}{1 + x}\right)^2 \tag{A}$$

INTEGRATION METHOD Equation (A) may be substituted in the design expression, Eq. (3-13), to evaluate the second-order assumption:

$$\frac{V}{F} = \frac{1}{k_2(p_t/R_g T)^2} \int_0^{x_1} \frac{dx}{[(1 - x)/(1 + x)]^2}$$

Integration yields

$$k_2 \left(\frac{p_t}{R_g T}\right)^2 \frac{V}{F} = \frac{4}{1 - x} + 4 \ln (1 - x) + x - 4 \tag{B}$$

which provides the relationship between V/F and x to be tested with the experimental data. Since

$$V = \pi \frac{3.3^2}{4} (80) = 684 \text{ cm}^3$$

the experimental data may be expressed in terms of V/F, as shown in Table 4-3. Equation (B) may be used to compute a value of k for each of the sets of x and V/F values given in the table. For example, at $x = 0.13$, substituting in Eq. (B), we obtain

$$k_2 \left[\frac{1}{0.082(518 + 273)}\right]^2 (2,160) = \frac{4}{1 - 0.13} + 4 \ln (1 - 0.13) + 0.13 - 4$$

$$k_2 = 0.33 \text{ liter/(g mole)(sec)}$$

Table 4-3

| Conversion | Feed rate | | V/F, (liters)(sec)/g mole | k_2, liters/(g mole)(sec) |
	g/hr	$g\ moles/sec$		
0.05	130	0.000825	828	0.32
0.13	50	0.000316	2,160	0.33
0.24	21	0.000131	5,210	0.32
0.35	10.8	0.0000680	10,000	0.33

Values of k_2 for the other three sets, shown in the last column of the table, are in good agreement with each other.

DIFFERENTIAL METHOD   The experimental V/F data are plotted against conversion in Fig. 4-1. According to Eq. (3-13), the slope of this curve at any conversion gives the rate of reaction at that point. The slopes of the curve at the experimental conversions are given in the second column of Table 4-4. These slopes may be compared with the assumed rate equation either by plotting ln r vs ln $[(1 - x)/(1 + x)]$ according to Eq. (A) and noting whether or not a straight line of slope 2 is obtained, or by computing values of k_2 directly from Eq. (A). Following this second procedure, at $x = 0.13$ we have

$$r = 4.9 \times 10^{-5} = k_2 \left[\frac{1}{0.082(791)} \right]^2 \left(\frac{1 - 0.13}{1 + 0.13} \right)^2$$

$$k_2 = 0.35 \text{ liter/(g mole)(sec)}$$

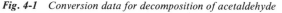

Fig. 4-1   Conversion data for decomposition of acetaldehyde

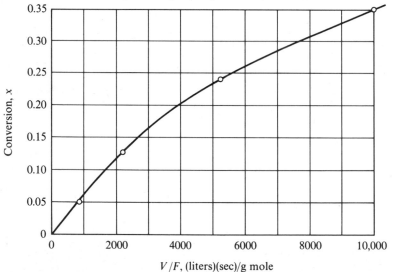

Table 4-4

Conversion	Slope from Fig. 4-1, g moles/(liter)(sec)	k_2, liters/(sec)(g mole)	k_1, sec^{-1}
0.05	6.2×10^{-5}	0.32	0.0045
0.13	4.9×10^{-5}	0.35	0.0041
0.24	2.8×10^{-5}	0.33	0.0030
0.35	2.0×10^{-5}	0.35	0.0027

The results at other conversions are shown in the third column of Table 4-4. Although there is some variation from point to point, there is no significant trend. Hence the differential method also confirms the validity of a second-order rate equation. The variation is due to errors associated with the measurement of slopes of the curve in Fig. 4-1.

FIRST-ORDER RATE EQUATION   For comparison, the differential method will be applied to a first-order mechanism. The rate expression replacing Eq. (A) is

$$\mathbf{r} = k_1[A] = k_1 \frac{p_t}{R_g T} \frac{1-x}{1+x} \tag{C}$$

This mechanism may be tested by computing k_1 values from the experimental rates of reaction according to Eq. (C). For example, at $x = 0.13$

$$\mathbf{r} = slope = 4.9 \times 10^{-5} = k_1 \frac{1}{0.082(791)} \frac{1 - 0.13}{1 + 0.13}$$

$$k_1 = 0.0041 \ sec^{-1}$$

The k_1 values for other conversions (fourth column of Table 4-4) show a distinct trend toward lower values at higher conversions. This indicates that the first-order rate is not a likely one.

Example 4-5   A flow reactor consisting of a 1-in. stainless-steel pipe 6 in. long and packed with inert rock salt is used to study the noncatalytic homogeneous reaction

$$\frac{4}{b} S_b + CH_4 \rightarrow 2H_2S + CS_2$$

The measurements are carried out at atmospheric pressure in the vapor phase at 600°C. From available data on the rate of dissociation of the sulfur species, it is reasonable to assume that the reactions

$$S_8 \rightleftharpoons 4S_2$$

$$S_6 \rightleftharpoons 3S_2$$

are very fast with respect to the combination of sulfur vapor with methane. Accordingly, assume that S_8, S_6, and S_2 are in equilibrium. The void volume, measured by benzene

Table 4-5

Run	Flow rate, g moles/hr			Reactants ratio, moles (S_2)/moles CH_4	Conversion
	$(CH_4)_f$	$(S_2)_f$	$(CS_2)_p$		
55	0.02975	0.0595	0.0079	2	0.268
58	0.0595	0.119	0.0086	2	0.144
57	0.119	0.238	0.0078	2	0.066
59	0.119	0.238	0.0072	2	0.060
56	0.238	0.476	0.0059	2	0.025
75	0.0595	0.119	0.0079	2	0.133
76	0.02975	0.0595	0.0080	2	0.269
77	0.119	0.238	0.0069	2	0.058
78	0.0893	0.0893	0.0087	1	0.0975
79	0.119	0.0595	0.0096	0.5	0.0807

displacement, was 35.2 ml. The conversion x of methane to carbon disulfide was measured for various flow rates and initial reactants ratios, and the results are shown in Table 4-5. (S_2) represents the total amount of sulfur vapor present, expressed as S_2. Hence

$$(S_2) = N_{S_2} + 3N_{S_6} + 4N_{S_8}$$

where the N values are the number of moles of each sulfur species.

(a) Assuming, first, that the only species of sulfur that exists is S_2, test the assumptions of first- and second-order rate equations with the experimental data. (b) Repeat part (a) with the assumption that S_2, S_6, and S_8 are in equilibrium but that only S_2 reacts with CH_4. (c) Using a second-order rate expression, determine which assumption regarding the reactive sulfur species best fits the experimental data.

Solution It is convenient to convert the data to a conversion x_t based on the total feed. For example, for run 55

$$F_t = 0.02975 + 0.0595 = 0.0892 \text{ g mole/hr}$$

$$\frac{V}{F_t} = \frac{35.2}{0.0892} = 395$$

$$x_t = \frac{0.0079}{0.0892} = 0.0890$$

where F_t is the molal feed rate. Table 4-6 shows the corresponding values of x_t and V/F_t. These results are also plotted in Fig. 4-2. Note that

$$x_t = \frac{\text{moles product}}{\text{moles total feed}} = x\frac{\text{moles } CH_4 \text{ in feed}}{\text{moles total feed}} = x\frac{1}{1 + a}$$

or

$$x = (a + 1)x_t$$

where a is the moles of sulfur in the feed per mole of methane.

Table 4-6

Run	Conversion		V/F_t, (ml)(hr)/g mole	$-\log[1-(a+1)(x_t)_1]$ or $-\log(1-x_1)$	$k_1 \times 10^4$	$(k_1)_s \times 10^4$
	x_1	$(x_t)_1$				
55	0.268	0.0890	395	0.135	7.87	3.93
58	0.144	0.0480	197	0.0675	7.90	3.95
57	0.066	0.022	98.6	0.0297	6.93	3.46
59	0.060	0.020	98.6	0.0269	6.29	3.15
56	0.025	0.00833	49.3	0.0110	5.14	2.57
75	0.133	0.0443	197	0.0620	7.25	3.62
76	0.269	0.0895	395	0.136	7.95	3.97
77	0.058	0.0193	98.6	0.0269	6.05	3.02
78	0.0975	0.0487	197	0.045	5.30	5.45
79	0.0807	0.0538	197	0.0368	4.31	9.90

(a) Two first-order mechanisms may be assumed, one with respect to methane and one with respect to sulfur vapor (S_2). Starting with the first-order assumption for methane and applying Eq. (3-13), we have

$$\frac{V}{F_t} = \int_0^{(x_t)_1} \frac{dx_t}{r} = \int_0^{(x_t)_1} \frac{dx_t}{k_1 p_{CH_4}} = \int_0^{(x_t)_1} \frac{dx_t}{k_1 y_{CH_4}} \tag{A}$$

$$p_{CH_4} = y_{CH_4} p_t = y_{CH_4}$$

The mole fraction of methane varies with the conversion. Hence to relate y_{CH_4} to x_t we must know the number of moles of each component as a function of conversion. Since we are assuming here that all the sulfur present is S_2, the number of moles of S_2

Fig. 4-2   Experimental conversion data for the sulfurization of methane

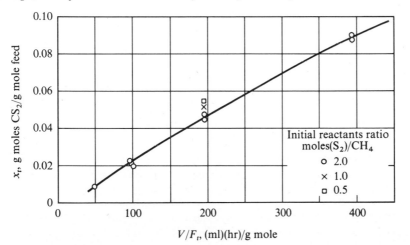

and CH_4 are simple to ascertain. For a mole ratio of S_2 to CH_4 of 2, the mole fraction entering the reactor is

$$y_{CH_4} = \frac{1}{2 + 1} = 0.333$$

If a basis of 1 mole of CH_4 and 2 moles of S_2 is chosen, at a point in the reactor where the conversion is x_t, the number of moles of each component will be

$$CS_2 = x = 3x_t$$

$$H_2S = 2x = 6x_t$$

$$CH_4 = 1 - x = 1 - 3x_t$$

$$S_2 = 2 - 6x_t$$

Total moles = 3

Thus the mole fractions are

$$y_{CH_4} = \frac{1 - 3x_t}{3} = 0.333 - x_t$$

$$y_{S_2} = \frac{2 - 6x_t}{3} = 0.667 - 2x_t$$

More generally, if the initial mole ratio of sulfur to CH_4 is a, then

$$\left. \begin{array}{l} y_{CH_4} = \dfrac{1 - (1 + a)x_t}{1 + a} = \dfrac{1}{1 + a} - x_t \\[3mm] y_{S_2} = \dfrac{a - 2(1 + a)x_t}{1 + a} = \dfrac{a}{1 + a} - 2x_t \end{array} \right\} \quad \text{(B)}$$

By the integral method, Eq. (A) can now be integrated with Eq. (B) to yield

$$\frac{V}{F_t} = \frac{1}{k_1} \int_0^{(x_t)_1} \frac{dx_t}{1/(1 + a) - x_t} = \frac{1}{k_1} \ln \frac{1/(1 + a) - (x_t)_1}{1/(1 + a)}$$

$$= -\frac{1}{k_1} \ln \left[1 - (a + 1)(x_t)_1 \right] = -\frac{1}{k_1} \ln (1 - x_1) \qquad \text{(C)}$$

where $(x_t)_1$ is the conversion leaving the reactor. The V/F_t-vs-$(x_t)_1$ data shown in Table 4-6 can now be used to test Eq. (C). Instead of plotting curves obtained from Eq. (C) with different k values for comparison with the experimental curve (Fig. 4-2), it is more effective to use one of two other approaches:

1. Plot the experimental V/F_t-vs-x data given in Table 4-6 as V/F_t vs $-\ln (1 - x_1)$. This method of plotting should result in a straight line if the assumed rate expression is correct. Figure 4-3 shows that this is not true if the rate is first order with respect to methane. At a constant reactants ratio of $a = 2.0$ the points fall on a straight line, but the data for other ratios deviate widely from the line.

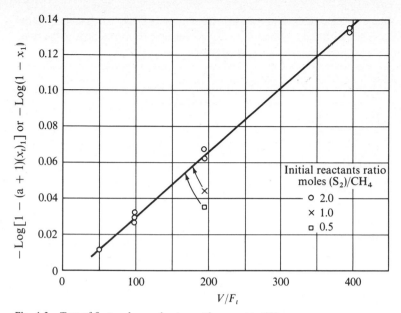

Fig. 4-3   *Test of first-order mechanism with respect to* CH_4

2.   From the data in Table 4-6, compute values of k_1 required by Eq. (C) for each of the runs. If these values are essentially constant, the assumed mechanism fits the data. For this case the resulting k_1 values are given in column 6 of Table 4-6. Again, the runs for different reactants ratios show that the assumed rate equation does not fit the data.

As we shall see later, the assumed rate equation may be complex enough to prevent analytical integration of the design equation (A). However, in such cases it is still possible to evaluate the integral $\int dx_t/r$ graphically and then apply either method 1 or method 2 to determine the suitability of the rate assumption. Both methods are general and hence applicable to any reaction.

The design equation for an assumed mechanism which is first order with respect to sulfur vapor S_2 will be similar to Eq. (C); for example,

$$\frac{V}{F_t} = \int_0^{(x_t)_1} \frac{dx_t}{(k_1)_S y_{S_2}} = \frac{1}{(k_1)_S} \int_0^{(x_t)_1} \frac{dx_t}{a/(1 + a) - 2x_t}$$

$$= -\frac{1}{2(k_1)_S} \ln \frac{a - 2(a + 1)(x_t)_1}{a} = \frac{-1}{2(k_1)_S} \ln\left(1 - \frac{2x_1}{a}\right) \qquad (D)$$

For $a = 2$, Eq. (D) becomes the same as Eq. (C) if $(k_1)_S = \frac{1}{2}k_1$. Hence the test of this mechanism is whether or not the two runs for $a = 1.0$ and $a = 0.5$ are either brought in line with the other data on a plot similar to Fig. 4-3 or else give the same value of

$(k_1)_S$. The values of $(k_1)_S$ are given in the last column of Table 4-6. It is clear that the results for the runs with $a = 1.0$ and 0.5 are far removed from the rest of the data. Hence neither first-order mechanism is satisfactory.

If a second-order mechanism is tried, again with the assumption that all the sulfur vapor is present as S_2, the design equation is

$$\frac{V}{F_t} = \int \frac{dx_t}{k_2 p_{CH_4} p_{S_2}} = \frac{1}{k_2} \int \frac{dx_t}{y_{CH_4} y_{S_2}}$$

$$= \frac{1}{k_2} \int \frac{dx_t}{\{[1/(1 + a)] - x_t\}\{[a/(1 + a)] - 2x_t\}}$$

This expression can be simplified and integrated to[1]

$$\frac{V}{F_t} = \frac{a + 1}{k_2} \int_0^{x_1} \frac{dx}{(1 - x)(a - 2x)}$$

$$= \frac{a + 1}{a - 2} \frac{1}{k_2} \left[\ln \left(1 - \frac{2}{a} x_1 \right) - \ln (1 - x_1) \right] \tag{E}$$

$$= \frac{3}{2k_2} \frac{x_1}{1 - x_1} \qquad \text{for } a = 2$$

Figure 4-4 shows a plot of V/F_t vs $[(a + 1)/(a - 2)] \{\ln [1 - (2/a)x_1] - \ln (1 - x_1)\}$, or $\frac{3}{2} x_1/(1 - x_1)$, for $a = 2$ for the experimental data. It is apparent that this correlation is an improvement over the previous ones, since the point for $a = 1.0$ is close to the straight line. However, the data point for $a = 0.5$ is still not satisfactorily correlated. From this preliminary investigation, neglecting the S_6 and S_8 species, we conclude that a second-order rate is more appropriate than a first-order one.

(b) The next step is to refine the analysis by considering the distribution of sulfur vapor between S_2 and S_6 and S_8. To take into account the presence of the S_6 and S_8 species we shall assume equilibrium among S_2, S_6, and S_8, utilizing the following equilibrium constants at 600°C:

$$\tfrac{1}{4} S_8 \rightarrow S_2 \qquad K_p = 0.930 \ (\text{atm})^{\tfrac{3}{4}}$$

$$\tfrac{1}{3} S_6 \rightarrow S_2 \qquad K_p = 0.669 \ (\text{atm})^{\tfrac{2}{3}}$$

This information, along with the stoichiometry of the reaction, can be used to develop expressions analogous to Eqs. (B) for the relation between y_{S_2}, or y_{CH_4}, and the conversion. As in part (a), s. ppose that there are a moles of sulfur vapor per mole of CH_4 entering the reactor. However, let a refer to the total moles of sulfur considered as S_2, not to the sum of the moles of S_2, S_6, and S_8. If α_0, β_0, and γ_0 represent the moles of S_2, S_6, and S_8 entering the reactor, the total moles of sulfur, considered as S_2, will be

$$a = \alpha_0 + 3\beta_0 + 4\gamma_0$$

[1] Equation (E) becomes indeterminate when $a = 2$. Hence it is necessary to substitute $a = 2$ prior to integration and obtain the particular solution.

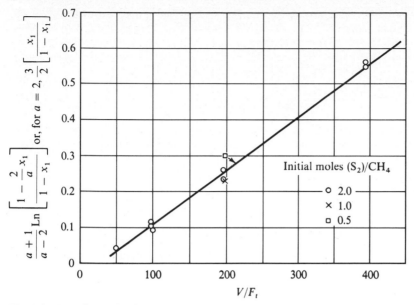

Fig. 4-4   *Test of second-order reaction mechanism*

At a point in the reactor where the conversion is x, the moles of each component will be

$$CS_2 = x$$

$$H_2S = 2x$$

$$CH_4 = 1 - x$$

$$S_2 = \alpha$$

$$S_6 = \beta$$

$$S_8 = \gamma$$

Total moles $= \alpha + \beta + \gamma + 2x + 1$

The conversion has been defined in terms of the methane reacted. It can also be written in terms of the sulfur used, providing a relationship between x, α, β, and γ. If (S_2) represents the total moles of sulfur (considered as S_2) present at any point in the reactor, then

$$x = \tfrac{1}{2}[a - (S_2)]$$

but $S_2 = \alpha + 3\beta + 4\gamma$

so that

$$x = \frac{a - \alpha - 3\beta - 4\gamma}{2} \tag{F}$$

Two additional relationships among x, α, β, and γ must exist because of the dissociation equilibria:

$$0.930 = \frac{p_{S_2}}{p_{S_8}^{\frac{1}{4}}} = \frac{y_{S_2}}{y_{S_8}^{\frac{1}{4}}} = \frac{\alpha}{\gamma^{\frac{1}{4}}(\alpha + \beta + \gamma + 2x + 1)^{\frac{3}{4}}} \qquad (G)$$

$$0.669 = \frac{p_{S_2}}{p_{S_6}^{\frac{1}{3}}} = \frac{y_{S_2}}{y_{S_6}^{\frac{1}{3}}} = \frac{\alpha}{\beta^{\frac{1}{3}}(\alpha + \beta + \gamma + 2x + 1)^{\frac{2}{3}}} \qquad (H)$$

Equations (F) to (H) permit the evaluation of α, β, and γ at any conversion and reactants ratio a. Then the mole fractions are immediately obtainable from the expressions

$$y_{S_2} = \frac{\alpha}{\alpha + \beta + \gamma + 2x + 1}$$

$$y_{CH_4} = \frac{1 - x}{\alpha + \beta + \gamma + 2x + 1}$$

To see the method of calculation let us take $a = 2$. Entering the reactor, $x = 0.0$, and Eqs. (F) to (H) become

$$2 = \alpha + 3\beta + 4\gamma$$

$$\gamma^{\frac{1}{4}}(\alpha + \beta + \gamma + 1)^{\frac{3}{4}} = \frac{\alpha}{0.93}$$

$$\beta^{\frac{1}{3}}(\alpha + \beta + \gamma + 1)^{\frac{2}{3}} = \frac{\alpha}{0.669}$$

Solving these by trial gives

$$\alpha = 0.782 \qquad y_{S_2} = \frac{0.782}{2.171} = 0.360$$

$$\beta = 0.340 \qquad y_{S_6} = \frac{0.340}{2.171} = 0.156$$

$$\gamma = 0.049 \qquad y_{S_8} = \frac{0.049}{2.171} = 0.023$$

$$CH_4 = 1.000 \qquad y_{CH_4} = \frac{1.000}{2.171} = 0.461$$

Total = 2.171 $\Sigma y = 1.000$

The mole fractions corresponding to any conversion can be evaluated in a similar manner. The results of such calculations are shown in Fig. 4-5, where composition of the reaction mixture is plotted against x. Information for other values of a are obtained in an analogous fashion.

The necessary data are now at hand for testing the second-order rate expression.

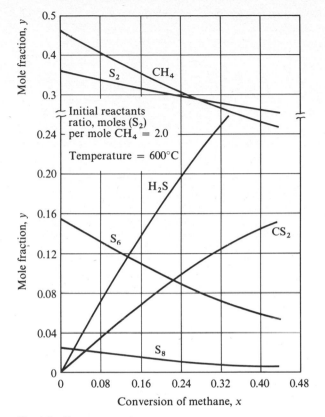

Fig. 4-5   *Composition of reaction mixture vs conversion for the sulfurization of methane*

However, the calculational procedure used in part (*a*) is not applicable because the design equation

$$\frac{V}{F_t} = \frac{1}{k_2'} \int_0^{(x_t)_1} \frac{dx_t}{y_{CH_4} y_{S_2}} \tag{I}$$

cannot be integrated analytically. Two alternatives are possible. The integral method entails graphical evaluation of the integral in Eq. (I) for each experimental point, with the measured value of the conversion as the upper limit in each case. This procedure leads to values of $\int_0^{(x_t)_1} dx_t/y_{CH_4} y_{S_2}$ for each value of V/F_t shown in Table 4-6. The results are given in the fourth column of Table 4-7 and are plotted in Fig. 4-6, with V/F_t as the abscissa and $\int_0^{(x_t)_1} dx_t/y_{CH_4} y_{S_2}$ as the ordinate. If the assumed mechanism is a satisfactory

Table 4-7   Test of reaction mechanisms for sulfurization of methane

Run	Reactants ratio	V/F_t	$\int_0^{(x_t)_1} dx_t/y_{CH_4}y_{S_b}$		
			$b = 2$	$b = 6$	$b = 8$
55	2	395	0.762	2.11	17.0
58	2	197	0.347	0.890	6.64
57	2	98.6	0.144	0.350	2.48
59	2	98.6	0.130	0.315	2.23
56	2	49.3	0.052	0.123	0.855
75	2	197	0.316	0.800	5.88
76	2	395	0.765	2.17	17.1
77	2	98.6	0.126	0.303	2.15
78	1	197	0.315	1.27	12.1
79	0.5	197	0.388	2.97	36.8

interpretation of the data, Fig. 4-6 should be a straight line with a slope equal to k_2'. This is apparent if Eq. (I) is rearranged as

$$\int_0^{(x_t)_1} \frac{dx_t}{y_{CH_4}y_{S_2}} = k_2' \frac{V}{F_t}$$

It is evident from the plot that this condition is satisfied.

The other approach, the differential method, entails graphical differentiation of the data for comparison with the differential form of Eq. (I). If the data are plotted

Fig. 4-6   Test of rate equation $\mathbf{r} = k_2' p_{CH_4} p_{S_2}$

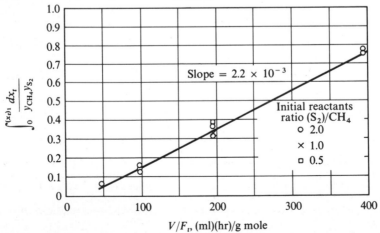

as x_t vs V/F_t, as in Fig. 4-2, the slope of the curve is equal to the rate of reaction. Thus, from Eq. (I),

$$d\left(\frac{V}{F_t}\right) = \frac{dx_t}{k_2' y_{CH_4} y_{S_2}}$$

or

$$\frac{dx_t}{d(V/F_t)} = \text{slope} = k_2' y_{CH_4} y_{S_2}$$

If the slope is taken at various values of the conversion from Fig. 4-2 and the corresponding values of y_{CH_4} and y_{S_2} are read from Fig. 4-5, and this is done for each reactants ratio, complete information on the effect of composition on the rate is available. The equation $\mathbf{r} = k_2' y_{CH_4} y_{S_2}$ can then be tested by observing the constancy in computed values of k_2'.

(c) The results of part (b) indicate that the assumption of S_2 as the active sulfur-vapor species in the second-order mechanism gives a satisfactory interpretation of the conversion data. However, it is important to know whether equally good correlations would be obtained by assuming that S_6 or S_8 is the reactive form. To do this, the design equations corresponding to (I) may be written for the other species

$$\frac{V}{F_t} = \frac{1}{k_6'} \int_0^{(x_t)_1} \frac{dx_t}{y_{CH_4} y_{S_6}} \tag{J}$$

$$\frac{V}{F_t} = \frac{1}{k_8'} \int_0^{(x_t)_1} \frac{dx_t}{y_{CH_4} y_{S_8}} \tag{K}$$

Figures 4-7 and 4-8, which show V/F_t plotted against the integrals in (J) and (K), were prepared by graphical integration, with Fig. 4-5 used for the relations of x_t to y_{S_6} and y_{S_8}. The values of the integrals are given in the last two columns of Table 4-7. We see that in contrast to Fig. 4-6, based on S_2, the data do not fall on a straight line as required by Eqs. (J) and (K). Also, the points for various reactants ratios are not in agreement. We conclude that a second-order mechanism based on either S_6 or S_8 as the reactive sulfur species does not agree with the experimental data.

The preferred equation for the rate of reaction is

$$\mathbf{r} = k_2' p_{CH_4} p_{S_2}$$

From Fig. 4-6, the slope of the straight line is 2.2×10^{-3} g mole/(hr)(ml)(atm²). Since this number is equal to k_2', the final expression for the rate of reaction at 600°C is[1]

$$\mathbf{r} = 2.2 \times 10^{-3} p_{CH_4} p_{S_2}$$

It is not advisable to make a firm statement regarding the mechanism of a reaction from data as limited as that in Example 4-5. It is perhaps best

---

[1] By evaluating the specific-reaction rate k_2' at other temperatures an activation energy can be determined. For this aspect of the problem see R. A. Fisher and J. M. Smith, *Ind. Eng. Chem.*, **42**, 704 (1950).

to consider that the final rate equation is a satisfactory interpretation of the kinetic data. The problem of interpreting laboratory kinetic data in terms of a rate equation for catalytic reactions will be considered in Chap. 10.

4-4 Design Procedure

Once a satisfactory expression has been established for the rate of reaction, the size of reactor as a function of operating variables can be found by combining and solving the rate and mass-balance (design) equations. The design equation assumes a simpler form when the volume of the reaction mixture (or density) is constant. If QC_0, where Q is the constant volumetric flow rate and C_0 is the feed concentration, is substituted for the reactant feed rate F and the conversion is expressed in terms of concentration according to Eq. (4-1), Eq. (3-13) may be written

$$\frac{V}{QC_0} = \int_0^x \frac{dx}{\mathbf{r}} = -\frac{1}{C_0} \int_{C_0}^C \frac{dC}{\mathbf{r}} \tag{4-4}$$

or, since V/Q is the residence or contact time,

$$\frac{V}{Q} = \theta = -\int_{C_0}^C \frac{dC}{\mathbf{r}} \tag{4-5}$$

Fig. 4-7 *Test of rate equation* $\mathbf{r} = k'_6 p_{CH_4} p_{S_6}$

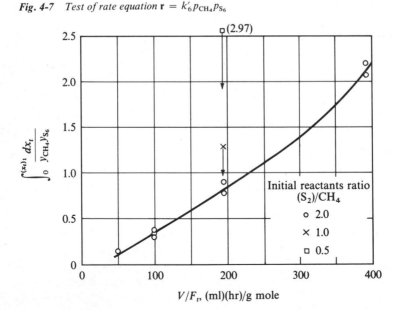

V/F_t, (ml)(hr)/g mole

Comparison of Eqs. (4-2) and (4-5) shows that the form of the design equations for ideal batch and tubular-flow reactors are identical if the real-time variable in the batch reactor is considered as the residence time in the flow case. The important point is that the integral $\int dC/\mathbf{r}$ is the same in both reactors. If this integral is evaluated for a given rate equation for an *ideal* batch reactor, the result is applicable for an *ideal* tubular-flow reactor; this conclusion will be utilized in Example 4-10.

When multiple reactions are involved, the yield and selectivity are important as well as the conversion. The following example illustrates the method of solving Eq. (4-4) for both single and consecutive reaction systems. The procedure is essentially the reverse of that for interpreting laboratory data on integral reactors (see Sec. 4-3).

Example 4-6   Hougen and Watson,[1] in an analysis of Kassell's data for the vapor-phase dehydrogenation of benzene in a homogeneous flow reactor, reported the reactions

1. $2C_6H_6(g) \rightarrow C_{12}H_{10}(g) + H_2(g)$

2. $C_6H_6(g) + C_{12}H_{10}(g) \rightarrow C_{18}H_{14}(g) + H_2(g)$

[1]O. A. Hougen and K. M. Watson, "Chemical Process Principles," vol. 3, p. 846, John Wiley & Sons, Inc., New York, 1947.

Fig. 4-8   *Test of rate equation* $\mathbf{r} = k_8' p_{CH_4} p_{S_8}$

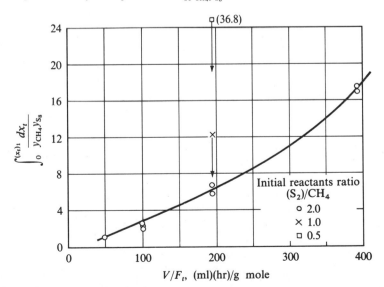

and the following rate equations:

$$\mathbf{r}_1 = 14.96 \times 10^6 e^{-15,200/T} \left(p_B{}^2 - \frac{p_D p_H}{K_1} \right) \qquad \text{lb moles benzene reacted/(hr)(ft}^3)$$

$$\mathbf{r}_2 = 8.67 \times 10^6 e^{-15,200/T} \left(p_B p_D - \frac{p_T p_H}{K_2} \right)$$

$$\text{lb moles triphenyl produced or diphenyl reacted/(hr)(ft}^3)$$

where p_B = partial pressure of benzene, atm
p_D = partial pressure of diphenyl, atm
p_T = partial pressure of triphenyl, atm
p_H = partial pressure of hydrogen, atm
T = temperature, °K
K_1, K_2 = equilibrium constants for the two reactions in terms of partial pressures

The data on which the rate equations are based were obtained at a total pressure of 1 atm and temperatures of 1265° and 1400°F in a 0.5-in. tube 3 ft long.

It is now proposed to design a tubular reactor which will operate at 1 atm pressure and 1400°F. (a) Determine the total conversion of benzene to di- and triphenyl as a function of space velocity. (b) Determine the reactor volume required to process 10,000 lb/hr of benzene as a function of the total conversion. First carry out the solution with the assumption that only reaction 1 occurs, and then proceed to the solution for the two consecutive reactions. Assume that the reactor will be operated isothermally and that no other reactions are significant.

Solution (a) Since the reactor is isothermal, the equilibrium constants K_1 and K_2 will have fixed values. They may be estimated at 1400°F from equations developed by Hougen and Watson, by methods analogous to those described in Chap. 1. The results are

$$K_1 = 0.312$$

$$K_2 = 0.480$$

As these values are not large, the reverse reactions may be important. At 1400°F (1033°K), in terms of the rate of disappearance of benzene, the two rates are

$$\mathbf{r}_1 = (14.96 \times 10^6)e^{-14.7} \left(p_B{}^2 - \frac{p_D p_H}{0.312} \right) = 6.23 \left(p_B{}^2 - \frac{p_D p_H}{0.312} \right) \qquad \text{(A)}$$

$$\mathbf{r}_2 = (8.67 \times 10^6)e^{-14.7} \left(p_B p_D - \frac{p_T p_H}{0.480} \right) = 3.61 \left(p_B p_D - \frac{p_T p_H}{0.480} \right) \qquad \text{(B)}$$

An equation of the form of Eq. (3-13) may be written for each reaction:

$$\frac{V}{F} = \int \frac{dx_1}{\mathbf{r}_1} \qquad \text{(C)}$$

$$\frac{V}{F} = \int \frac{dx_2}{\mathbf{r}_2} \qquad \text{(D)}$$

where the conversion x_1 is the pound moles of benzene disappearing by reaction 1 per pound mole of feed, and the conversion x_2 is the pound moles of benzene disappearing by reaction 2 per pound mole of feed.

If we take as a basis 1.0 mole of entering benzene, the moles of each component are

$H_2 = \frac{1}{2}x_1 + x_2$

$C_{12}H_{10} = \frac{1}{2}x_1 - x_2$

$C_6H_6 = 1 - x_1 - x_2$

$C_{18}H_{14} = x_2$

Total moles $= 1.0$

Since the total moles equals 1.0 regardless of x_1 and x_2, the mole fractions of each component are also given by these quantities. If the components are assumed to behave as ideal gases, the partial pressures are

$p_H = 1.0(\frac{1}{2}x_1 + x_2) = \frac{1}{2}x_1 + x_2$

$p_D = \frac{1}{2}x_1 - x_2$

$p_B = 1 - x_1 - x_2$

$p_T = x_2$

With these relationships the rate equations (A) and (B) can be expressed in terms of x_1 and x_2 as

$$r_1 = 6.23\left[(1 - x_1 - x_2)^2 - \frac{(\frac{1}{2}x_1 - x_2)(\frac{1}{2}x_1 + x_2)}{0.312}\right] \tag{E}$$

$$r_2 = 3.61\left[(1 - x_1 - x_2)(\frac{1}{2}x_1 - x_2) - \frac{x_2(\frac{1}{2}x_1 + x_2)}{0.480}\right] \tag{F}$$

In principle, Eqs. (E) and (F) can be substituted in the design equations (C) and (D) and values of exit conversions x_1 and x_2 computed for various values of V/F. If only the first reaction is considered, then only x_1 is involved and direct integration is possible. Let us carry out the solution first with this assumption.

SIMPLIFIED SOLUTION (REACTION 2 ASSUMED UNIMPORTANT) Since $x_2 = 0$, Eq. (E) simplifies to

$$r = 6.23[(1 - x_1)^2 - 0.801(x_1)^2] \tag{G}$$

This may be substituted in design Eq. (C) and integrated directly:

$$\frac{V}{F} = \frac{1}{6.23}\int_0^{x_1} \frac{dx_1}{(1 - x_1)^2 - 0.801(x_1)^2} = \frac{1}{6.23}\int_0^{x_1} \frac{dx_1}{(1 - 0.105x_1)(1 - 1.895x_1)}$$

$$= \frac{0.559}{6.23}\ln\frac{1 - 0.105x_1}{1 - 1.895x_1} \tag{H}$$

Table 4-8

x_1, lb moles C_6H_6 decomposed/ lb mole feed	V/F, ft³ (hr)/ (lb mole feed)	Space velocity, hr⁻¹
0.0	0	∞
0.1	0.0179	21,200
0.2	0.0409	9,280
0.3	0.0725	5,230
0.4	0.123	3,080
0.5	0.260	1,460
0.52	0.382	990
0.528*	∞	0

*Equilibrium conditions.

Table 4-8 gives the values of V/F corresponding to successive values of x_1 obtained by substituting in Eq. (H). A particular V/F value represents the ratio of volume of reactor to feed rate necessary to give the corresponding conversion. Figure 4-9 is a plot of the data given in the table.

The equilibrium conversion is 52.8%, as that is the value corresponding to infinite time. The same result could have been obtained by using the equilibrium constant $K_1 = 0.312$ to compute the equilibrium yield by the method described in Chap. 1.

Fig. 4-9   *Dehydrogenation of benzene (simplified solution)*

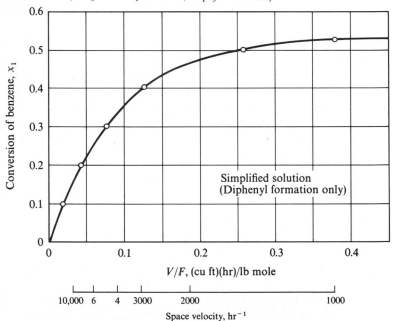

GENERAL SOLUTION (BOTH REACTIONS 1 AND 2 CONSIDERED)   The rate equations (E) and (F) and the design equations (C) and (D) provide the means for determining x_1 and x_2 as a function of V/F. If (C) and (D) are written in difference form, the integration can be accomplished by a stepwise numerical approach. In a small element of volume ΔV the conversions Δx_1 and Δx_2 are

$$\Delta x_1 = \frac{\Delta V}{F}\, \bar{r}_1 = \Delta\left(\frac{V}{F}\right)\bar{r}_1 \tag{C'}$$

$$\Delta x_2 = \frac{\Delta V}{F}\, \bar{r}_2 = \Delta\left(\frac{V}{F}\right)\bar{r}_2 \tag{D'}$$

where $\bar{r}_1$ and $\bar{r}_2$ are average values of the rate for the conversion range x_1 to $x_1 + \Delta x_1$ and $x_2 + \Delta x_2$.

To illustrate the method of solving Eqs. (C') and (D') let us start at the entrance to the reactor and carry out several stepwise calculations. The rates at the entrance are given by Eqs. (E) and (F) with $x_1 = x_2 = 0$,

$(r_1)_0 = 6.23(1 - 0) = 6.23$ lb moles C_6H_6 converted/(hr)(ft^3)

$(r_2)_0 = 3.61(0) = 0$

Choose an interval of $\Delta(V/F) = 0.005$ ft^3/(lb mole feed)(hr). If the initial rates are constant over this interval, the conversions at the end of the interval are, according to Eqs. (C') and (D'),

$x_1 = 0 + 0.005(6.23) = 0.03115$

$x_2 = 0 + 0.005(0) = 0$

The rates at the end of the interval are

$$(r_1)_1 = 6.23\left[(1 - 0.03115)^2 - \frac{1}{4}\frac{0.03115^2}{0.312} \right] = 6.23(0.938) = 5.84$$

$$(r_2)_1 = 3.61\left[(1 - 0.03115)\frac{0.03115}{2} - 0 \right] = 0.0544$$

The average values of the rates in the first increment of $\Delta(V/F)$ may be taken as the arithmetic average of the rates entering and leaving the increment; for example,

$$\bar{r}_1 = \frac{(r_1)_0 + (r_1)_1}{2} = \frac{6.23 + 5.84}{2} = 6.04$$

$$\bar{r}_2 = \frac{(r_2)_0 + (r_2)_1}{2} = \frac{0 + 0.0544}{2} = 0.0272$$

With these revised values the conversions at the end of the first increment are, again according to Eqs. (C') and (D'),

$(x_1)_1 = 0 + 0.005(6.04) = 0.0302$

$(x_2)_1 = 0 + 0.005(0.0272) = 0.000136$

Proceeding to the second increment of $\Delta(V/F)$, we can speed up the computations by anticipating the average rates in the increment instead of using the values at the beginning of the increment. Estimate that the average rates will be

$$\bar{r}_1 = 6.04 - (6.23 - 5.84) = 5.65$$

$$\bar{r}_2 = 0.0272 + (0.0544 - 0) = 0.0816$$

With these values the conversion occurring within the second increment is given by Eqs. (C′) and (D′). Thus

$$\Delta x_1 = 0.005(5.65) = 0.0282$$

$$(x_1)_2 = x_{1_1} + \Delta x_1 = 0.0302 + 0.0282 = 0.0584$$

and

$$\Delta x_2 = 0.005(0.0816) = 0.000408$$

$$(x_2)_2 = 0.000136 + 0.000408 = 0.000544$$

The estimated values of the average rates can be checked by evaluating rates at the beginning and end of the increment. At the end of the increment

$$(r_1)_2 = 6.23\left[(1 - 0.0584 - 0.0005)^2 - \left(\frac{0.0584}{2} - 0.0005\right)\right.$$
$$\left.\left(\frac{0.0584}{2} + 0.0005\right)\frac{1}{0.312}\right]$$

$$= 6.23(0.883) = 5.50$$

$$(r_2)_2 = 3.61\left[(1 - 0.0584 - 0.0005)\left(\frac{0.0584}{2} - 0.0005\right) - 0.000544\right.$$
$$\left.\left(\frac{0.0584}{2} + 0.0005\right)\frac{1}{0.480}\right]$$

$$= 3.61(0.0270) = 0.0975$$

At the beginning of the increment

$$(r_1)_1 = 6.23\left[(1 - 0.0302 - 0.0001)^2 - \left(\frac{0.0302}{2} - 0.0001\right)\right.$$
$$\left.\left(\frac{0.0302}{2} + 0.001\right)\left(\frac{1}{0.312}\right)\right]$$

$$= 5.86$$

$$(r_2)_1 = 3.61\left[(1 - 0.0302 - 0.0001)\left(\frac{0.0302}{2} - 0.0001\right) - 0.000136\right.$$
$$\left.\left(\frac{0.0302}{2} + 0.0001\right)\left(\frac{1}{0.480}\right)\right]$$

$$= 0.0529$$

Thus the average values are

$$\bar{r}_1 = \frac{r_{1_1} + r_{1_2}}{2} = \frac{5.86 + 5.50}{2} = 5.68 \qquad \text{vs 5.65 estimated}$$

$$\bar{r}_2 = \frac{r_{2_1} + r_{2_2}}{2} = \frac{0.0529 + 0.0975}{2} = 0.0752 \qquad \text{vs 0.0816 estimated}$$

The agreement with the estimated average rates is sufficiently close that a second calculation is unnecessary. The final values for the conversion leaving the second increment ($V/F = 0.005 + 0.005 = 0.01$) will be

$$(x_1)_2 = 0.0302 + 0.005(5.68) = 0.0586$$

$$(x_2)_2 = 0.000136 + 0.005(0.0752) = 0.000512$$

This stepwise procedure may be repeated until any desired value of $x_1 + x_2$ is reached. Such iterative calculations are most easily carried out by machine computation. The results from zero to equilibrium are shown in Table 4-9 and Fig. 4-10. Space velocities $v_F F/V$, based on $v_F = 379$ ft³/mole (60°F, 1 atm), are also shown.

(b) The reactor volume required to process 10,000 lb/hr of benzene may be computed from the V/F data in Table 4-9. For a total conversion of 41.7%, for example, $V/F = 0.12$ and the reactor volume is

$$V = 0.12F = 0.12\,\frac{10,000}{78} = 15.4 \text{ ft}^3$$

Fig. 4-10   *Dehydrogenation of benzene*

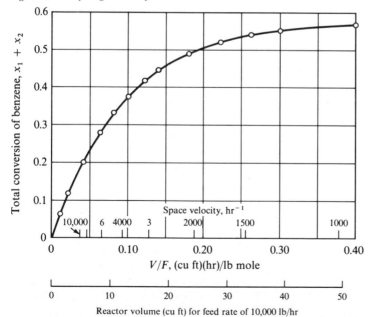

Table 4-9   Conversion vs V/F for the dehydrogenation of benzene

| V/F | Space velocity, hr^{-1} | Conversion | | | Composition of mixture (molal) | | | |
		x_1	x_2	x_t	C_6H_6	$C_{12}H_{10}$	$C_{18}H_{14}$	H_2
0	∞	0	0	0	1.000	0	0	0
0.005	75,800	0.0302	0.00014	0.0303	0.970	0.0150	0.00014	0.0152
0.010	37,900	0.0586	0.00051	0.0591	0.941	0.0288	0.00051	0.0298
0.020	18,950	0.1105	0.00184	0.112	0.888	0.0534	0.00184	0.0571
0.040	9,500	0.197	0.0062	0.203	0.797	0.0923	0.0062	0.1047
0.060	6,320	0.264	0.0119	0.276	0.724	0.1201	0.0119	0.1439
0.080	4,740	0.316	0.0180	0.334	0.666	0.140	0.0180	0.176
0.100	3,790	0.356	0.0242	0.380	0.620	0.154	0.0242	0.202
0.120	3,160	0.387	0.0302	0.417	0.583	0.163	0.0302	0.224
0.140	2,710	0.411	0.0359	0.447	0.533	0.170	0.0359	0.241
0.180	2,110	0.445	0.0459	0.491	0.509	0.177	0.0459	0.268
0.220	1,720	0.468	0.0549	0.523	0.477	0.179	0.0549	0.289
0.260	1,460	0.478	0.0617	0.540	0.460	0.177	0.0617	0.301
0.300	1,260	0.485	0.0673	0.552	0.448	0.175	0.0673	0.310
0.400	950	0.494	0.0773	0.571	0.429	0.170	0.0773	0.324
∞*	0	0.496	0.091	0.587	0.413	0.157	0.091	0.339

*Equilibrium conditions.

If a 4-in.-ID tubular reactor were used, a length of about 175 ft would be required. The complete curve of volume vs total conversion is also shown in Fig. 4-10.

The equilibrium total conversion of 49.6 + 9.1 = 58.7% and the equilibrium composition are shown as the values for V/F at infinity in Table 4-9. These results could also have been obtained from Eqs. (E) and (F). Thus at equilibrium, $r_1 = r_2 = 0$, and

$$(1 - x_1 - x_2)^2 = \frac{(\frac{1}{2}x_1 - x_2)(\frac{1}{2}x_1 + x_2)}{0.312}$$

$$(1 - x_1 - x_2)(\tfrac{1}{2}x_1 - x_2) = \frac{x_2(\frac{1}{2}x_1 + x_2)}{0.480}$$

Simultaneous solution of these two equations leads to values of $x_1 = 0.496$ and $x_2 = 0.091$, as shown.

The rates of each reaction as a function of V/F are shown in Fig. 4-11. In both cases the rates fall off toward zero as V/F approaches high values. The rate of reaction 2 also approaches zero at very low values of V/F because diphenyl is not present in the benzene entering the reactor.

The last four columns in Table 4-9 give the composition of the reaction mixture. With a fixed feed rate the increasing V/F values correspond to moving through the reactor, i.e., increasing reactor volume. Hence these columns show how the composition varies with position in the reactor. Note that the mole fraction of benzene decreases

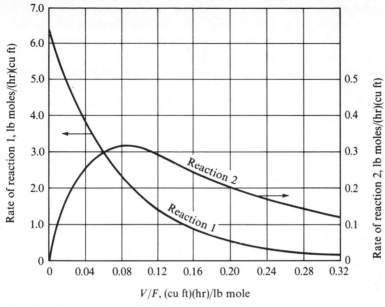

Fig. 4-11   *Rates of reaction for dehydrogenation of benzene*

continuously as the reaction mixture proceeds through the reactor, while the diphenyl content reaches a maximum at a V/F value of about 0.22, typical of the intermediate product in consecutive reactions.

Example 4-6 illustrates the design procedure for a homogeneous flow reactor operating at constant temperature and pressure. When the pressure is not constant and the rate of reaction varies significantly with the pressure (gaseous components), the pressure drop in the reactor must be taken into account. The method of accomplishing this involves no new principles. In comparison with the case in Example 4-6, the differences are as follows:

1.   The effect of pressure on the rate of reaction must be accounted for; i.e., rate equations (A) and (B) would be used instead of (E) and (F), which are based upon 1 atm pressure.
2.   The same stepwise numerical procedure is applicable, except that the pressure at any point in the reactor must be computed and the appropriate value used in Eqs. (A) and (B).

The changes in pressure are due to friction, potential-energy changes, and kinetic-energy changes in the flow process in the reactor. Hence the pressure drop may be computed from the Bernoulli equation, with suitable auxiliary relationships, such as the Fanning equation for the friction. Since the pressure drop is rather small for flow in empty pipes, which is the usual case for homogeneous reactors, the effect of pressure may sometimes be neglected. However, this is not invariably the case. Thermal cracking of hydrocarbons is accomplished by passing the fluid through pipes which may be several hundred feet long, and in such instances the pressure drop may be of the order of 100 psi at an absolute pressure of 500 psia.

IDEAL STIRRED-TANK REACTORS (FLOW)

As we saw in Chap. 3, the relation between conversion and flow rate in a continuous-flow, stirred-tank reactor [Eq. (3-3)] is an algebraic one, in contrast to the integral relations for batch reactors [Eq. (3-10)] and plug-flow reactors [Eq. (3-13)]. If the volume (density) is constant, the simplified form, for the continuous-flow type analogous to Eqs. (4-2) and (4-5) for ideal batch and plug-flow reactors, is[1]

$$\frac{V}{Q} = \bar{\theta} = \frac{C_0 - C}{\mathbf{r}} \tag{4-6}$$

Here C and $\mathbf{r}$ refer to conditions in the product stream (or in the reactor) and $\bar{\theta}$ is the average residence time. It is equal to V/Q and also equal to the average of the distribution of residence times for an ideal stirred-tank reactor. This distribution function is developed in Chap. 6.

Since the temperature must be the same in all parts of this type of reactor, isothermal operation is always achieved as long as steady-state conditions prevail. However, the reactor temperature may be different from that of the feed stream, because of either the heat of reaction or the energy exchange with the surroundings. Hence the treatment in this chapter is restricted to cases where the feed and reactor temperatures are the same. The more general case will be considered in Chap. 5, along with noniso-thermal behavior.

4-5 Single-stirred-tank Reactors

The algebraic nature of Eq. (4-6) simplifies the analytical treatment for the case of multiple reactions to the solution of an assembly of algebraic equations. This is, in general, a simpler problem than treatment of an assembly of integral equations, applicable for plug-flow or batch reactors.

---

[1] The rate in this equation, as in Eqs. (3-10) and (3-13), is a positive quantity. That is, it is a rate of disappearance of reactant or production of product.

The following two examples illustrate the procedure for single and multiple reactions.

Example 4-7   Eldridge and Piret[1] have investigated the hydrolysis of acetic anhydride in stirred-tank flow reactors of about 1,800 cm^3 volume at temperatures from 15 to 40°C. Their experimental results for several volumetric feed rates are shown in Table 4-10. Independent determination by these investigators from other experiments in

Table 4-10

Entering anhydride concentration, moles/cm^3	Volumetric feed rate, cm^3/min	% hydrolysis of anhydride	T, °C
2.1×10^{-4}	378	25.8	15
1.4×10^{-4}	582	33.1	25
1.37×10^{-4}	395	40.8	25
1.08×10^{-4}	555	15.3	10
0.52×10^{-4}	490	16.4	10
0.95×10^{-4}	575	55.0	40
0.925×10^{-4}	540	55.7	40
1.87×10^{-4}	500	58.3	40
2.02×10^{-4}	88.5	88.2	40

batch-operated reactors resulted in the following first-order equations for the rate of reaction, in g moles/(cm^3)(min):

$$r = \begin{cases} 0.0567C & \text{at } 10°C \\ 0.0806C & \text{at } 15°C \\ 0.1580C & \text{at } 25°C \\ 0.380C & \text{at } 40°C \end{cases}$$

where C is the concentration of acetic anhydride, in gram moles per cubic centimeter.

For each run compute the percent of hydrolysis and compare it with the observed value in the table. In all cases the feed temperature is the same as the temperature of the reaction mixture. Since the concentrations are low, the density of the solution may be assumed constant.

Solution   The hydrolysis reaction,

$$(CH_3CO)_2O + H_2O \rightarrow 2CH_3COOH$$

would be expected to be second order. In this case, since dilute aqueous solutions were employed, the water concentration is essentially constant, so that a first-order equation is satisfactory. Equation (4-6) is applicable because the reaction mixture is a constant-density liquid.

[1]J. W. Eldridge and E. L. Piret, *Chem. Eng. Progr.*, **46**, 290 (1950).

Substituting the rate equation $\mathbf{r} = kC$ into Eq. (4-6) and solving the resulting expression for the exit concentration C_1, we have

$$C_1 = \frac{C_0}{1 + k(V/Q)} = \frac{C_0}{1 + k\bar{\theta}}$$

The fraction of the anhydride hydrolyzed, y, is equal to $(C_0 - C_1)/C_0$, and so

$$y = 1 - \frac{C_1}{C_0} = 1 - \frac{1}{1 + k\bar{\theta}} \tag{A}$$

Equation (A) can be employed to determine the percent hydrolyzed from $\bar{\theta}$ and the reaction-rate constant. For the first set of data in the table, at 15°C,

$$k = 0.0806 \text{ min}^{-1} \qquad \bar{\theta} = \frac{1,800}{378} \text{ min}$$

$$k\bar{\theta} = 0.0806 \frac{1,800}{378} = 0.384 \text{ (dimensionless)}$$

Substitution in Eq. (A) gives

$$y = 1 - \frac{1}{1 + 0.384} = 0.277 \qquad \text{or } 27.7\%$$

This agrees reasonably well with the experimental value of 25.8%.

Table 4-11 shows the results of similar calculations for the other sets of data. The computed values are based on the concept of complete mixing and hence on the

Table 4-11

	% hydrolysis	
Volumetric feed rate, cm³/min	Experimental	Calculated from Eq. (A)
378	25.8	27.7
582	33.1	32.8
395	40.8	41.8
555	15.3	15.5
490	16.4	17.2
575	55.0	54.4
540	55.7	55.9
500	58.3	57.8
88.5	88.2	88.5

assumption of uniform composition and temperature throughout the mass. The agreement between the calculated results and the experimental ones is a measure of the validity of the complete mixing assumption. Tests with successively less stirring can be used to study the level of agitation required for complete mixing. MacDonald and Piret[1] have made such studies for a first-order reaction system.

[1] R. W. MacDonald and E. L. Piret, *Chem. Eng. Progr.*, **47**, 363 (1951).

Example 4-8 An ideal continuous stirred-tank reactor is used for the homogeneous polymerization of monomer M. The volumetric flow rate is Q, the volume of the reactor is V, and the density of the reaction solution is invariant with composition. The concentration of monomer in the feed is $[M]_0$. The polymer product is produced by an initiation step and a consecutive series of propagation reactions. The reaction mechanism and rate equations may be described as follows, where P_1 is the activated monomer and $P_2, \ldots, P_n$ are polymer molecules containing n monomer units:

INITIATION

$$M \rightarrow P_1 \qquad\qquad \mathbf{r} = k_i[M] \qquad \text{moles/(time)(vol)}$$

PROPAGATION

$$P_1 + M \rightarrow P_2$$
$$P_2 + M \rightarrow P_3$$
$$\cdots\cdots\cdots\cdots\cdots \qquad \mathbf{r} = k_p[P_n][M] \qquad \text{moles/(time)(vol)}$$
$$P_n + M \rightarrow P_{n+1}$$

All the propagation steps have the same second-order rate constant k_p.

If k_i, k_p, $[M]_0$, Q, and V are known, derive expressions for (a) the conversion of monomer, (b) the concentration of polymer P_n in the product stream, and (c) the weight fraction w_n of the polymer product (this is a form of the molecular weight distribution of the product).

Solution (a) Equation (4-6) is valid and may be applied to each polymer product and to the monomer. Solving Eq. (4-6) for the rate and applying it to the monomer, we obtain

$$\mathbf{r}_m = \frac{M_0 - M}{V/Q} = \frac{M_0 - M}{\bar{\theta}} \tag{A}$$

The monomer is consumed by the initiation step and all the propagation reactions, and so its rate of disappearance is

$$\mathbf{r}_m = k_i[M] + k_p[M] \sum_{n=1}^{\infty} [P_n] \tag{B}$$

Combining Eqs. (A) and (B) gives

$$\frac{M_0 - M}{\bar{\theta}} = k_i[M] + k_p[M] \sum_{n=1}^{\infty} [P_n] \tag{C}$$

Writing Eq. (A) for P_1, noting that there is no P_1 in the feed, and inserting the appropriate expressions for the rate, we obtain

$$\mathbf{r}_1 = -k_i[M] + k_p[P_1][M] = \frac{0 - [P_1]}{\bar{\theta}} = -\frac{[P_1]}{\bar{\theta}}$$

This may be written

$$k_i[M] = \frac{[P_1]}{\bar{\theta}} + k_p[P_1][M] \tag{D}$$

or

$$[P_1] = \frac{k_i[M]}{(1/\bar{\theta}) + k_p[M]} \tag{E}$$

Similarly, for P_2 and P_n the mass balances are

$$k_p[M][P_1] = \frac{[P_2]}{\bar{\theta}} + k_p[P_2][M] \tag{F}$$

$$k_p[M][P_{n-1}] = \frac{[P_n]}{\bar{\theta}} + k_p[P_n][M] \tag{G}$$

Solving for P_2 from Eq. (F) and P_n from Eq. (G) gives

$$[P_2] = \frac{k_p[M][P_1]}{(1/\bar{\theta}) + k_p[M]} \tag{H}$$

$$[P_n] = \frac{k_p[M][P_{n-1}]}{(1/\bar{\theta}) + k_p[M]} \tag{I}$$

Inserting Eq. (E) in Eq. (H) to eliminate $[P_1]$ and proceeding in a similar way for $[P_2]$, $[P_3]$, ..., $[P_{n-1}]$, we can eliminate all intermediate concentrations to obtain

$$[P_n] = \frac{k_i([M])^n(k_p)^{n-1}}{\{k_p[M] + 1/\bar{\theta}\}^n} = \frac{k_i}{k_p} \frac{1}{\{1 + 1/\bar{\theta} k_p[M]\}^n} \tag{J}$$

Summing up all the mass balances for the individual species written in the form of Eqs. (D), (F), and (G) gives a simple relationship between the concentration of unreacted monomer and the concentrations of all the polymer products,

$$k_i[M] = \frac{P_1}{\bar{\theta}} + \frac{P_2}{\bar{\theta}} + \cdots = \frac{1}{\bar{\theta}} \sum_{n=1}^{\infty} P_n \tag{K}$$

Combining this result with Eq. (C) provides a relationship between $[M]$ and $[M]_0$,

$$\frac{[M]_0 - [M]}{\bar{\theta}} = k_i[M] + k_i k_p[M]^2 \bar{\theta} \tag{L}$$

Since the conversion of monomer is

$$x = \frac{[M]_0 - [M]}{[M]_0} \tag{M}$$

Eq. (L) may be written

$$x = k_i\bar{\theta}(1 - x)\{1 + k_p\bar{\theta}[M]_0(1 - x)\} \tag{N}$$

This is the answer to the first part of the problem. Note that the conversion depends on the initial monomer concentration as well as on the rate constants and $\bar{\theta}$.

(b) If we know x, we can find $[M]$ from Eq. (M) and then calculate $[P_n]$ from Eq. (J) for any value of n.

(c) The weight fraction of P_n in the polymer product is the weight of polymer

P_n divided by the weight of total polymer product. The latter is equal to the weight of monomer that has reacted. If W_0 is the molecular weight of monomer, the weight fraction of P_n is

$$w_n = \frac{[P_n]nW_0}{([M]_0 - [M])W_0} = \frac{n[P_n]}{[M]_0 - [M]} \tag{P}$$

Substituting Eq. (J) for $[P_n]$ in Eq. (P) gives

$$w_n = \frac{k_i n}{k_p([M]_0 - [M])} \frac{1}{\{1 + 1/(\bar{\theta}k_p[M])\}^n} \tag{Q}$$

As an illustration of these results Eqs. (N) and (Q) have been used to calculate the conversion and molecular weight distribution for the case

$$k_i = 10^{-2} \text{ sec}^{-1}$$

$$k_p = 10^2 \text{ cm}^3/(\text{g mole})(\text{sec})$$

$$[M]_0 = 1.0 \text{ g moles/(liter)}$$

Calculations were carried out for average residence times ($\bar{\theta} = V/Q$) from 3 to 1,000 sec. Figure 4-12 shows the conversion and w_n for several products. In this range of $\bar{\theta}$ the conversion increases from 3.7 to 97.3%. The weight fraction of P_1 decreases continuously with residence time. This is expected, since w refers to the weight fraction of product (not of total effluent from the reactor), and at low $\bar{\theta}$ the product will be nearly all P_1. At the other extreme, w_{10} is low but increases continuously, since higher

Fig. 4-12   Distribution of products in stirred-tank reactor for polymerization

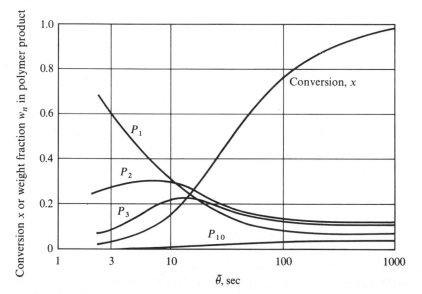

products ($m > 10$) are not formed in significant amounts. Curves are also shown for two products of intermediate molecular weight, P_2 and P_3. These curves show maxima at intermediate $\bar{\theta}$.

The curves in Fig. 4-12 are helpful in choosing a reactor size appropriate for producing a specified product distribution. For example, if the objective is to make the maximum amount of trimer (P_3), the reactor volume should be such that $\bar{\theta}$ is about 13 sec. This problem does not allow for destruction of active polymer P_n by termination reactions. This is often the case and leads to a more symmetrical distribution of molecular weights. Denbigh[1] has considered this and similar polymerization problems.

It was possible to solve this polymerization problem in closed form because we were dealing with algebraic equations, that is, with an ideal stirred-tank reactor. The same problem in a tubular-flow reactor would require the solution of a series of integral equations, which is possible only by numerical methods.

4-6 Stirred-tank Reactors in Series

In some cases it may be desirable to use a series of stirred-tank reactors, with the exit stream from the first serving as the feed to the second, and so on. For constant density the exit concentration or conversion can be solved by consecutive application of Eq. (4-6) to each reactor. MacDonald and Piret[1] have derived solutions for a number of rate expressions and for systems of reversible, consecutive, and simultaneous reactions. Graphical procedures have also been developed. The kinds of calculations involved are illustrated for the simple case of a first-order reaction in Example 4-9.

Example 4-9   Acetic anhydride is to be hydrolyzed in three stirred-tank reactors operated in series. Suppose that each has a volume of 1,800 cm^3, that the temperature is constant and equal to 25°C, and that the feed rate to the first reactor is 582 cm^3/min. Compute the percent of hydrolysis accomplished in the three reactors.

Solution   From the results of Example 4-7, the fraction hydrolyzed in the stream leaving the first reactor is 0.328. If the anhydride concentration leaving the first reactor is designated as C_1, and that leaving the second is C_2, Eq. (4-6) applied to the second reactor is

$$C_2 = \frac{C_1}{1 + k\bar{\theta}_2}$$

or

$$\frac{C_1 - C_2}{C_1} = \text{fraction hydrolyzed in reactor 2} = 1 - \frac{1}{1 + k\bar{\theta}_2}$$

[1] K. G. Denbigh, *Trans. Faraday Soc.*, **43**, 648 (1947).
[1] R. W. MacDonald and E. L. Piret, *Chem. Eng. Progr.*, **47**, 363 (1951).

Since $\bar{\theta}_2 = \bar{\theta}_1$, the fraction hydrolyzed in the second reactor will be the same as in the first. Therefore the following equations may be written for the three reactors:

$$\frac{C_0 - C_1}{C_0} = 0.328 \quad \text{or } C_1 = (1 - 0.328)C_0 = 0.672C_0$$

$$\frac{C_1 - C_2}{C_1} = 0.328 \quad \text{or } C_2 = 0.672C_1 = (0.672)^2\, C_0$$

$$\frac{C_2 - C_3}{C_2} = 0.328 \quad \text{or } C_3 = 0.672C_2 = (0.672)^3\, C_0$$

The equation for C_3 may be generalized to the form

$$C_n = (0.672)^n C_0 = \left(\frac{1}{1 + k\bar{\theta}_i}\right)^n C_0$$

or

$$1 - \frac{C_n}{C_0} = 1 - \frac{1}{(1 + k\bar{\theta}_i)^n} = \text{fraction hydrolyzed in } n \text{ reactors} \qquad \text{(A)}$$

where n is the number of reactors in series, and $\bar{\theta}_i$ is the residence time in each stage. For $n = 3$, the fraction hydrolyzed is $1 - 0.672^3 = 0.697$, or 69.7%.

If the restriction of equal residence times in each reactor is removed, Eq. (A) of Example 4-9 becomes

$$x_n = 1 - \frac{1}{(1 + k\bar{\theta}_1) \ldots (1 + k\bar{\theta}_i) \ldots (1 + k\bar{\theta}_n)} \qquad (4\text{-}6A)$$

where x_n is the conversion in the effluent from the nth reactor and $\bar{\theta}_i = V_i/Q$ is the mean residence time in the ith reactor.

Graphical methods can be used to obtain the conversion from a series of reactors and have the advantage of displaying the concentration in each reactor. Moreover, no additional complications are introduced when the rate equation is not first order. As an illustration of the procedure consider three stirred-tank reactors in series, each with a different volume, operating as shown in Fig. 4-13a. The density is constant, so that at steady state the volumetric flow rate to each reactor is the same. The flow rate and reactant concentration of the feed (Q and C_0) are known, as are the volumes of each reactor. We construct a graph of the rate of reaction $\mathbf{r}$ vs reactant composition. The curved line in Fig. 4-13b shows how the rate varies with C according to the rate equation, which may be of any order.

Equation (4-6) applied to the first stage is

$$\frac{V_1}{Q} = \bar{\theta}_1 = \frac{C_0 - C_1}{\mathbf{r}_1}$$

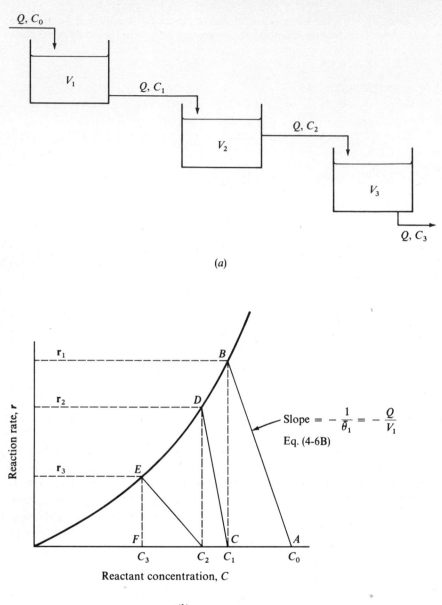

Fig. 4-13 *(a) Steady operation of three stirred-tank reactors in series.*
(b) Graphical solution for stirred-tank reactors in series.

or

$$\frac{1}{\bar{\theta}_1} = \frac{r_1}{C_0 - C_1}$$

For a stirred-tank reactor, r_1 and C_1 will represent a point B on the rate line in Fig. 4-13b. If C_0 is located on the abscissa of this figure (point A), a straight line from that point to B will have a slope of $r_1/(C_1 - C_0) = -r_1/(C_0 - C_1)$. From the equation above, this slope is equal to $-Q/V_1$, which is known. Hence the conditions in the effluent stream from the first reactor can be found by constructing a straight line from point A with a slope of $-Q/V_1$ and noting the point at which it intersects the rate curve.

The second stage is constructed by locating C_1 on the abscissa (point C) and then drawing a straight line from C with a slope of $-Q/V_2$. The intersection of this line with the reaction curve (point D) establishes the effluent concentration C_2 from the second reactor. Similar construction for the third reactor is shown in Fig. 4-13b.

4-7 Comparison of Stirred-tank and Tubular-flow Reactors

The stirred-tank reactor has certain advantages because of the uniform temperature, pressure, and composition attained as a result of mixing. As mentioned, it is possible to operate such a reactor under isothermal conditions even when the heat of reaction is high—an impossibility in the long tubular type. When operation within a small temperature range is desired, for example, to minimize side reactions or avoid unfavorable rates, the opportunity for isothermal operation at the optimum temperature is a distinct advantage. Stirred-tank reactors, by virtue of their large volumes (and hence their large V/F values) provide a long residence time, which, combined with the isothermal nature of the reactor, permits operation at the optimum temperature for a long reaction time. For rate equations of certain types the selectivity in multiple reactions may be greater in tank reactors than in tubular-flow reactors for the same residence time. For other forms of the rate equations the reverse is true. Examples later in this section illustrate this point.

Efficient stirred-tank reactors are difficult to construct for gaseous systems because of the mixing problem. Fixed baffles and mechanical stirrers can be used, but these do not ensure complete mixing. Hence stirred-tank equipment is generally restricted to liquid systems.

For high-pressure reactions it is usually necessary, because of cost considerations, to use small-diameter tubular reactors rather than tank types. Tank reactors that are to be operated at high pressures require a large wall thickness and complex sealing arrangements for the mixer shaft, factors which increase both initial and maintenance costs. Stirred-tank performance may be achieved in a recycle form of tubular-flow reactor, as illustrated in Fig. 1-3a. The diameter may be minimized for high-pressure operation by constructing the reactor in the form of a closed loop of tubing with entrance and exit connections and with a recycle pump in the loop.

The rate of heat transfer per unit mass of reaction mixture is generally lower in the conventional tank type than in a small-diameter tubular reactor, chiefly because of the lower ratio of surface area (available for heat transfer) to volume in the tank reactors and their lower heat-transfer coefficients. So, in instances where the heat of reaction is high it may be desirable to use a tubular reactor. For example, various thermal reactions of hydrocarbons require significant amounts of thermal energy at an elevated temperature level. This would be difficult to accomplish with a large-diameter reactor because of the limited external heat-transfer surface (per unit mass of reaction mixture) and the low coefficient of heat transfer from the oil in the tank to the tank wall. In the tubular reactors (pipe stills) used in industry the coefficient of heat transfer can be increased by forcing the oil through the tubes at a high speed. It is also apparent that severe difficulties would arise in attempting to provide for efficient stirring under reaction conditions (800 to 1200°F, 300 to 800 psia). The tubular-loop reactor operated at high circulation rates can give complete mixing in a small-diameter tube and high rates of heat transfer. By introducing small solid particles that are free to move, a type is obtained in which there is considerable mixing. Such fluidized-bed reactors also give improved heat-transfer coefficients between the fluid and the wall.

In general, then, stirred-tank reactors have been used mainly for liquid-phase reaction systems at low or medium pressures. Stirred-tank reactors can be used when the heat of reaction is high, but only if the temperature level obtained in their isothermal operation is satisfactory from other standpoints. If the reactions involved are endothermic and a high temperature is required, tubular reactors are usually indicated. However, a tank type may be employed for a highly exothermic reaction. For example, the production of hexamethylenetetramine by reacting ammonia and formaldehyde (in aqueous solution) is highly exothermic, but the rate of reaction is rapid and 100% conversion is possible over a range of temperature of at least 80 to 100°C. By adjusting the rate of feed and reactor volume, it is possible to add the feed at 20°C and remove enough heat to keep the reaction mixture below 100°C.

Denbigh and coworkers[1] have discussed the advantages and disadvantages of continuous-stirred-tank reactors, especially in comparison with batch-operated tank reactors. Stead, Page, and Denbigh[2] describe experimental techniques for evaluating rate equations from stirred-tank data. In stirred-tank equipment the reaction occurs at a rate determined by the composition of the exit stream from the reactor. Since the rate generally decreases with the extent of conversion, the tank reactor operates at the lowest point in the range between the high rate corresponding to the composition in the reactor feed and the low rate corresponding to the exit composition. In the tubular type maximum advantage is taken of the high rates corresponding to low conversions in the first part of the reactor. This means that the tank reactor must have a larger volume for a given feed rate (larger V/F value). Of course, this reasoning does not take into account the effects of side reactions or temperature variations; these may offset this disadvantage of the tank reactor, as illustrated in Example 5-3. Also, the total volume required in a tank-flow reactor can be reduced by using several small units in series.

The relation between volumes required in stirred-tank and tubular-flow reactors can be illustrated by reference to a constant-density first-order reaction. Equation (4-6) is applicable for the stirred-tank reactor and gives

$$\frac{V}{Q} = \frac{C_0 - C}{\mathbf{r}} = \frac{C_0 - C}{kC}$$

or

$$x = \frac{C_0 - C}{C_0} = \frac{k(V_s/Q)}{1 + k(V_s/Q)} \tag{4-7}$$

For the tubular-flow case Eq. (4-4) can be used:

$$\frac{V_p}{Q} = C_0 \int \frac{dx}{\mathbf{r}} = C_0 \int_0^x \frac{dx}{kC_0(1 - x)} = \frac{-1}{k} \ln(1 - x)$$

or

$$x = 1 - e^{-k(V_p/Q)} \tag{4-8}$$

[1] K. G. Denbigh, *Trans. Faraday Soc.*, **40**, 352 (1944), **43**, 648 (1947); K. G. Denbigh, M. Hicks, and F. M. Page, *Trans. Faraday Soc.*, **44**, 479 (1948).
[2] B. Stead, F. M. Page, and K. G. Denbigh, *Disc. Faraday Soc.*, **2**, 263 (1947).

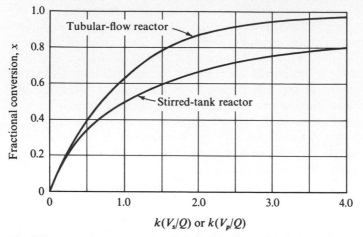

Fig. 4-14   *Conversion in stirred-tank and tubular-flow (plug-flow) reactors*

Equations (4-7) and (4-8) are plotted in Fig. 4-14 as conversion vs $k(V/Q)$. For equal flow rates $k(V/Q)$ is proportional to the volume. It is clear that for any conversion the volume required is largest for the tank reactor and that the difference increases with residence time. We can obtain a direct measure of the ratio of volume V_s of the stirred-tank reactor to volume V_p of the tubular-flow (plug-flow) reactor at the same conversion by equating Eqs. (4-7) and (4-8).

Fig. 4-15   *Ratio of volumes required for stirred-tank and tubular-flow (plug-flow) reactors*

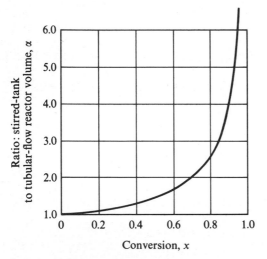

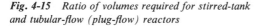

$$\frac{k(V_s/Q)}{1 + k(V_s/Q)} = 1 - e^{-k(V_p/Q)}$$

If α is the ratio of volumes, $\alpha = V_s/V_p$, then the previous equation can be written in terms of V_p and α.

$$\frac{\alpha k(V_p/Q)}{1 + \alpha k(V_p/Q)} = 1 - e^{-k(V_p/Q)} \tag{4-9}$$

If we now replace $k(V_p/Q)$ in Eq. (4-9) with the function of x from Eq. (4-8) and solve for α, we have the ratio of volumes as a function of conversion,

$$\alpha = \frac{x}{(x - 1) \ln (1 - x)} \tag{4-10}$$

This result is plotted in Fig. 4-15 and shows that at low conversions there is little to be gained in using a tubular-flow reactor, but at conversions of 70% or larger more than twice as much volume is required for a stirred-tank unit.

Selectivity may also be different in stirred-tank and tubular-flow reactors. It has been shown[1] that, depending on the kinetics and nature of the multiple reactions, selectivity obtained in a stirred-tank reactor may be less, the same as, or greater than that for a tubular-flow reactor. Examples of reaction systems for each result are given in Table 4-12. The order of

Table 4-12 Effect of mixing on selectivity for various types of reaction systems

Reaction system	Reaction type	Overall selectivity*
1. $A \rightarrow B$ $\;\;\;\;B \rightarrow D$	Consecutive (first order)	$S_S < S_p$ (selectivity of B with respect to D)
2. $A + B \rightarrow R$ $\;\;\;\;A + R \rightarrow S$	Consecutive (second order)	$S_S < S_p$ (selectivity of R with respect to S)
3. $A \rightarrow B$ $\;\;\;\;A \rightarrow C$	Parallel (equal, first order)	$S_p = S_S$ (selectivity of B with respect to C)
4. $A + B \rightarrow C$ $\;\;\;\;A + B \rightarrow D$	Parallel (equal, second order)	$S_p = S_S$ (selectivity of C with respect to D)
5. $A + B \rightarrow C$ $\;\;\;\;2A \rightarrow D$	Parallel (unequal order with respect to A)	$S_S > S_p$ (selectivity of C with respect to D)

*Selectivity, as defined in Chap. 2, is the ratio of the yield of one product to that of another
[1] T. E. Corrigan, G. A. Lessells, and M. J. Dean, *Ind. Eng. Chem.*, **60**, 62 (1968).

the rate equation is assumed to follow the stoichiometry for each reaction. Since the stirred-tank reactor corresponds to complete mixing and the tubular-flow unit to no axial mixing, the table shows the effect of mixing upon selectivity. The following example illustrates the method of establishing the conclusions given in Table 4-12.

Example 4-10   Develop equations for the selectivity of product B with respect to D for reaction system 1 of Table 4-12 for stirred-tank and tubular-flow reactors. Assume isothermal conditions and constant density. In the feed $[A] = [A]_0$ and $[B] = [D] = 0$.

Solution   The reaction sequence is

$$A \xrightarrow{k_1} B$$

$$B \xrightarrow{k_3} D$$

In Sec. 2-10 this system was analyzed for a constant-density batch reactor. Since the tubular-flow reactor will have the same performance (see Sec. 4-4), the results in Sec. 2-10 can be applied here if t is replaced with the residence time V/Q. Hence the overall selectivity for a *tubular-flow* reactor is given by the ratio of Eqs. (2-106) and (2-107). Since $[A]/[A]_0 = 1 - x_t$, this ratio is

$$S_p = \frac{x_B}{x_D} = \frac{[k_1/(k_1 - k_3)][(1 - x_t)^{k_3/k_1} - (1 - x_t)]}{[k_1/(k_1 - k_3)][1 - (1 - x_t)^{k_3/k_1}] - [k_3/(k_1 - k_3)]x_t} \tag{A}$$

For the *stirred-tank* case Eq. (4-6) may be written for components A, B, and D as follows:

$$\bar{\theta} = \frac{V}{Q} = \frac{[A]_0 - [A]}{k_1[A]} \quad \text{or} \quad [A] = \frac{[A]_0}{1 + k_1\bar{\theta}} \tag{B}$$

$$\bar{\theta} = \frac{V}{Q} = \frac{0 - [B]}{k_3[B] - k_1[A]} \quad \text{or} \quad [B] = \frac{k_1\bar{\theta}[A]}{1 + k_3\bar{\theta}} \tag{C}$$

$$\bar{\theta} = \frac{V}{Q} = \frac{0 - [D]}{-k_3[B]} \quad \text{or} \quad [D] = k_3\bar{\theta}[B] \tag{D}$$

From Eqs. (B) and (C),

$$\frac{[B]}{[A]_0} = x_B = \frac{k_1\bar{\theta}}{(1 + k_1\bar{\theta})(1 + k_3\bar{\theta})} \tag{E}$$

Using this result in Eq. (D) gives

$$\frac{[D]}{[A]_0} = x_D = \frac{k_1\bar{\theta}k_3\bar{\theta}}{(1 + k_1\theta)(1 + k_3\theta)} \tag{F}$$

Then the selectivity S_s in the stirred-tank reactor will be

$$S_s = \frac{x_B}{x_D} = \frac{1}{k_3\bar{\theta}} \tag{G}$$

This result may be expressed in terms of the total conversion of A by noting, from Eq. (B), that

$$k_1 \bar{\theta} = \frac{[A]_0}{[A]} - 1 = \frac{x_t}{1 - x_t}$$

Using this result in Eq. (G) to eliminate $\bar{\theta}$ gives

$$S_s = \frac{k_1}{k_3} \frac{1 - x_t}{x_t} \tag{H}$$

Equations (A) and (H) can be employed to calculate S_p and S_s for any conversion. The results are shown in Fig. 4-16 for $k_1/k_3 = 2$. The *relative* position of the two curves would be the same for other values of k_1/k_3. Note that the selectivity of B with respect to D is greater in the tubular-flow reactor for all conversions, although the difference approaches zero as the conversion approaches zero.

Table 4-12 compares selectivities for *single* reactors. For some reaction systems a combination of stirred-tank and tubular-flow units may give higher selectivities than a single reactor of the same total volume. The combinations of number and arrangement of reactors and reaction systems are huge. However, the approach to selectivity evaluation is always the same and follows the methods described in Example 4-10. A simple illus-

Fig. 4-16 *Selectivity for consecutive reactions in stirred-tank and tubular-flow reactors*

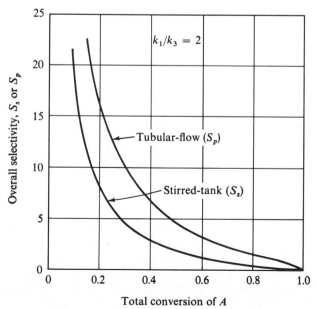

tration of the conversion obtained in a combination of reactors is given in the following example.

Example 4-11   A dilute aqueous solution of acetic anhydride is to be hydrolyzed continuously at 25°C. At this temperature the rate equation is

$$r = 0.158C, \text{ g mole/(cm}^3)(\text{min})$$

where C is the concentration of anhydride, in gram moles per cubic centimeter. The feed rate to be treated is 500 cm^3/min of solution, with an anhydride concentration of 1.5×10^{-4} g mole/cm^3.

There are two 2.5-liter and one 5-liter reaction vessels available, with excellent agitation devices.

(a) Would the conversion be greater if the one 5-liter vessel were used as a steady-flow tank reactor or if the two 2.5-liter vessels were used as reactors in series? In the latter case all the feed would be sent to the first reactor and the product from that would be the feed to the second reactor.

(b) Would a higher conversion be obtained if the two 2.5-liter vessels were operated in parallel; that is, if 250 cm^3/min of feed were fed to each reactor and then the effluent streams from each reactor joined to form the final product?

(c) Compare the conversions calculated in parts (a) and (b) with that obtainable in a tubular-flow reactor of 5-liter volume.

(d) Would the conversion be increased if a tank-flow reactor of 2.5 liters were followed with a 2.5-liter tubular-flow reactor?

Assume that the density of the solutions is constant and that operation is steady state.

Solution   Since this is a first-order constant-density reaction, Eqs. (4-7) and (4-8) give the conversions for single-stirred-tank and ideal tubular-flow reactors in terms of residence time V/Q. For multiple-stirred-tank reactors Eq. (A) of Example (4-9) is applicable.

(a) For a single 5-liter vessel, $\bar{\theta} = 5{,}000/500 = 10$ min. From Eq. (4-7),

$$x = \frac{0.158(10)}{1 + 0.158(10)} = 0.612$$

For two 2.5-liter reactors in series,

$$\bar{\theta}_1 = \bar{\theta}_2 = \frac{2{,}500}{500} = 5 \text{ min}$$

Substituting in Eq. (A) of Example 4-9 gives

$$x = 1 - \frac{1}{(1 + k\bar{\theta})^2} = 1 - \frac{1}{[1 + 0.158(5)]^2} = 0.688$$

(b) For a 2.5-liter reactor with a feed rate of 250 cm^3/min, $\bar{\theta} = 10$ min. Hence the conversion will be the same as for the 5-liter reactor with $Q = 500$ cm^3/min, that is, 0.612.

(c) For a single tubular-flow reactor, from Eq. (4-8),

$$x = 1 - e^{-0.158(5,000/500)} = 1 - 0.206 = 0.794$$

(d) In the first reactor $\bar{\theta}_1 = 2,500/500 = 5$ min. Hence the conversion in the product stream from the first reactor will be

$$x_1 = \frac{k\bar{\theta}_1}{1 + k\bar{\theta}_1} = \frac{0.158(5)}{1 + 0.158(5)} = 0.442$$

When the feed stream to a tubular-flow reactor is x_1 rather than zero, integration of Eq. (4-5) gives

$$\theta = C_0 \int_{x_1}^{x_2} \frac{dx}{\mathbf{r}} = C_0 \int \frac{dx}{kC_0(1 - x)} = -\frac{1}{k}[\ln(1 - x_2) - \ln(1 - x_1)]$$

or

$$x_2 = 1 - (1 - x_1)e^{-k\theta}$$

The residence time in the second tubular-flow reactor is also 5 min. With a feed of conversion $x_1 = 0.442$, the final conversion would be

$$x_2 = 1 - (1 - 0.442)e^{-0.158(5)} = 1 - 0.254 = 0.746$$

The various results, arranged in order or increasing conversion, are shown in Table 4-13.

Table 4-13

Type	Conversion, %
Single-stirred tank (5 liters)	61.2
Two stirred tanks in parallel (each 2.5 liters)	61.2
Two stirred tanks in series (each 2.5 liters)	68.8
Stirred tank followed by tubular-flow reactor (each 2.5 liters)	74.6
Single tubular-flow reactor (5 liters)	79.4

In the previous example increasing the number of stirred-tank reactors from one to two (with the same total residence time) caused an increase in conversion from 61.2% to 68.8%. Further increase in number of tank reactors in series would lead to a maximum conversion of 79.4%, the value for a tubular-flow reactor with the same residence time. Thus an infinite number of stirred-tank reactors in series is equivalent to a tubular-flow reactor, provided the total residence time is the same. This may be demonstrated by comparing Eq. (4-8) with Eq. (A) of Example 4-9, which is applicable for equal residence time in each stirred tank. For a total residence time of $\bar{\theta}_t$, Eq. (A) becomes

$$x = 1 - \frac{1}{(1 + k\bar{\theta}_t/n)^n} \qquad (4\text{-}11)$$

It is known that

$$\lim_{n \to \infty} \left(1 + \frac{\alpha}{n}\right)^n = e^\alpha$$

If we use this result with $\alpha = k\bar{\theta}_t$, Eq. (4-11) becomes

$$x = 1 - e^{-k\theta_t}$$

which is the same as Eq. (4-8).

4-8 Non-steady-flow (Semibatch) Reactors

The semibatch reactor was defined in Chap. 3 (Fig. 3-1c) as a tank type operated on a non-steady-flow basis. Semibatch behavior occurs when a tank-flow reactor is started up, when its operating conditions are changed from one steady state to another, or when it is shut down. Purging processes in which an inert material is added to the reactor can also be classified as semibatch operation.

In addition to applications arising from short-period deviations from steady state, the semibatch reactor often is used for its own particular characteristics. For example, it is sometimes advantageous to add all of one reactant initially and then add the other reactant continuously. When the heat of reaction is large, the energy evolution can be controlled by regulating the rate of addition of one of the reactants. In this way the poor heat-transfer characteristic of tank reactors can be partially eliminated. This form of operation also allows for a degree of control of concentration of the reaction mixture, and hence rate of reaction, that is not possible in batch or continuous-flow reactors. Another example is the case in which the reactants are all added initially to the vessel but one of the products is removed continuously, as in the removal of water by boiling in esterification reactions. The advantage here is an increase in rate, owing to the removal of one of the products of a reversible reaction and to increased concentrations of reactants.

The mass-balance equations for semibatch operation may include all four of the terms in the general balance, Eq. (3-1). The feed and withdrawal streams from the reactor cause changes in the composition and volume of the reaction mixture, in addition to the changes due to the reaction itself. Many operating alternatives exist. Frequently numerical solution is required because of the complicated form of the mass balance. After a discussion of general equations, this class of design problem will be illustrated with two specific examples. Here again ideal reactor performance (complete mixing) is assumed.

In general the conversion of reactant is not a very useful term for semibatch operation because when reactant is present initially in the reactor and is added and deleted in feed and exit streams, there may be ambiguity about the total amount upon which to define x. Instead we shall formulate Eq. (3-1) in terms of the mass fraction w of reactant. If F_0 is the total mass feed rate and F_1 is the withdrawal rate, the mass balance for the reactant is

$$F_0 w_0 - F_1 w_1 - \mathbf{r}V = M \frac{d(VC_1)}{dt} \tag{4-12}$$

where C_1 is the molal concentration of reactant in the reactor and M is the molecular weight of reactant. Alternately, the term on the right side for the accumulation could be expressed as $d(m_t w_1)/dt$ where m_t is the total mass in the reactor.

If the feed-stream conditions and the initial state in the reactor are known, Eq. (4-12) can always be integrated, although numerical procedures may be required. An important case in which analytical integration is possible is when the feed and exit flow rates, feed composition, and density are all constant and the reaction is first order. Piret and Mason[1] have analyzed single and cascades (reactors in series) of stirred-tank reactors operating under these restrictions. The results are a reasonable representation of the behavior for many systems under startup and shutdown periods. With constant density, the concentration accounts fully for changes in amount of reactant. Also, constant density along with constant flow rates means that the reactor volume V will remain constant. Under these restrictions Eq. (4-12) may be written

$$QC_0 - QC_1 - \mathbf{r}V = V \frac{dC_1}{dt}$$

or

$$C_0 - C_1 - \mathbf{r}\bar{\theta} = \bar{\theta} \frac{dC_1}{dt} \tag{4-13}$$

If the rate is first order and the temperature is constant, Eq. (4-13)

[1] D. R. Mason and E. L. Piret, *Ind. Eng. Chem.*, **42**, 817 (1950); *Ind. Eng. Chem.*, **43**, 1210 (1951).

is a linear differential equation which can be integrated analytically. In terms of $\bar{\theta}$, it may be written

$$\frac{dC_1}{dt} + \left(\frac{1}{\bar{\theta}} + k\right)C_1 = \frac{C_0}{\bar{\theta}} \tag{4-14}$$

where k is the first-order rate constant.

Example 4-12 Acetic anhydride is hydrolyzed at 40°C in a semibatch system operated by initially charging the stirred-tank reactor with 10 liters of an aqueous solution containing 0.50×10^{-4} g mole anhydride/cm^3. The vessel is heated to 40°C, and at that time a feed solution containing 3.0×10^{-4} g mole anhydride/cm^3 is added at the rate of 2 liters/min. Product is withdrawn at the same rate. The density may be assumed constant, and the rate is

$$\mathbf{r} = kC \qquad \text{g moles/(cm}^3\text{)(min)}$$

$$k = 0.380 \text{ min}^{-1}$$

Determine the concentration of the solution leaving the reactor as a function of time.

Solution Equation (4-14) is applicable, and in this case $C_0/\bar{\theta}$ is a constant. Thus

$$\bar{\theta} = \frac{V}{Q} = \frac{10,000}{2,000} = 5 \text{ min}$$

$$\frac{C_0}{\bar{\theta}} = \frac{3.00 \times 10^{-4}}{5} = 6 \times 10^{-5} \text{ g mole/(cm}^3\text{)(min)}$$

$$\frac{1}{\bar{\theta}} + k = \frac{1}{5} + 0.380 = 0.580 \text{ min}^{-1}$$

The integrated solution of Eq. (4-14) is

$$C_1 = \frac{C_0}{\bar{\theta}} \frac{1}{(1/\bar{\theta}) + k} + Ie^{-(1/\bar{\theta} + k)t}$$

The constant of integration I may be obtained by noting that at $t = 0$ and $C_1 = 0.50 \times 10^{-4}$

$$I = 0.50 \times 10^{-4} - \frac{C_0}{\bar{\theta}} \frac{1}{(1/\bar{\theta}) + k}$$

$$= 0.50 \times 10^{-4} - \frac{6 \times 10^{-5}}{0.580} = -5.34 \times 10^{-5}$$

Then the final expression for the product concentration is

$$C_1 = \frac{6 \times 10^{-5}}{0.580} - 5.34 \times 10^{-5}e^{-0.58t}$$

$$= 10.3 \times 10^{-5} - 5.34 \times 10^{-5}e^{-0.58t}$$

Table 4-14 shows values of C_1 for increasing time, measured from the instant of addition of the feed stream. These results indicate that after 5 min the reactor is operating at very nearly steady-state conditions. Hence, the concentration at this time could also be obtained directly from Eq. (4-6) for steady-state operation.

Table 4-14

t, min	C_1, g moles/cm^3
0	5.00×10^{-5}
1	7.35×10^{-5}
2	8.67×10^{-5}
3	9.40×10^{-5}
5	10.0×10^{-5}
∞	10.3×10^{-5}

Figure 4-17 shows a special type of semibatch reactor system in which there is a continuous feed, no withdrawal of product, and a mass $(m_A)_i$ of component A initially in the reactor. The application of Eq. (4-12) will be considered for a reactor of this type when the rate equation is not first order and when the volume of the reaction mixture varies. Since there is no exit stream, Eq. (4-12) takes the form

$$F_0(w_A)_0 - \mathbf{r}V = M_A \frac{d(V[A]_1)}{dt} \tag{4-15}$$

The rate $\mathbf{r}$ could be expressed in terms of the concentrations of the reactants and this equation integrated numerically for any rate function. However, in this case conversion has meaning, and concentration can be expressed in terms of this variable. Define x at any time t as the ratio of the amount of reactant A converted to the total amount of A added up to that time. Then the concentration of A in the reactor will be related to x by

$$[A]_1 = \frac{[(m_A)_i + F_0(w_A)_0 t](1 - x)}{M_A V} \tag{4-16}$$

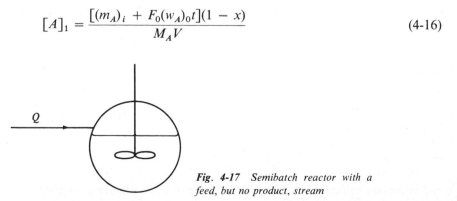

Fig. 4-17 *Semibatch reactor with a feed, but no product, stream*

and

$$d(V[A]_1) = \frac{1}{M_A} \{-[(m_A)_i + F_0(w_A)_0 t] \, dx + (1 - x)F_0(w_A)_0 \, dt\}$$

Substituting this expression into Eq. (4-15) and simplifying leads to the result

$$F_0(w_A)_0 x \, dt + [(m_A)_i + F_0(w_A)_0 t] \, dx = \mathbf{r}V \, dt \qquad (4\text{-}17)$$

This is the mass balance from which the conversion may be computed at any time.

Note that if reactant A were not present in both the initial charge and the feed stream, one of the two terms on the left side of the equation would disappear. If there were no reactant A in the feed, the batch-reactor equation would be obtained. If there were no A in the initial charge to the reactor, the expression would become

$$F_0(w_A)_0(x \, dt + t \, dx) = V\mathbf{r} \, dt \qquad (4\text{-}18)$$

In general, the volume will vary with changes in both the mass and the density of the reaction mixture. However, for most systems, density changes will be small. With this restriction,

$$V = V_0 + \frac{F_0}{\rho} t \qquad (4\text{-}19)$$

The rate of reaction is the other quantity that must be expressed in terms of x and t before the mass balance can be integrated. This may be accomplished by substituting expressions for the concentration, such as Eq. (4-16), into the normal form, $\mathbf{r} = f([A], [B], \ldots)$, of the rate equation.

The general procedure for designing the reactor is as follows:

1. The mass balance is first written in difference form,

$$[(m_A)_i + F_0(w_A)_0 t] \Delta x = [\mathbf{r}V - F_0(w_A)_0 x] \Delta t \qquad (4\text{-}20)$$

and the value of Δx is computed for a chosen time increment Δt. The average values of $\mathbf{r}V$ and x for the first Δt can be taken equal to the known values at $t = 0$ as a first estimate.
2. The reactor volume at time Δt is computed from Eq. (4-19).
3. With the conversion computed in step 1 and the reactor volume V from step 2, the concentrations at the end of the first time increment are determined from equations such as Eq. (4-16).
4. The rate at the end of the first increment is evaluated from the concentrations established in step 2.
5. A second estimate of the average value of $\mathbf{r}V$ is determined from the

known values at the beginning and end of the increment.

6. Equation (4-20) is employed to calculate a second estimate of Δx.
7. This procedure is continued until the average values of rV do not change. Generally the second estimate is sufficient.
8. Subsequent time increments are chosen, and steps 1 to 7 are repeated.

This procedure follows the same pattern as that for stepwise solutions of design equations for simple batch and flow reactors. In each case all the quantities in the differential mass balance are functions of two variables, conversion and reactor volume for the flow case and conversion and time for the batch case.

Example 4-13   The esterification of acetic acid and ethyl alcohol is to be carried out in a semibatch stirred-tank reactor at a constant temperature of $100°C$. The alcohol is added initially as 400 lb of pure C_2H_5OH. An aqueous solution of acetic acid is then added at a rate of 3.92 lb/min for 120 min. The solution contains 42.6% by weight acid. Assume that the density is constant and equal to that of water.

The reaction is reversible, and the specific reaction rates may be taken as the same as those used in Example 4-2:

$$CH_3COOH + C_2H_5OH \rightleftharpoons CH_3COOC_2H_5 + H_2O$$

$$k = 4.76 \times 10^{-4} \text{ liter/(g mole)(min)}$$

$$k' = 1.63 \times 10^{-4} \text{ liter/(g mole)(min)}$$

Compute the conversion of acetic acid to ester as a function of time from $t = 0$ until the last amount of acid is added (120 min).

Solution   As the conversion is to be based on the acetic acid and there is no acid in the initial charge to the reactor, the proper design equation is Eq. (4-18). Since the rate constants are in molal units, the concentrations will also be expressed in those units.

If $[H]$, $[OH]$, $[W]$, and $[E]$ represent acid, alcohol, water, and ester concentrations, in pound moles per cubic foot, equations similar to Eq. (4-16) for each component are

$$[H] = \frac{F_0(w_H)_0 t(1 - x)}{M_H V} \qquad \text{[analogous to Eq. (4-16), with } (m_A)_0 = 0]$$

$$[OH] = \frac{(1/M_{OH})(m_{OH})_i - xF_0(w_H)_0 t(1/M_H)}{V}$$

$$[W] = \frac{(1/M_W)F_0(w_W)_0 t + xF_0(w_H)_0 t(1/M_H)}{V}$$

$$[E] = \frac{xF_0(w_H)_0 t(1/M_H)}{V}$$

Substituting these expressions into the rate equations gives a relationship for the net rate of conversion of the acid in terms of x, t, and V,

$$\mathbf{r} = \frac{k\{F_0(w_H)_0 t(1 - x)[(m_{OH})_i/(M_{OH}) - xF_0(w_H)_0 t/M_H]\}}{M_H V^2}$$

$$-k' \frac{xF_0(w_H)_0 t[F_0(w_W)_0 t/M_W + xF_0(w_H)_0 t/M_H]}{M_H V^2} \quad \text{(A)}$$

where $F_0 = 3.92$ lb/min

$(w_H)_0 = 0.426$

$(m_{OH})_i = 400$ lb

$(w_W)_0 = 1 - 0.426 = 0.574$

M_H = molecular weight acetic acid = 60

M_{OH} = molecular weight ethyl alcohol = 46

M_W = molecular weight water = 18

The rate constants, expressed as $\text{ft}^3/(\text{lb mole})(\text{min})$, become

$$k = 4.76 \times 10^{-4} \frac{1}{28.32} (454) = 7.63 \times 10^{-3}$$

$$k' = 2.62 \times 10^{-3}$$

Substituting these results in Eq. (A) to obtain a working expression for the rate, in lb moles/$(\text{ft}^3)(\text{min})$, we have

$$\mathbf{r} = \frac{1.28 \times 10^{-2} t(1 - x)(8.68 - 0.0278xt)}{60 V^2}$$

$$-\frac{4.37 \times 10^{-3} t^2 x(0.125 + 0.0278x)}{60 V^2}$$

or

$$\mathbf{r} = \frac{1}{V^2}[21.3 \times 10^{-5} t(1 - x)(8.68 - 0.0278xt)$$

$$- 7.13 \times 10^{-5} t^2 x(0.125 + 0.0278x)] \quad \text{(B)}$$

Since the density is assumed constant, Eq. (4-19) is applicable and the reactor volume, in cubic feet, is

$$V = V_0 + \frac{F_0}{\rho} t = \frac{m_0 + F_0 t}{\rho} = \frac{400 + 3.92t}{59.8} = 6.69 + 0.0655t \quad \text{(C)}$$

Equation (4-18) may be written

$$F_0(w_H)_0 t \, dx = \mathbf{r} V M_H \, dt - F_0(w_H)_0 x \, dt$$

or

$$3.92(0.426)t \, dx = \mathbf{r} V(60) \, dt - 3.92(0.426)x \, dt$$

$$t \, dx = 35.9 \mathbf{r} V \, dt - x \, dt$$

In difference form this expression becomes

$$\Delta(xt) = 35.9\,(\mathbf{r}\,V)_a\,\Delta t \tag{D}$$

Equations (B), (C), and (D) may now be employed to carry out a stepwise solution. Since the concentration equations have been combined with the rate equation in expression (B), step 3 in the general procedure may be omitted. Let us carry out calculations for the first time increment in detail.

STEP 1   Choose an initial time increment of 5 min. The initial volume V is 6.69 ft^3 (based on the density of water at 100°C). The initial rate as computed from Eq. (B) is zero, since at $t = 0$ there is no acid in the reactor. To obtain a reasonable estimate of $(\mathbf{r}V)_a$ with which to start the computations, calculate $\mathbf{r}$ for an average $t = \frac{5}{2} = 2.5$ min:

$$\mathbf{r} = \left(\frac{1}{6.69}\right)^2 [21.3 \times 10^{-5}\,(2.5)(1 - 0)(8.68 - 0) - 0]$$

$$= 1.04 \times 10^{-4}\ \text{lb mole/(ft}^3)(\text{min})$$

Hence the first estimate of $(\mathbf{r}V)_a$ is

$$(\mathbf{r}V)_a = 1.04 \times 10^{-4}\,(6.69) = 6.94 \times 10^{-4}$$

From Eq. (D),

$$\Delta(tx) = 35.9(6.94 \times 10^{-4})(5) = 0.1250$$

$$t_1 x_1 - 0 = 0.1250$$

or $x_1 = 0 + 0.0250 = 0.0250$

STEP 2   The reactor volume at the end of the first increment is given by Eq. (C) as

$$V_1 = 6.69 + 0.0655(5) = 7.02\ ft^3$$

STEP 4   The rate at the end of the increment is obtained from Eq. (B), with $x = 0.0250$, $t = 5$, and $V_1 = 7.02$:

$$\mathbf{r} = \left(\frac{1}{7.02}\right)^2 [21.3 \times 10^{-5}(5)\,0.975(8.68 - 0.003)$$

$$- 7.13 \times 10^{-5}(0.625)(0.125 + 0.0007)]$$

$$= \frac{0.898 \times 10^{-2} - 0.56 \times 10^{-5}}{7.02^2} = 1.82 \times 10^{-4}$$

STEP 5   At the end of the increment $\mathbf{r}V = (1.82 \times 10^{-4})(7.02) = 12.8 \times 10^{-4}$. At the beginning of the increment $(t = 0)$ $\mathbf{r}V = 0(6.69) = 0$. Hence the second estimate of $(\mathbf{r}V)_a$ is

$$(\mathbf{r}V)_a = \frac{12.8 \times 10^{-4} + 0}{2} = 6.40 \times 10^{-4}$$

STEP 6   The second estimate of Δx is given by Eq. (D),

$$\Delta(tx) = 35.9(6.40 \times 10^{-4})(5) = 0.115$$

STEP 7   Recalculation of $(rV)_a$ will not change the value of 6.40×10^{-4} significantly. Hence at the end of the first increment

$$x_1 = \frac{0.115}{5} = 0.0230$$

$$V_1 = 7.02 \text{ ft}^3$$

$$r_1 = 1.82 \times 10^{-4} \text{ lb mole/(min)(ft}^3)$$

$$(rV)_a = 12.8 \times 10^{-4}$$

For the second increment of 5 min, estimate $(rV)_a = 18.0 \times 10^{-4}$. Then the first estimate of Δx is

$$\Delta(tx) = 35.9(18.0 \times 10^{-4})(5) = 0.323$$

$$t_2 x_2 - t_1 x_1 = t_2 x_2 - 0.115 = 0.323$$

$$x_2 = 0.0115 + 0.0323 = 0.0438$$

$$V_2 = 6.69 + 0.0655 \times 10 = 7.34 \text{ ft}^3$$

$$r_2 = \frac{1}{7.34^2}(1.71 \times 10^{-2} - 0.004 \times 10^{-2}) = 3.18 \times 10^{-4}$$

$$(rV)_2 = 23.3 \times 10^{-4}$$

$$(rV)_a = \frac{(12.8 + 23.3) \times 10^{-4}}{2} = 18.1 \times 10^{-4}$$

Since this result is close to the first estimate of 18.0×10^{-4}, further calculations of the rate are unnecessary. At the end of the second increment ($t = 10$ min)

$$x_2 = 0.044$$

$$V_2 = 7.34 \text{ ft}^3$$

$$r_2 = 3.18 \times 10^{-4}$$

$$(rV)_2 = 23.3 \times 10^{-4}$$

The results for the entire time interval 0 to 120 min, determined by continuing these stepwise calculations,[1] are summarized in Table 4-15.

As has been mentioned, when reverse reactions are important, continuous removal of one or more of the reaction products will increase the conversion obtainable in a given time. Thus one reactant could be charged to the reactor, a second reactant could be added continuously, and one of

[1] This solution of the problem is approximate because of the assumption of constant density of the reaction mixture. However, if density-vs-composition data were available, the same stepwise method of calculation could be employed to obtain a more precise solution. Equation (C) would have to be modified to take into account density differences.

Table 4-15   Semibatch-reactor design for esterification of acetic acid with ethyl alcohol

t, min	V, ft^3	Conversion of acid	r, lb moles acid reacted/(min)(ft^3)
0	6.69	0	0
5	7.02	0.023	1.82×10^{-4}
10	7.34	0.044	3.2×10^{-4}
20	8.00	0.081	5.2×10^{-4}
40	9.31	0.138	7.0×10^{-4}
60	10.6	0.178	7.3×10^{-4}
80	11.9	0.205	6.9×10^{-4}
100	13.2	0.224	6.3×10^{-4}
120	14.5	0.237	5.7×10^{-4}

the products could be withdrawn continuously.[1] This form of semibatch reactor can be treated by modifying the design equation (4-17), the volume equation (4-19), and the expressions for the concentration to take into account the withdrawal of material.

Problems

4-1.   Nitrous oxide decomposes approximately according to a second-order rate equation. The specific reaction rate in the forward direction,

$$2N_2O \rightarrow 2N_2 + O_2$$

is $k = 977$ cm^3/(g mole)(sec) at 895°C. Calculate the fraction decomposed at 1.0 and 10 sec. and at 10 min in a constant-volume batch reactor. The rate of the reverse reaction is negligible and the initial pressure (all N_2O) is 1 atm.

4-2.   An aqueous solution of ethyl acetate is to be saponified with sodium hydroxide. The initial concentration of ethyl acetate is 5.0 g/liter and that of caustic soda is 0.10 normal. Values of the second-order rate constant, in liters/(g mole)(min), are

$$k = \begin{cases} 23.5 & \text{at } 0°C \\ 92.4 & \text{at } 20°C \end{cases}$$

The reaction is essentially irreversible. Estimate the time required to saponify 95% of the ester at 40°C.

4-3.   It has been mentioned that gaseous reactions are more suitably carried out on a commercial scale in flow equipment than in batch reactors. Consider the following example. Watson[2] has studied the thermal (noncatalytic) cracking of

[1] This withdrawal might be accomplished by distillation, for example.
[2] K. M. Watson, Chem. Eng. Progr., 44, 229 (1948).

butenes at 1 atm pressure in a flow reactor. The rate equation determined from his experimental data is

$$\log k_1 = -\frac{60,000}{4.575T} + 15.27$$

where k_1 = butenes cracked, g moles/(hr)(liter)(atm), and T is in degrees Kelvin. Although the feed consists of a number of different butenes and the products vary from coke to butadiene, the irreversible reaction may be considered as a first-order one,

$$C_4H_8 \rightarrow C_4H_6 + H_2$$

It is desired to crack butenes in a batch-type reactor which will operate at 1200°F and will be equipped for efficient agitation. The initial charge to the reactor will consist of 1 lb mole of butenes and 10 lb moles of steam. Under these conditions the change in total moles during the course of the reaction can be neglected. (a) Determine the time required for a conversion of 30% of the butenes. (b) Determine the reactor volume required.

(c) Suppose that the feed consists of 10 moles of steam per mole of hydrocarbon, as before, but this time the hydrocarbon fraction contains 60 mole % butenes and 40% butadiene. The butadiene may undergo two reactions: cracking and polymerization to the dimer. Assuming that the rates of these reactions are known, outline a method of determining the conversion of butenes and of butadiene for a given reaction time.

4-4.   The decomposition of phosphine is irreversible and first order at 650°C,

$$4PH_3(g) \rightarrow P_4(g) + 6H_2(g)$$

The rate constant (sec^{-1}) is reported as

$$\log k = -\frac{18,963}{T} + 2 \log T + 12.130$$

where T is in degrees Kelvin. In a closed vessel (constant volume) initially containing phosphine at 1 atm pressure, what will be the pressure after 50, 100, and 500 sec. The temperature is maintained at 650°C.

4-5.   A first-order homogeneous gas-phase reaction, $A \rightarrow 3R$, is first studied in a constant-pressure batch reactor. At a pressure of 2 atm and starting with pure A, the volume increases by 75% in 15 min. If the same reaction is carried out in a constant-volume reactor, and the initial pressure is 2 atm, how long is required for the pressure to reach 3 atm?

4-6.   Suppose the initial mixture in Prob. 4-5 consisted of 70 mole % A and 30 mole % helium (inert) at a total pressure of 2 atm. In a constant-pressure reactor, what would be the increase in volume in 15 min?

4-7.[1]   A stage in the production of propionic acid, C_2H_5COOH, is the acidification of a water solution of the sodium salt according to the reaction

---

[1] From W. F. Stevens, "An Undergraduate Course in Homogeneous Reaction Kinetics," presented at Fourth Summer School for Chemical Engineering Teachers, Pennsylvania State University, June 27, 1955.

$$C_2H_5COONa + HCl \rightarrow C_2H_5COOH + NaCl$$

The reaction rate may be represented by a second-order reversible equation. Laboratory data on the rate of reaction are obtained by taking 10-ml samples of the reaction solution at varying times and neutralizing the unreacted HCl with 0.515-normal NaOH. The original amount of acid is determined from a sample taken at zero time. The temperature is 50°C, and the initial moles of HCl and C_2H_5COONa are the same. The data are as follows:

t, min	0	10	20	30	50	∞
NaOH required, ml	52.5	32.1	23.5	18.9	14.4	10.5

Determine the size of an agitated batch reactor to produce propionic acid at an average rate of 1,000 lb/hr. Twenty minutes is required for charging the reactor and heating to 50°C, and 10 min is necessary for cooling and removing the products. The final conversion is to be 75% of the sodium propionate. The initial charge to the reactor will contain 256 lb of C_2H_5COONa and 97.5 lb of HCl per 100 gal. Assume that the density of the reaction mixture is 9.9 lb/gal and remains constant.

4-8. The production of carbon disulfide from methane and sulfur vapor can be carried out homogeneously or with a solid catalyst. Also, some solid materials act as a poison, retarding the reaction. The following data were obtained on a flow basis at a constant temperature of 625°C and with an initial reactants ratio of 1 mole of CH_4 to 2 moles of sulfur vapor (considered as S_2). The first set of data was obtained with the reactor empty (effective volume 67.0 ml), and the second set was obtained after packing the reactor with a granular material (7 mesh) which reduced the void volume to 35.2 ml.

Set	Run	Feed rate, g moles/hr		Production of CS_2, g moles/hr	Conversion of methane
		CH_4	S_2		
1	1	0.417	0.834	0.0531	0.127
	2	0.238	0.476	0.0391	0.164
	3	0.119	0.238	0.0312	0.262
2	1	0.119	0.238	0.0204	0.171
	2	0.178	0.357	0.0220	0.123

Was the granular material acting as a catalyst or as a poison in this case?

4-9. Butadiene and steam (0.5 mole steam/mole butadiene) are fed to a tubular-flow reactor which operates at 1180°F and a constant pressure of 1 atm. The reactor is noncatalytic. Considering only the reversible polymerization reaction to the dimer, determine (a) the length of 4-in.-ID reactor required to obtain a conversion of 40% of the butadiene with a total feed rate of 20 lb moles/hr and (b) the space velocity, measured as (liters feed gas)/(hr)(liters reactor volume) (at 1180°F and 1 atm), required to obtain a conversion of 40%.

The polymerization reaction is second order and has a specific reaction-rate constant given by

$$\log k = -\frac{5{,}470}{T} + 8.063$$

where k is C_4H_6 polymerized, in g moles/(liter)(hr)(atm^2), and T is in degrees Kelvin. The reverse (depolymerization) reaction is first order. At 1180°F (911°K) the equilibrium constant for the reaction is 1.27.

4-10.   A mixture of butenes and steam is to be thermally (noncatalytically) cracked in a tubular-flow reactor at a constant temperature of 1200°F and a constant pressure of 1.0 atm. Although the feed consists of a number of different butenes[1] and the products vary from coke to butadiene, the rate of reaction may be adequately represented by the first-order mechanism

$$\mathbf{r}_1 = \frac{\epsilon}{\rho} k_1 p_4$$

$$\log k_1 = -\frac{60{,}000}{4.575T} + 15.27$$

The rate was determined experimentally in a reactor packed with inert quartz chips, and the reactor to be designed in this problem will also be so packed. The data and notation are as follows:

$\mathbf{r}_1$ = g moles butenes cracked/(g quartz chips)(hr)
ϵ = void fraction = 0.40
ρ = bulk density of bed packed with quartz chips = 1,100 g/liter
p_4 = partial pressure of butenes, atm
T = temperature, °K

The ratio of steam to butenes entering the reactor will be 10:1.0 on a molal basis. Under these conditions the change in total number of moles during the course of the reaction can be neglected.

(a) Determine the conversion as a function of size of reactor. Also prepare a plot of conversion of butenes versus two abscissas: (1) lb quartz chips/lb mole butene feed per hr, covering a range of values from 0 to 3,000, and (2) space velocity, defined as (ft^3 feed)/(hr)(ft^3 void volume) at 1200°F. What total volume of reactor would be required for a 20% conversion with a butenes feed rate of 5 lb moles/hr?

(b) Suppose that the feed consists of 10 moles of steam per mole of total hydrocarbons. The hydrocarbon fraction is 60 mole % butenes and 40 mole % butadiene. Consider that the butenes react as in (a) and that the butadiene may undergo two reactions, cracking and polymerization to the dimer. The rate for cracking is

$$\mathbf{r}_2 = \frac{\epsilon}{\rho} k_2 p_4''$$

$$\log k_2 = -\frac{30{,}000}{4.575T} + 7.241$$

[1] See *Chem. Eng. Progr.*, **44**, 229 (1948).

where r_2 is butadiene cracked, in g moles/(g quartz chips)(hr), and p_4'' is partial pressure of butadiene, in atmospheres; the rate for polymerization to the dimer is

$$r_3 = \frac{\epsilon}{\rho} k_3 (p_4'')^2$$

$$\log k_3 = -\frac{25,000}{4.575 T} + 8.063$$

where r_3 is butadiene polymerized, in g moles/(g quartz chips)(hr).

Determine the conversion of butenes and of butadiene as a function of W/F from 0 to 3,000 lb chips/(lb mole feed per hour). Assume that the total number of moles is constant and neglect all reactions except those mentioned.

4-11. The following conversion data were obtained in a tubular-flow reactor for the gaseous pyrolysis of acetone at 520°C and 1 atmosphere. The reaction is

$$CH_3COCH_3 \rightarrow CH_2=C=O + CH_4$$

Flow rate, g/hr	130.0	50.0	21.0	10.8
Conversion of acetone	0.05	0.13	0.24	0.35

The reactor was 80 cm long and had an inside diameter of 3.3 cm. What rate equation is suggested by these data?

4-12. A small pilot plant for the photochlorination of hydrocarbons consists of an ideal tubular-flow reactor which is irradiated, and a recycle system, as shown in the sketch. The HCl produced is separated at the top of the reactor, and the liquid stream is recycled. The Cl_2 is dissolved in the hydrocarbon (designated as RH_3) before it enters the reactor. It is desired to predict what effect the type of reactor operation will have on the ratio $[RH_2Cl]/[RHCl_2]$ in the product stream. Determine this ratio, as a function of total conversion of RH_3, for two

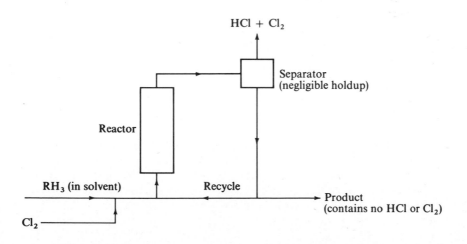

extremes: zero reflux ratio (ratio of recycle flow rate to product-stream flow rate) and infinite reflux ratio. The concentration of hydrocarbon in the feed entering the reactor is $[RH_3]_0$. The hydrocarbon is dissolved in an inert solvent in which Cl_2 is also soluble. There is a stoichiometric excess of Cl_2 fed to the reactor. The reactor operates isothermally, and the HCl product remains in solution until it reaches the separator at the top of the reactor.

Plot curves of $[RH_2Cl]/[RHCl_2]$ vs conversion for each of the conditions and for three types of kinetics. The reactions are

1. $RH_3 + Cl_2 \rightarrow RH_2Cl + HCl$

2. $RH_2Cl + Cl_2 \rightarrow RHCl_2 + HCl$

The three types of kinetics are as follows:

(a) First order (with equal rate constants), so that the rates of production, in g moles/(vol)(time), are

$$\mathbf{r}_{RH_3} = -k_1[RH_3]$$

$$\mathbf{r}_{RH_2Cl} = k_1[RH_3] - k_1[RH_2Cl]$$

(b) Second order (with equal rate constants),

$$\mathbf{r}_{RH_3} = -k_2[RH_3][Cl_2]$$

$$\mathbf{r}_{RH_2Cl} = k_2[RH_3][Cl_2] - k_2[RH_2Cl][Cl_2]$$

(c) Chain kinetics, for which the *elementary* steps are

INITIATION

$$Cl_2 + hv \rightarrow 2Cl^* \qquad \mathbf{r} = \varphi\alpha I[Cl_2] = k_1[Cl_2]$$

PROPAGATION

$$Cl^* + RH_3 \overset{k_2}{\rightarrow} RH_2^* + HCl$$

$$RH_2^* + Cl_2 \overset{k_3}{\rightarrow} RH_2Cl + Cl^*$$

$$Cl^* + RH_2Cl \overset{k_2}{\rightarrow} RHCl^* + HCl$$

$$RHCl^* + Cl_2 \overset{k_3}{\rightarrow} RHCl_2 + Cl^*$$

TERMINATION

$$RH_2^* \overset{k_6}{\rightarrow} \text{end product}$$

$$RHCl^* \overset{k_6}{\rightarrow} \text{end product}$$

In solving the problem for this case, use the stationary-state hypothesis for the intermediates (free radicals Cl^*, RH_2^*, $RHCl^*$) to obtain rate equations for $\mathbf{r}_{RH_3}$ and $\mathbf{r}_{RH_2Cl}$ analogous to the rate equations for the first-order and second-order cases. Note that the rate constant for the termination steps is usually much less than that for the propagation steps.

4-13.   A homogeneous liquid-phase polymerization is carried out in a completely mixed stirred-tank reactor which has an average residence time of 33.6 sec. The concentration of monomer in the feed stream is 1.0×10^{-3} g mole/cm^3. The polymerization reactions follow a two-step process:

1. An initiation reaction producing an active form of the monomer, P_1. This reaction has a rate constant $k_i = 0.10$ sec^{-1}.
2. Propagation reactions, where the monomer reacts with successive polymers of the form P_n. These are all second order with the same rate constant, $k_p = 500$ cm^3/(g mole)(sec).

(a) Of the total polymer content of the exit stream, what is the weight-fraction distribution of polymer molecules from P_1 to P_{10}? (b) For the same reactor and flow rates but another reaction, the initiation rate constant is the same and monomer feed concentration is again 1.0×10^{-3} g mole/cm^3. The weight-fraction distribution of the total polymer in the exit stream in this case is

Polymer	P_1	P_2	P_3	P_4	P_7	P_{10}	P_{20}
Weight fraction	0.0180	0.0314	0.0409	0.0470	0.0546	0.0503	0.0250

What rate constant for the propagation reactions is indicated by these data?

4-14.   The following irreversible first-order reactions occur at constant density:

$$A \xrightarrow{k_1} B \xrightarrow{k_2} C$$

$$k_1 = 0.15 \text{ min}^{-1} \qquad k_2 = 0.05 \text{ min}^{-1}$$

This reaction system is to be analyzed in continuous-flow reactors with a volumetric feed rate of 5 ft^3/min and feed composition $[A] = [A]_0$ and $[B] = [C] = 0$. For the highest production rate of B, which of the following reactors is preferable?

(a) A single-stirred tank of volume $V = 10$ ft^3
(b) Two stirred tanks in series, each with a volume of 5 ft^3
(c) Two stirred tanks in parallel, each of 5-ft^3 volume and with the feed stream split equally between them
(d) A plug-flow (ideal tubular-flow) reactor with a volume of 10 ft^3

4-15.   Acetic anhydride is hydrolyzed in three stirred-tank reactors operated in series. The feed flows to the first reactor ($V = 1$ liter) at a rate of 400 cm^3/min. The second and third reactors have volumes of 2 and 1.5 liters, respectively. The temperature is 25°C, and at this condition the first-order irreversible rate constant is 0.158 min^{-1}. Use a graphical method to calculate the fraction hydrolyzed in the effluent from the third reactor.

4-16.   Suppose in Prob. 4-15 that the first reactor is operated at 10°C, the second at 40°C, and the third at 25°C. The additional rate constants are 0.0567 min^{-1} (at 10°C) and 0.380 min^{-1} (at 40°C). Determine the fraction hydrolyzed in the effluent from the third reactor.

4-17.   Benzene is to be chlorinated in the liquid phase in a kettle-type reactor operated on a steady-state basis. Liquid benzene is added continuously, and the liquid product and gaseous hydrogen chloride are removed continuously. The chlorine gas is bubbled continuously into the liquid reaction mixture in the kettle. The rate of reaction may be assumed large enough that there is no unreacted chlorine in the reaction products. Also, the concentrations of chlorine and HCl in the reaction mixture will be small. The density of the liquid mixture may be assumed constant.

     At the constant operating temperature of 55°C the significant reactions are the three substitution ones leading to mono-, di-, and trichlorobenzene. Each reaction is second order and irreversible. The three reactions are

1. $C_6H_6 + Cl_2 \xrightarrow{k_1} C_6H_5Cl + HCl$

2. $C_6H_5Cl + Cl_2 \xrightarrow{k_2} C_6H_4Cl_2 + HCl$

3. $C_6H_4Cl_2 + Cl_2 \xrightarrow{k_3} C_6H_3Cl_3 + HCl$

It was noted in Chap. 2 that at 55°C the ratios of the rate constants are

$$\frac{k_1}{k_2} = 8.0$$

$$\frac{k_2}{k_3} = 30$$

     Under the proposed operating conditions the composition of the liquid product will be constant for any one run. Different products will be obtained for different ratios of benzene and chlorine fed to the reactor. Compute the composition of the liquid product for the case where 1.4 moles chlorine/mole benzene are fed to the reactor.

4-18.   Reconsider Prob. 4-17 for the case where, instead of a single reactor, a two-reactor system is used. The liquid stream enters the first reactor (as pure benzene) and flows from the first reactor to the second, and finally the product is withdrawn from the second reactor. Gaseous hydrogen chloride is withdrawn from each reactor. Plot the composition of the product vs moles of total chlorine added per mole of benzene. Cover a range of the latter variable from 0 to 2.5. One-half the total chlorine is added to each reactor.

4-19.   The successive irreversible reactions

$$A \xrightarrow{k_1} B \xrightarrow{k_2} C$$

are first order. They are carried out in a series of identical stirred-tank reactors operating at the same temperature and at constant density. Derive an expression for the number of reactors needed to give the maximum concentration of B in the effluent in terms of the total average residence time and the rate constants. The feed stream contains no B or C. What is the number for the specific case where $k_2 = 0.1$ hr^{-1}, $k_1 = 0.05$ hr^{-1}, and the total residence time is 1.5 hr.

4-20.   A constant-density reaction is carried out in a stirred-tank reactor of volume V.

The reaction is first order and irreversible, with a rate constant k_1. The volumetric feed rate is Q. Under these conditions the conversion in the product stream is x_1 [given by Eq. (4-7)].

(a) If one-half of the product stream is recycled and the makeup feed rate is reduced to $Q/2$, what will be the change in conversion in the product stream? What will be the change in the production rate of the product? (b) If one-half of the product stream is recycled, but the fresh-feed rate is maintained equal to Q, what will be the effects on the conversion and production rate?

4-21. Two stirred tanks of volume V_1 and $2V_1$ are available for carrying out a first-order irreversible reaction at constant density and temperature. If the flow rate of the feed stream is Q, which of the following arrangements would give the highest production rate of product?

(a) Parallel operation of the two reactors, with equal average residence times in each
(b) Parallel operation, with different average residence times
(c) Series operation, with the feed stream entering the larger reactor
(d) Series operation, with the feed entering the smaller reactor

4-22. Repeat Example 4-12 with the modification that the effluent from the first reactor is fed to a second reactor. The second reactor originally contains 10 liters of an anhydride solution of concentration 0.50×10^{-4} g mole/cm³. Product is withdrawn from reactor 2 at a constant rate of 2 liters/min. Temperatures in both are 40°C, and all other conditions are the same as in Example 4-12.

(a) Determine the concentration of anhydride in the solution leaving the second reactor from zero time until steady-state conditions are reached. (b) Suppose that reactor 2 was originally empty and that its capacity is 10 liters. After it is filled, product is withdrawn at the rate of 2 liters/min. What would be the concentration of the first anhydride solution leaving the second reactor?

4-23. Ethyl acetate is to be produced in a 100-gal reactor. The reactor originally holds 100 gal of solution containing 20% by weight ethanol and 35% by weight acetic acid. Its density is 8.7 lb/gal; assume that this value remains constant for all compositions. The reactor is heated to 100°C and this temperature is maintained. Pure ethanol is added at a rate of 2 gpm (17.4 lb/min). Solution is withdrawn at the same volumetric rate. At this level the rate data are

$$\mathbf{r} = k[H][OH] \qquad k = 4.76 \times 10^{-4} \text{ liter/(g mole)(min)}$$

$$\mathbf{r}' = k'[E][W] \qquad k' = 1.63 \times 10^{-4} \text{ liter/(g mole)(min)}$$

where H, OH, W, and E refer to acid, alcohol, water, and ester. The concentrations are expressed in gram moles per liter.

Determine the concentration of ester in the product stream for time values from 0 to 1 hr. What is the percent conversion of the total amount of ethanol added? Assume that no water is vaporized in the reactor.

4-24. Ethyl acetate is to be saponified by adding a 0.05-normal solution of sodium hydroxide continuously to a kettle containing the ethyl acetate. The reactor is initially charged with 100 gal of an aqueous solution containing 10 g ethyl acetate/liter. The sodium hydroxide solution is added at a rate of 1.0 gpm until stoichiometric amounts are present. The reaction is relatively fast and irreversible, with a specific reaction rate of 92 liters/(g mole)(min) at 20°C. Assuming that the contents of the kettle are well mixed, determine the concentration of unreacted ethyl acetate as a function of time. At what time will this concentration be a maximum?

5

TEMPERATURE EFFECTS IN HOMOGENEOUS REACTORS

Nonuniform temperatures, or a temperature level different from that of the surroundings, are common in operating reactors. The temperature may be varied deliberately to achieve optimum rates of reaction, or high heats of reaction and limited heat-transfer rates may cause unintended nonisothermal conditions. Reactor design is usually sensitive to small temperature changes because of the exponential effect of temperature on the rate (the Arrhenius equation). The temperature profile, or history, in a reactor is established by an energy balance such as those presented in Chap. 3 for ideal batch and flow reactors.

In an endothermic reaction the temperature decreases as the conversion increases, unless energy is added to the system in excess of that absorbed by the reaction. Both the reduction in concentration of reactants, due to increasing conversion, and the reduction in temperature cause the rate to fall (Fig. 5-1a). Hence the conversion in a nonisothermal reactor will normally be less than that for an isothermal one. Adding energy to reduce the temperature drop will limit the reduction in conversion. If the

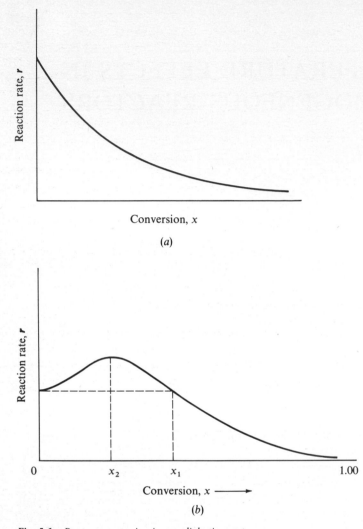

Fig. 5-1   *Rate vs conversion in an adiabatic reactor:*
(a) endothermic reaction (b) exothermic reaction

reaction is reversible, adding energy provides the further advantage of increasing the maximum (equilibrium) conversion. A practical example is the dehydrogenation of butenes to butadiene. Only at an elevated temperature is the equilibrium conversion high enough for the process to be profitable. Also, only at high temperatures will the reaction rate be high enough to assure a close approach to the equilibrium conversion.

Successful designs for such processes must provide for the addition of energy to counteract the heat of reaction. Energy can be added in many ways: by passing a hot fluid through a jacket around the reactor; by adding a large amount of diluent to the feed, which will absorb the heat of reaction and so prevent a large temperature drop; by a simultaneous exothermic reaction involving part of the reactants or other components added to the feed; or by inserting heaters in the reactor.

In an exothermic reaction the temperature increases as the conversion increases. At low conversions the rising temperature increases the rate more than it is reduced by the fall in reactants concentration. Normally the conversion will be greater than for isothermal operation. However, undesirable side reactions and other factors may limit the permissible temperatures. In these cases successful design depends on effective removal of the heat of reaction to prevent excessive temperatures (hot spots). In general, the same methods are employed as for adding energy in endothermic reactions.

The increase in rate as an exothermic reaction progresses is hindered by the limit on the conversion. The limit for an irreversible reaction is 100%. As the limit is approached, the concentration of reactants and the rate both approach zero, regardless of the temperature level. Hence the curve of reaction rate vs conversion for an exothermic reaction in an adiabatic reactor has a maximum, as shown in Fig. 5-1b. At low conversions the relative change in reactants concentration with conversion is small, and the rate increases exponentially because of the dominant temperature effect. At high conversions the reactants concentration approaches zero, and so does the rate.

In Examples 5-2 and 5-3 we shall see a quantitative comparison of the efficiency of tubular-flow and stirred-tank reactors for converting reactant into product. Some conclusions can be given here on the basis of the curves in Fig. 5-1a and b. In a tubular-flow reactor the conversion in the product stream will be the sum of the conversions in each volume element. The total conversion thus depends on the whole curve of rate vs conversion, as shown in Fig. 5-1a and b. In fact, the *average* rate over the conversion range will be a measure of the total conversion in the product stream. In a stirred-tank reactor, however, the conversion is determined solely by the rate in the *exit* stream, that is, by the *point* at the exit conversion in Fig. 5-1a and b. If the rate-vs-conversion curve is a falling one, as in Fig. 5-1a, the average rate will be higher than the rate at the end point. Hence a tubular-flow reactor is better for this case, which is descriptive of an endothermic reaction system. For an exothermic reaction the answer is not as clear cut because of the maximum in the curve in Fig. 5-1b. At low conversions a stirred-tank unit is most efficient, while at high conversions a tubular-flow unit is preferable. At intermediate conversion levels the choice depends on whether the average rate is greater or less than the rate at the exit conversion. At any conversion greater than x_1 (the conversion where the rate is the same as that at zero conversion) the average rate will be greater than that at the end point, so that a tubular-flow reactor is best. At some exit conversion less than x_1, but greater than x_2, the stirred tank will become more efficient and will remain so down to zero conversion.

We have been discussing adiabatic reactors in Fig. 5-1a and b, but heat-transfer rates are usually insufficient to change the form of the curves. Hence the same conclusions apply to almost all nonisothermal conditions.

The shape of the curve in Fig. 5-1b introduces the possibility of multiple, stable operating conditions for exothermic reactions in stirred-tank reactors. This is discussed in Sec. 5-4.

5-1 Ideal Batch Reactors

The energy balance developed in Chap. 3 for batch reactors, Eq. (3-11), expressed the time rate of change of temperature in terms of the rate of reaction and the energy exchange with the surroundings. This equation may be used with the mass balance, Eq. (3-9) [or Eq. (4-2) if the density is constant], to calculate the temperature and composition of the reaction mixture at any time. Also required is a rate equation giving **r** as a function of temperature and composition. Normally a stepwise numerical solution is necessary. This is accomplished by writing the balances in difference form and solving them for successive time increments.

The energy balance can be expressed in terms of the conversion rather than the rate by combining Eqs. (3-11) and (3-9) to eliminate **r**. Thus

$$m_t c_v \frac{dT}{dt} = \frac{-\Delta H}{M} m \frac{dx}{dt} + UA_h(T_s - T) \tag{5-1}$$

where m = mass of limiting reactant (corresponding to zero conversion)

M = molecular weight of reactant

A_h = heat-transfer area

T_s = temperature of surroundings

This form is advantageous for adiabatic operation, for then time is not involved; that is, the equation reduces to

$$m_t c_v \, dT = \frac{-\Delta H}{M} m \, dx \tag{5-2}$$

All the coefficients except c_v and ΔH are constant during the course of the reaction, and the variation in these is usually small. If they are taken as constants, Eq. (5-2) can be immediately integrated to give the following relationship between temperature and conversion:

$$T - T_0 = \frac{-\Delta H}{Mc_v} \frac{m}{m_t} (x - x_0)$$

or

$$T - T_0 = \frac{-\Delta H}{Mc_v} w(x - x_0) \tag{5-3}$$

where w is the weight fraction of limiting reactant, corresponding to zero conversion. If the conversion initially is zero ($x_0 = 0$), w is the weight fraction of reactant at initial conditions. For adiabatic operation Eq. (5-3) eliminates the need for simultaneous stepwise solution of the mass and energy balances. Both adiabatic and nonadiabatic design calculations are illustrated in the following example.

Example 5-1 In a study of the production of drying oils by the decomposition of acetylated castor oil, Grummitt and Fleming[1] were able to correlate decomposition data on the basis of a first-order reaction written as

Acetylated castor oil (l) → $CH_3COOH(g)$ + drying oil (l)

$\mathbf{r} = kC$

where $\mathbf{r}$ is rate of decomposition, in grams of acetic acid produced per minute per milliliter, and C is concentration of acetic acid, in grams per milliliter, equivalent to acetylated castor oil. Data obtained over the temperature range 295 to 340°C indicated an activation energy of 44,500 cal/g mole in accordance with the following expression for the specific-reaction-rate constant k:

$$\ln k = \frac{-44,500}{R_g T} + 35.2$$

where T is in degrees Kelvin.

If a batch reactor initially contains 500 lb of acetylated castor oil at 340°C (density 0.90) and the operation is adiabatic, plot curves of conversion (fraction of the acetylated oil that is decomposed) and temperature vs time. It is estimated that the endothermic heat effect for this reaction is 15,000 cal/g mole of acetic acid vapor. The acetylated oil charged to the reactor contains 0.156 g of equivalent acetic acid per gram of oil; i.e., complete decomposition of 1 g of the oil would yield 0.156 g of acetic acid. Assume that the specific heat of the liquid reaction mixture is constant and equal to 0.6 Btu/(lb)(°F). Also assume that the acetic acid vapor produced leaves the reactor at the temperature of the reaction mixture.

[1] *Ind. Eng. Chem.*, **37**, 485 (1945).

Solution   The mass balance, assuming no change in volume during the course of the reaction, is Eq. (4-2),

$$t_1 = C_0 \int_0^{x_1} \frac{dx}{\mathbf{r}}$$

where C_0 and $\mathbf{r}$ are measured in terms of equivalent acetic acid. If we replace $\mathbf{r}$ with the rate equation, this expression becomes

$$t_1 = \int_0^{x_1} \frac{C_0\, dx}{C e^{35.2 - 44,500/R_g T}} \tag{A}$$

The relationship between the concentration and the conversion is

$$C = C_0(1 - x)$$

In terms of x, Eq. (A) may be written

$$t_1 = \int_0^{x_1} \frac{dx}{e^{35.2 - 44,500/R_g T}(1 - x)} \tag{B}$$

Since the reactor operates adiabatically, Eq. (5-3) is applicable. In this example $w = 0.156$ g of equivalent acetic acid per gram of oil. Hence

$$T - T_0 = -\frac{15,000(0.156)}{60(0.6)}(x - 0) \tag{C}$$

$$T - (340 + 273) = -65x$$

or

$$T = 613 - 65x \tag{D}$$

There are two methods of obtaining a curve of t vs x from Eqs. (B) and (D). The first approach is to write Eq. (B) in difference form for a small change in conversion, Δx, and solve by stepwise numerical integration. As an illustration let us follow through three incremental calculations using the modified Euler method. We write Eq. (B) as

$$\Delta t = \left[\frac{1}{(1 - x)e^{35.2 - 44,500/R_g T}} \right]_{av} \Delta x = \left(\frac{1}{R} \right)_{av} \Delta x \tag{B'}$$

Note that it is the multiplier of Δx that must be averaged. This means that the reciprocal of the rate, not the rate itself, is averaged over the increment. Entering the reactor,

$$T_0 = 613°K$$

$$x_0 = 0$$

$$\frac{1}{R_0} = \frac{1}{(1 - 0)e^{35.2 - 44,500/1.98(613)}} = 4.15$$

For the first increment choose $\Delta x = 0.1$. Then at $x = 0.1$, from Eq. (D) we have

$$T_1 = 613 - 65(0.10) = 606.5°K$$

$$\frac{1}{R_1} = \frac{1}{(1 - 0.1)e^{35.2 - 44,500/1.98(606.5)}} = 6.15$$

For the first increment of Δx the average value of $1/R$ will be

$$\left(\frac{1}{R}\right)_{av} = \frac{4.15 + 6.15}{2} = 5.15$$

Substituting this result in Eq. (B') gives the time at the end of the first increment,

$$\Delta t = t_1 - 0 = 5.2(0.1) = 0.52 \text{ min}$$

The procedure for the second increment would be similar. A value of $1/R_2$ is calculated at $x_2 = 0.2$ and $T_2 = 606.5 - 6.5 = 600°K$. This is averaged with $1/R_1$, and Eq. (B') is used to compute Δt_2 and t_2. The results for three increments are shown in Table 5-1. The smaller the size of the increment of conversion, the more appropriate the use of an arithmetic average value of $1/R$, and the more accurate the final results.

Table 5-1

Conversion	T, °K	t, min
0	613	0
0.10	606.5	0.52
0.20	600.0	1.40
0.30	593.6	3.00

An alternate approach, instead of stepwise numerical integration, is to plot temperature vs conversion from Eq. (D), and then from this information plot $1/r$ vs conversion. Graphical integration under this second curve up to any conversion x corresponds to the value of the integral of dx/r, which, multiplied by C_0, equals the time required for that conversion. Figure 5-2 represents the temperature-conversion relationship corresponding to Eq. (D). At any conversion the temperature and $1/r$ can be evaluated. For example, at $x = 0.2$, $T = 600°K$ from Fig. 5-2, and

$$\mathbf{r} = C_0(1 - x)e^{35.2 - 44,500/R_g T} = 0.8e^{-2.2}C_0 = 0.0886C_0$$

$$\frac{1}{\mathbf{r}} = \frac{11.3}{C_0}$$

Also shown in Fig. 5-2 is a plot of the $1/r$ values obtained in this manner. Graphical integration under this curve yields the time t for any x, as shown by Eq. (4-2). The results are summarized in Table 5-2. The conversion-vs-residence-time values obtained by the numerical-integration process are approximately equivalent to those listed in Table 5-1.

The final curves of temperature and conversion vs time (Fig. 5-3) show the necessity of supplying energy (as heat) to a highly endothermic reaction if large conversions are desired. In the present case, where no energy was supplied, the temperature decreased

Table 5-2

Conversion	0	0.10	0.20	0.30	0.40
θ, min	0	0.54	1.43	2.93	5.53

so rapidly that the reaction essentially stopped after a conversion of 50% was reached. If, instead of operating adiabatically, a constant rate of energy Q_s Btu/min had been *added* to the reactor, the energy balance, according to Eq. (5-1), would have been

$$500(0.6)(1.8)\frac{dT}{dt} = \frac{-15,000}{60}0.156(500)(1.8)\frac{dT}{dt} + Q_s$$

Integration gives the desired relation between T and x_1, but in this case the time is also involved:

$$\Delta T = 0.00185 Q_s \, \Delta t - 65 \, \Delta x \tag{D'}$$

Equation (D') is analogous to Eq. (D) for the adiabatic case. The graphical solution cannot be employed in this case because curves of T and $1/\mathbf{r}$ vs conversion cannot be easily determined independently of the time. However, the numerical-solution approach is satisfactory. The equations involved would be (D') and

$$\Delta t = \left[\frac{1}{(1-x)e^{35.2-44,500/R_gT}}\right]_{av} \Delta x \tag{B''}$$

These twin expressions can be solved by choosing an increment of conversion, assuming a corresponding time interval Δt, and evaluating T_1 at $x_1 = 0 + \Delta x$ from Eq. (D'); the assumed values of Δt may then be checked from Eq. (B''). For example, let $Q_s =$

Fig. 5-2   *Temperature vs conversion for decomposition of acetylated castor oil*

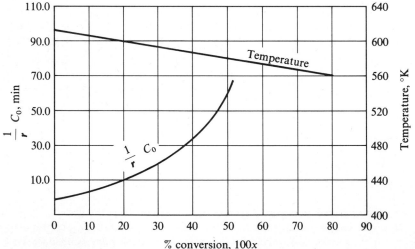

3,000 Btu/min and choose Δx equal to 0.10. As a first trial, assume $\Delta t = 0.50$ min, which is somewhat less than the adiabatic case. Then from Eq. (D′)

$$T_1 = 613 + \Delta T_1 = 613 - 65(0.1) + 0.00185(3,000)0.50$$

$$= 613 - 6.5 + 2.8 = 609.3°K$$

Now Δt can be obtained from Eq. (B″). First the average value of $1/R$ over the increment must be determined. At $x = 0.0$

$$R_0 = (1 - 0)e^{35.2 - 44,500/R_g(613)} = e^{-1.4}$$

$$\frac{1}{R_0} = 4.15$$

At $x_1 = 0.1$

$$R_1 = (1 - 0.1)e^{35.2 - 44,500/R_g(609.3)} = 0.9e^{-1.6}$$

$$\frac{1}{R_1} = 5.50$$

$$\left(\frac{1}{R}\right)_{av} = \frac{4.15 + 5.50}{2} = 4.83$$

From Eq. (B″)

$$\Delta t = 4.83(0.10) = 0.48 \text{ min}$$

Fig. 5-3   *Temperature and conversion vs time*

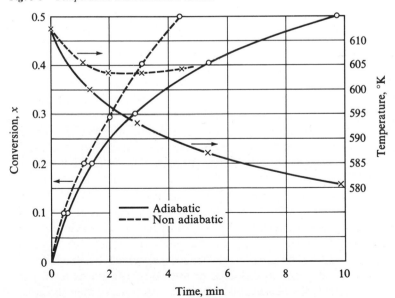

This is close enough to the assumed value of $\Delta t = 0.50$ to be satisfactory. Values of Δt for successive increments can be obtained by repeating the previous calculations. The results are summarized in Table 5-3. For comparison, the adiabatic operating conditions are also shown.

Table 5-3

	Nonadiabatic		Adiabatic	
Conversion	$T, °K$	t, min	$T, °K$	t, min
0	613	0	613	0
0.10	608.7	0.48	606.5	0.54
0.20	605.5	1.08	600.0	1.43
0.30	604.0	1.97	593.5	2.93
0.40	603.9	3.10	587.0	5.53
0.50	604.3	4.39	580.5	9.85

These results show that 30% conversion is obtained in 30% less time when 3000 Btu/min, or $3,000/500 = 6$ Btu/lb of charge, is added to the reactor. Temperature and conversion curves for this nonadiabatic case are also indicated in Fig. 5-3.

5-2 *Ideal Tubular-flow Reactors*

The energy balance for tubular-flow reactors, Eq. (3-18), is valid for steady-state conditions and when there are no radial temperature gradients and no axial dispersion of energy. The last two requirements are satisfied by the limitation to ideal reactors. As mentioned earlier, deviations from ideal conditions are most likely to be significant for heterogeneous reactions. Therefore more general equations for tubular reactors are presented in Chap. 13, although some consideration is given to isothermal cases in Chap. 6.

Simultaneous application of Eq. (3-18), the mass balance, Eq. (3-12), and the rate equation establishes the temperature and composition as a function of reactor volume. The calculations are the same as those for batch reactors (illustrated in Example 5-1), except that the reactor volume replaces time as the independent variable. The procedure is not repeated here for single reactions; instead the treatment for multiple-reaction systems is illustrated. The most significant point is the effect of temperature on the product distribution (selectivity) in the effluent stream from the reactor. Example 5-2 concerns a nonisothermal reactor in which two parallel reactions occur, and the desired product is produced by the reaction with the higher activation energy. A case of consecutive reactions is then discussed in which the desirable product is produced by the reaction with the lower activation energy. Some conclusions about favorable operating temperatures for systems of reactions are immediately apparent. For example, a high temperature is indicated for parallel (simultaneous) reactions when the desirable product is formed by the reaction with the highest activation energy. If the desirable reaction has the lowest activation energy, a low temperature is favored. The same conclusions are true, in general, for consecutive reactions of the form

1. $A + B \overset{k_1}{\rightarrow} R$

2. $R + B \overset{k_2}{\rightarrow} S$

where R is the desirable product. That is, high selectivity of R with respect to S is favored by high temperature if $E_1 > E_2$ and low temperature if $E_1 < E_2$. Numerical substantiation of some of these conclusions is shown in Example 5-2 and the subsequent discussion. The literature may be consulted for more thorough treatment of optimum temperatures in reactors.[1]

For multiple-reaction systems the maximum selectivity for a given product will require operation at a different temperature for each location in the reactor. However, it is rarely of value to find this optimum temperature-vs-position relationship because of the practical difficulty in achieving a specified temperature profile. It is important to be able to predict the general type of profile that will give the optimum yield, for it may be possible to design the reactor to conform to this general trend. These comments apply equally to batch reactors, where the temperature-time relationship rather than the temperature-position profile is pertinent.

For adiabatic operation Eq. (3-18) reduces to an integrable form,

$$F_t c_p \, dT = F \frac{-\Delta H}{M} \, dx \tag{5-4}$$

If c_p and ΔH are constant, this may be integrated to give an expression analogous to Eq. (5-3) for batch reactors,

$$T - T_0 = \frac{-\Delta H}{M c_p} w(x - x_0) \tag{5-5}$$

where $w = F/F_t$ is the weight fraction of reactant in the feed and the subscript zero denotes feed conditions.

[1] Octave Levenspiel, "Chemical Reaction Engineering," chap. 8, John Wiley & Sons, Inc., New York, 1962; K. G. Denbigh, "Chemical Reactor Theory," chaps. 5 and 6, Cambridge University Press, New York, 1965; K. G. Denbigh, *Chem. Eng. Sci.*, **8**, 125 (1958); O. Bilous and N. R. Amundson, *Chem. Eng. Sci.*, **5**, 81 (1956).

Example 5-2 It is proposed to design a pilot plant for the production of allyl chloride. The reactants consist of 4 moles propylene/mole chlorine and enter the reactor at 200°C. The reactor will be a vertical tube of 2 in. ID. If the combined feed rate is 0.85 lb mole/hr, determine the conversion to allyl chloride as a function of tube length. The pressure may be assumed constant and equal to 29.4 psia.

The reactants will be preheated separately to 200°C and mixed at the entrance to the reactor. At this low temperature explosion difficulties on mixing are not serious. The reactor will be jacketed with boiling ethylene glycol, so that the inside-wall temperature will be constant and equal to 200°C. The inside-heat-transfer coefficient may be taken as 5.0 Btu/(hr)(ft^2)(°F).

ADDITIONAL DATA AND NOTES The basic development of the allyl chloride process has been reported by Groll and Hearne[1] and Fairbairn, Cheney, and Cherniavsky.[2] It was found that the three chief reactions were allyl chloride formation,

1. $Cl_2 + C_3H_6 \rightarrow CH_2 = CH—CH_2Cl + HCl$

the parallel addition reaction giving 1,2-dichloropropane,

2. $Cl_2 + C_3H_6 \rightarrow CH_2Cl—CHCl—CH_3$

and consecutive chlorination of allyl chloride to give 1,3-dichloro-1-propene

3. $Cl_2 + CH_2 = CH—CH_2Cl \rightarrow CHCl = CH—CH_2Cl + HCl$

To simplify the kinetic treatment of the problem we shall consider only the first two reactions. The heats of reaction, in calories per gram mole, are shown in Table 5-4. The molal heat capacities C_p will be assumed constant and equal to the values given in Table 5-5.

No information is published for the rate-of-reaction equations, although it is known that reaction 2 takes place at temperatures as low as 100°C, while reaction 1 has an insignificant rate below 200°C. As the temperature increases above 200°C, the rate of reaction 1 increases rapidly, until at 500°C it is several times as fast as that of reaction 2. From this general information and some published data on the effect of residence time and temperature on the conversion, the proposed rate equations are

$$\mathbf{r}_1 = 206,000e^{-27,200/R_gT} p_{C_3H_6}p_{Cl_2}$$

$$\mathbf{r}_2 = 11.7e^{-6,860/R_gT} p_{C_3H_6}p_{Cl_2}$$

Table 5-4

ΔH	273°K	355°K
Reaction 1	−26,800	−26,700
Reaction 2	−44,100	−44,000

[1] H. P. A. Groll and G. Hearne, *Ind. Eng. Chem.,* **31**, 1530 (1939).

[2] A. W. Fairbairn, H. A. Cheney, and A. J. Cherniavsky, *Chem. Eng. Progr.,* **43**, 280 (1947).

Table 5-5

Component	C_p, Btu/(lb mole)(°R)
Propylene (*g*)	25.3
Chlorine (*g*)	8.6
Hydrogen chloride (*g*)	7.2
Allyl chloride (*g*)	28.0
1,2-Dichloropropane (*g*)	30.7

where r_1 and r_2 are in lb moles of Cl_2 converted per hour per cubic foot, T is in degrees Rankine, and the partial pressure p is in atmospheres.

Solution   Since there will be heat transfer from the reaction gases to the glycol jacket, the temperature in the reactor will depend on the length z, and the design calculations have to be carried out by the stepwise integration of the rate and energy equations. The conversion resulting from each of the simultaneous reactions must be obtained separately. It is convenient in this problem to define the conversion as moles of chlorine reacted per mole of *total* feed rather than per mole of chlorine in the feed. If this conversion is $(x_t)_1$ for reaction 1 and $(x_t)_2$ for 2, the mass-balance equation, Eq. (3-12), are

$$r_1 \, dV = F_t \, d(x_t)_1$$

$$r_2 \, dV = F_t \, d(x_t)_2$$

or, in difference form,

$$\Delta(x_t)_1 = \bar{r}_1 \, \Delta\left(\frac{V}{F_t}\right) \tag{A}$$

$$\Delta(x_t)_2 = \bar{r}_2 \, \Delta\left(\frac{V}{F_t}\right) \tag{B}$$

where the overbar indicates average value and F_t is the total *molal* feed rate.

The energy balance defining the temperature is given by Eq. (3-18). It is convenient for numerical solution to write this in terms of the rates of reaction instead of the conversion. This can be done by substituting for dx/dV the quantities in Eqs. (A) and (B). If we take both reactions into account and note that $F_t \, dx_t = F \, dx$, Eq. (3-18) becomes[1]

$$F_t C_p \frac{dT}{dV} = U(T_s - T) \frac{dA_h}{dV} - (r_1 \, \Delta H_1 + r_2 \, \Delta H_2)$$

If z is the reactor length and d is its diameter,

$$dV = \frac{\pi d^2}{4} \, dz$$

$$dA_h = \pi d \, dz$$

[1] The molecular weight M is not necessary as it is in Eq. (3-18), because the rates are expressed in moles.

Using these relations and writing the result in difference form gives

$$F_t C_p \, \Delta T = \pi d \Delta z \, U (T_s - T)_{av} - (\bar{r}_1 \, \Delta H_1 + \bar{r}_2 \, \Delta H_2) \frac{\pi d^2}{4} \Delta z \qquad \text{(C)}$$

Equations (A) to (C) and the rate expressions can be solved for $(x_t)_1$, $(x_t)_2$, and T as a function of reactor length. Equations (A) and (B) may be expressed in terms of Δz by noting

$$\Delta \left(\frac{V}{F_t} \right) = \frac{\pi d^2}{4 F_t} \Delta z \qquad \text{(D)}$$

Then they may be written

$$\Delta(x_t)_1 = \bar{r}_1 \frac{\pi d^2}{4 F_t} \Delta z \qquad \text{(A')}$$

$$\Delta(x_t)_2 = \bar{r}_2 \frac{\pi d^2}{4 F_t} \Delta z \qquad \text{(B')}$$

The solution procedure is as follows:

1. From the known initial conversion $[(x_t)_1 = 0$ and $(x_t)_2 = 0]$ and temperature (200°C), r_1 and r_2 entering the reactor are computed.
2. An arbitrary increment of reactor length Δz is chosen. The smaller this increment, the more accurate the solution (and the more time consuming the calculations).
3. For the chosen Δz, first estimates of the conversion occurring within the increment are obtained from Eqs. (A') and (B'). It is convenient to assume that the average values of r_1 and r_2 are equal to the initial values evaluated in step 1.
4. The change in temperature within the increment ΔT is determined from Eq. (C). An estimate of the mean temperature difference $(T_s - T)_{av}$ in the increment is required in order to evaluate the heat-loss term. Once ΔT has been computed, the estimate of $(T_s - T)_{av}$ can be checked. Hence a trial-and-error calculation is necessary to evaluate ΔT.
5. From the conversion and temperature at the end of the first increment, as determined in steps 3 and 4, the rate of reaction is computed at this position in the reactor. Then steps 3 and 4 are repeated, using for $\bar{r}_1$ and $\bar{r}_2$ the arithmetic average of the values at the beginning and end of the increment. This, in turn, will give more precise values of the conversion and temperature at the end of the first increment and permit a third estimate of the average values of the rates. If this third estimate agrees with the second, the next increment of reactor length is chosen and the procedure is repeated.

The steps will be illustrated by carrying out the numerical calculations for the first two increments. Note that these kinds of calculations are well suited for digital-machine computation.

The rate equations should be converted to a form involving the conversions rather than partial pressures. If 4 moles propylene/mole chlorine enter the reactor, then at a point where the conversions are $(x_t)_1$ and $(x_t)_2$, the moles of each component will be as follows:

Chlorine[1] $= 1 - 5(x_t)_1 - 5(x_t)_2$

Propylene $= 4 - 5(x_t)_1 - 5(x_t)_2$

Allyl chloride $= 5(x_t)_1$

Dichloropropane $= 5(x_t)_2$

Hydrogen chloride $= 5(x_t)_1$

Total moles $= 5[1 - (x_t)_2]$

If we assume that all the components behave as ideal gases at $p_t = 29.4$ psia, the partial pressures of chlorine and propylene, in atmospheres, are given by

$$p_{C_3H_6} = \frac{29.4}{14.7}\left[\frac{4 - 5(x_t)_1 - 5(x_t)_2}{5[1 - (x_t)_2]}\right] = 2\,\frac{0.8 - (x_t)_1 - (x_t)_2}{1 - (x_t)_2}$$

$$p_{Cl} = \frac{29.4}{14.7}\left[\frac{1 - 5(x_t)_1 - 5(x_t)_2}{5[1 - (x_t)_2]}\right] = 2\,\frac{0.2 - (x_t)_1 - (x_t)_2}{1 - (x_t)_2}$$

These expressions can be substituted in the original proposed rate equations to relate the rates to the conversions and the temperature. The results are

$$\mathbf{r}_1 = 824{,}000e^{-13{,}700/T}\,\frac{[0.8 - (x_t)_1 - (x_t)_2][0.2 - (x_t)_1 - (x_t)_2]}{[1 - (x_t)_2]^2} \tag{E}$$

$$\mathbf{r}_2 = 46.8e^{-3{,}460/T}\,\frac{[0.8 - (x_t)_1 - (x_t)_2][0.2 - (x_t)_1 - (x_t)_2]}{[1 - (x_t)_2]^2} \tag{F}$$

Equations (A′) to (F) can now be employed to carry out the steps in the solution.

STEP 1   The rates of reaction at the reactor entrance are given by Eqs. (E) and (F), with $T = (200 + 273)(1.8) = 852°R$ and $(x_t)_1 = (x_t)_2 = 0$:

$$\mathbf{r}_1 = 824{,}000e^{-16.1}(0.16) = 0.0135 \text{ lb mole/(hr)(ft}^3)$$

$$\mathbf{r}_2 = 46.8e^{-4.06}(0.16) = 0.129 \text{ lb mole/(hr)(ft}^3)$$

STEP 2   An increment of reactor length is chosen as $\Delta z = 4.0$ ft.

STEP 3   Assuming that the rates computed in step 1 are average values for the increment, the first estimates of the conversion in the increment are given by Eqs. (A′) and (B′):

$$\frac{\pi d^2}{4} = \frac{\pi}{4}\left(\frac{2}{12}\right)^2 = 0.0218 \text{ ft}^2$$

$$\frac{\pi d^2}{4F_t} = \frac{0.0218}{0.85} = 0.0257$$

[1] The coefficient 5 appears in this equation because x_t is the conversion based on 1 mole of *total* feed. Note that the maximum value for the sum of $(x_t)_1$ and $(x_t)_2$ is 0.2, which corresponds to complete conversion of the chlorine to either allyl chloride or dichloropropane.

hence

$$\Delta(x_t)_1 = 0.0135(0.0257)(4.0) = 0.0014$$

$$(x_t)_1 = 0 + 0.0014 = 0.0014$$

Similarly, for the second reaction we obtain

$$\Delta(x_t)_2 = 0.129(0.0257)(4.0) = 0.0133$$

$$(x_t)_2 = 0 + 0.0133 = 0.0133$$

STEP 4   To use Eq. (C) for estimating ΔT for the increment we must know ΔH_1, ΔH_2, T_s, U, F_t, C_p, and $(T_s - T)_{av}$: The heat-of-reaction data in Table 5-4 are at low temperatures, but they show little change with temperature; hence ΔH_1 and ΔH_2 will be assumed constant:

$$\Delta H_1 = -26,700(1.8) = -48,000 \text{ Btu/lb mole}$$

$$\Delta H_2 = -44,000(1.8) = -79,200 \text{ Btu/lb mole}$$

$$T_s = (200 + 273)(1.8) = 852°R$$

$$h = 5 \text{ Btu/(hr)(ft}^2)(°F)^1$$

Since the conversion is small during the first increment, the heat capacity per unit time of the total reaction mixture is essentially that of the chlorine and propylene in the feed to the reactor,

$$F_t C_p = \frac{0.85}{5}(8.6) + \frac{4}{5}(0.85)(25.3)$$

$$= 0.85(21.8) = 18.5 \text{ Btu/(hr)(°F)}$$

If the mean temperature difference $(T_s - T)_{av}$ for the first increment is *estimated* to be $-20°F$, substitution in Eq. (C) gives

$$18.5 \, \Delta T = \pi \tfrac{2}{12}(5)(-20)(4) - [0.0135(-48,000) + 0.129(-79,200)](0.0218)(4)$$

$$\Delta T = \frac{946 - 210}{18.5} = 40°F$$

$$T_1 = 852 + 40 = 892°R \qquad \text{temperature at end of first increment}$$

$$T_s - T \text{ at entrance} = 0$$

$$T_s - T \text{ at end of increment} = 852 - 892 = -40°F$$

$$(T_s - T)_{av} = \frac{0 + (-40)}{2} = -20° \qquad \text{vs } -20° \text{ assumed}$$

[1] In the expression for h, since the inside-wall temperature is known, the heat transfer with the surroundings can be evaluated from $h(T_s - T)_{av}\pi d \, \Delta z$, where h is the inside-film coefficient and T_s is the inside-wall-surface temperature. The overall coefficient U is not necessary in this case.

STEP 5   At the end of the first increment the first estimate of the conversions and temperature is

$$(x_t)_1 = 0.0014$$

$$(x_t)_2 = 0.0133$$

$$T_1 = 852 + 40 = 892°R$$

Substitution of these conditions in rate equations (E) and (F) yields

$$\mathbf{r}_1 = 0.0256 \qquad \mathbf{r}_2 = 0.143$$

A second, and more accurate, estimate of the average values of the rates for the first increment can now be made:

$$\bar{\mathbf{r}}_1 = \frac{0.0135 + 0.0256}{2} = 0.0195$$

$$\bar{\mathbf{r}}_2 = \frac{0.129 + 0.143}{2} = 0.136$$

The second estimate of the conversion in the increment, obtained from these rates and Eqs. (A′) and (B′), is

$$\Delta(x_t)_1 = 0.0195(0.0257)(4.0) = 0.0020$$

$$(x_t)_1 = 0 + 0.0020 = 0.0020$$

$$\Delta(x_t)_2 = 0.136(0.0257)(4.0) = 0.0140$$

$$(x_t)_2 = 0 + 0.0140 = 0.0140$$

The second estimate of ΔT, determined from Eq. (C), is

$$\Delta T = 41°F$$

$$T_1 = 852 + 41 = 893°R$$

At the revised values of $(x_t)_1$, $(x_t)_2$, and T (at the end of the first increment) the rates are

$$\mathbf{r}_1 = 0.0258$$

$$\mathbf{r}_2 = 0.144$$

Since these values are essentially unchanged from the previous estimate, the average rates will also be the same, and no further calculations for this increment are necessary.

For the second increment Δz is again chosen as 4 ft, and the average rates are estimated to be

$$\bar{\mathbf{r}}_1 = 0.032$$

$$\bar{\mathbf{r}}_2 = 0.15$$

Then from Eqs. (A′) and (B′),

$$\Delta(x_t)_1 = 0.032(0.0257)(4) = 0.0033$$

$$(x_t)_1 = 0.0020 + 0.0033 = 0.0053$$

$$\Delta(x_t)_2 = 0.15(0.0257)(4.0) = 0.0154$$

$$(x_t)_2 = 0.0140 + 0.0154 = 0.0294$$

From Eq. (C), estimating $(T_s - T)_{av} = -55°F$, we have

$$18.5 \, \Delta T = \pi \tfrac{2}{12}(5)(-20)(4) - [0.032(-48,000) + 0.15(-79,200)](0.0218)(4)$$

$$\Delta T = 32°F$$

$$T_2 = 893 + 32 = 925°R$$

The estimated $(T_s - T)_{av}$ for the second increment can be checked by evaluating the temperature differences at the beginning and end of the increment:

$$T_s - T = 852 - 893 = -41°F \qquad \text{beginning of increment}$$

$$T_s - T = 852 - 925 = -73°F \qquad \text{end of increment}$$

$$(T_s - T)_{av} = -57°F \qquad \text{vs} -55°F \text{ assumed}$$

According to the first estimate, the conversions and temperature at the end of the second increment are

$$(x_t)_1 = 0.0053$$

$$(x_t)_2 = 0.0294$$

$$T_2 = 925°R$$

Using these quantities in the rate equations, we obtain

$$\mathbf{r}_1 = 0.0400$$

$$\mathbf{r}_2 = 0.145$$

A second estimate of the average rates during the increment is given by the expressions

$$\bar{\mathbf{r}}_1 = \frac{0.0258 + 0.0400}{2} = 0.0329$$

$$\bar{\mathbf{r}}_2 = \frac{0.144 + 0.145}{2} = 0.144$$

Since these values are close to the first estimates, additional rate calculations for the increment are unnecessary. The revised values for conversion at the end of the increment are

$$(x_t)_1 = 0.0020 + 0.0329(0.0257)(4.0) = 0.0054$$

$$(x_t)_2 = 0.0140 + 0.144(0.0257)(4.0) = 0.0288$$

The revised temperature is given by

$$F_t C_p \Delta T = -[0.0329(-48,000) + 0.144(-79,200)](0.0218)(4)$$

$$+ 5(-57)(0.524)(4) = 536$$

Since the total conversion is becoming significant at the end of the second increment, it may be necessary to evaluate $F_t C_p$ in terms of the actual mixture in the reactor instead of assuming that it equals that for the feed. This may be approximated by noting that $F_t C_p = \Sigma_i F_i (C_p)_i$. To evaluate this summation molal units will be used. The rate of flow of each component at the end of the second increment will be

Allyl chloride $= F_t(x_t)_1 = 0.85(0.0054) = 0.0046$ mole/hr

Dichloropropane $= F_t(x_t)_2 = 0.85(0.0288) = 0.0245$

Chlorine $= F_t[0.2 - (x_t)_1 - (x_t)_2] = 0.85(0.1658) = 0.1410$

Propylene $= F_t[0.8 - (x_t)_1 - (x_t)_2] = 0.85(0.7658) = 0.6510$

Hydrogen chloride $= F_t(x_t)_1 = 0.85(0.0054) = 0.0046$

Thus

$$\Sigma F_i (C_p)_i = 0.0046(28) + 0.0245(30.7) + 0.1410(8.6)$$

$$+ 0.6510(25.3) + 0.0046(7.2)$$

$$= 18.6 \text{ Btu/(hr)(°R)}$$

This result turns out not to be appreciably different from that based on the feed, but it will become so when the conversion reaches larger values. Hence, from the energy balance

$$\Delta T = \frac{536}{18.6} = 29°\text{F}$$

and the revised temperature at the end of the increment is

$$T_2 = 893 + 29 = 922°\text{R}$$

The results of these calculations indicate two trends. First, the rate of reaction 1 is relatively low with respect to reaction 2 at low temperatures but increases rapidly with temperature, while the rate of reaction 2 is not very sensitive to temperature changes. This is because the activation energy for reaction 1 is greater than that for reaction 2. The second trend is the increase in rate of heat transfer to the surroundings with an increase in reactor length. This effect, which is a consequence of the increase in $(T - T_s)_{av}$, offsets the rise in temperature due to the exothermic nature of the reactions, and results ultimately in a decrease in temperature with reactor length. This point is reached when the heat transferred from the reactor tube to the surroundings is greater than the heat evolved from the reactions. Additional incremental calculations (Figs. 5-4, 5-5, and Table 5-6) indicate that the point at which the temperature starts to decrease is about 18 ft from the entrance to the reactor. This point of maximum temperature occurs at too low a level for the rate of reaction 1 to become large with respect to that of reaction 2. Therefore the conversion to allyl chloride never reaches a high

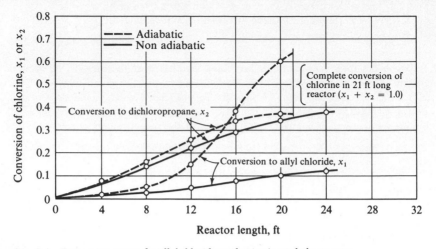

Fig. 5-4   *Conversion curves for allyl chloride production in a tubular reactor*

Fig. 5-5   *Temperature profiles for allyl chloride production in a tubular reactor*

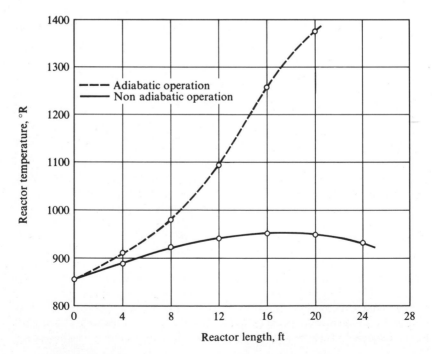

Table 5-6   Conversion vs reactor length for chlorination of propylene

NON ADIABATIC OPERATION

Reactor volume, ft^3	Reactor length, ft^3	Conversion per mole feed			T, °R	Conversion of chlorine in feed			Selectivity $S_o = x_1/x_2$
		$(x_t)_1$	$(x_t)_2$	Total		x_1	x_2	Total	
0	0	0.0	0.0	0.0	852	0.0	0.0	0.0	. . .
0.087	4	0.0020	0.0140	0.0160	893	0.0100	0.070	0.080	0.14
0.174	8	0.0054	0.0288	0.0342	922	0.0270	0.145	0.172	0.19
0.261	12	0.0098	0.0435	0.0533	940	0.0490	0.217	0.266	0.23
0.349	16	0.0147	0.0575	0.0723	949	0.0735	0.287	0.361	0.36
0.436	20	0.0193	0.0698	0.0891	949	0.0964	0.349	0.445	0.28

ADIABATIC OPERATION

0	0	0.0	0.0	0.0	852	0.0	0.0	0.0	. . .
0.087	4	0.0023	0.0144	0.0167	905	0.0115	0.0720	0.0835	0.16
0.174	8	0.0088	0.0315	0.0403	981	0.0440	0.157	0.201	0.28
0.261	12	0.0283	0.0505	0.0788	1,093	0.142	0.252	0.394	0.56
0.349	16	0.0779	0.0667	0.1446	1,261	0.389	0.334	0.723	1.16
0.436	20	0.1204	0.0739	0.1943	1,380	0.602	0.369	0.971	1.63

value (most of the reaction is to the dichloropropane), and the selectivity for allyl chloride is always less than unity, as noted in the last column of Table 5-6. A reactor of this type is not well suited for the production of allyl chloride. An adiabatic reactor would eliminate the maximum in the temperature and result in higher temperatures, which in turn would favor the production of allyl chloride. The only change in the calculations for such an adiabatic case is that the term $U(T_s - T)_{av} \, \pi d \, \Delta z$ would be zero. For comparison purposes the temperature and conversion results for such an adiabatic reactor are also shown in Figs. 5-4 and 5-5 and Table 5-6. The selectivity now is greater than unity at high conversions.

Even adiabatic operation results in the formation of considerable amounts of the undesirable dichloropropane. This occurs in the first part of the reactor, where the temperature of the flowing mixture is low. This is an illustration of the discussion at the beginning of the chapter with respect to Fig. 5-1a and b. The conditions correspond to the low-conversion range of Fig. 5-1b before the maximum rate is reached. A tubular-flow reactor is less desirable for these conditions than a stirred-tank unit. The same reaction system is illustrated in Example 5-3 for a stirred-tank unit.

Consider next an illustration of consecutive reactions where the intermediate component is the desired product. Suppose A and B produce products X, Y, and Z according to the following homogeneous reactions, which follow second-order kinetics:

1. $A + B \rightarrow X + Z$

2. $X + B \rightarrow Y + Z$

The system has a constant density and the reactions are irreversible. The specific reaction rates for each reaction are

$$k_1 = A_1 \, e^{-E_1/R_g T} \tag{5-6}$$

$$k_2 = A_2 \, e^{-E_2/R_g T} \tag{5-7}$$

    The problem is to find the constant operating temperature in a tubular-flow reactor for which the yield of X will be a maximum. We would also like to know the value of the yield and the overall selectivity of X with respect to Y at this condition. For these objectives it is not necessary to know the individual values of A_1, A_2, E_1, and E_2 in Eqs. (5-6) and (5-7), but only the ratio k_2/k_1. Suppose this is defined by

$$\frac{A_2}{A_1} = 7.95 \times 10^{11} \tag{5-8}$$

$$E_2 - E_1 = 16,700 \; \text{cal/g mole} \tag{5-9}$$

and that we take the permissible range of operating temperatures as 5 to 45°C. This problem involves consecutive reactions for which the activation energy of the reaction producing the desired product is less than that of the other reaction; that is, $E_1 < E_2$. According to the discussion at the beginning of the section, the lowest temperature, 5°C, would give the best yield of X. We wish to verify this and calculate the maximum yield and the selectivity.

    The reactions represent a common type of substitution reaction; for example, the successive chlorination of benzene considered in Example 2-8 is of this form. In that example yields of primary and secondary products were obtained for an isothermal batch reactor. As pointed out in Sec. 4-4, the results for batch reactors may be used for tubular-flow reactors if the time is replaced by the residence time. Actually, in Example 2-8 the yield equations were expressed in terms of fraction of benzene unreacted rather than time. Therefore identical results apply for the tubular-flow reactor. Hence the yield of X is given by Eq. (F) of Example 2-8,[1] with $[A]/[A]_0$ replacing $[B]/[B]_0$; that is,

$$x_X = \frac{[X]}{[A]_0} = \frac{1}{1 - \alpha} \left[\left(\frac{[A]}{[A_0]} \right)^{\alpha} - \frac{[A]}{[A_0]} \right] \tag{5-10}$$

---

[1] Actually, by the equation immediately preceding Eq. (F). The same result is also given by Eq. (2-106), which was developed for consecutive first-order reactions. As shown in Example 2-8, in this case the results are the same for first- or second-order kinetics.

where

$$\alpha = \frac{k_2}{k_1} = \frac{A_2 e^{-E_2/R_g T}}{A_1 e^{-E_1/R_g T}} = 7.95 \times 10^{11} e^{-16,700/R_g T} \tag{5-11}$$

The yield of X, described by Eq. (5-10), has a maximum value at an intermediate value of $[A]/[A]_0$, that is, at an intermediate value of the conversion $(1 - [A]/[A]_0)$. We first find this local maximum, and then express it as a function of α. Once this is done, we can find the value of α, and hence the temperature, that gives the overall maximum within the allowable temperature range.

To determine the local maximum let us set the derivative of x_X with respect to $[A]/[A]_0$ (call this z) equal to zero; that is,

$$\frac{dx_X}{dz} = 0 = \frac{1}{1 - \alpha} \left[\alpha(z)^{\alpha - 1} - 1 \right]$$

or

$$z = \frac{[A]}{[A]_0} = \left(\frac{1}{\alpha} \right)^{1/(\alpha - 1)} \tag{5-12}$$

This is the value of $[A]/[A]_0$ for the local maximum of x_X. Substituting this value in Eq. (5-10) gives the local maximum in terms of α,[1]

$$(x_X)_{max} = \frac{1}{1 - \alpha} \left(\frac{1}{\alpha} \right)^{1/(\alpha - 1)} \left(\frac{1}{\alpha} - 1 \right) = \left(\frac{1}{\alpha} \right)^{\alpha/(\alpha - 1)} \tag{5-13}$$

From Eq. (5-11) we can find the value of α for any temperature over the range 5 to 45°C. Substituting these results for α in Eq. (5-13) gives the corresponding local-maximum yields of X. The results are shown in Fig. 5-6, plotted as $(x_X)_{max}$ vs temperature. As expected, the overall maximum occurs at 5°C and is about 0.84. The corresponding value of α is 0.06.

The total conversion at maximum yield conditions is, from Eq. (5-12),

$$x_t = 1 - \frac{[A]}{[A]_0} = 1 - \left(\frac{1}{\alpha} \right)^{1/(\alpha - 1)}$$

$$= 1 - \left(\frac{1}{0.06} \right)^{1/(0.06 - 1)} = 1 - 0.05 = 0.95 \quad \text{or } 95\%$$

From the stoichiometry of the reactions, the mass balance of A is

$$[A] = [A]_0 - [X] - [Y]$$

[1] This same result was obtained in Example 2-7 for a batch reactor.

or

$$\frac{[A]}{[A]_0} = 1 - \frac{[X]}{[A]_0} - \frac{[Y]}{[A]_0}$$

$$x_Y = \frac{[Y]}{[A]_0} = 1 - \frac{[A]}{[A]_0} - \frac{[X]}{[A]_0} = x_t - x_X \tag{5-14}$$

Hence

$$x_Y = 0.95 - 0.84 = 0.11$$

Of the total conversion of 95%, 84% is to X and 11% is to Y. The overall selectivity of X to Y is 0.84/0.11 = 7.6 at the maximum yield of X. The selectivity is not necessarily a maximum at the conditions of maximum yield. If the separation of X from Y in the product stream requires an expensive process, it might be more profitable to operate the reactor at a conversion level at which x_X is less than 84% but the selectivity is higher than 7.6. The selectivity may be evaluated for any total conversion and α from Eqs. (5-10) and (5-14); thus the selectivity is

$$S_o = \frac{x_X}{x_Y} = \frac{1/(1-\alpha)[(1-x_t)^\alpha - (1-x_t)]}{x_t - 1/(1-\alpha)[(1-x_t)^\alpha - (1-x_t)]} \tag{5-15}$$

Equation (5-15) indicates that S_o decreases sharply with x_t. The relationship is shown for 5°C by the dotted curve in Fig. 5-7. This curve suggests that

Fig. 5-6 *Maximum yield of intermediate product (X) for consecutive reactions*

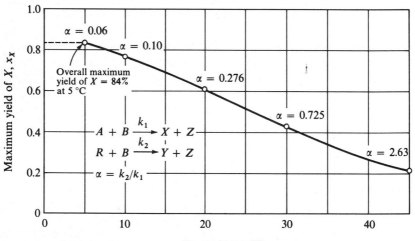

if a selectivity of 20 is required to reduce separation costs, the reactor should be designed to give a conversion of about 73%, at which the yield of X is only 70%. In this case the maximum yield occurs at a conversion of 95%, while the maximum selectivity is at a conversion approaching zero. Figure 5-7 also includes the curves for yield vs total conversion for several temperatures.

One final point is of significance. As conditions have been attained for the maximum yield of X, the reactor has become relatively large. At 5°C the reaction rate will be low and the total conversion is approaching 100%. Whether it is advisable to operate at these conditions for maximum conversion of X depends on the economics of reactor costs, separation costs, and price of product X.

5-3 Ideal Stirred-tank Reactors (Flow)

The ideal stirred-tank reactor operates isothermally and hence at a constant rate. However, an energy balance is needed to predict the constant temperature when the heat of reaction is sufficient (or the heat exchange between the surroundings and reactor is insufficient) to cause a difference between

Fig. 5-7 *Yield of intermediate product* (X) *vs total conversion for consecutive reactions*

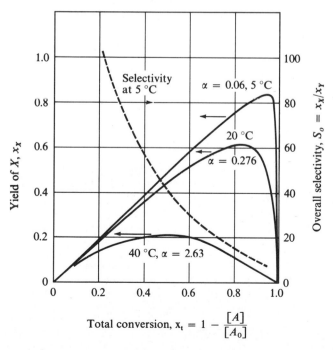

feed and reactor temperatures. The proper energy balance is the algebraic relation presented as Eq. (3-6). This expression, the mass balance [Eq. (3-3)], and the rate equation are sufficient to fix the temperature and composition of the reaction mixture leaving the reactor. Sometimes simultaneous solution of the equations is required. For example, if a reactor of known volume is to be used, the calculation of the conversion and the temperature requires a simultaneous trial-and-error solution of Eqs. (3-3), (3-6), and the rate equation expressing the temperature dependency of r. In contrast, if the objective is the reactor volume needed to obtain a specified conversion, Eq. (3-6) can be solved independently for the reactor exit temperature T_e. The rate at T_e can be found from the rate equation and used in Eq. (3-3) to obtain the reactor volume. When more than one reaction occurs, a separate mass balance is necessary for each reaction.

A stirred tank may give better or worse selectivities than a tubular-flow unit in multiple-reaction systems. As usual, the key point is the relative values of the activation energies for the reactions. In particular, for a set of parallel reactions, where the desired product is formed by the reaction with the higher activation energy, the stirred tank is advantageous. The production of allyl chloride considered in Example 5-2 is a case in point. The performance of a stirred-tank reactor for this system is discussed next, and the results are compared with the performance of the tubular-flow reactor.

Example 5-3 Consider the design of a continuous stirred-tank reactor for the production of allyl chloride from propylene, using the reaction-rate data given in Example 5-2. So that we may compare the two types of reactors, the same feed condition will be employed:

$F_t = 0.85$ mole/hr
$T = 200°C$
$p = 29.4$ psia
Molal ratio of propylene to chlorine $= 4.0$

The operation is adiabatic. Also, suitable baffles and entrance nozzles will be used so that, although the contents of the reactor are gaseous, they will be of uniform temperature, pressure, and composition.

Using the heats of reaction and heat capacities given in Example 5-2, determine the conversion of chlorine to allyl chloride expected for a range of sizes of reactors (i.e., reactor volumes).

Solution The rate of each reaction (allyl chloride and dichloropropane formation) will be a constant and should be evaluated at the temperature and composition of the stream leaving the reactor. The temperature is determined by Eq. (3-6). For adiabatic operation and zero conversion in the feed this becomes, using molal units for F,

$$F_t C_p (T_F - T_e) - F[(x_t)_1 \Delta H_1 + (x_t)_2 \Delta H_2] = 0 \tag{A}$$

where the subscript e is omitted from the conversion for convenience. Thus $(x_t)_1$ and $(x_t)_2$ refer to conversion in the product, based on total feed (as in Example 5-2).

If only one reaction were involved, this energy balance could be used to calculate T_e for a series of arbitarily chosen conversions $(x_t)_1$. Then each value of $(x_t)_1$ would fix a rate $\mathbf{r}_1$. Substitution of these rates in the mass balance, Eq. (3-3), would give the reactor volumes corresponding to the conversions. However, in this case two equations of the form of (3-3) must be satisfied,

$$(x_t)_1 - 0 = \mathbf{r}_1 \frac{V}{F_t} \qquad\qquad (B)$$

$$(x_t)_2 - 0 = \mathbf{r}_2 \frac{V}{F_t} \qquad\qquad (C)$$

The rate expressions are obtained by substituting T_e for the temperature in Eqs. (E) and (F) of Example 5-2. Thus

$$\mathbf{r}_1 = 824,000\, e^{-13,700/T_e} \frac{[0.8 - (x_t)_1 - (x_t)_2][0.2 - (x_t)_1 - (x_t)_2]}{[1 - (x_t)_2]^2} \qquad (D)$$

$$\mathbf{r}_2 = 46.8\, e^{-3,460/T_e} \frac{[0.8 - (x_t)_1 - (x_t)_2][0.2 - (x_t)_1 - (x_t)_2]}{[1 - (x_t)_2]^2} \qquad (E)$$

The problem is solution of Eqs. (A) to (E) for the five unknowns $(x_t)_1$, $(x_t)_2$, T_e, $\mathbf{r}_1$, and $\mathbf{r}_2$ at different values of the reactor volume V. One procedure which is not tedious is first to choose a value of T_e. Then, from the ratio of Eqs. (B) and (C), using (D) and (E) for $\mathbf{r}_1$ and $\mathbf{r}_2$, we obtain the ratio $(x_t)_1/(x_t)_2$. Employing this ratio in Eq. (A) will give separate values for each conversion. Finally, the corresponding reactor volume can be obtained from either Eq. (B) or Eq. (C). This approach will be illustrated by including the numerical calculations for an exit temperature of $1302°R$ ($450°C$),

$$\frac{(x_t)_1}{(x_t)_2} = \frac{\mathbf{r}_1}{\mathbf{r}_2} = \frac{824,000\, e^{-13,700/1,302}}{46.8\, e^{-3,460/1,302}} = 6.77$$

Using this ratio in Eq. (A) and noting that the heat capacity of the feed $F_t C_p$ was determined in Example 5-2 as 18.5 Btu/(hr)(°F), we have

$$18.5(T_F - T_e) + 0.85[6.77(x_t)_2\,(48,000) + (x_t)_2(79,200)]$$

$$= 18.5(852 - 1,302) + 343,000(x_t)_2 = 0$$

$$(x_t)_2 = 0.0243$$

$$(x_t)_1 = 6.77(0.0243) = 0.164$$

The reactor volume required for these conversions is, from Eq. (B),

$$0.164 = \frac{V}{0.85}(824,000\, e^{-13,700/1,302}) \frac{(0.8 - 0.164 - 0.024)(0.2 - 0.164 - 0.024)}{(1 - 0.0243)^2}$$

$$V = \frac{0.164(0.85)}{0.167} = 0.83\ \text{ft}^3$$

Table 5-7   Conversion vs reactor volume for adiabatic tank reactor: allyl chloride production

Reactor (or exit) temperature		Conversion per mole feed		Conversion of chlorine in feed		Reactor volume,	Selectivity
°R	°C	$(x_t)_1$	$(x_t)_2$	x_1	x_2	ft³	$S_o = x_1/x_2$
960	260	0.0098	0.0237	0.049	0.119	0.12	0.41
1032	300	0.0282	0.0324	0.14	0.162	0.15	0.86
1122	350	0.0660	0.0341	0.33	0.171	0.18	1.93
1212	400	0.114	0.0298	0.57	0.149	0.24	3.82
1257	425	0.138	0.0273	0.69	0.136	0.34	5.07
1302	450	0.164	0.0243	0.82	0.121	0.83	6.78

The corresponding values of the conversions and volume for other temperatures are summarized in Table 5-7. Comparing the results with those of Example 5-2, we see that the adiabatic stirred-tank reactor gives much higher yields and selectivies for allyl chloride than the tubular-flow type for the same reactor volume. In the tubular-flow equipment considerable dichloropropane is formed in the initial sections of the reactor, where the temperature is relatively low. This is avoided in the adiabatic tank reactor by operation at a constant temperature high enough to favor allyl chloride formation. For example, if the adiabatic tank reactor is operated at 450°C, 82% of the chlorine is converted to allyl chloride and 12% is converted to dichloropropane; the total conversion is 94%. In the adiabatic tubular reactor of Example 5-2 the products contained much greater amounts of dichloropropane for all reactor volumes. These conclusions are summarized in Fig. 5-8, where the fraction of chlorine converted to each product is shown plotted against reactor volume for the tubular and the tank reactors.

5-4 Stable Operating Conditions in Stirred-tank Reactors[1]

In Example 5-3 the temperature and conversion leaving the reactor were obtained by simultaneous solution of the mass and energy balances. The results for each temperature in Table 5-7 represented such a solution and corresponded to a different reactor, i.e., a different reactor volume. However, the numerical trial-and-error solution required for this multiple-reaction system hid important features of reactor behavior. Let us therefore reconsider the performance of a stirred-tank reactor for a simple single-reaction system.

Suppose an exothermic irreversible reaction with first-order kinetics is carried out in an adiabatic stirred-tank reactor, as shown in Fig. 5-9. The

[1] For more complete discussions of this subject see C. van Heerden, *Ind. Eng. Chem.*, **45**, 1242 (1953); K. G. Denbigh, *Chem. Eng. Sci.*, **8**, 125 (1958).

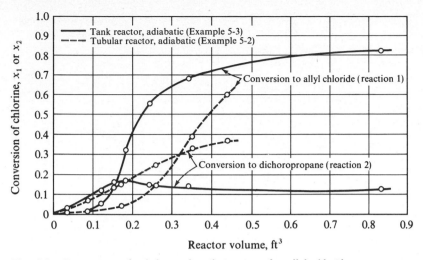

Fig. 5-8 *Comparison of tubular and tank reactors for allyl chloride production*

permissible, or stable, operating temperatures and conversions can be shown analytically and graphically for this system by combining the mass and energy balances. If the density is constant, Eq. (4-7) represents the mass balance for this case. Since V/Q is the average residence time, Eq. (4-7) may be written

$$x = \frac{k\bar{\theta}}{1 + k\bar{\theta}} \tag{5-16}$$

In terms of temperature, Eq. (5-16) takes the form

$$x = \frac{\bar{\theta}Ae^{-E/R_gT}}{1 + \bar{\theta}Ae^{-E/R_gT}} \tag{5-17}$$

where E is the activation energy and A is the frequency factor. At a fixed $\bar{\theta}$ (that is, for a given reactor), Eq. (5-17) expresses the result of the combined effects of temperature and reactant concentration on the rate as referred to in Fig. 5-1b. At low conversion levels the conversion increases approximately exponentially with temperature, since the exponential term in the denominator is small with respect to unity. At high temperatures the reactant composition, and hence the rate, approaches zero; then the exponential term dominates the denominator and the conversion approaches a constant value. For an irreversible reaction Eq. (5-17) shows that this value is unity (100% conversion). The S-shaped curve representing this T-vs-x relationship is sketched in Fig. 5-9.

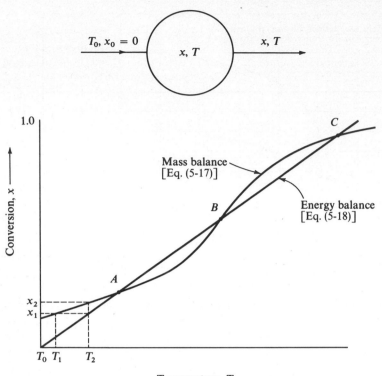

Fig.5-9   *Temperature vs conversion for a first-order irreversible reaction in an adiabatic stirred-tank reactor*

The energy balance [Eq. (3-6)] for adiabatic operation and zero conversion in the feed is

$$x - 0 = \frac{MF_t c_p (T - T_0)}{F(-\Delta H)} = \frac{M}{w_0} \frac{c_p (T - T_0)}{-\Delta H}$$

where w_0 is the weight fraction of reactant in the feed. For constant-density systems it is convenient to express M/w_0 as ρ/C_0, so that

$$x = \frac{\rho c_p}{C_0(-\Delta H)}(T - T_0) \tag{5-18}$$

Since the heat of reaction usually varies little with temperature, Eq. (5-18) shows a linear relationship between $T - T_0$ and conversion and is represented by the straight line in Fig. 5-9.

For a given reactor and kinetics the operating temperature and conversion in the effluent stream are established by the simultaneous solution of Eqs. (5-17) and (5-18). The intersections shown in Fig. 5-9 indicate that such solutions are possible at three points, A, B, and C. Stable steady-state operation cannot exist at other temperatures. For example, suppose that the initial temperature is below the temperature at point A, say, at T_1. Figure 5-9 shows that the conversion required by the mass balance is greater than that corresponding to the energy balance. Thus the energy evolved at the conversion x_1, corresponding to T_1 from Eq. (5-17), will raise the temperature to T_2. This is the temperature corresponding to x_1 in Eq. (5-18). The conversion at T_2 is x_2, from Eq. (5-17). Hence a further heating of the reaction mixture occurs. This transient heating will continue until point A is reached. At initial temperatures between points A and B the reaction rate is too small to justify steady-state operation, and the reaction mixture cools to the point A. Initial temperatures between B and C would be similar to temperatures below point A, so that transient heating will occur until point C is reached. At initial temperatures above C transient cooling will take place until the temperature drops to C. The temperature-vs-time relationship may be evaluated by writing and solving the *transient* energy balance.

Point B is different from A and C. After small initial displacements from B the system does not return to B, whereas disturbances from points A and C are followed by a return to these stable points.

The relative position of the mass-balance curve and the energy-balance line in Fig. 5-9 depends on the chemical properties (A, E, and ΔH) and physical properties (ρ and c_p) of the system and the operating conditions ($\bar{\theta}$ and C_0), according to Eqs. (5-17) and (5-18). These properties and conditions determine whether or not stable operating conditions are possible and how many stable operating points exist. For example, consider a series of reactions with the same A and E at constant $\bar{\theta}$. Then the relation between x and T determined by Eq. (5-17) is fixed, as shown by the S-shaped curve in Fig. 5-10. Suppose also that ρ, c_p, and C_0 are constant, and that the reactions are distinguished by different heats of reaction. For reaction 1, which has the lowest $-\Delta H$, the energy-balance line will be steep and will intersect the mass-balance curve at a small value of $T - T_0$. This point is the only stable operating condition. The reactor will operate at a low conversion and at a temperature only slightly above the feed temperature T_0. For reaction 2, with an intermediate value of $-\Delta H$, there will be two stable operating conditions, points A and C in Fig. 5-10, and a metastable point, B. This is the situation described in Fig. 5-9. Reaction 3 has a high $-\Delta H$; there is just one intersection, and this occurs at nearly complete conversion. As shown in Fig. 5-10, point D, the reactor temperature is far above T_0.

Stability behavior in heterogeneous reactions is similar. For example, the interaction of chemical and physical processes (rates of chemical reaction and mass and heat transfer) in reactions between gases and solid particles causes results analogous to those discussed here. We shall return to this subject in Chap. 10.

Calculations of stable operating conditions are illustrated in the following example.

Example 5-4　A first-order homogeneous (liquid-phase) reaction is carried out in an ideal stirred-tank reactor. The concentration of reactant in the feed is 3.0 g moles/liter and the volumetric flow rate is 60 cm^3/sec. The density and specific heat of the reaction mixture are constant at 1.0 g/cm^3 and 1.0 cal/(g)(°C), respectively. The reactor volume is 18 liters. There is no product in the feed stream and the reactor operates adiabatically. The heat and rate of the irreversible reaction are

$$\Delta H = -50,000 \text{ cal/g mole}$$

$$\mathbf{r} = 4.48 \times 10^6 \, e^{-15,000/R_g T} \, C \qquad \text{g moles/(sec)(cm}^3)$$

where C is the reactant concentration, in gram moles per cubic centimeter, and T is in degrees Kelvin. If the feed stream is at 25°C, what are the steady-state conversions and temperatures in the product stream?

Solution　Stable operating conditions correspond to the solutions of Eqs. (5-17) and (5-18). Starting with Eq. (5-17),

$$\bar{\theta} A = \frac{V}{Q} A = \frac{18,000}{60} (4.48 \times 10^6) = 1.34 \times 10^9$$

so that Eq. (5-17) becomes

Fig. 5-10　*Temperature rise vs conversion as a function of heat of reaction in an adiabatic stirred-tank reactor*

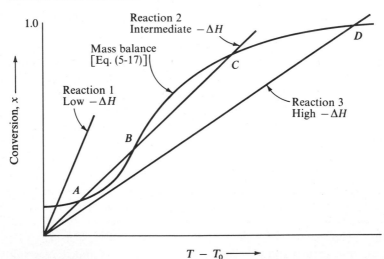

$$x = \frac{1.34 \times 10^9 \, e^{-15,000/R_g T}}{1 + 1.34 \times 10^9 \, e^{-15,000/R_g T}} \tag{A}$$

This relation between x and T is shown as the S-shaped curve in Fig. 5-11. With the given numerical values, Eq. (5-18) takes the form

$$x = \frac{1.0(1.0)}{3.0 \times 10^{-3} \, (50,000)} (T - T_0) = \frac{1}{150} (T - 298) \tag{B}$$

This energy balance is the straight line shown in Fig. 5-11.

Table 5-8

Intersection	$T - T_0$, °C	T, °K	Conversion, %
A	3	301	1.5
C	146	444	98.0

The intersections of the two curves in Fig. 5-11 give the possible operating temperatures and conversions. The intersections at both ends of the curves are stable operating points and give the results shown in Table 5-8. Point B at $T = 298 + 48 = 346$°K$(73$°C$)$ represents a metastable point, as described in connection with Fig. 5-9. If the reaction had been started in the usual way by adding feed at 25°C, point A would be the steady-state operating condition at $25 + 3 = 28$°C and the conversion would be only 1.5%. To obtain a high conversion the initial temperature would have to be above 73°C (point B). Then the reaction temperature would increase to $25 + 146 = 171$°C and the conversion would be 98%.

Fig. 5-11   Temperature rise vs conversion

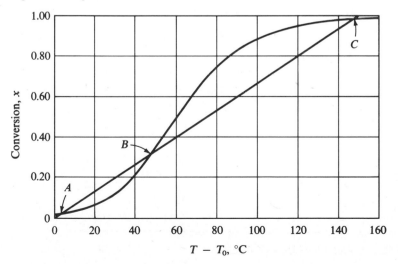

5-5 Semibatch Reactors

The performance of semibatch reactors under isothermal conditions was studied in Sec. 4-8. When the temperature is not constant, an energy balance must be solved simultaneously with the mass-balance equation. In general, the energy balance for a semibatch reactor (Fig. 3-1c) will include all four items of Eq. (3-2). Following the nomenclature of Sec. 4-8, let F_0 and F_1 be the total mass-flow rates of feed and product streams and H_0 and H_1 the corresponding enthalpies above a reference state. Then, following Eq. (3-2) term by term, the energy balance over an element of time Δt is

$$
\begin{array}{cccc}
\text{Energy in} & \text{energy in} & \text{energy from} & \text{energy accumulated} \\
\text{feed stream} - & \text{product stream} + & \text{surroundings} & = \text{in reactor}
\end{array}
$$

(5-19)

$$
F_0 H_0 \, \Delta t \quad - F_1 H_1 \, \Delta t \quad + UA_h(T_s - T_1) \, \Delta t = \frac{\Delta H}{M} \mathbf{r} V \, \Delta t + m_t c_v \, \Delta T_1
$$

The accumulation term represents the energy increase due to the change in composition and the change in temperature of the reactor contents. This involves the heat of reaction ΔH and the specific heat c_v, as described in the development of Eq. (3-11) for batch reactors. Similarly, the enthalpies H_0 and H_1 depend on both temperature and composition. Let us take the reference temperature as T_1 and the reference composition as that corresponding to the composition of the feed (zero conversion). Then, following the development of Eq. (3-5),

$$
H_0 = c_p(T_0 - T_1) + 0 \frac{\Delta H}{M} w_0 = c_p(T_0 - T_1) \tag{5-20}
$$

$$
H_1 = c_p(T_1 - T_1) + x_1 \frac{\Delta H}{M} w_1 = x_1 w_1 \frac{\Delta H}{M} \tag{5-21}
$$

where w = weight fraction of reactant
M = molecular weight of reactant
x_1 = conversion

Substituting Eqs. (5-20) and (5-21) in Eq. (5-19), simplifying, and taking the limit at $\Delta t \to 0$ gives

$$
F_0 c_p(T_0 - T_1) + F_1 x_1 w_1 \frac{\Delta H}{M} + UA_h(T_s - T_1) =
$$

$$
\frac{\Delta H}{M} \mathbf{r} V + m_t c_v \frac{dT_1}{dt} \tag{5-22}
$$

This expression, along with the mass balance [Eq. (4-12)] and the rate equation, can be solved for the temperature and composition in the reactor (or product stream) as a function of time.

There are a multitude of variations for semibatch operation. Equation (5-22) already includes restrictions that limit its application to specific operating conditions; for example, constant mass-flow rates. A frequently encountered case for nonisothermal operation is one in which there is *no product stream*, one reactant is present in the reactor, and the temperature is controlled by the flow rate of the feed stream containing the second reactant. Figure 4-17 shows this type, and Example 4-13 illustrates the design calculations for isothermal operation. The energy balance for this situation reduces to

$$F_0 c_p (T_0 - T_1) + UA_h(T_s - T_1) = \frac{\Delta H}{M} \mathbf{r} V + m_t c_v \frac{dT_1}{dt} \tag{5-23}$$

When temperature control is critical the reaction rate is often very fast with respect to the rate of heat transfer. Then the factors that determine the design are the rate of energy exchange with the surroundings, the feed temperature, and the feed rate. Under these circumstances only the energy balance is required. The reaction may be assumed to be at thermodynamic equilibrium, so that the chemical kinetics equation (and the mass balance) is unnecessary. The rate of reaction is equal to the rate of addition of reactant in the feed stream. Hence the quantity $\mathbf{r}V$ is equal to $F_0 w_0$, and Eq. (5-23) becomes

$$m_t c_v \frac{dT_1}{dt} = F_0 c_p (T_0 - T_1) + UA_h(T_s - T_1) - \frac{\Delta H}{M} F_0 w_0 \tag{5-24}$$

Solution of this expression shows how the temperature varies with time for various combinations of feed rate, feed temperature, and heat-exchange rate. The conversion obtained under such conditions is always the equilibrium value corresponding to the temperature at the end of the process. Its application to a practical problem is illustrated in Example 5-5.

Example 5-5 Hexamethylenetetramine (HMT) is to be produced in a semibatch reactor by adding an aqueous ammonia solution (25 wt % NH_3) at the rate of 2 gpm to an initial charge of 238 gal (at 25°C) of formalin solution containing 42% by weight formaldehyde. The original temperature of the formalin solution is raised to 50°C in order to start the reaction. The temperature of the NH_4OH solution is 25°C. The heat of reaction in the liquid phase may be assumed independent of temperature and concentration and taken as -960 Btu/lb of HMT. If the reactor can be operated at a temperature of 100°C, the rate of reaction is very fast in comparison with the rate of heat transfer with the surroundings. Temperatures higher than 100°C are not desirable because of vaporization and increase in pressure.

It is proposed to cool the reactor by internal coils through which water is passed. The overall heat-transfer coefficient between the stirred reaction mixture and the cooling water will be 85 Btu/(hr)(ft^2)(°F). The water rate through the coils is such that its temperature varies little, and an average value of 25°C may be used. Given the following data, calculate the length of 1-in.-OD tubing required for the cooling coils.

Density of ammonia solution = 0.91 g/cm³
Density of formalin (42%) at 25°C = 1.10 g/cm³
Specific heat of reaction mixture (assume constant), c_v = 1.0 Btu/(lb)(°F)
Specific heat of 25 wt % NH_3 solution, c_p = 1.0 Btu/(lb)(°F)

The rate of the reverse reaction is negligible.

Solution   Since the rate is very rapid and the reaction is irreversible, the ammonia in the inlet stream will be completely converted to HMT just as soon as it is added to the reactor according to the reaction

$$4NH_3 + 6HCHO \rightarrow N_4(CH_2)_6 + 6H_2O$$

Since 4 moles of ammonia are required for 6 moles of formaldehyde, the total amount of ammonia required to react with all the charge of formalin solution will be

$$(NH_3)_t = \frac{238(8.33)(1.10)(0.42)}{30} \frac{4}{6}(17) = 346 \text{ lb}$$

From the ammonia feed rate of 2 gpm, the total time of reaction will be

$$t_t = \frac{346}{2(8.33)(0.91)(0.25)} = 91.3 \text{ min}$$

The heat-transfer surface is to be sufficient to prevent the temperature from exceeding 100°C. Hence at 100°C, dT/dt in Eq. (5-24) will be zero. At temperatures below 100°C the driving force, $T - T_s$, will be insufficient to transfer enough energy to the cooling coils to maintain a constant temperature. Thus at the start of the addition of ammonia the last term in Eq. (5-24), which is positive for an exothermic reaction, will be greater than the sum of the first and second terms, and the temperature of the reaction mixture will increase. From a practical standpoint this heating period would be reduced to a minimum by shutting off the flow of cooling water until the temperature reaches 100°C.

To determine the required heat-transfer area, Eq. (5-24) may be used when the temperature is 100°C and $dT_1/dt = 0$. Thus

$$UA_h(T_1 - T_s) = \frac{-\Delta H}{M_{NH_3}} F_0 w_0 + F_0 c_p(T_0 - T_1)$$

$$85A_h(100 - 25)(1.8) = \frac{-\Delta H}{M_{NH_3}} F_0 w_0 + F_0(1.0)(25 - 100)(1.8) \tag{A}$$

The heat of reaction is -960 Btu/lb HMT. On the basis of NH_3,

$$\frac{\Delta H}{M_{NH_3}} = -\frac{960(140)}{4(17)} = -1,975 \text{ Btu/lb } NH_3$$

The feed rate is

$F_0 = 2(60)(8.33)(0.91) = 910 \text{ lb/hr}$

$w_0 = 0.25$

Substituting these values in Eq. (A) and solving for the heat-transfer area, we obtain

$$A_h = \frac{910(1{,}975)(0.25) - 1(75)(1.8)}{85(75)(1.8)} = 28.3 \text{ ft}^2$$

If the heat-transfer coefficient of 85 is based on the outside area of the tubes, the length L of 1-in.-OD coil is

$$L = \frac{28.3}{\pi D} = \frac{28.3(12)}{\pi} = 108 \text{ ft}$$

An approximate size of the reactor can be obtained by noting that the total mass of mixture at the end of the process will be

$$910\left(\frac{91.3}{60}\right) + 238(8.33)(1.10) = 3{,}560 \text{ lb}$$

If we assume the density of the HMT solution to be 72 lb/ft^3, the minimum reactor volume is 50 ft^3. A cylindrical vessel 4 ft in diameter and 6 ft in height would provide 33% excess capacity. If the 1-in. tubing were wound into a 3-ft-diameter coil, approximately 12 loops would be needed.

The length of time necessary to raise the reaction temperature from its initial value of 50°C to 100°C can be obtained by integrating Eq. (5-24). With the water rate shut off $UA_h(T - T_s)$ is negligible, and the expression becomes

$$\int_{(T_1)_0}^{T_1} \frac{dT_1}{(-\Delta H/M_{\text{NH}_3})F_0 w_0 + F_0 c_p(T_0 - T)} = \int_0^t \frac{dt}{(m_0 + F_0 t)c_v}$$

where m_t has been replaced by $m_0 + F_0 t$. If ΔH and c_v do not vary with temperature, this equation may be integrated to yield

$$-\frac{1}{F_0 c_p} \ln \frac{(-\Delta H/M_{\text{NH}_3})F_0 w_0 + F_0 c_p(T_0 - T_1)}{(-\Delta H/M_{\text{NH}_3})F_0 w_0 + F_0 c_p[T_0 - (T_1)_0]} = \frac{1}{F_0 c_v} \ln \frac{m_0 + F_0 t}{m_0} \quad \text{(B)}$$

Equation (B) expresses the temperature as a function of time during the heating period. Taking $(T_1)_0 = 50° + 273°$ and $T_0 = 25° + 273°$ and expressing t in hours, the time required for T_1 to reach 100°C is

$$-\frac{1}{1(910)} \ln \frac{449{,}000 + (910)(1)(25 - 100)(1.8)}{449{,}000 + (910)(1)(25 - 50)(1.8)} = \frac{1}{910(1)} \ln \frac{2{,}180 + 910t}{2{,}180}$$

$$\ln(1 + 0.418t) = -\ln \frac{326{,}000}{408{,}000}$$

$$1 + 0.418t = 1.25$$

$$t = 0.60 \text{ hr} \qquad \text{or } 36 \text{ min}$$

In summary, the reaction temperature would rise from 50 to 100°C in 36 min after the ammonia feed is started, provided water is not run through the cooling coil. After 36 min the water flow would be started in order to maintain the reactor temperature at 100°C. After a total time of 93 min sufficient ammonia would have been added to convert all the formaldehyde to HMT.

Problems

5-1.   The liquid-phase hydrolysis of dilute aqueous acetic anhydride solutions is second order and irreversible, as indicated by the reaction

$$(CH_3CO)_2O + H_2O \rightarrow 2CH_3COOH$$

A batch reactor for carrying out the hydrolysis is charged with 200 liters of anhydride solution at 15°C and a concentration of 2.16×10^{-4} g mole/cm^3. The specific heat and density of the reaction mixture are essentially constant and equal to 0.9 cal/(g)(°C) and 1.05 g/cm^3, respectively. The heat of reaction may be assumed constant and equal to $-50,000$ cal/g mole. The rate has been investigated over a range of temperatures, of which the following results are typical:

t, °C	10	15	25	40
r, g mole/(cm^3)(min)	$0.0567C$	$0.0806C$	$0.1580C$	$0.380C$

where C is acetic anhydride concentration, in gram moles per cubic centimeter. (*a*) Explain why the rate expression can be written as shown in the table even though the reaction is second-order. (*b*) If the reactor is cooled so that operation is isothermal at 15°C, what time would be required to obtain a conversion of 70% of the anhydride? (*c*) Determine an analytical expression for the rate of reaction in terms of temperature and concentration. (*d*) What time is required for a conversion of 70% if the reactor is operated adiabatically?

5-2.   A reactor for the production of drying oils by the decomposition of acetylated castor oil is to be designed for a conversion of 70%. The initial charge will be 500 lb and the initial temperature 340°C, as in Example 5-1. In fact, all the conditions of Example 5-1 apply, except instead of adiabatic operation, heat will be supplied electrically with a cal-rod unit in the form of a 1-in.-OD coil immersed in the reaction mixture. The power input and the stirring in the reactor will be such that the surface temperature of the heater is maintained constant at 700°K. The heat-transfer coefficient may be taken equal to 60 Btu/(hr)(ft^2)(°F). What length of heater will be required if the conversion of 70% is to be obtained in 20 min?

5-3.   Reconsider the pilot plant discussed in Example 5-2 for the production of allyl chloride. It has been proposed to reduce the extent of the side reaction to dichloropropane by preheating the feed to 300°C. To obtain a more uniform axial temperature profile, inert nitrogen will be added to the feed to give a composition corresponding to 5 moles of N_2, 4 moles of C_3H_6, and 1 mole of Cl_2. The total feed rate will be 0.85 lb mole/hr. If all other conditions are

also the same, calculate the temperature and conversion profiles for a tube
length of 20 ft. Comment on this method of operation in comparison with that in
Example 5-2 with respect to selectivity and production rate of allyl chloride.

5-4.   Consider the irreversible constant-density reaction sequence

$$A + B \xrightarrow{k_2} C \qquad r_2 = k_2[A][B]$$

$$C \xrightarrow{k_1} D \qquad r_1 = k_1[C]$$

The rate constants are

$$k_2 = A_2\, e^{-E_2/R_gT}$$

$$k_1 = A_1\, e^{-E_1/R_gT}$$

Component C is the desired product. The feed contains no C or D, and $[A] = [A]_0$, $[B] = [A]_0$.

(a) If an isothermal tubular-flow reactor is employed, develop equations for
calculating the conversion of A for which the yield of C is a maximum and
for the constant temperature at which the reactor should operate to give the
highest value of the maximum yield of C. (b) Develop equations for predicting
the conversion of A for which the selectivity of C with respect to D is a maximum.

5-5.   Repeat Prob. 5-4 for a stirred-tank reactor.

5-6.   Reconsider Example 5-3 for the case where the feed composition is $N_2 = 50$
mole %, $C_3H_6 = 40$ mole %, $Cl_2 = 10$ mole %. The feed temperature will be
300°C, with all other conditions the same as in Example 5-3. Determine conversions and temperatures for reactor volumes up to 0.5 ft³.

5-7.   A first-order irreversible (liquid-phase) reaction is carried out in a stirred-tank
flow reactor. The density is 1.2 g/cm³ and the specific heat is 0.9 cal/(g)(°C).
The volumetric flow rate is 200 cm³/sec and the reactor volume is 10 liters.
The rate constant is

$$k = 1.8 \times 10^5\, e^{-12,000/R_gT} \qquad sec^{-1}$$

where T is in degrees Kelvin. If the heat of reaction is $\Delta H = -46,000$ cal/g
mole and the feed temperature is 20°C, what are the possible temperatures and
conversions for stable, adiabatic operation at a feed concentration of 4.0 g
mole/liter?

5-8.   For irreversible reactions the maximum conversion is unity, regardless of
temperature level. Since the rate normally increases with temperature, the
maximum conversion is obtained in a given reactor if the temperature is at its
highest permissible level at every location in the reactor. The permissible
temperature limit is determined by the possibility of undesirable side reactions
and strength or corrosion resistance of construction materials. For endothermic
reversible reactions both rate and maximum conversion increase with temperature. Here again the highest permissible temperature gives the maximum
conversion. For reversible exothermic reactions increasing the temperature
reduces the maximum (equilibrium) conversion but increases the forward rate.
To obtain the maximum conversion a high temperature is needed at low conversions (where the reverse reaction is unimportant) and a lower temperature at
higher conversions.

Consider a reversible first-order reaction $A \rightleftharpoons B$ for which, at 298°K,
$\Delta F°_{298} = -2500$ cal/g mole

$\Delta H°_{298} = -20,000$ cal/g mole

The reaction mixture is an ideal liquid solution (constant density) at all temperatures. (a) Assuming that $\Delta H°$ is constant, plot a curve of the equilibrium conversion vs temperature from 0 to 100°C. (b) If the forward-rate constant is

$$k = 5 \times 10^8 \, e^{-12,500/R_g T} \qquad \text{min}^{-1}$$

determine the conversion in the effluent from an isothermal tubular-flow reactor for which the volumetric feed rate is 100 liters/min and the reactor volume is 1,500 liters. Calculate the conversion for a series of temperatures from 0 to 100°C and plot the results on the figure prepared for part (a). (c) Suppose that the maximum permissible temperature is 100°C and the concentration of A in the feed stream is 2 g moles/liter (the feed contains no B). Determine the maximum conversion obtainable in the reactor of part (b) if the temperature can be varied along the length of the reactor. First prepare curves of net rate vs temperature at constant conversion for several conversion levels. Note that each curve will show an increase in rate with temperature increase at low temperatures. Then as the temperature corresponding to equilibrium for the fixed conversion is approached the curve will go through a maximum, and it will finally fall to zero rate when this temperature is reached. The maximum conversion in the reactor effluent will be obtained when the net rate is a maximum at every position (or conversion) in the reactor. Hence the locus of the maxima in the curves for net rate vs temperature will determine the optimum temperature profile. Plot this profile, first as temperature vs conversion and then as temperature vs V/Q.

5-9.   Consider the same reaction system, rate equation, feed rate, and composition as in Prob. 5-8. The reaction is to be carried out in two stirred-tank reactors, each with a volume of 750 liters. What should be the temperature (within the range 0 to 100°C) in each reactor in order to obtain the maximum conversion in the effluent from the second reactor? The graphical method for multiple-stirred-tank reactors (described in Sec. 4-6) may be helpful.

5-10.   The HMT reactor described in Example 5-5 is to be redesigned so that the reaction temperature rises from 50 to 100°C as uniformly as possible as the ammonia solution is added. To accomplish this the rate of addition of ammonia must be varied. The cooling water will flow through the coils throughout the run. Determine the feed rate as a function of time, and the time required to add all the ammonia, to meet these operating requirements. All other conditions are the same as in Example 5-5.

6
DEVIATIONS FROM IDEAL REACTOR PERFORMANCE

In Chap. 3 equations were developed for calculating the conversion for two extreme mixing states: the ideal stirred-tank reactor, corresponding to complete mixing, and the ideal tubular-flow reactor, corresponding to no axial mixing and uniform velocity in the direction of flow. The reasons for deviations from ideal forms in actual reactors were also pointed out (see Fig. 3-4). The objective in this chapter is to evaluate quantitatively the effect of deviations on the conversion. If the velocity and local rate of mixing (micromixing) of every element of fluid in the reactor were known and the differential mass balance could be integrated, an exact solution for the conversion could be obtained. Since such complete information is unavailable for actual reactors, approximate methods, using readily obtainable data and models of mixing, are necessary. It is difficult to measure velocities and concentrations within a reactor, but data can be obtained on the feed and effluent streams. Such end-effect, or response, data consist of measuring the effect observed in the effluent stream when the concentration of an inert component in the feed is changed. In this chapter the treatment is restricted to a single reaction in a homogeneous reactor operated isothermally. Nonideal behavior in heterogeneous reactors will be discussed in Chap. 13.

6-1 Mixing Concepts and Models

Deviations from ideal flow can be classified in two types. In one type of deviation elements of fluid may move through the reactor at different velocities, causing channeling and "dead spots." For such behavior to occur, the elements of fluid must not completely mix locally, but remain at least partially *segregated* as they move through the reactor. The other deviation refers to the extent of the local or micromixing. For example, there may be some mixing or diffusion in the direction of flow in a tubular reactor.

We shall see in Sec. 6-3 that response measurements can be used to determine the residence-time distribution (RTD) of the elements of fluid in a reactor. However, the RTD is not enough to determine both the extent of segregated flow in the reactor and the extent of micromixing. Hence residence-time information is not sufficient, in general, to evaluate the conversion in a nonideal reactor. Put another way, the RTD is affected by both types of deviations. There can be a large number of different mixing states, i.e., different extents of segregated flow, which give the same RTD.[1] However, for first-order kinetics the distinction between mixing states has no effect on the conversion (Sec. 6-7). In this case the RTD provides all the information necessary to calculate a correct conversion. Any mixing model which gives the actual RTD may be used to calculate the conversion for first-order kinetics.

We shall consider three methods of estimating deviations from ideal reactor performance. The first method is to determine the actual RTD from experimental response data and then calculate the conversion by assuming the flow to be wholly segregated (Sec. 6-8). This model should be a good approximation, for example, for a tubular-flow reactor, where the flow is streamline. It would not describe a nearly ideal stirred-tank reactor, for here the fluid is nearly completely mixed when it enters the reactor. In this case no error is introduced by an approximation of the RTD, since the actual

[1] Much of the development of mixing and residence-time-distribution concepts in reactors is due to P. V. Danckwerts, *Chem. Eng. Sci.*, **8**, 93 (1958), and T. N. Zwietering, *Chem. Eng. Sci.*, **11**, (1959). Even though the flow is wholly segregated within the reactor, complete micromixing will occur at the exit. The conversion and other properties of the effluent stream are average values, based on a completely mixed stream. Zwietering has used the concept of the location in a reactor where complete mixing occurs as a parameter for describing the mixing state. For example, in an ideal stirred tank complete micromixing occurs as the feed enters. At the other extreme, in completely segregated flow micromixing occurs at the exit of the reactor.

RTD is used. An error does arise from the assumption of segregated flow when there may be some micromixing; as noted, this error disappears for first-order kinetics.

The other two methods are subject to both these errors, since both the form of the RTD and the extent of micromixing are assumed. Their advantage is that they permit analytical solution for the conversion. In the *axial-dispersion model* the reactor is represented by allowing for axial diffusion in an otherwise ideal tubular-flow reactor. In this case the RTD for the actual reactor is used to calculate the best axial diffusivity for the model (Sec. 6-5), and this diffusivity is then employed to predict the conversion (Sec. 6-9). This is a good approximation for most tubular reactors with turbulent flow, since the deviations from plug-flow performance are small. In the third model the reactor is represented by a series of ideal stirred tanks of equal volume. Response data from the actual reactor are used to determine the number of tanks in series (Sec. 6-6). Then the conversion can be evaluated by the method for multiple stirred tanks in series (Sec. 6-10).

Another model, which will not be analyzed, is the plug-flow reactor with recycle shown in Fig. 6-1. The reactor itself behaves as an ideal tubular type, but mixing is introduced by the recycle stream. When the recycle rate becomes very large, ideal stirred-tank performance is obtained, and when the recycle is zero, plug-flow operation results. The response data on the actual reactor are used to evaluate the recycle rate, and then the conversion is estimated for a plug-flow reactor with this recycle rate.

In all these models except the segregated-flow approach, which employs the actual experimental response data, a single parameter is used to describe the data. One parameter does not always provide an adequate description. If the experimental response data cannot be described, for example, by a single axial diffusivity, the dispersion model may not give an accurate con-

Fig. 6-1 *Ideal tubular-flow reactor with recycle*

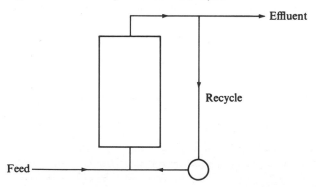

version. Many other, usually more elaborate models have been proposed.[1] Some employ more than one parameter in order to describe more accurately both the RTD and the extent of micromixing in the reactor.

In the next section RTD functions are discussed, and in Secs. 6-3 and 6-4 the RTD is evaluated from response data, with illustrations for ideal reactors. Later sections deal with application of the single-parameter axial-dispersion and number-of-tanks in series models. Finally, conversions are evaluated by the various methods and compared with results for ideal reactors to determine the magnitude of the deviations.

6-2 The Residence-time Distribution Function

The time it takes a molecule to pass through a reactor is called its *residence time* θ. Two properties of θ are important: the time elapsed since the molecule entered the reactor (its age) and the remaining time it will spend in the reactor (its residual lifetime). We are concerned mainly with the sum of these times, which is θ, but it is important to note that micromixing can occur only between molecules that have the same residual lifetime; molecules cannot mix at some point in the reactor and then unmix at a later point in order to have different residual lifetimes. A convenient definition of *residence-time distribution function* is the fraction $J(\theta)$ of the effluent stream that has a residence time less than θ. None of the fluid can have passed through the reactor in zero time, so $J = 0$ at $\theta = 0$. Similarly, none of the fluid can remain in the reactor indefinitely, so that J approaches 1 as θ approaches infinity. A plot of $J(\theta)$ vs θ has the characteristics shown in Fig. 6-2a.

Variations in density, such as those due to temperature and pressure gradients, can effect the residence time and are superimposed on effects due to velocity variations and micromixing. We are concerned with micromixing in this chapter, and we shall therefore suppose that the density of each element of fluid remains constant as it passes through the reactor. Under these conditions the *mean residence time*, averaged for all the elements of fluid, is given by Eq. (3-21) or by

$$\bar{\theta} = \theta_F = \frac{V}{F_t/\rho} = \frac{V}{Q}$$

where Q is the volumetric flow rate. For constant density, Q is the same for the feed as for the effluent stream.

From the definition of $J(\theta)$ we can also say that $dJ(\theta)$ is the volume

[1]L. A. Spillman and O. Levenspiel, *Chem. Eng. Sci.*, **20**, 247 (1965); R. L. Curl, *AIChE J.*, **9**, 175 (1963); S. A. Shain, *AIChE J.*, **12**, 806 (1966); H. Weinstein and R. J. Adler, *Chem. Eng. Sci.*, **22**, 65 (1967); D. Y. Ng and D. W. T. Rippin, *Chem. Eng. Sci.*, **22**, 3, 247 (1967).

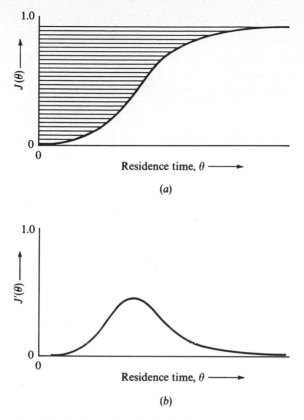

(a)

(b)

Fig. 6-2 *Residence-time distribution functions*

fraction of the effluent stream that has a residence time between θ and $\theta + d\theta$. Hence the mean residence time is also given by

$$\bar{\theta} = \frac{\int_0^1 \theta \, dJ(\theta)}{\int_0^1 dJ(\theta)} = \int_0^1 \theta \, dJ(\theta) = \frac{V}{Q} \tag{6-1}$$

The shaded area in Fig. 6-2a represents $\bar{\theta}$. The RTD can also be described in terms of the slope of the curve in Fig. 6-2a. This function, $J'(\theta) = dJ(\theta)/d\theta$, will have the shape usually associated with distribution curves, as noted in Fig. 6-2b. The quantity $J'(\theta) \, d\theta$ represents the fraction of the effluent stream with a residence time between θ and $\theta + d\theta$. Substituting $J'(\theta) \, d\theta$ for $dJ(\theta)$ in Eq. (6-1) gives an expression for $\bar{\theta}$ in terms of $J'(\theta)$,

$$\bar{\theta} = \int_0^\infty \theta \, J'(\theta) \, d\theta \tag{6-2}$$

6-3 Residence-time Distributions from Response Measurements

The RTD for a given reactor and flow rate can be established from response-type experiments. In these experiments the concentration of an inert tracer is disturbed in the feed stream and its effect on the effluent stream is measured. The three most common perturbations are a step function, a pulse (square wave), and a sinusoidal wave. The relationships between the observed concentration-vs-time curves and the RTD are examined here for step functions and pulses. The analysis of sinusoidal perturbations is more complex but is available in the literature.[1]

Step-function input   Suppose a stream with a molecular concentration C_0 flows through a reactor at a constant volumetric rate Q. Imagine that at $\theta = 0$ all the molecules entering the feed are marked to distinguish them from the molecules which have entered prior to $\theta = 0$. Since the total concentration is not changed by the marking process, its value will be C_0 at any θ. However, the concentration C of marked molecules in the effluent will change with θ, since some marked molecules would spend longer than others in the reactor. The input and output (response) concentration ratios C/C_0 are shown generally for this situation in Fig. 6-3b and c. The exact shape of the response curve depends on the mixing state of the system.

At a time θ when the concentration of marked molecules in the effluent is C the rate of flow of these molecules will be CQ. All the marked molecules have entered the reactor in a time less than θ. By definition, $J(\theta)$ is the fraction of the total molecules that have this residence time range. Since the total flow rate of molecules is C_0Q, the product $C_0QJ(\theta)$ will also describe the flow rate of marked molecules. Equating these two expressions gives

$$CQ = C_0QJ(\theta)$$

or

$$J(\theta) = \left(\frac{C}{C_0}\right)_{\text{step}} \tag{6-3}$$

In practice it is not possible to start marking the instant molecules start entering the reactor. Instead, a quantity of inert miscible substance is added to the feed stream in the apparatus shown in Fig. 6-3a. Provided that the tracer can be added quickly, that the tracer molecules move through the reactor in the same way as the feed stream, and that the detection of tracer in the effluent is rapid, the experimental results are satisfactory. In such an experiment the concentration of tracer in the feed takes the place of C_0

[1] H. Kramers and G. Alberda, *Chem. Eng. Sci.*, **2**, 173 (1953).

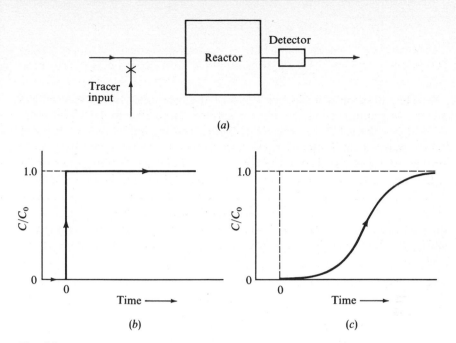

Fig. 6-3 *Response to a step-function input: (a) apparatus for input-response studies, (b) step-function input, (c) response in effluent*

in Eq. (6-3), and C is the concentration of tracer molecules in the effluent. The mathematical description of this step-function input is

$$C = \begin{cases} 0 & \text{for } t < 0 \\ C_0 & \text{for } t > 0 \end{cases} \qquad \text{at } z = 0 \qquad (6\text{-}4)$$

Equation (6-3) shows that the C/C_0 response to a step-function input gives the RTD function $J(\theta)$ directly. This approach provides a simple experimental procedure for measuring the RTD for an actual reactor.

Pulse (square-wave) input Suppose the marking experiment is conducted in a different way. This time let all the molecules be marked only for a very short time interval at $\theta = 0$. The total molecules marked would be

$$M = C_0 Q\, \Delta t_0 \qquad (6\text{-}5)$$

where Δt_0 is the marking interval and C_0 is again the total concentration of molecules. This input is shown graphically in Fig. 6-4a and is defined mathematically by

$$C = \begin{cases} 0 & \text{for } t < 0 \\ C_0 & \text{for } 0 < t < \Delta t_0 \qquad \text{at } z = 0 \\ 0 & \text{for } t > \Delta t_0 \end{cases} \qquad (6\text{-}6)$$

The variation in residence times of the molecules in the reactor will disperse the input pulse, giving a response curve like that in Fig. 6-4b. To evaluate the RTD we proceed as before and formulate two expressions for the marked molecules at some time θ. Since C is the concentration of marked molecules at θ, the number of such molecules leaving the reactor in the time period θ to $\theta + d\theta$ will be $CQ \, d\theta$. All the marked molecules in the effluent will have a residence time θ to $\theta + d\theta$ because they were added only at $\theta = 0$. By definition, the fraction of the effluent stream consisting of such molecules will be $dJ(\theta)$ or $J'(\theta) \, d\theta$. The number of such molecules will be $MJ'(\theta) \, d\theta$. Equating the two expressions for the number of marked molecules gives

$$CQ \, d\theta = MJ'(\theta) \, d\theta$$

or

$$J'(\theta) = \frac{(C)_{\text{pulse}} \, Q}{M} \qquad (6\text{-}7)$$

Equation (6-7) shows that $J'(\theta)$ can be obtained from the measured response curve for a pulse input. In practice, instead of marking molecules we obtain this curve by introducing, as quickly as possible, a pulse of miscible tracer in the feed and detecting its concentration in the effluent (see Fig. 6-3a). The volumetric flow rate would be known, and M can be calculated. Since it is difficult to measure Δt_0 and C_0 accurately, M is better determined from the area of the response curve; that is,

Fig. 6-4   *Response to a pulse input: (a) pulse input to feed, (b) response in effluent*

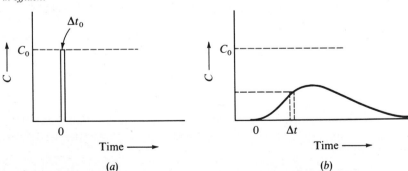

$$M = Q \int_0^\infty C_{\text{pulse}} \, d\theta \tag{6-8}$$

This follows from the fact that all the tracer molecules will ultimately appear in the effluent stream. With this formulation for M, Eq. (6-7) becomes

$$J'(\theta) = \frac{C_{\text{pulse}}}{\int_0^\infty C_{\text{pulse}} \, d\theta} \tag{6-9}$$

With this result we can evaluate $J'(\theta)$ for any reactor, using only the measured response curve to a pulse input. Since $dJ(\theta) = J'(\theta) \, d\theta$, the RTD in terms of $J(\theta)$ is given by

$$J(\theta) = \frac{\int_0^\theta C_{\text{pulse}} \, d\theta}{\int_0^\infty C_{\text{pulse}} \, d\theta} \tag{6-10}$$

6-4 Residence-time Distributions for Reactors with Known Mixing States

For reactors with known mixing characteristics the response curve and the RTD can be predicted; no experiments are necessary. As an illustration let us develop the RTD for the plug-flow reactor, a single ideal stirred-tank reactor, and a tubular reactor with laminar flow.

The characteristics of uniform velocity profile and no axial mixing in a *plug-flow reactor* require that the residence time be a constant, $\theta = V/Q$. The curve for response to a step-function input is as shown in Fig. 6-5. From Eq. (6-3), the response curve is equal to $J(\theta)$. Then $J(\theta) = 0$ for $\theta < V/Q$ and $J(\theta) = 1$ for $\theta \geq V/Q$. The input and response curve for a pulse input would correspond to narrow peaks at $\theta = 0$ and $\theta = V/Q$, as shown in Fig. 6-6 (solid lines). The response curve, according to Eq. (6-7), is proportional to $J'(\theta)$.

Fig. 6-5 *Response curves to a step input for ideal reactors*

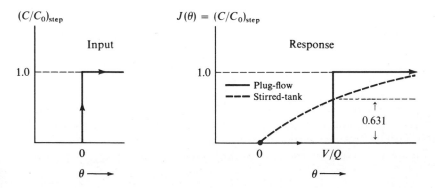

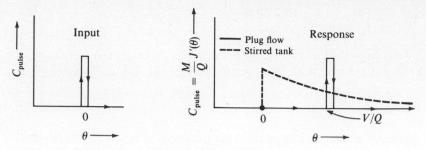

Fig. 6-6　*Response curves to a pulse input for ideal reactors*

For an ideal *stirred-tank reactor* $(C/C_0)_{step}$ can be calculated by applying Eq. (3-1) for a step-function input. The third term is zero, since there is no reaction. At a time θ after the tracer concentration in the feed is increased to C_0 the other terms in Eq. (3-1) give

$$C_0 Q\,\Delta\theta - CQ\,\Delta\theta = V\,\Delta C$$

where C is the effluent concentration at θ and ΔC is the change in concentration of tracer in the reactor during $\Delta\theta$. Dividing by $\Delta\theta$ and taking the limit as $\Delta\theta \to 0$ gives

$$\frac{dC}{d\theta} = \frac{Q}{V}(C_0 - C) = \frac{1}{\bar\theta}(C_0 - C) \tag{6-11}$$

With the initial condition $C = 0$ at $\theta \le 0$, the solution of Eq. (6-11) is

$$\left(\frac{C}{C_0}\right)_{step} = J(\theta) = 1 - e^{-(\theta/\bar\theta)} \tag{6-12}$$

The curve described by Eq. (6-12), the dotted line in Fig. 6-5, indicates a large spread in residence times. The asymptotic tail shows that a few molecules remain in the reactor for a very long time. At the mean residence time $\bar\theta = V/Q$, $J(\theta) = 0.631$; or 63.1% of the effluent stream has a residence time less than the mean value.

A similar analysis for a pulse input would give a response curve of C_{pulse} vs θ, but this can be obtained more easily by differentiating the $J(\theta)$ curve in Fig. 6-5. From Eq. (6-7), the derivative $J'(\theta)$ is proportional to

The derivative of the dashed line in Fig. 6-5 will be largest at $\theta = 0$ and will continually decrease toward zero as θ increases. Such a distribution curve, given as the dashed line in Fig. 6-6, shows that the most probable [largest $J'(\theta)\,d\theta$] residence time is at $\theta = 0$ for a stirred-tank reactor.

Figure 6-5 is useful for determining how closely an actual reactor fits one of the two ideal forms. The measured response curve may be super-imposed on Fig. 6-5 to show the extent of the deviations from either plug-flow or stirred-tank behavior.

A tubular reactor with *laminar flow* has been mentioned as a good approximation to segregated flow. If the dispersion due to molecular diffusion is neglected, the approximation is exact. Since the flow is segregated and the velocity profile is known, the RTD can be calculated.[1] It is instructive to compare the calculated results with those for the ideal forms given in Fig. 6-5. The velocity in the axial direction for laminar flow is parabolic,

$$u(r) = 2 \frac{Q}{\pi r_0{}^2} \left[1 - \left(\frac{r}{r_0} \right)^2 \right]$$ (6-13)

where r is the radial position and r_0 is the tube radius.[2] Since u is a function of r, the residence time also varies with r. If the reactor length is L, θ at any r is

$$\theta = \frac{L}{u} = \frac{\pi r_0{}^2}{2Q} \frac{L}{[1 - (r/r_0)^2]}$$ (6-14)

Since $L\pi r_0{}^2 = V$, this expression becomes

$$\theta = \frac{V/Q}{2[1 - (r/r_0)^2]}$$ (6-15)

We first find $J(r)$ and then use Eq. (6-15) to convert $J(r)$ to $J(\theta)$. The fraction of the effluent stream of radius between r and $r + dr$ will be

$$dJ(r) = \frac{u(2\pi r \, dr)}{Q}$$ (6-16)

Substituting Eq. (6-13) for u and simplifying gives

$$dJ(r) = \frac{4}{r_0{}^2} \left[1 - \left(\frac{r}{r_0} \right)^2 \right] r \, dr$$ (6-17)

To replace r with θ in this expression we first differentiate Eq. (6-15) and then solve for $r \, dr$ to obtain

$$r \, dr = \frac{V}{4Q} \frac{r_0{}^2}{\theta^2} d\theta$$ (6-18)

[1] The behavior of this type of reactor was first considered by K. G. Denbigh, *J. Appl. Chem.*, **1**, 227 (1951); and H. Kramers and K. R. Westerterp "Elements of Chemical Reactor Design and Operation," p. 85, Academic Press, Inc., New York, 1963.

[2] In this discussion we temporarily abandon our convention that subscript zero indicates initial conditions.

Substituting $1 - (r/r_0)^2$ from Eq. (6-15) and $r\,dr$ from Eq. (6-18) in Eq. (6-17) yields

$$dJ(\theta) = \frac{1}{2}\left(\frac{V}{Q}\right)^2 \frac{d\theta}{\theta^3} = \frac{1}{2}\bar{\theta}^2 \frac{d\theta}{\theta^3} \tag{6-19}$$

or

$$\frac{dJ(\theta)}{d\theta} = J'(\theta) = \frac{1}{2}\frac{\bar{\theta}^2}{\theta^3} \tag{6-20}$$

Integration of Eq. (6-19) will give $J(\theta)$. The minimum residence time is not zero, but corresponds to the maximum velocity at the center of the tube. From Eq. (6-15), this is

$$\theta_{min} = \frac{1}{2}\frac{V}{Q} = \frac{1}{2}\bar{\theta} \tag{6-21}$$

Integration of Eq. (6-19) from θ_{min} to θ gives

$$J(\theta) = \frac{1}{2}\bar{\theta}^2 \int_{\frac{1}{2}\bar{\theta}}^{\theta} \frac{d\theta}{\theta^3} = 1 - \frac{1}{4}\left(\frac{\theta}{\bar{\theta}}\right)^{-2} \tag{6-22}$$

Equation (6-20) or Eq. (6-22) gives the desired RTD function for a tubular reactor with laminar flow.

$J(\theta)$ is plotted against $\theta/\bar{\theta}$ in Fig. 6-7; also shown are the curves for the two ideal reactors, taken from Fig. 6-5. The comparison brings out pertinent points about reactor behavior. Although the plug-flow reactor might be expected to be a better representation of the laminar case than the stirred-tank reactor, the RTD for the latter more closely follows the laminar-reactor curve for $\theta/\bar{\theta}$ from about 0.6 to 1.5. However, there is no possibility for θ to be less than 0.5 in the laminar-flow case. Hence the stirred-tank form is not applicable at all in the low θ region. At high θ the three curves approach coincidence. Conversions for these reactors are compared in Sec. 6-7.

6-5 Interpretation of Response Data by the Dispersion Model

Here we wish to determine the axial diffusivity from response measurements, in preparation for using the dispersion model for conversion calculations. According to this model, the actual reactor can be represented by a tubular-flow reactor in which axial dispersion takes place according to the effective diffusivity D_L. It is supposed that the axial velocity u and the concentration are uniform across the diameter, as in a plug-flow reactor.

    Imagine that a step function of inert tracer concentration C_0 is introduced into the feed at $\theta = 0$. By solving the transient mass balance, we can

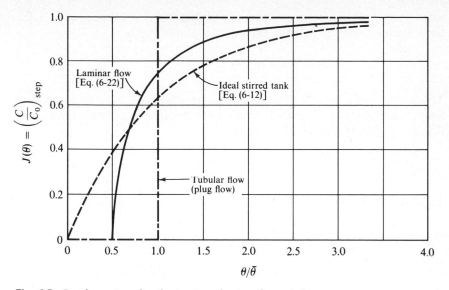

Fig. 6-7 *Residence-time distribution in a laminar-flow tubular reactor (segregated flow)*

determine the response C/C_0 vs θ as a function of D_L. In writing Eq. (3-1) for the mass balance we omit the term involving reaction but include axial dispersion in the input and output terms. The result is

$$\left[\left(-D_L \frac{\partial C}{\partial z} + uC\right)\pi r_0{}^2\right]_z \Delta\theta - \left[\left(-D_L \frac{\partial C}{\partial z} + uC\right)\pi r_0{}^2\right]_{z+\Delta z} \Delta\theta$$
$$= \pi r_0{}^2 \, \Delta z \, \Delta C$$

Canceling $\pi r_0{}^2$, dividing by $\Delta z \, \Delta\theta$, and taking the limit as $\Delta z \to 0$ gives

$$D_L \frac{\partial^2 C}{\partial z^2} - u \frac{\partial C}{\partial z} = \frac{\partial C}{\partial \theta} \tag{6-23}$$

The initial condition is

$$C = \begin{cases} 0 & \text{at } z > 0 \text{ for } \theta = 0 \\ C_0 & \text{at } z < 0 \text{ for } \theta = 0 \end{cases} \tag{6-24}$$

If no axial dispersion occurs in the feed line, the proper boundary condition at $z = 0$ is

$$-D_L \left(\frac{\partial C}{\partial z}\right)_{>0} + u(C)_{>0} = uC_0 \qquad \text{at } z = 0 \text{ for } \theta \geq 0 \tag{6-25}$$

where >0 designates the position just inside the reactor at $z = 0$. It has been shown[1] that at the exit the correct condition is

$$\frac{dC}{dz} = 0 \qquad \text{at } z = L \text{ for } \theta \geq 0 \tag{6-26}$$

The solution with these boundary conditions is difficult, but a good approximation, particularly when D_L is small, may be obtained by substituting for Eqs. (6-25) and (6-26) the expression

$$C = \begin{cases} C_0 & \text{at } z = -\infty \text{ for } \theta \geq 0 \\ 0 & \text{at } z = \infty \text{ for } \theta \geq 0 \end{cases} \tag{6-27}$$

The solution is most easily obtained by making the substitution

$$\alpha = \frac{z - u\theta}{\sqrt{4D_L\theta}} \tag{6-28}$$

Then the partial differential equation (6-23) becomes the ordinary differential equation

$$\frac{d^2 C^*}{d\alpha^2} + 2\alpha \frac{dC^*}{d\alpha} = 0 \tag{6-29}$$

where C^* is the dimensionless concentration C/C_0. The boundary conditions now are

$$C^* = \begin{cases} 1 & \text{for } \alpha = -\infty \\ 0 & \text{for } \alpha = \infty \end{cases} \tag{6-30}$$

Equations (6-29) and (6-30) are readily solvable to give C^* as a function of α, or z and θ, through Eq. (6-28). Substituting $z = L$ gives the response at the end of the reactor as a function of θ; this is[2]

[1] P. V. Danckwerts, *Chem. Eng. Sci.*, **2**, 1 (1953); J. F. Wehner and R. H. Wilhelm, *Chem. Eng. Sci.*, **6**, 89 (1959). The boundary conditions for this type of problem have been the subject of numerous reports. For an analysis of the general problem see G. Standart, *Chem. Eng. Sci.*, **23**, 645 (1968).

[2] The *error function* erf is defined as

$$\text{erf}(y) = \frac{2}{\sqrt{\pi}} \int_0^y e^{-x^2} dx$$

$$\text{erf}(\pm\infty) = \pm 1 \qquad \text{erf}(0) = 0 \qquad \text{erf}(-y) = -\text{erf}(y).$$

values of erf (y) are given in standard mathematical tables; see, for example, "Handbook of Chemistry and Physics," 46th ed., p. A-113, Chemical Rubber Publishing Company, Cleveland, Ohio, 1964.

$$C^*_{z=L} = \left(\frac{C}{C_0}\right)_{\text{step}} = \frac{1}{2}\left[1 - \text{erf}\left(\frac{1}{2}\sqrt{\frac{uL}{D_L}}\frac{1 - \theta/(L/u)}{\sqrt{\theta/(L/u)}}\right)\right] \qquad (6\text{-}31)$$

The mean residence time $\bar{\theta} = V/Q$, or L/u. Hence, Eq. (6-31) can be written so that C/C_0 is a function of D_L/uL and $\theta/\bar{\theta}$,

$$\left(\frac{C}{C_0}\right)_{\text{step}} = \frac{1}{2}\left[1 - \text{erf}\left(\frac{1}{2}\sqrt{\frac{uL}{D_L}}\frac{1 - \theta/\bar{\theta}}{\sqrt{\theta/\bar{\theta}}}\right)\right] \qquad (6\text{-}32)$$

Equation (6-32) is plotted in Fig. 6-8 with curves for various values of D_L/uL. When $D_L/uL = 0$ there is no axial diffusion and the reactor fulfills the requirements of plug-flow behavior. The curve for this case in Fig. 6-8 is the expected step-function response for a plug-flow reactor (see Fig. 6-5 or Fig. 6-7). The other extreme, $D_L/uL = \infty$, corresponds to an infinite diffusivity (complete mixing), and stirred-tank performance is obtained. The curve for $D_L/uL = \infty$ in Fig. 6-8 is identical to that described by Eq. (6-12) and shown in Figs. 6-5 and 6-7. The curves between the two extremes allow for intermediate degrees of axial mixing. The use of the dispersion model presupposes that in an actual reactor the mixing state, perhaps consisting of some segregated flow and some axial mixing, can be represented by a particular value of D_L/uL.

In Example 6-1 D_L/uL is evaluated with the response curve for the laminar-flow reactor described in Sec. 6-4 (Fig. 6-7). Example 6-2 treats the same problem, but with other response data.

Fig. 6-8   Response curves for dispersion model

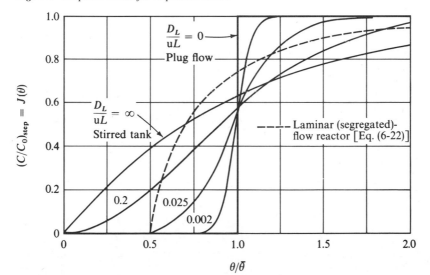

Example 6-1 Determine how well the dispersion model fits the segregated-flow (laminar-flow) reactor of Sec. 6-4.

Solution Equation (6-22) gives the RTD, or $(C/C_0)_{step}$, for the laminar-flow tubular reactor. This equation, shown as a dashed line in Fig. 6-8, is in marked contrast to the dispersion curves. It is evident that introducing axial diffusion cannot account for the RTD given by segregated flow. The shape of the dashed curve is so different from those for the dispersion model that a value of D_L/uL cannot be chosen that will be even approximately correct. The effect on conversion of using a model which does not predict the correct RTD is illustrated in Sec. 6-7.

Example 6-2 Experimental response measurements on a continuous-flow tubular reactor give the following RTD:

$\theta/\bar{\theta}$	0	0.5	0.70	0.875	1.0	1.5	2.0	2.5	3.0
$J(\theta)$	0	0.10	0.22	0.40	0.57	0.84	0.94	0.98	0.99

Determine how well the dispersion model can be fitted to the RTD for this reactor and evaluate an appropriate D_L/uL.

Solution Figure 6-9 shows the given RTD data and curves for two values of D_L/uL as computed from Eq. (6-32). In contrast to Example 6-1, the dispersion model fits the given RTD reasonably well. For $D_L/uL = 0.085$ the agreement is good for large $\theta/\bar{\theta}$, while the curve for $D_L/uL = 0.15$ fits well for small $\theta/\bar{\theta}$. An average value of 0.117 will be used in Sec. 6-9 for conversion calculations.

6-6 Interpretation of Response Data by the Series-of-stirred-tanks Model

In the series-of-stirred-tanks model the actual reactor is simulated by n ideal stirred tanks in series. The total volume of the tanks is the same as the volume of the actual reactor. Thus for a given flow rate the total mean residence time is also the same. The mean residence time per tank is $\bar{\theta}_t/n$. Figure 6-10a describes the situation. The objective is to find the value of n for which the response curve of the model would best fit the response curve for the actual reactor. To do this the relation between $(C/C_0)_{step}$ and n should be developed.

In Sec. 6-4 we found the desired result for $n = 1$; it was Eq. (6-12). The same method of writing a differential mass balance can be employed to find $(C/C_0)_{step}$ for any value of n. The result (see Example 6-3) is the series

$$(C_n/C_0)_{step} = J_n(\theta) = 1 - e^{-n\theta/\bar{\theta}_t} \left[1 + \frac{n\theta}{\bar{\theta}_t} + \frac{1}{2!} \left(\frac{n\theta}{\bar{\theta}_t} \right)^2 \right.$$

$$\left. + \cdots + \frac{1}{(n-1)!} \left(\frac{n\theta}{\bar{\theta}_t} \right)^{n-1} \right] \quad (6-33)$$

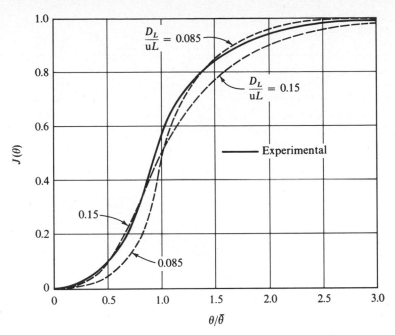

Fig. 6-9 *Fit of dispersion model to RTD data (Example 6-2)*

The number of terms in brackets depends on n, the last term being the one with a power of $n\theta/\bar{\theta}_t$ equal to $n - 1$. For example, if $n = 1$, the last term would be 1, or the first term in brackets. This result is the same as Eq. (6-12).

Figure 6-10b is a plot of Eq. (6-33) for various values of n. The similarity between Figs. 6-8 and 6-10b indicates that the axial-dispersion and series-of-stirred-tanks models give the same general shape of response curve. The analogy is exact for $n = 1$, for this curve in Fig. 6-10b agrees exactly with that in Fig. 6-8 for infinite dispersion, $D_L/uL = \infty$; both represent the behavior of an ideal stirred-tank reactor. Agreement is exact also at the other extreme, the plug-flow reactor ($n = \infty$ in Fig. 6-10b and $D_L/uL = 0$ in Fig. 6-8). The shapes of the curves for the two models are more nearly the same the larger the value of n.

The series-of-stirred-tanks model could not represent the RTD for the laminar-flow reactor shown in Fig. 6-7. However, the RTD data given in Example 6-2 can be simulated approximately. The dashed curve in Fig. 6-10b is a plot of this RTD. While no integer value[1] of n coincides with this curve for all $\theta/\bar{\theta}$, the curve for $n = 5$ gives approximately the correct shape. Comparison of the fit in Figs. 6-9 and 6-10b indicates that about the same

---

[1] B. A. Buffham and L. G. Gibilaro [*AIChE J.*, **14**, 805 (1968)] have shown how the stirred-tanks-in-series model, with noninteger values of n, can be used to fit RTD data.

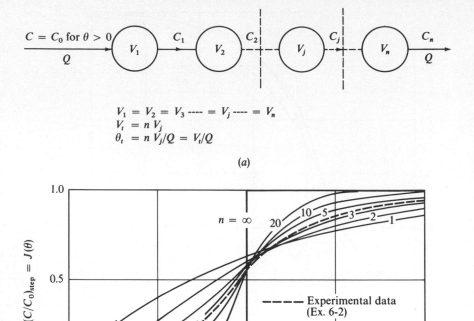

$$V_1 = V_2 = V_3 \cdots = V_j \cdots = V_n$$
$$V_t = n V_j$$
$$\theta_t = n V_j/Q = V_t/Q$$

(a)

(b)

Fig. 6-10   (a) *Series of ideal stirred-tank reactors; (b) response curves for series of stirred tanks*

deviation exists between experimental and predicted RTD curves for either model.

Example 6-3   Develop Eq. (6-33) by deriving $(C_n/C_0)_{\text{step}}$ for $n = 2$ and 3 and then generalizing the results.

Solution   Figure 6-10a describes a step-function input of tracer of concentration C_0 for a series of ideal stirred-tank reactors. A mass balance on the j reactor in a series of n is, according to Eq. (3-1),

$$C_{j-1}Q - C_jQ = V_j \frac{dC_j}{d\theta}$$

or

$$\frac{dC_j}{d\theta} + \frac{n}{\bar{\theta}_t} C_j = \frac{n}{\bar{\theta}_t} C_{j-1} \tag{A}$$

since $\bar{\theta}_t = nV_j/Q$. The initial condition is $C_j = 0$ at $\theta = 0$. Integrating Eq. (A) formally with this initial condition gives

$$C_j = \frac{n}{\bar{\theta}_t} e^{-n\theta/\bar{\theta}_t} \int_0^\theta C_{j-1} e^{n\theta/\bar{\theta}_t} \, d\theta \tag{B}$$

Equation (B) may be integrated for successive stages in an n-stage system. For the first stage $C_{j-1} = C_0$, so that

$$C_1 = C_0 \frac{n}{\bar{\theta}_t} e^{-n\theta/\bar{\theta}_t} \int_0^\theta e^{n\theta/\bar{\theta}_t} \, d\theta$$

or

$$\frac{C_1}{C_0} = 1 - e^{-n\theta/\bar{\theta}_t} \tag{C}$$

When there is just one stage ($n = 1$) this becomes the familiar result for a single tank. In general Eq. (C) gives the response in the effluent from the first stage of an n-stage system.

Continuing to the second stage, we note that $C_{j-1} = C_1$ and that this is given by Eq. (C). Hence Eq. (B) becomes

$$C_2 = \frac{n}{\bar{\theta}_t} e^{-n\theta/\bar{\theta}_t} \int_0^\theta C_0(1 - e^{-n\theta/\bar{\theta}_t}) e^{n\theta/\bar{\theta}_t} \, d\theta \tag{D}$$

Integrating gives the response for the effluent from the second stage of an n-stage unit,

$$\frac{C_2}{C_0} = 1 - e^{-n\theta/\bar{\theta}_t} \left(1 + \frac{n\theta}{\bar{\theta}_t}\right) \tag{E}$$

For the third stage ($j = 3$) C_{j-1} in Eq. (B) becomes C_2, as given by Eq. (E). Proceeding as before, we find that the response in the effluent from the third stage is

$$\frac{C_3}{C_0} = 1 - e^{-n\theta/\bar{\theta}_t} \left[1 + \frac{n\theta}{\bar{\theta}_t} + \frac{1}{2}\left(\frac{n\theta}{\bar{\theta}_t}\right)^2\right] \tag{F}$$

By analogy with Eqs. (C), (E), and (F), the response from the fourth stage will be

$$\frac{C_4}{C_0} = 1 - e^{-n\theta/\bar{\theta}_t} \left[1 + \frac{n\theta}{\bar{\theta}_t} + \frac{1}{2}\left(\frac{n\theta}{\bar{\theta}_t}\right)^2 + \frac{1}{3!}\left(\frac{n\theta}{\bar{\theta}_t}\right)^3\right] \tag{G}$$

Hence the response from the nth stage of an n-stage system is described by Eq. (6-33).

6-7 Conversion in Nonideal Reactors

The maximum effect of the RTD on conversion is evident from the comparison of tubular-flow and stirred-tank reactors. Figure 4-14 shows such a comparison for first-order kinetics, and it is apparent that the differences are sizable at high conversion levels. For example, when $k_1\bar{\theta} = 4.0$ the conversion in the stirred-tank reactor is 80%, while in the tubular-flow unit it is 98%. The differences would be larger for second-order kinetics and smaller for half-order kinetics. These differences are for the extremes of RTD described by the two ideal reactors. For a reactor which followed neither ideal form, but showed an intermediate RTD, the conversion would be between the two extremes.

The other effect associated with nonideal flow depends on the extent of micromixing. As mentioned in Sec. 6-1, this error does not exist for first-order kinetics. This can be shown in a simple way. Consider two elements of fluid, of equal volume, moving through a reactor. A single reaction occurs for which the rate is given by kC^α. Suppose the reactant concentration is C_1 in one element and C_2 in the other. Two extremes of micromixing may be imagined. In one, the elements are adjacent to each other and mix completely, giving a concentration $(C_1 + C_2)/2$. Afterward these elements flow on through the reactor in a completely mixed state. In the other case the two elements do not mix at all, but flow through the reactor with the same residence time, retaining their individual concentrations. The rate of reaction for complete micromixing is, per unit volume

$$\mathbf{r}_m = k\left(\frac{C_1 + C_2}{2}\right)^\alpha$$

For no micromixing (segregated flow) it is

$$\mathbf{r}_s = \tfrac{1}{2}(kC_1{}^\alpha + kC_2{}^\alpha) \tag{6-34}$$

For the same residence time the ratio of the rates is equal to the ratio of the conversions. Thus

$$\frac{x_m}{x_s} = \frac{\mathbf{r}_m}{\mathbf{r}_s} = \frac{[(C_1 + C_2)/2]^\alpha}{(C_1{}^\alpha + C_2{}^\alpha)/2} \tag{6-35}$$

Evaluation of Eq. (6-35) for different values of α shows that

$$\frac{x_m}{x_s} = 1 \qquad \text{for } \alpha = 1 \text{ (first order)} \tag{6-36}$$

$$\frac{x_m}{x_s} > 1 \qquad \text{for } \alpha < 1 \tag{6-37}$$

$$\frac{x_m}{x_s} < 1 \qquad \text{for } \alpha > 1 \tag{6-38}$$

Thus for first-order kinetics the conversion will be the same regardless of the extent of micromixing. If the kinetics are second order or larger, then $x_m < x_s$, so that micromixing reduces the conversion. To obtain the maximum conversion for a given residence time for second-order kinetics a reactor with segregated flow is desirable. In contrast, for half-order reactions, a reactor with micromixing would give the higher conversion. The effect of micromixing on the conversion for non-first-order kinetics is generally less than the effect of the RTD, although exceptions can occur. Also, it should be remembered that unaccounted-for temperature differences in reactors can affect the isothermal conversion much more than deviations due either to the RTD or to micromixing. The magnitudes of these latter effects for representative cases are illustrated in later examples.

6-8   Conversion According to the Segregated-flow Model

In segregated flow the conversion in each element is determined by the conversion-vs-time relationship $x(\theta)$ as the element moves through the reactor. This relationship is given by the batch-reactor equations of Chap. 2 (see, for example, Table 2-5) according to the appropriate kinetics. Since the conversion in each element depends on its residence time, to obtain the average conversion we must divide the effluent into elements according to their residence time, that is, according to the RTD. The conversion in element i is $x(\theta_i)\, dJ(\theta_i)$ or $x(\theta_i)J'(\theta_i)\, d\theta$. Therefore the average conversion in the effluent from the reactor will be

$$\bar{x} = \int_0^1 x(\theta)\, dJ(\theta) = \int_0^\infty x(\theta)J'(\theta)\, d\theta \tag{6-39}$$

To illustrate the use of Eq. (6-39) suppose we use the RTD for a stirred-tank reactor [Eq. (6-12)] but assume segregated flow. We wish to calculate the conversion for an irreversible first-order reaction. The function $x(\theta)$ is, from the second entry of Table 2-5,

$$x(\theta) = 1 - \frac{C}{C_0} = 1 - e^{-k_1\theta}$$

Using this function in Eq. (6-39), the mean conversion is

$$\bar{x} = \int_0^\infty (1 - e^{-k_1\theta})\frac{1}{\bar{\theta}} e^{-\theta/\bar{\theta}}\, d\theta = \frac{k_1\bar{\theta}}{1 + k_1\bar{\theta}} \tag{6-40}$$

In contrast, suppose complete micromixing is assumed. Then, for the same RTD, an ideal stirred-tank reactor results. The conversion for a first-order reaction in this case is given by Eq. (4-7), which is identical to the above expression for segregated flow. This verifies the conclusion of Eq. (6-36) that the extent of micromixing does not affect conversion for first-order kinetics (as long as the correct RTD is used). The same development has been carried out for half-order and second-order kinetics,[1] and the results verify Eqs. (6-37) and (6-38). For example, the conversion was about 78% for a second-order reaction in segregated flow and 72% for complete micromixing, both based on the RTD of an ideal stirred-tank reactor. These figures are for $k_2 C_0 \bar{\theta} = 10$, a value that gives about the maximum effect of micromixing. For half-order reactions the maximum effect is about the same, but reversed. These results give an idea of the magnitude of deviations due to micromixing.

Example 6-4 Consider the laminar-flow reactor described in Sec. 6-4 and calculate the conversion for a first-order reaction for which $k_1 = 0.1 \text{ sec}^{-1}$ and $\bar{\theta} = 10 \text{ sec}$.

Solution   The RTD for this reactor is given by Eq. (6-20) or Eq. (6-22). The flow is segregated, so that Eq. (6-39) is valid for calculating the conversion. Since $x(\theta)$ for a first-order reaction is

$$x(\theta) = 1 - e^{-k_1\theta}$$

Eq. (6-39) becomes

$$\bar{x} = \int_{\frac{1}{2}\bar{\theta}}^{\infty} (1 - e^{-k_1\theta}) \, \frac{1}{2} \frac{\bar{\theta}^2}{\theta^3} \, d\theta$$

Note that the lower limit for residence time is not zero for this reactor, but is instead $\frac{1}{2}\bar{\theta} = 5 \text{ sec}$. Inserting numerical values and integrating numerically, we obtain

$$\bar{x} = 50 \int_{5 \text{ sec}}^{\infty} (1 - e^{-0.1\theta}) \frac{d\theta}{\theta^3}$$

$$= 0.52$$

For the same $\bar{\theta}$ and k_1 the plug-flow result [Eq. (4-8)] is

$$\bar{x} = 1 - e^{k_1\bar{\theta}} = 0.63$$

and the stirred-tank result [Eq. (4-7)] is

$$\bar{x} = \frac{k_1\bar{\theta}}{1 + k_1\bar{\theta}} = 0.50$$

For calculating conversion in this laminar-flow reactor the RTD for a stirred-tank reactor is more appropriate than that for a plug-flow reactor. This is not apparent from a comparison of the three RTDs shown in Fig. 6-7.

Example 6-5 Calculate the conversion for a first-order reaction ($k = 0.1 \text{ sec}^{-1}$, $\bar{\theta} = 10 \text{ sec}$), using the RTD given in Example 6-2 and assuming the flow to be segregated.

Solution   Equation (6-39) for this application is

---

[1]H. Kramers and K. R. Westerterp, "Elements of Chemical Reactor Design and Operation," p. 87, Academic Press, Inc., New York, 1963; K. G. Denbigh, *J. Appl. Chem.*, **1**, 227 (1951).

$$\bar{x} = \int_0^1 (1 - e^{-0.1\theta})\, dJ(\theta)$$

where $J(\theta)$ is given by the data in Example 6-2. A plot of $1 - e^{-0.1\theta}$ vs $J(\theta)$ is shown in Fig. 6-11. According to the above equation for $\bar{x}$, the area under the curve, from $J(\theta) = 0$ to $J(\theta) = 1$, is the conversion. Evaluation of this area gives $\bar{x} = 0.61$. This is much closer to the plug-flow result of 0.63 than to the stirred-tank result of 0.50. The RTD for this case is shown in Fig. 6-10b. This figure shows that not one, but five stirred tanks in series would best fit the RTD data. Thus the RTD comparison confirms that the single stirred tank would not provide a close simulation of the actual reactor. From this and the preceding example we have a good idea of how the RTD effects the conversion for first-order kinetics. The results are summarized in Table 6-1

Table 6-1   *Conversion vs RTD for first-order kinetics ($k = 0.1$ sec^{-1}, $\bar{\theta} = 10$ sec)*

Type of reactor	Conversion	RTD
Plug-flow reactor	0.63	$\theta = 10$ sec, Fig. 6-7, dashed line
Actual tubular reactor	0.61	Fig. 6-10b
Tubular reactor, laminar flow	0.52	Fig. 6-7, solid curve
Stirred-tank reactor	0.50	Fig. 6-7, dashed curve

Fig. 6-11   *Graph for conversion calculation according to segregated-flow model*

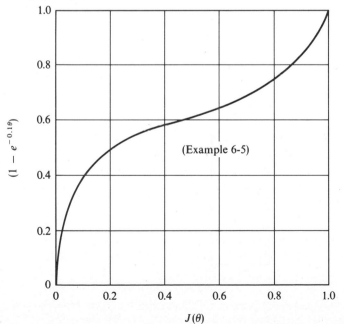

(Example 6-5)

6-9   Conversion According to the Dispersion Model

In Sec. 6-5 a non-steady-state mass balance for a tubular-flow reactor (plug flow except for axial dispersion) was used to evaluate an effective diffusivity. Now we consider the problem of calculating the conversion when a reaction occurs in a dispersion-model reactor operated at steady-state conditions. Again a mass balance is written, this time for steady state and including reaction and axial-dispersion terms. It is considered now that the axial diffusivity is known.

Figure 6-12 shows a section Δz of the reactor with upward flow at a uniform velocity u. From Eq. (3-1), a mass balance around this section will be

$$A\left(-D_L\frac{dC}{dz} + uC\right)_z - A\left(-D_L\frac{dC}{dz} + uC\right)_{z+\Delta z} - \mathbf{r}A\,\Delta z = 0$$

where A is the cross-sectional area and $\mathbf{r}$ is the rate of reaction. Dividing by Δz and taking the limit as $\Delta z \to 0$ gives

$$D_L\frac{d^2C}{dz^2} - u\frac{dC}{dz} - \mathbf{r} = 0 \tag{6-41}$$

Note the similarities and differences between this steady-state equation and the transient one, Eq. (6-23). The boundary conditions for the two expressions are the same, that is, Eqs. (6-25) and (6-26). For first-order kinetics the solution is straightforward. Introducing $\mathbf{r} = k_1C$ and using a dimensionless concentration $C^* = C/C_0$ and reactor length $z^* = z/L$, we obtain for the differential equation and boundary conditions

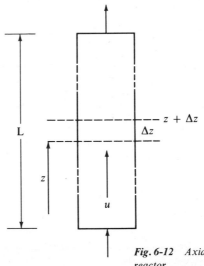

Fig. 6-12 *Axial section of tubular-flow reactor*

$$\frac{D_L}{uL}\frac{d^2C^*}{dz^{*2}} - \frac{dC^*}{dz^*} - k_1\bar{\theta}C^* = 0 \tag{6-42}$$

$$C^* - \frac{D_L}{uL}\frac{dC^*}{dz^*} = 1 \qquad \text{at } z^* = 0 \tag{6-43}$$

$$\frac{dC^*}{dz^*} = 0 \qquad \text{at } z^* = 1 \tag{6-44}$$

where $\bar{\theta} = L/u$. The solution of Eqs. (6-42) to (6-44) gives C as a function of z. Evaluating this function at the end of the reactor, $z = L$, we find

$$\frac{C}{C_0} = 1 - x = \frac{4\beta}{(1 + \beta)^2 \, e^{-\frac{1}{2}(uL/D_L)(1-\beta)} - (1 - \beta)^2 \, e^{-\frac{1}{2}(uL/D_L)(1+\beta)}} \tag{6-45}$$

where

$$\beta = \left(1 + 4k_1\bar{\theta}\frac{D_L}{uL}\right)^{\frac{1}{2}} \tag{6-46}$$

If D_L/uL, and of course $k\bar{\theta}$, are known, Eq. (6-45) gives the conversion predicted by the dispersion model, provided the reaction is first order. For most other kinetics numerical solution of the differential equation is necessary.[1]

As mentioned earlier, the dispersion model, like other models, is subject to two errors: inadequate representation of the RTD and improper allowance for the extent of micromixing. We can evaluate the first error for a specific case by using the dispersion model to obtain an alternate solution for Example 6-5. The second error does not exist for first-order kinetics. However, the maximum value of this error for second- and half-order kinetics was indicated in Sec. 6-8.

Example 6-6   In Example 6-2, $D_L/uL = 0.117$ was found to provide the best fit of the dispersion model to the reported RTD. Use this result to evaluate the conversion for a first-order reaction, with $k_1 = 0.1 \text{ sec}^{-1}$ and $\bar{\theta} = 10 \text{ sec}$.

Solution   From Eq. (6-46),

$$\beta = [1 + 4(0.1)(10)(0.117)]^{\frac{1}{2}} = 1.21$$

Then, from Eq. (6-45),

$$1 - x = \frac{4(1.21)}{4.88e^{0.90} - 0.0445e^{-9.45}} = 0.40$$

[1] A. Burghardt and T. Zaleski [*Chem. Eng. Sci.,* **23**, 575 (1968)] have reviewed solutions of Eq. (6-41) for various forms of the rate equation.

or

$$x = 0.60$$

This result differs little from the value $x = 0.61$ obtained with the actual RTD (Example 6-5). We may conclude that the fit of the dispersion model to the actual RTD is adequate for conversion calculations.

6-10 Conversion According to the Series-of-stirred-tanks Model

The mass balance of reacting component for the first of a series of ideal stirred-tank reactors (see Secs. 4-6 and 4-7) is

$$C_0 - k\bar{\theta}_i C_1{}^\alpha = C_1 \tag{6-47}$$

where $\bar{\theta}_i$ = mean residence time (V/Q) in the first reactor

$\quad C_1$ = concentration of reactant in stream leaving the first reactor

$\quad \alpha$ = order of rate equation

If additional reactors with the same mean residence time are in series with the first one, the balance for the second reactor would be

$$C_1 - k\bar{\theta}_i C_2{}^\alpha = C_2 \tag{6-48}$$

and for the last reactor

$$C_{n-1} - k\bar{\theta}_i C_n{}^\alpha = C_n \tag{6-49}$$

Regardless of the value of α, the series of mass balances can be solved numerically for C_n and the conversion, provided $k\bar{\theta}_i$ and n are known. To use the stirred-tank model, n is found from response data, as illustrated in Sec. 6-6.

For first-order kinetics an analytical solution of Eqs. (6-47) to (6-49) is easily obtained; in fact, Eq. (4-11) gives the solution in terms of the total residence time $(\bar{\theta}_t = \bar{\theta}_i n)$:

$$x = 1 - \frac{1}{(1 + k\bar{\theta}_t/n)^n} \tag{4-11}$$

This model of reactor behavior is illustrated in Example 6-7 by predicting the conversion from Eq. (4-11) for the same conditions as used in Examples 6-5 and 6-6.

Example 6-7   Using the series-of-stirred-tanks model to simulate the RTD of the reactor described in Example 6-2, predict the conversion for a first-order reaction for which $k = 0.1 \text{ sec}^{-1}$ and $\bar{\theta} = 10 \text{ sec}$.

Solution   In Sec. 6-6 we found that five stirred tanks in series fit the RTD of the reactor. With $n = 5$, the conversion from Eq. (4-11) is

$$x = 1 - \frac{1}{[1 + 0.1(10)/5]^5} = 1 - \frac{1}{(1.2)^5} = 1 - 0.40 = 0.60$$

This result is the same as that obtained with the dispersion model. Also, it deviates little from the result $x = 0.61$ obtained with the actual RTD (Example 6-5). In this case the estimated RTD obtained with either the dispersion or tanks-in-series model is a satisfactory representation for conversion calculations.

The results of Examples 6-6 and 6-7 are for one case, but they are representative of the situation for many reactors. We saw in Secs. 6-7 and 6-8 that extremes of RTD can have large effects on the conversion, particularly at high conversion levels. However, with relatively simple models to estimate the RTD, little error need be involved. Put differently, if an engineer were to use an ideal stirred-tank reactor to simulate a nearly ideal tubular-flow unit, the predicted conversion would be seriously in error. However, if the measured RTD or a reasonable model were employed, the result would be approximately correct. The residual error will be due to uncertainty in the extent of micromixing.

In practice an extreme RTD is likely to be encountered in reactors of specialized design for very viscous fluids. For example, consider a stirred-tank reactor for polymerizing a viscous fluid, with two concentric cooling coils, as shown in Figure 6-13a. In such a situation the fluid between the coils, or between the coils and the reactor wall, may be nearly stagnant, while the fluid in the central section is well stirred. Here neither stirred-tanks-in-series nor dispersion models may fit the actual RTD. However, if response data are obtained, it is usually possible to develop a special model to fit the RTD. Thus a combination plug-flow and ideal stirred-tank reactor, as in Fig. 6-13b, might simulate the actual reactor. The flow rates to each reactor and their volumes provide adjustable parameters which can be chosen to fit the response data. The conversion in the exit stream from the simulation unit may be calculated by combining the results for each reactor as obtained from Eqs. (4-7) and (4-8). If this model is inadequate, additional plug-flow and stirred-tank units, of adjustable volumes, may be added.

Although the influence of nonideal flow on conversion may be only a small percentage, if the conversion level is high, the effect on subsequent separation units may be significant. Suppose that the conversion by an approximate model is 99% but proper accounting for nonideal flow would give 95%. Suppose also that the product must have a purity of 99.5%. The separation systems following the reactor would be considerably different in the two cases. One system would be designed to reduce the impurity from 1.0 to 0.5%, while the other would be for a reduction from 5.0 to 0.5%. Of course, errors in conversion estimates from any source, such as temperature variations, would have the same effect.

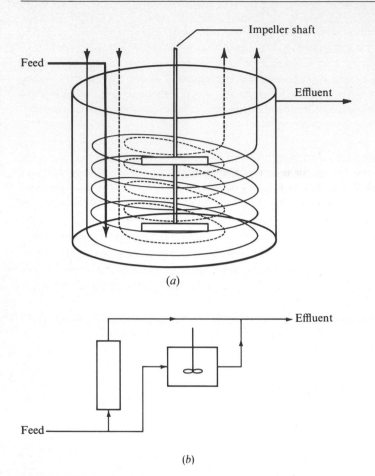

Fig. 6-13 *Polymerization tank reactor and simulation model: (a) polymerization reactor, with two concentric cooling coils, (b) reactor model of parallel plug-flow and stirred-tank units*

Denbigh[1] has provided useful guidelines for deciding when deviations (in conversion) from ideal tubular-flow performance are significant. In laminar flow, molecular diffusion in the axial direction causes little deviation if the reactor is reasonably long with respect to its diameter. Molecular diffusion in the radial direction may be important, particularly for gases, but it serves to offset the deviation from ideal performance caused by the velocity distribution. That is, radial diffusion tends to make the reactor

[1]K. G. Denbigh, "Chemical Reactor Theory," pp. 50–51, Cambridge University Press, New York, 1965.

behave more like an ideal tubular-flow unit. For turbulent flow, axial diffusion (which now includes eddy diffusion) can be significant. Its importance may be neglected if the Reynolds number is greater than 10^4 and the length is at least 50 times the diameter of the reactor. In turbulent flow, as in laminar flow, radial diffusion tends to improve the accuracy of the plug-flow model.

Problems

6-1. The RTD for a reactor is to be determined by tracer-concentration measurements. With a constant-density isothermal system, the following effluent concentration data are obtained in response to a pulse of tracer added to the feed:

t, min	0	5	10	15	20	25	30	35
Tracer concentration, g/cm³	0	3.0	5.0	5.0	4.0	2.0	1.0	0

(a) Plot both $J(\theta)$ and $J'(\theta)$ vs time on the same figure. (b) What is the mean residence time for the flow rate used?

6-2. Response measurements to a step-function input are made for a reaction vessel. The following data are obtained for a given volumetric flow rate:

t, sec	0	15	25	35	45	55	65	75	95
Tracer concentration, g/cm³	0	0.5	1.0	2.0	4.0	5.5	6.5	7.0	7.7

(a) Plot the RTD expressed as $J(\theta)$ vs time. (b) On the same graph plot $J'(\theta)$ vs time. (c) What is $\bar{\theta}$ for this flow rate?

6-3. For the RTD of Prob. 6-1 determine (a) the value of D_L/uL which gives the most accurate fit of the dispersion model to the data and (b) the most accurate integer value of n if the stirred-tanks-in-series model is used.

6-4. Repeat Prob. 6-3 for the RTD determined for Prob. 6-2.

6-5. Determine the RTD for an isothermal tubular-flow reactor in which the liquid is in laminar flow in an annulus of inner radius r_1 and outer radius $r_2 (r_1/r_2 = \alpha)$. Neglect molecular diffusion. The velocity in the axial direction at any radius r (between r_1 and r_2) is given by

$$u(r) = A\left[1 - \left(\frac{r}{r_2}\right)^2 + \frac{1 - \alpha^2}{\ln(1/\alpha)}\ln\frac{r}{r_2}\right]$$

Express the RTD, as $J(\theta)$, in terms of the mean residence time $\bar{\theta}$.

6-6. An ideal stirred-tank reactor followed by a plug-flow reactor is proposed as a model for the RTD of the reactor system in Prob. 6-1. The volumetric flow rate and combined volume of the two reactors will be the same as in Prob. 6-1. What ratio of the volume of the plug-flow and stirred-tank vessels would best represent the RTD? Comment on the suitability of the model.

6-7.   Calculate the conversion in a laminar-flow tubular reactor for a second-order reaction $A + B \rightarrow C$ for which $k_2 = 100 \text{ cm}^3/(\text{g mole})(\text{sec})$ and $\bar{\theta} = 10$ sec. The feed concentration of both reactants is the same, 10^{-3} g mole/cm^3. Neglect molecular diffusion, so that the flow is segregated.

6-8.   Calculate the conversion for the laminar-flow reactor of Prob. 6-7, using the dispersion model to represent the actual RTD.

6-9.   Calculate the conversion for the laminar-flow reactor of Prob. 6-7, using the stirred-tanks-in-series model to represent the RTD.

6-10. Reconsider the system in Prob. 6-7, with all conditions the same, except that volumes of different reactors required to obtain the same conversions (for the same flow rate) will be compared. Calculate the ratio of the volumes required for a laminar-flow reactor and a plug-flow reactor for several conversion levels between 0 and 100%. Do the results depend on the feed concentrations and the rate constant?

6-11. Repeat Prob. 6-10 for a first-order reaction for which $k_1 = 0.1 \text{ sec}^{-1}$.

6-12. The flow characteristics of a continuous reactor are studied by suddenly introducing a quantity of miscible tracer into the feed stream. The concentrations of tracer in the effluent at various times after the instant of addition are:

t, min	0.1	0.2	1.0	2.0	5.0	10.	30
Tracer concentration, mg/liter	0.20	0.17	0.15	0.125	0.07	0.02	0.001

(a) What type of ideal reactor does the actual vessel most closely approach? (b) If an isothermal first-order reaction ($k_1 = 0.15 \text{ min}^{-1}$) occurs in the vessel, what conversion may be expected?

6-13. Another first-order reaction is to be studied at the same flow rate in the vessel characterized in Prob. 6-12. Measurements for this reaction in a plug-flow reactor, operating at the same mean residence time, gave a conversion of 75.2%. (a) Calculate the value of the rate constant for this reaction. (b) Calculate the conversion expected for this reaction in the reactor of Prob. 6-12. (c) Are any assumptions made in obtaining the answer to part (b)? If the reaction were second order, would additional assumptions be necessary to use the same procedure for part (b)?

7

HETEROGENEOUS REACTORS

The kinetics and reactor designs we have considered thus far are for homogeneous reactions. The remainder of the book is devoted to heterogeneous reactions. Many systems are heterogeneous because a catalyst is necessary, and this substance is commonly (but not always) in a phase different from that of the reactants and products. Accordingly, our first objective will be a study (Chaps. 8 and 9) of heterogeneous catalysis and kinetics of heterogeneous catalytic reactions.

The fact that phase boundaries are inherent in heterogeneous systems introduces the need to deal with physical processes (mass and energy transfer) between the bulk fluid and the catalyst. Thus physical processes affect reactor design in a more intrinsic way for heterogeneous reactions than for homogeneous ones. It is common practice to write mass and energy balances for heterogeneous and homogeneous systems in the same way; e.g., as given in Chap. 3. When this is done, the rate equation used for heterogeneous systems will include the effects of physical processes.

To illustrate, in Eq. (3-1) the rate is expressed per unit volume. If the volume element includes a heterogeneous mixture of reaction fluid and solid catalyst particles, the proper rate to use in Eq. (3-1) will include the effects of mass- and energy-transfer processes from fluid to solid surface and within the solid particle. Such rates are sometimes called *global*, or *overall*, *rates* of reaction. The advantage of using the global rate is that the design equations for heterogeneous systems are identical to those for homogeneous systems, as presented in Chaps. 3 to 5. However, expressions for the global reaction rate must be formulated in terms of properties of the bulk fluid. This is accomplished by writing expressions for the rate of each step in the overall process. The sequence of steps for converting reactants to products is as follows: *global steps*

1. Transport of reactants from the bulk fluid to the fluid-solid interface (external surface of catalyst particle)
2. Intraparticle transport of reactants into the catalyst particle (if it is porous)
3. Adsorption of reactants at interior sites of the catalyst particle
4. Chemical reaction of adsorbed reactants to adsorbed products (surface reaction)
5. Desorption of adsorbed products
6. Transport of products from the interior sites to the outer surface of the catalyst particle
7. Transport of products from the fluid-solid interface into the bulk-fluid stream

At steady state the rates of the individual steps will be identical. This equality can be used to develop a global rate equation in terms of the concentrations and temperatures of the bulk fluid. The derivation of such equations will be considered in detail in Chaps. 10 to 12, but a very simple treatment is given next to illustrate the nature of the problem.

7-1 Global Rates of Reaction

Consider an irreversible gas-phase reaction

$$A(g) \rightarrow B(g)$$

which requires a solid catalyst C. Suppose that the temperature is constant and that the reaction is carried out by passing the gas over a bed of *non-porous* particles of C. Since the catalyst is nonporous, steps 2 and 6 are not involved. The problem is to formulate the rate of reaction per unit volume[1]

---

[1] As defined, the volume element on which r_v is based must include at least one catalyst particle; otherwise the interphase physical effects cannot be considered. Yet this rate is used in Eqs. (3-1) and (3-2) as a rate applicable to a differential volume, i.e., as a point rate. This is an approximation which arises from treating the discrete nature of a bed of catalyst particles

of the bed—that is, the *global rate for a catalyst particle*, $\mathbf{r}_v$—in terms of the temperature and concentration of A in the bulk-gas stream. Note that these are the quantities that are measurable or can be specified (as design requirements), rather than the temperature and concentration at the gas-particle interface. Global rates of catalytic reactions are usually expressed per unit mass of catalyst, i.e., as $\mathbf{r}_p$. These are easily converted to rates per unit volume by multiplying them by the bulk density ρ of the bed of catalyst particles.

The overall conversion of A to B in the bulk gas occurs according to steps 1, 3 to 5, and 7 in series. Let us further simplify the problem by supposing that steps 3 to 5 may be represented by a single first-order rate equation. Then the overall reaction process may be described in three steps: gas A is transported from the bulk gas to the solid surface, the reaction occurs at the interface, and finally, product B is transported from the catalyst surface to the bulk gas. Since the reaction is irreversible, the concentration of B at the catalyst surface does not influence the rate. This means that $\mathbf{r}_p$ can be formulated by considering only the first two steps involving A. Since the rates of these two steps will be the same at steady state, the disappearance of A can be expressed in two ways: either as the rate of transport of A to the catalyst surface,

$$\mathbf{r}_p = k_m a_m (C_b - C_s) \tag{7-1}$$

or as the rate of reaction at the catalyst surface,

$$\mathbf{r}_p = k a_m C_s \tag{7-2}$$

In Eq. (7-1) k_m is the usual mass-transfer coefficient based on a unit of transfer surface, i.e., a unit of external area of the catalyst particle. In order to express the rate per unit mass[1] of catalyst, we multiply k_m by the external area per unit mass, a_m. In Eq. (7-2) k is the *reaction-rate constant* per unit surface. Since a positive concentration difference between bulk gas and solid surface is necessary to transport A to the catalyst, the surface concentration C_s will be less than the bulk-gas concentration C_b. Hence Eq. (7-2) shows that the rate is less than it would be for $C_s = C_b$. Here the effect of the mass-transfer resistance is to reduce the rate. Figure 7-1 shows schematically how the concentration varies between bulk gas and catalyst surface.

---

as a continuum. It is a necessary approximation because it is not yet possible to take into account variations in heat- and mass-transfer coefficients with position on the surface of a single catalyst particle. Thus average values of these coefficients over the particle surface are employed in formulating global rates. This is the value denoted by k_m in Eq. (7-1).

[1] Instead of rate per unit mass of catalyst, Eqs. (7-1) and (7-2) could be expressed as rates per unit external surface, in which case a_m would not be needed in Eq. (7-1). Alternately, a rate per particle could be used. We shall generally use the rate per unit mass.

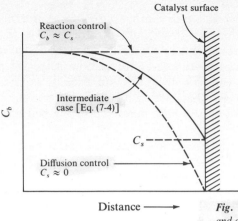

Fig. 7-1 Mass transfer between fluid and catalyst surface

The global rate can be expressed in terms of C_b by first solving for C_s from Eqs. (7-1) and (7-2); thus

$$C_s = \frac{k_m}{k_m + k} C_b \qquad (7\text{-}3)$$

Then this result is substituted in either Eq. (7-1) or Eq. (7-2) to give

$$\mathbf{r}_p = \frac{k k_m a_m}{k_m + k} C_b = \frac{a_m}{1/k + 1/k_m} C_b \qquad (7\text{-}4)$$

This is the expression for the global rate in terms of the bulk-reactant concentration. The concentration profile in this case is shown by the solid line in Fig. 7-1. It is a very restricted illustration of a global rate, since heat-transfer resistances were not considered (constant temperature was assumed) and only external mass transfer is involved (the catalyst particle is non-porous). These restrictions are removed in the detailed treatment in Chaps. 10 and 11, but this simple example illustrates the meaning of global rates of reaction for heterogeneous systems.

A hypothetical reaction has been used to develop the previous results. It is of interest to know the magnitude of the global rate for real situations. Fortunately, considerable experimental data are available. Measurements[1] for the oxidation of SO_2 with air on a platinum catalyst, deposited on 1/8 × 1/8-in. cylindrical pellets,[2] gave a global rate of 0.0956 g mole/(hr)(g catalyst). The bulk temperature of the gas was 465°C and the gas velocity in the catalyst bed was 350 lb/(hr) (ft²). The bulk composition corresponded to a 10% con-

[1]R. W. Olson, R. W. Schuler, and J. M. Smith, *Chem. Eng. Progr.*, **42**, 614 (1950).

[2]The particles were of porous alumina, but the platinum was deposited on the outer surface, so that, in effect, a nonporous catalyst was used.

version of a 6.5 mole % SO_2–93.5% air mixture. At these conditions both mass- and heat-transfer resistances between the bulk gas and the surface of the catalyst pellets were important. Calculations with Eq. (7-1) indicated that the partial pressure of SO_2 dropped to 0.040 atm at the solid surface from the value of 0.06 atm in the bulk gas. This relatively large difference means that the diffusion resistance was large. If the rate were evaluated without consideration of this resistance—that is, if it were evaluated for $p_{SO_2} = 0.06$ atm at 465°C—it would be 0.333 g moles SO_2 reacted/(hr) (g catalyst). At $p_{SO_2} = 0.04$ atm it would be 0.0730. Hence the error in neglecting the mass-transfer resistance would be large; that is, the rate so computed would be (0.333/0.0730) (100), or 350% higher than the global rate.

Actually, the temperature at the catalyst surface is about 15°C above the bulk-gas temperature (see Example 10-2) because of heat-transfer resistance. Hence the reaction at the catalyst surface occurred at 480°C. According to the activation energy for this reaction, a 15°C temperature rise would increase the rate about 31%. Hence neglecting the heat-transfer resistance leads to a rate 31% less than the global value. If both thermal and diffusion resistances were neglected, the rate would be (0.333/0.0956) (100), or 250% higher than the global rate. For this exothermic reaction the diffusion and thermal resistances have opposite effects on the rate. The example is extreme in that mass-transfer resistance was relatively large in comparison with reaction resistance. At lower temperatures and/or higher gas velocities past the catalyst pellet (higher k_m) the mass-transfer effect would be less.

Frequently heat-transfer resistances are much larger than mass-transfer resistances, in contrast to the preceding example. As an illustration, consider the data of Maymo and Smith[1] for the oxidation of hydrogen with oxygen on a platinum-on-alumina catalyst. Rates and temperatures were measured for a single porous catalyst pellet (1.86 cm diameter) suspended in a well-mixed gas containing primarily hydrogen with a small percentage of oxygen and water vapor. Because of the turbulence in the gas, the concentration difference between bulk gas and pellet surface was negligible. However, there was a significant thermal resistance, so that the pellet-surface temperature was greater than the bulk-gas temperature. The pellet was porous with uniform distribution of platinum throughout. Hence there were internal resistances, both to mass and to heat transfer. In one particular run the measured rate of water production was 49.8×10^{-6} mole/(g catalyst) (sec). This is the global value and includes the effects of both internal and external physical resistances. The bulk-gas, pellet-surface, and pellet-center temperatures were 89.9, 101, and 147.7°C, respectively. The rate of the chemical step on the surface was also measured by experiments with fine particles of catalyst for which there were no temperature or concentration differences between bulk

[1]J. A. Maymo and J. M. Smith, *AIChE J.*, **12**, 845 (1966).

gas and surface or interior of particle. This intrinsic rate was correlated by the equation

$$\mathbf{r} = 0.655 \, p_{O_2}^{0.804} \, e^{-5,230/R_g T} \tag{7-5}$$

where p is in atmospheres and T is in degrees Kelvin. We can evaluate the effect of the external resistance in the following way. First calculate the rate at the surface temperature $(101 + 273°K)$ and $p_{O_2} = 0.0527$ atm, which is the oxygen partial pressure, either in the bulk gas or at the pellet surface. Substituting these quantities in Eq. (7-5) gives $\mathbf{r} = 54.3 \times 10^{-6}$ mole/(g catalyst)(sec). If the rate were evaluated at bulk-gas conditions $(89.9°C$ and $p_{O_2} = 0.0527$ atm), the result from Eq. (7-5) would be 43.6×10^{-6} mole/(g catalyst) (sec). Hence neglecting the external-temperature difference would give a rate $[(54.3 - 43.6)/54.3]$ (100), or 20%, less than the correct value.

This example also shows the effects of mass- and energy-transfer resistances within the catalyst pellet. The temperature increases toward the center of the pellet and increases the rate, but the oxygen concentration goes down, tending to reduce the rate. The global value of 49.8×10^{-6} is the resultant balance of both factors. Hence the net error in using the bulk conditions to evaluate the rate would be $[(49.8 - 43.6)/49.8]$ (100), 12.5%. In this case the rate increase due to external and internal thermal effects more than balances the adverse effect of internal mass-transfer resistance. The procedure for calculating the effects of internal gradients on the rate is presented in Chap. 11.

One additional illustration is of interest. This refers to the hydro-fluorination of UO_2 pellets with $HF(g)$ according to the reaction

$$UO_2(s) + 4HF(g) \rightarrow 4UF_4(s) + 2H_2O(g)$$

This is a different kind of heterogeneous reaction—a gas-solid noncatalytic one. Let us examine the process at initial conditions $(t \rightarrow 0)$, so that there has been no opportunity for a layer of $UF_4(s)$ to be formed around the UO_2 pellet. The process is much like that for gas-solid catalytic reactions. Hydrogen fluoride gas is transferred from the bulk gas to the surface of the UO_2 pellets and reacts at the pellet-gas interface, and H_2O diffuses out into the bulk gas. If the pellet is nonporous, all the reaction occurs at the outer surface of the UO_2 pellet, and only an external transport process is possible. Costa[1] studied this system by suspending spherical pellets 2 cm in diameter in a stirred-tank reactor. In one run, at a bulk-gas temperature of 377°C, the surface temperature was 462°C and the observed rate was $-\mathbf{r}_{UO_2} = 6.9 \times 10^{-6}$ mole UO_2/(sec) (cm² reaction surface). At these conditions the concentrations of

---

[1] E. C. Costa and J. M. Smith, *Proc. Fourth European Symp. Chem. Reaction Eng.*, Brussels, September, 1968.

HF gas were 1.12×10^{-5} g mole/cm^3 at the surface of the pellet and 1.38×10^{-5} g mole/cm^3 in the bulk gas. The intrinsic rate of reaction was found to be given by

$$\mathbf{r} = 40\ C_{HF}\ e^{-6,070/R_g T}\ \text{g mole/(cm}^2)\ \text{(sec)} \tag{7-6}$$

where T is in degrees Kelvin and C_{HF} is in gram moles per cubic centimeter. If the rate is evaluated at bulk-gas conditions, Eq. (7-6) gives 5.0×10^{-6} g mole/(sec)(cm^2). Then the combined effects of an external temperature and concentration difference serve to increase the rate from 5.0×10^{-6} to 6.9×10^{-6} g mole/(cm^2)(sec), or 38%. In this case the temperature at the catalyst surface is 95°C higher than the bulk-gas temperature, and this has a dominating effect on the rate of reaction. The effect of external mass-transfer resistance is to reduce the rate by the ratio of concentrations of HF, 1.12:1.38, or 19%.

7-2 Types of Heterogeneous Reactions

All the examples in Sec. 7-1 were of the gas-solid form. This is, perhaps, the most important type of heterogeneous system because of its utilization in the chemical industry. Most salable chemicals are prepared by converting raw materials via chemical reactions. Such reactions usually require catalysts, and these are normally solids. Since the temperatures must be high for rapid rates, the reacting fluid commonly is in the gas phase. Examples of large-scale gas-solid catalytic reactions are the major hydrocarbon transformations: cracking, reforming, dehydrogenation (for example, butadiene and butenes from butane), isomerization, desulfurization, etc. Kinetics and reactor design for such systems have naturally received major attention, and most of our emphasis on design is directed to this type (Chap. 13).

Gas-solid heterogeneous reactions may be noncatalytic. An example is the hydrofluorination of uranium dioxide pellets referred to in Sec. 7-1. Since one reactant is in the solid phase and is consumed, the rate of reaction varies with time. Hence such processes are basically transient, in comparison with the steady-state operation of gas-solid catalytic reactors. The process for smelting ores such as zinc sulfide,

$$ZnS(s) + \tfrac{3}{2}O_2(g) \rightarrow ZnO(s) + SO_2(g)$$

is of this type. The conversion of $CaCO_3$ to CaO in the line kiln is another example. Yet another is the process for making HCl in a transport reactor[1] from salt particles; the reaction is

$$2NaCl(s) + SO_3(g) + H_2O(g) \rightarrow Na_2SO_4(s) + 2HCl(g)$$

[1] A transport reactor is like a fluidized-bed reactor, except that the fluidized particles move through and out of the reactor with the gas phase.

In many noncatalytic types a solid product builds up around the reacting core [for example, $Na_2SO_4(s)$ is deposited around the NaCl particles in the last illustration]. This introduces the additional physical processes of heat and mass transfer through a product layer around the solid reactant. A somewhat different form of noncatalytic gas-solid reaction is the regeneration of catalysts which have been deactivated by the deposition of a substance on the interior surface. The most common is the burning of carbon (with air) which has been gradually deposited on catalyst particles used in hydrocarbon reactions. Many of the physical and chemical steps involved here are the same as those for gas-solid catalytic reactions. The chief difference is the transient nature of the noncatalytic reaction. This type of heterogeneous reaction will be considered in Chap. 14.

Liquid-solid reactions, where the solid phase is a catalyst, is another type of heterogeneous system that is common in the chemical and petroleum industries. Alkylation with $AlCl_3(s)$ catalyst is an example. In these systems the catalyst frequently forms complexes with reactants and/or products and becomes a poorly defined solid-liquid mixture, best described as a sludge. Analytical treatment in these cases is difficult. An important modification of the liquid-solid catalytic type arises when one reactant is gaseous. Hydrogenation of liquids normally is of this type. A slurry is formed of solid catalyst particles and hydrogen is bubbled into the slurry. These reactions are usually carried out in stirred-tank reactors, where the high heat capacity of the tank contents simplifies temperature control of the exothermic hydrogenation reaction. Such gas-liquid-solid systems involve several physical steps, and indeed, resistance to diffusion of dissolved hydrogen to the catalyst particle is frequently significant. The quantitative treatment of combined physical and chemical steps in such slurry reactors is considered in Chaps. 10 and 13. Polymerization systems are often of this type. For example, ethylene is normally polymerized by dissolving it in a solvent containing suspended catalyst particles. The trickle-bed reactor is a somewhat different form of gas-liquid-solid system. Here gas and liquid, usually in cocurrent flow, pass over a bed of catalyst particles in a fixed bed. This form is used when the volatility of the liquid is so low that complete gasification is not practical. Sulfur compounds are removed from heavy hydrocarbon liquids in such reactors. The gas phase here is primarily hydrogen, which is necessary for the desulfurization reactions.

Liquid-liquid reactions are sometimes encountered. Alkylation of hydrocarbons with aqueous solution of sulfuric acid as a catalyst is an example. As in liquid-solid systems, definition of the liquid phase containing the catalyst may be difficult, reducing the effectiveness of a fundamental analysis in terms of chemical and physical steps.

Solid-solid noncatalytic reactions are important in ceramics manufacture.[1] It appears that diffusion resistances may be important to some extent in all such systems. The diffusion process itself is hard to define in solid-solid systems, since at least two possibilities exist: volume diffusion in the solid and surface diffusion along interfaces and crystal boundaries.[2] Little is known about the kinetics of solid-solid reactions at the reacting interface because most measurements include diffusion effects.

Problems

7-1. In a sketch similar to Fig. 7-1 show schematically the concentration profiles for a first-order *reversible* catalytic reaction. Consider three cases: (*a*) reaction control, (*b*) diffusion control, and (*c*) an intermediate case.

7-2. Explain why mass-transfer resistance reduces the global rate more at higher temperatures than at lower temperatures. Assume no heat-transfer resistance.

7-3. For endothermic reactions, do mass- and heat-transfer resistances have complementary or counterbalancing effects on the global rate?

7-4. A gaseous reaction with a solid catalyst is carried out in a flow reactor. The system is isothermal, but it is believed that mass-transfer resistances are important. (*a*) Would increasing the turbulence in the gas region next to the catalyst surface increase or decrease the global rate? (*b*) If the system is not isothermal and the reaction is exothermic, would increasing the turbulence increase or decrease the global rate?

[1] W. Kingery "Kinetics of High Temperature Processes," John Wiley & Sons, Inc., New York, 1959; G. Cohn, *Chem. Rev.*, **42**, 527 (1948).

[2] R. J. Arrowsmith and J. M. Smith, *Ind. Eng. Chem., Fund. Quart.*, **5**, 327 (1966).

8

HETEROGENEOUS CATALYSIS

As kinetic information began to accumulate during the last century, it appeared that the rates of a number of reactions were influenced by the presence of a material which itself was unchanged during the process. In 1836 Berzelius[1] reviewed the evidence and concluded that a "catalytic" force was in operation. Among the cases he studied were the conversion of starch into sugar in the presence of acids, the decomposition of hydrogen peroxide in alkaline solutions, and the combination of hydrogen and oxygen on the surface of spongy platinum. In these three examples the acids, the alkaline ions, and the spongy platinum were the materials which increased the rate and yet were virtually unchanged by the reaction. Although the concept of a catalytic force proposed by Berzelius has now been discarded, the term "catalysis" is retained to describe all processes in which the rate of a reaction is influenced by a substance that remains chemically unaffected.

In this chapter we shall consider first the general characteristics of heterogeneous catalysis and adsorption (physical and chemical) and then physical properties of solid catalysts and methods of preparation. Kinetics

[1] J. J. Berzelius, *Jahresber. Chem.*, **15**, 237 (1836).

and mechanism of adsorption and fluid-solid catalytic reactions are taken up in Chap. 9.

GENERAL CHARACTERISTICS[*]

8-1 The Nature of Catalytic Reactions

Although the catalyst remains unchanged at the end of the process, there is no requirement that the material not take part in the reaction. In fact, present theories of catalyst activity postulate that the material does actively participate in the reaction. From the concept of the energy of activation developed in Chap. 2, the mechanism of catalysis would have to be such that the free energy of activation is lowered by the presence of the catalytic material. A catalyst is effective in increasing the rate of a reaction because it makes possible an alternative mechanism, each step of which has a lower free energy of activation than that for the uncatalyzed process. Consider the reaction between hydrogen and oxygen in the presence of spongy platinum. According to the proposed concept, hydrogen combines with the spongy platinum to form an intermediate substance, which then reacts with oxygen to provide the final product and reproduce the catalyst. It is postulated that the steps involving the platinum surface occur at a faster rate than the homogeneous reaction between hydrogen and oxygen.

The combination or complexing of reactant and catalyst is a widely accepted basis for explaining catalysis. For example, suppose the overall reaction

$$A + B \rightleftharpoons C$$

is catalyzed via two active centers, or catalytic sites, X_1 and X_2, which form complexes with A and B. The reaction is truly catalytic if the sequence of steps is such that the centers X_1 and X_2 are regenerated after they have caused the formation of C. In a general way the process may be written

1. $A + X_1 \rightleftharpoons AX_1$

2. $B + X_2 \rightleftharpoons BX_2$

3. $AX_1 + BX_2 \rightleftharpoons C + X_1 + X_2$

Note that whereas X_1 and X_2 are combined and regenerated a number of times, it does not necessarily follow that their catalyzing ability and/or number remain constant forever. For example, poisons can intervene to slowly remove X_1 and/or X_2 from the system, arresting the catalytic rate. What distinguishes this decline in catalytic activity from that of a non-

[*]This material was written jointly with Professor J. J. Carberry.

catalytic reaction in which X_1 and X_2 are not regenerated is that the complexing-regenerating sequence occurs a great many times before X_1 and X_2 become inactive. In the noncatalytic sequence no regeneration of X occurs. Hence, while catalysts can deteriorate, their active lifetime is far greater than the time required for reaction.

A relatively small amount of catalyst can cause conversion of a large amount of reactant. For example, Glasstone[1] points out that cupric ions in the concentration of 10^{-9} mole/liter appreciably increase the rate of the oxidation of sodium sulfide by oxygen. However, the idea that a small amount of the catalyst can cause a large amount of reaction does not mean that the catalyst concentration is unimportant. In fact, when the reaction does not entail a chain mechanism, the rate of the reaction is usually proportional to the concentration of the catalyst. This is perhaps most readily understood by considering the case of surface catalytic reactions. In the reaction of hydrogen and oxygen with platinum catalyst the rate is found to be directly proportional to the platinum surface. Here a simple proportionality exists between platinum surface area and the number of centers X which catalyze the oxidation of hydrogen. While a simple relationship may not often exist in solid-catalyzed reactions, in homogeneous catalysis there is often a direct proportionality between rate and catalyst concentration. For example, the hydrolysis of esters in an acid solution will depend on the concentration of hydrogen ion acting as a catalyst.

The position of equilibrium in a reversible reaction is not changed by the presence of the catalyst. This conclusion has been verified experimentally in several instances. For example, the oxidation of sulfur dioxide by oxygen has been studied with three catalysts: platinum, ferric oxide, and vanadium pentoxide. In all three cases the *equilibrium compositions* were the same.

An important characteristic of a catalyst is its effect on selectivity when several reactions are possible. A good illustration is the decomposition of ethanol. Thermal decomposition gives water, acetaldehyde, ethylene, and hydrogen. If, however, ethanol vapor is suitably contacted with alumina particles, ethylene and water are the only products. In contrast, dehydrogenation to acetaldehyde is virtually the sole reaction when ethanol is reacted over a copper catalyst.

The general characteristics of catalysis may be summarized as follows:

1. A catalyst accelerates reaction by providing alternate paths to products, the activation energy of each catalytic step being less than that for the homogeneous (noncatalytic) reaction.
2. In the reaction cycle active centers of catalysis are combined with at

[1]S. Glasstone, "Textbook of Physical Chemistry," p. 1104, D. Van Nostrand Company, Inc., New York, 1940.

least one reactant and then freed with the appearance of product. The freed center then recombines with reactant to produce another cycle, and so on.

3. Comparatively small quantities of catalytic centers are required to produce large amounts of product.

4. Equilibrium conversion is not altered by catalysis. A catalyst which accelerates the forward reaction in an equilibrium system is a catalyst for the reverse reaction.

5. The catalyst can radically alter selectivity.

Examples have been observed of negative catalysis, where the rate is decreased by the catalyst. Perhaps the most reasonable theory is that developed for chain reactions. In these cases it is postulated that the catalyst breaks the reaction chains, or sequence of steps, in the mechanism. For example, nitric oxide reduces the rate of decomposition of acetaldehyde and ethyl ether. Apparently nitric oxide has the characteristic of combining with the free radicals involved in the reaction mechanism. The halogens, particularly iodine, also act as negative catalysts in certain gaseous reactions. In the combination of hydrogen and oxygen, where a chain mechanism is probably involved, iodine presumably destroys the radicals necessary for the propagation of the chains.

8-2 The Mechanism of Catalytic Reactions

The concept that a catalyst provides an alternate mechanism for accomplishing a reaction, and that this alternate path is a more rapid one, has been developed in many individual cases. The basis of this idea is that the catalyst and one or more of the reactants form an intermediate complex, a loosely bound compound which is unstable, and that this complex then takes part in subsequent reactions which result in the final products and the regenerated catalyst. Homogeneous catalysis can frequently be explained in terms of this concept. For example, consider catalysis by acids and bases. In aqueous solutions acids and bases can increase the rate of hydrolysis of sugars, starches, and esters. The kinetics of the hydrolysis of ethyl acetate catalyzed by hydrochloric acid can be explained by the following mechanism:

1. $CH_3COOC_2H_5 + H^+ \rightleftharpoons CH_3COOC_2H_5[H^+]$

2. $CH_3COOC_2H_5[H^+] + H_2O \rightleftharpoons C_2H_5OH + H^+ + CH_3COOH$

For this catalytic sequence to be rapid with respect to noncatalytic hydrolysis, the free energy of activation of reaction steps 1 and 2 must each be less than the free energy of activation for the noncatalytic reaction,

$$CH_3COOC_2H_5 + H_2O \rightleftharpoons CH_3COOH + C_2H_5OH$$

Similarly, the heterogeneous catalytic hydrogenation of ethylene on a solid catalyst might be represented by the steps

1. $C_2H_4 + X_1 \overset{\Delta F_1}{\rightleftharpoons} C_2H_4X_1$

2. $H_2 + X_1C_2H_4 \overset{\Delta F_2}{\rightleftharpoons} C_2H_4[X_1]H_2$

3. $C_2H_4[X_1]H_2 \overset{\Delta F_3}{\rightleftharpoons} C_2H_6 + X_1$

where $[X_1]$ is the solid catalyst and $C_2H_4[X_1]H_2$ represents the complex formed between the reactants and the catalyst. The homogeneous reaction, according to the absolute theory of reaction rates discussed in Chap. 2, would be written

$$C_2H_4 + H_2 \overset{\Delta F^*}{\rightleftharpoons} \overset{\text{Activated complex}}{C_2H_4 \cdot H_2} \rightarrow C_2H_6$$

where the free-energy change for the formation of the activated complex ΔF^* is the free energy of activation for the homogeneous reaction. The effectiveness of the catalyst is explained on the basis that the free energy of activation of each of the steps in the catalytic mechanism is less than ΔF^*.

These illustrations, particularly that for ethylene hydrogenation, are grossly oversimplified. They must be considered phenomenological models, not mechanisms. The actual mechanism of ethylene hydrogenation is quite complex. In spite of the considerable effort focused on this reaction, a mechanism satisfactory to all investigators has yet to be offered. The system does, however, provide an opportunity to compare homogeneous and heterogeneous rates. Using published data, Boudart[1] found that the homogeneous and catalytic rates can be expressed as

$$r_{hom} = 10^{27} e^{-43,000/R_g T}$$

$$r_{cat} = 2 \times 10^{27} e^{-13,000/R_g T} \qquad \text{(CuO-MgO catalyst)}$$

At $600°K$ the relative rates are

$$\frac{r_{cat}}{r_{hom}} = e^{(43,000-13,000)/600 R_g} \simeq 10^{11}$$

In this case the catalyst has caused a radical reduction in overall activation energy, presumably by replacing a difficult homogeneous step by a more easily executed surface reaction involving adsorbed ethylene. The results lead to the kinetics observed by Wynkoop and Wilhelm,[2] a reaction first order in H_2 and zero order in strongly adsorbed ethylene.

[1] M. Boudart, *Ind. Chim. Belg.*, **23**, 383 (1958).

[2] R. Wynkoop and R. H. Wilhelm, *Chem. Eng. Progr.*, **46**, 300 (1950).

The three steps postulated for the catalytic hydrogenation of ethylene indicate that the rate may be influenced by both adsorption and desorption (steps 1 and 3) and the surface reaction (step 2). Two extreme cases can be imagined: that step 1 or step 3 is slow with respect to step 2 or that step 2 is relatively slow. In the first situation rates of adsorption or desorption are of interest, while in the second the surface concentration of the adsorbed species corresponding to equilibrium with respect to steps 1 and 3 is needed. In any case we should like to know the number of sites on the catalyst surface, or at least the surface area of the catalyst. These questions require a study of adsorption. More is known about adsorption of gases, and this will be emphasized in the sections that follow.

ADSORPTION ON SOLID SURFACES

8-3 Surface Chemistry and Adsorption

Even the most carefully polished surfaces are not smooth in a microscopic sense, but are irregular, with valleys and peaks alternating over the area. The regions of irregularity are particularly susceptible to residual force fields. At these locations the surface atoms of the solid may attract other atoms or molecules in the surrounding gas or liquid phase. Similarly, the surfaces of pure crystals have nonuniform force fields because of the atomic structure in the crystal. Such surfaces also have sites or active centers where adsorption is enhanced. Two types of adsorption may occur.

Physical Adsorption[1] The first type of adsorption is nonspecific and somewhat similar to the process of condensation. The forces attracting the fluid molecules to the solid surface are relatively weak, and the heat evolved during the adsorption process is of the same order of magnitude as the heat of condensation, 0.5 to 5 kcal/g mole. Equilibrium between the solid surface and the gas molecules is usually rapidly attained and easily reversible, because the energy requirements are small. The energy of activation for physical adsorption is usually no more than 1 kcal/g mole. This is a direct consequence of the fact that the forces involved in physical adsorption are weak. Physical adsorption cannot explain the catalytic activity of solids for reactions between relatively stable molecules, because there is no possibility of large reductions in activation energy. Reactions of atoms and free radicals at surfaces sometimes involve small activation energies, and in these cases physical adsorption may play a

[1]For a detailed treatment of physical adsorption see D. M. Young and A. D. Crowell, "Physical Adsorption of Gases," Butterworths & Co. (Publishers), London, 1962.

role. Also, physical adsorption serves to concentrate the molecules of a substance at a surface. This can be of importance in cases involving reaction between a chemisorbed reactant and a coreactant which can be physically adsorbed. In such a system the catalytic reaction would occur between chemisorbed and physically adsorbed reactants. Catalysis cannot be attributed solely to physical adsorption. Thus *all* solids will physically adsorb gases under suitable conditions, and yet all solids are not catalysts.

The amount of physical adsorption decreases rapidly as the temperature is raised and is generally very small above the critical temperatures of the adsorbed component. This is further evidence that physical adsorption is not responsible for catalysis. For example, the rate of oxidation of sulfur dioxide on a platinum catalyst becomes appreciable only above 300°C; yet this is considerably above the critical temperature of sulfur dioxide (157°C) or of oxygen (-119°C). Physical adsorption is not highly dependent on the irregularities in the nature of the surface, but is usually directly proportional to the amount of surface. However, the extent of adsorption is not limited to a monomolecular layer on the solid surface, especially near the condensation temperature. As the layers of molecules build up on the solid surface, the process becomes progressively more like one of condensation.

Physical-adsorption studies are valuable in determining the physical properties of solid catalysts. Thus the questions of surface area and pore-size distribution in porous catalysts can be answered from physical-adsorption measurements. These aspects of physical adsorption are considered in Secs. 8-5 and 8-7.

Chemisorption[1]   The second type of adsorption is specific and involves forces much stronger than in physical adsorption. According to Langmuir's pioneer work,[2] the adsorbed molecules are held to the surface by valence forces of the same type as those occurring between atoms in molecules. He observed that a stable oxide film was formed on the surface of tungsten wires in the presence of oxygen. This material was not the normal oxide WO_3, because it exhibited different chemical properties. However, analysis of the walls of the vessel holding the wire indicated that WO_3 was given off from the surface upon desorption. This suggested a process of the type

$$3O_2 + 2W \rightarrow 2[W \cdot O_3]$$

$$2[W \cdot O_3] \rightarrow 2WO_3$$

[1] For a detailed treatment of chemisorption see D. O. Hayward and B. M. W. Trapnell, "Chemisorption," 2d ed., Butterworths & Co. (Publishers), London, 1964.

[2] I. Langmuir, *J. Am. Chem. Soc.*, **38**, 221 (1916).

where $[W \cdot O_3]$ represents the adsorbed compound. Further evidence for the theory that such adsorption involves valence bonds is found in the large heats of adsorption. Observed values are of the same magnitude as the heat of chemical reactions, 5 to 100 kcal/g mole.

Taylor[1] suggested the name *chemisorption* for describing this second type of combination of gas molecules with solid surfaces. Because of the high heat of adsorption, the energy possessed by chemisorbed molecules can be substantially different from that of the molecules alone. Hence the energy of activation for reactions involving chemisorbed molecules can be considerably less than that for reactions involving the molecules alone. It is on this basis that chemisorption offers an explanation for the catalytic effect of solid surfaces. *energy from absorption helps reaction*

Two kinds of chemisorption are encountered: activated and, less frequently, nonactivated. *Activated chemisorption* means that the rate varies with temperature according to a finite activation energy in the Arrhenius equation. However, in some systems chemisorption occurs very rapidly, suggesting an activation energy near zero. This is termed *nonactivated chemisorption.*[2] It is often found that for a given gas and solid the initial chemisorption is nonactivated, while later stages of the process are slow and temperature dependent (activated adsorption).

With respect to adsorption *equilibrium*, the relationship between temperature and quantity adsorbed (both physically and chemically) is shown in Fig. 8-1. Chemisorption is assumed to be activated in this case. As the critical temperature of the component is exceeded, physical adsorption approaches a very low equilibrium value. As the temperature is raised, the amount of activated adsorption becomes important because the rate is high enough for significant quantities to be adsorbed in a reasonable amount of time. In an ordinary adsorption experiment involving the usual time periods the adsorption curve actually rises with increasing temperatures from the minimum value, as shown by the solid line in Fig. 8-1. When the temperature is increased still further, the decreasing equilibrium value for activated adsorption retards the process, and the quantity adsorbed passes through a maximum. At these high temperatures even the rate of the relatively slow activated process may be sufficient to give results closely approaching equilibrium. Hence the solid curve representing the amount adsorbed approaches the dashed equilibrium value for the activated adsorption process.

It has been explained that the effectiveness of solid catalysts for reactions of stable molecules is dependent upon chemisorption. Granting

[1] H. S. Taylor, *J. Am. Chem. Soc.*, **53**, 578 (1931).

[2] As an illustration, chemisorption of hydrogen on nickel at low temperatures is nonactivated; see G. Padberg and J. M. Smith, *J. Catalysis*, **12**, 111 (1968).

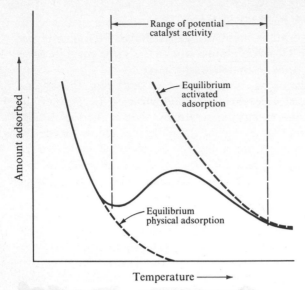

Fig. 8-1 *Effect of temperature on physical and activated adsorption.*

this, the temperature range over which a given catalyst is effective must coincide with the range where chemisorption of one or more of the reactants is appreciable. This is indicated on Fig. 8-1 by the dashed vertical lines. There is a relationship between the extent of chemisorption of a gas on a solid and the effectiveness of the solid as a catalyst. For example, many metallic and metal-oxide surfaces adsorb oxygen easily, and these materials are also found to be good catalysts for oxidation reactions. When reactions proceed catalytically at low temperatures, Fig. 8-1 does not apply. In these cases catalysis is due to nonactivated chemisorption. Thus ethylene is hydrogenated on nickel at $-78°C$, at which temperature there would surely exist physical adsorption of the ethylene.

An important feature of chemisorption is that its magnitude will not exceed that corresponding to a monomolecular layer. This limitation is due to the fact that the valence forces holding the molecules on the surface diminish rapidly with distance. These forces become too small to form the adsorption compound when the distance from the surface is much greater than the usual bond distances.

The differences between chemisorption and physical adsorption are summarized in Table 8-1.

The key concept for quantitative treatment of both physical and chemical adsorption is that formulated by Langmuir.[1] While his concern

[1]I. Langmuir, *J. Am. Chem. Soc.*, **40**, 1361 (1918).

Table 8-1   Physical vs chemical adsorption

Parameter	Physical adsorption	Chemisorption
Adsorbent	All solids	Some solids
Adsorbate	All gases below critical temperature	Some chemically reactive gases
Temperature range	Low temperature	Generally high temperature
Heat of adsorption	Low ($\approx \Delta H_{cond}$)	High, order of heat of reaction
Rate (activation energy)	Very rapid, low E	Nonactivated, low E; activated, high E
Coverage	Multilayer possible	Monolayer
Reversibility	Highly reversible	Often irreversible
Importance	For determination of surface area and pore size	For determination of active-center area and elucidation of surface-reaction kinetics

was with chemisorption, Brunauer, Emmett, and Teller gainfully employed the concepts to derive a valuable relationship between the volume of a gas physically adsorbed and total surface area of the adsorbent (see Sec. 8-5). Also, the Langmuir treatment can be extended to develop useful relations for chemisorption rates and rate of catalytic reactions, even for surfaces which do not obey the basic postulates of the Langmuir theory. This second application is given in Chap. 9.

8-4 The Langmuir Treatment of Adsorption

The derivations may be carried out by using as a measure of the amount adsorbed either the fraction of the surface covered or the concentration of the gas adsorbed on the surface. Both procedures will be illustrated, although the second is the more useful for kinetic developments (Chap. 9). The important assumptions are as follows:[1]

1. All the surface of the catalyst has the same activity for adsorption; i.e., it is energetically uniform. The concept of nonuniform surface with active centers can be employed if it is assumed that all the active centers have the same activity for adsorption and that the rest of the surface has none, or that an average activity can be used.
2. There is no interaction between adsorbed molecules. This means that the amount adsorbed has no effect on the rate of adsorption per site.

[1] It is also tacitly supposed that each site can accommodate only one adsorbed particle.

3. All the adsorption occurs by the same mechanism, and each adsorbed complex has the same structure.
4. The extent of adsorption is less than one complete monomolecular layer on the surface.

In the system of solid surface and gas, the molecules of gas will be continually striking the surface and a fraction of these will adhere. However, because of their kinetic, rotational, and vibrational energy, the more energetic molecules will be continually leaving the surface. An equilibrium will be established such that the rate at which molecules strike the surface, and remain for an appreciable length of time, will be exactly balanced by the rate at which molecules leave the surface.

The rate of adsorption $\mathbf{r}_a$ will be equal to the rate of collision $\mathbf{r}_c$ of molecules with the surface multiplied by a factor F representing the fraction of the colliding molecules that adhere. At a fixed temperature the number of collisions will be proportional to the pressure p of the gas, and the fraction F will be constant. Hence the rate of adsorption per unit of bare surface will be $\mathbf{r}_c F$. This is equal to kp, where k is a constant involving the fraction F and the proportionality between $\mathbf{r}_c$ and p.

Since the adsorption is limited to complete coverage by a monomolecular layer, the surface may be divided into two parts: the fraction θ covered by the adsorbed molecules and the fraction $1 - \theta$, which is bare. Since only those molecules striking the uncovered part of the surface can be adsorbed, the rate of adsorption per unit of total surface will be proportional to $1 - \theta$; that is,

$$\mathbf{r}_a = kp(1 - \theta) \tag{8-1}$$

The rate of desorption will be proportional to the fraction of covered surface

$$\mathbf{r}_d = k'\theta \tag{8-2}$$

The amount adsorbed at equilibrium is obtained by equating $\mathbf{r}_a$ and $\mathbf{r}_d$ and solving for θ. The result, called the *Langmuir isotherm*, is

$$\theta = \frac{kp}{k' + kp} = \frac{Kp}{1 + Kp} = \frac{v}{v_m} \tag{8-3}$$

where $K = k/k'$ is the adsorption equilibrium constant, expressed in units of (pressure)$^{-1}$. The fraction θ is proportional to volume of gas adsorbed, v, since the adsorption is less than a monomolecular layer. Hence Eq. (8-3) may be regarded as a relationship between the pressure of the gas and the volume adsorbed. This is indicated by writing $\theta = v/v_m$, where v_m is the volume adsorbed when all the active sites are covered, i.e., when there is a complete monomolecular layer.

The concentration form of Eq. (8-3) can be obtained by introducing the concept of an adsorbed concentration $\overline{C}$, expressed in moles per gram of catalyst. If $\overline{C}_m$ represents the concentration corresponding to a complete monomolecular layer on the catalyst, then the rate of adsorption, moles/(sec) (g catalyst) is, by analogy with Eq. (8-1),

$$\mathbf{r}_a = k_c C_g (\overline{C}_m - \overline{C}) \tag{8-4}$$

where k_c is the rate constant for the catalyst and C_g is the concentration of adsorbable component in the gas. Similarly, Eq. (8-2) becomes

$$\mathbf{r}_d = k_c' \overline{C} \tag{8-5}$$

At equilibrium the rates given by Eqs. (8-4) and (8-5) are equal, so that

$$\overline{C} = \frac{K_c \overline{C}_m C_g}{1 + K_c C_g} \tag{8-6}$$

where now the equilibrium constant K_c is equal to k_c/k_c' and is expressed in cubic centimeters per gram mole. Since $\overline{C}/\overline{C}_m = \theta$, Eq. (8-6) may also be written

$$\theta = \frac{K_c C_g}{1 + K_c C_g} \tag{8-7}$$

which is a form analogous to Eq. (8-3), since C_g is proportional to p.

Equation (8-6) predicts that adsorption data should have the general form shown in Fig. 8-2. Note that at low values of C_g (or low surface coverages θ) the expression becomes a straight line with a slope equal to $K_c \overline{C}_m$. The data points in Fig. 8-2 are for the physical adsorption of n-butane on silica gel ($S_g = 832$ m^2/g) at 50°C.[1] The solid line represents Eq. (8-6), where

$$\overline{C}_m = 0.85 \times 10^{-3} \text{ g mole/g silica gel}$$

and the equilibrium constant of adsorption is

$$K_c = 4.1 \times 10^5 \text{ cm}^3/\text{g mole}$$

For this instance of physical adsorption Eq. (8-6) fits the data rather well. The measurements were made on mixtures of n-butane in helium (at 1 atm total pressure) to vary C_g. The percentage of n-butane in the gas corresponding to the concentration C_g is shown as a second abscissa in the figure. Also, θ is shown as a second ordinate. This was calculated from $\theta = \overline{C}/\overline{C}_m$. It is interesting to note that the isotherm is linear up to about $\theta = 0.10$, or 10% surface coverage.

[1] Data from R. L. Cerro, "Adsorption Studies by Chromatography," master's thesis, University of California, Davis, Calif., September, 1968.

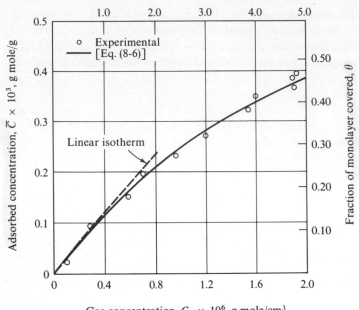

Fig. 8-2   *Adsorption-equilibrium data for n-butane on silica gel (surface area 832 m²/g) at 50°C*

Chemisorption data often do not fit Eq. (8-6). However, the basic concepts on which the Langmuir isotherm is based, the ideas of a dynamic equilibrium between rates of adsorption and desorption and a finite adsorption time, are sound and of great value in developing the kinetics of fluid-solid catalytic reactions. Equations (8-4) to (8-6) form the basis for the rate equations presented in Chap. 9.

PHYSICAL PROPERTIES OF CATALYSTS

The surface area of a solid has a pronounced effect on the amount of gas adsorbed and on its activity as a catalyst. For example, if a sample of fresh Raney nickel, which is highly porous and has a large surface, is held in the hand, the heat due to adsorption of oxygen can be felt immediately. No heat is apparent in the same mass of nonporous nickel. This relationship between surface area and extent of adsorption has led to the development of highly porous materials with areas as high as 1,500 m²/g. Sometimes the catalytic material itself can be prepared in a form with large surface area. When this is not possible, materials which can be so prepared may be used as a carrier or support on which the catalytic substance is dispersed. Silica gel and alumina are widely used as supports.

The dependence of rates of adsorption and catalytic reactions on surface makes it imperative to have a reliable method of measuring surface area. Otherwise it would not be possible to compare different catalysts (whose areas are different) to ascertain the intrinsic activity per unit surface. For surface areas in the range of hundreds of square meters per gram a porous material with equivalent cylindrical pore radii (see Sec. 8-5) in the range of 10 to 100 Å is needed. The following example shows that such areas are not possible with nonporous particles of the size which can be economically manufactured.

Example 8-1   Spray drying and other procedures for manufacturing small particles can produce particles as small as 2 to 5 microns. Calculate the external surface area of nonporous spherical particles of 2 microns diameter. What size particles would be necessary if the external surface is to be 100 m^2/g? The density of the particles is 2.0 g/cm^3.

Solution   The external surface area per unit volume of a spherical particle of diameter d_p is

$$\frac{\pi d_p^2}{\pi d_p^3/6} = \frac{6}{d_p}$$

If the particle density is ρ_P, the surface area, per gram of particles, would be

$$S_g = \frac{6}{\rho_p d_p}$$

For $d_p = 2$ microns (2×10^{-4} cm) and $\rho_p = 2.0$ g/cm^3

$$S_g = \frac{6}{2(2 \times 10^{-4})} = 1.5 \times 10^4 \; cm^2/g$$

or

$$S_g = 1.5 \; m^2/g$$

This is about the largest surface area to be expected for nonporous particles. If a surface of 100 m^2/g were required, the spherical particles would have a diameter of

$$d_p = \frac{6}{\rho_p S_g} = \frac{6}{2.0(100 \times 10^4)} = 0.02 \times 10^{-4} \; cm$$

or

$$d_p = 0.02 \; micron$$

Particles as small as this cannot, at present, be produced on a commercial scale. It may be noted that the smaller particles found in a fluidized-bed reactor are retained on 400 mesh size, which has a sieve opening of 37 microns.

When the major catalytic surface is in the interior of a solid particle, the resistance to transport of mass and energy from the external surface to the interior can have a significant effect on the global rate of reaction. Quantitative treatment of this problem is the objective in Chap. 11. It is sufficient here to note that this treatment rests on a geometric model for the extent and distribution of void spaces within the complex porous structure of the particle. It would be best to know the size and shape of each void space in the particle. In the absence of this information the parameters in the model should be evaluated from reliable and readily obtainable geometric properties. In addition to the surface area, three other properties fall into this classification: void volume, the density of the solid material in the particle, and the distribution of void volume according to void size (pore-volume distribution). The methods of measurement of these four properties are considered in Secs. 8-5 to 8-7.

8-5 Determination of Surface Area

The standard method for measuring catalyst areas is based on the physical adsorption of a gas on the solid surface. Usually the amount of nitrogen adsorbed at equilibrium at the normal boiling point ($-195.8°C$) is measured over a range of nitrogen pressures below 1 atm. Under these conditions several layers of molecules may be adsorbed on top of each other on the surface. The amount adsorbed when one molecular layer is attained must be identified in order to determine the area. The historical steps in the development of the Brunauer-Emmett-Teller method[1] are clearly explained by Emmett.[2] There may be some uncertainty as to whether the values given by this method correspond exactly to the surface area. However, this is relatively unimportant, since the procedure is standardized and the results are reproducible. It should be noted that the surface area so measured may not be the area effective for catalysis. For example, only certain parts of the surface, the active centers, may be active for chemisorption, while nitrogen may be physically adsorbed on much more of the surface. Also, when the catalyst is dispersed on a large-area support, only part of the support area may be covered by catalytically active atoms. For example, a nickel-on-kieselguhr catalyst was found to have a surface of 205 m^2/g as measured by nitrogen adsorption.[3] To determine the area covered by nickel atoms, hydrogen was chemisorbed on the catalyst at 25°C. From the amount of hydrogen chemisorbed, the surface area of nickel atoms was calculated to be about 40 m^2/g. It would be most useful to know surface areas for

[1] S. Brunauer, P. H. Emmett, and E. Teller, *J. Am. Chem. Soc.*, **60**, 309 (1938).

[2] P. H. Emmett (ed.), "Catalysis," vol. I, chap. 2, Reinhold Publishing Corporation, New York, 1954.

[3] G. Padberg and J. M. Smith, *J. Catalysis*, **12**, 111 (1968).

chemisorption of the reactant at reaction conditions. However, this would require measurement of relatively small amounts of chemisorption at different, and often troublesome, conditions (high temperature and/or pressure), for each reaction system. In contrast, nitrogen can be adsorbed easily and rapidly in a routine fashion with standard equipment.

In the classical method of determining surface area an all-glass apparatus is used to measure the volume of gas adsorbed on a sample of the solid material.[1] The apparatus operates at a low pressure which can be varied from near zero up to about 1 atm. The operating temperature is in the range of the normal boiling point. The data obtained are gas volumes at a series of pressures in the adsorption chamber. The observed volumes are normally corrected to cubic centimeters at 0°C and 1 atm (standard temperature and pressure) and plotted against the pressure in millimeters, or as the ratio of the pressure to the vapor pressure at the operating temperature. Typical results from Brunauer and Emmett's work[2] are shown in Fig. 8-3 for the adsorption of several gases on a 0.606-g sample of silica gel. To simplify the classical experimental procedure a flow method has been developed in which a mixture of helium and the gas to be adsorbed is passed continuously over the sample of solid.[3] The operating total pressure is constant, and the partial pressure of adsorbable gas is varied by changing the composition of the mixture. The procedure[4] is to pass a mixture of known composition over the sample until equilibrium is reached, that is, until the solid has adsorbed an amount of adsorbable component corresponding to equilibrium at its partial pressure in the mixture. Then the gas is desorbed by heating the sample while a stream of pure helium flows over it. The amount desorbed is measured with a thermal-conductivity cell or other detector. This gives one point on an isotherm, such as shown in Fig. 8-3. Then the process is repeated at successively different compositions of the mixture until the whole isotherm is obtained.

The curves in Fig. 8-3 are similar to the extent that at low pressures they rise more or less steeply and then flatten out for a linear section at intermediate pressures. After careful analysis of much data it was concluded that the lower part of the linear region corresponded to complete monomolecular adsorption. If this point could be located with precision, the

[1] For a complete description of apparatus and techniques see L. G. Joyner, "Scientific and Industrial Glass Blowing and Laboratory Techniques," Instruments Publishing Company, Pittsburgh, 1949; see also S. Brunauer, "The Adsorption of Gases and Vapors," vol. 1, Princeton University Press, Princeton, N.J., 1943.

[2] S. Brunauer and P. H. Emmett, J. Am. Chem. Soc., 59, 2682 (1937).

[3] F. M. Nelson and F. T. Eggertsen, Anal. Chem., 30, 1387 (1958).

[4] A description of the operating procedure and the data obtained are given by S. Masamune and J. M. Smith [AIChE J., 10, 246 (1964)] for the adsorption of nitrogen on Vycor (porous glass).

volume of one monomolecular layer of gas, v_m, could then be read from the curve and the surface area evaluated. The Brunauer-Emmett-Teller method locates this point from an equation obtained by extending the Langmuir isotherm to apply to multilayer adsorption. The development is briefly summarized as follows: Equation (8-3) can be rearranged to the form

$$\frac{p}{v} = \frac{1}{Kv_m} + \frac{p}{v_m} \tag{8-8}$$

Brunauer, Emmett, and Teller adapted this equation for multilayer adsorption and arrived at the result

$$\frac{p}{v(p_0 - p)} = \frac{1}{v_m c} + \frac{(c - 1)p}{c v_m p_0} \tag{8-9}$$

where p_0 is the saturation or vapor pressure and c is a constant for the particular temperature and gas-solid system.

According to Eq. (8-9), a plot of $p/v(p_0 - p)$ vs p/p_0 should give a straight line. The data of Fig. 8-3 are replotted in this fashion in Fig. 8-4.

Fig. 8-3 *Adsorption isotherms for various gases on a 0.606-g sample of silica gel [by permission from P. H. Emmett (ed.), "Catalysis," vol. I, Reinhold Publishing Corporation, New York, 1954]*

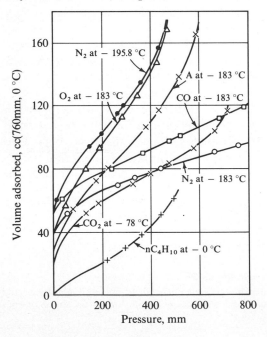

Of additional significance is the fact that such straight lines can be safely extrapolated to $p/p_0 = 0$. The intercept I obtained from this extrapolation, along with the slope s of the straight line, gives two equations from which v_m can be obtained,

$$I = \frac{1}{v_m c} \quad \text{at } p/p_0 = 0 \tag{8-10}$$

$$s = \frac{c - 1}{v_m c} \tag{8-11}$$

Solving these equations for the volume of adsorbed gas corresponding to a monomolecular layer gives

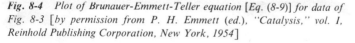

$$v_m = \frac{1}{I + s} \tag{8-12}$$

The volume v_m can be readily converted to the number of molecules adsorbed. However, to determine the absolute surface area it is necessary to select a value for the area covered by one adsorbed molecule. If this is α, the total surface area is given by

Fig. 8-4 *Plot of Brunauer-Emmett-Teller equation [Eq. (8-9)] for data of Fig. 8-3 [by permission from P. H. Emmett (ed.), "Catalysis," vol. 1, Reinhold Publishing Corporation, New York, 1954]*

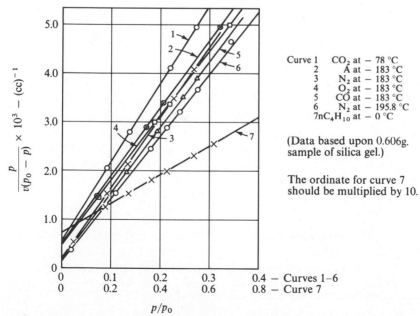

Curve 1 CO_2 at $-78\,°C$
      2   A at $-183\,°C$
      3   N_2 at $-183\,°C$
      4   O_2 at $-183\,°C$
      5   CO at $-183\,°C$
      6   N_2 at $-195.8\,°C$
7 nC_4H_{10} at $-0\,°C$

(Data based upon 0.606g. sample of silica gel.)

The ordinate for curve 7 should be multiplied by 10.

$$S_g = \left[\frac{v_m N_0}{V} \right] \alpha \tag{8-13}$$

where N_0 is Avogadro's number, 6.02×10^{23} molecules/mole, and V is the volume per mole of gas at conditions of v_m. Since v_m is recorded at standard temperature and pressure, $V = 22,400$ cm^3/g mole. The term in brackets represents the number of molecules adsorbed. If v_m is based on a 1.0 g sample, then S_g is the total surface per gram of solid adsorbent.

The value of α has been the subject of considerable investigation. Emmett and Brunauer[1] proposed that α is the projected area of a molecule on the surface when the molecules are arranged in close two-dimensional packing. This value is slightly larger than that obtained by assuming that the adsorbed molecules are spherical and their projected area on the surface is circular. The proposed equation is

$$\alpha = 1.09 \left[\frac{M}{N_0 \rho} \right]^{\frac{2}{3}} \tag{8-14}$$

where M is molecular weight and ρ is the density of the adsorbed molecules. The term in brackets represents the volume of one adsorbed molecule. The density is normally taken as that of the pure liquid at the temperature of the adsorption experiment. For example, for N_2 at $-195.8°C$, $\rho = 0.808$ g/cm^3.

In theory, the adsorption measurements can be made with a number of different gases. However, it has been found that even when values of α are calculated from Eq. (8-14) for each gas the results are somewhat different (see Example 8-3). Therefore it has become standard procedure to employ N_2 at its normal boiling point ($-195.8°C$). The reason for the variation in areas obtained with different gases is not well understood. Nevertheless, if the measurements are carried out with one gas at one temperature, the results for different catalysts may be compared with confidence.

With the value of ρ for N_2 at $-195.8°C$, the area per molecule from Eq. (8-14) is 16.2×10^{-16} cm^2, or 16.2 Å^2. If this result is used in Eq. (8-13), along with the known values of N_0 and V, the surface area per gram is

$$S_g = 4.35 \times 10^4 v_m \qquad \text{cm}^2/\text{g solid adsorbent} \tag{8-15}$$

Remember in using Eq. (8-15) that it is based on adsorption measurements with N_2 at $-195.8°C$.

Table 8-2 shows surface areas determined by the Brunauer-Emmett-

[1]P. H. Emmett and S. Brunauer, *J. Am. Chem. Soc.*, **59**, 1553 (1937).

Table 8-2   Surface area, pore volume, and mean pore radii for typical solid catalysts

Catalyst	Surface area, m^2/g	Pore volume, cm^3/g	Mean pore radius, Å
Activated carbons	500–1,500	0.6–0.8	10–20
Silica gels	200–600	0.4	15–100
SiO-Al$_2$O$_3$ cracking catalysts	200–500	0.2–0.7	33–150
Activated clays	150–225	0.4–0.52	100
Activated alumina	175	0.39	45
Celite (Kieselguhr)	4.2	1.1	11,000
Synthetic ammonia catalysts, Fe	. . .	0.12	200–1,000
Pumice	0.38		
Fused copper	0.23		

SOURCE: In part from A. Wheeler, "Advances in Catalysis," vol. III, pp. 250–326, Academic Press, Inc., New York, 1950.

Teller method for a number of common catalysts and carriers. Calculations of surface areas from adsorption data are illustrated in Examples 8-2 and 8-3.

Example 8-2   From the Brunauer-Emmett-Teller plot in Fig. 8-4 estimate the surface area per gram of the silica gel. Use the data for adsorption of nitrogen at $-195.8°C$.

Solution   From curve 6 of Fig. 8-4, the intercept on the ordinate is

$$I = 0.1 \times 10^{-3} \text{ cm}^{-3}$$

The slope of the curve is

$$s = \frac{(5.3 - 0.1) \times 10^{-3}}{0.4 - 0} = 13 \times 10^{-3} \text{ cm}^{-3}$$

These values of s and I may be substituted in Eq. (8-12) to obtain v_m,

$$v_m = \frac{10^3}{0.1 + 13} \frac{1}{0.606} = 126 \text{ cm}^3/\text{g catalyst}$$

The factor 0.606 is introduced because the data in Fig. 8-4 are for a silica gel sample of 0.606 g, and v_m is the monomolecular volume per gram. For nitrogen at $-195.8°C$, the application of Eq. (8-15) yields

$$S_g = 4.35(126) = 550 \text{ m}^2/\text{g}$$

Example 8-3   For comparison, estimate the surface area of the silica gel by using the adsorption data for oxygen at $-183°C$. The density of the liquefied oxygen at $-183°C$, from the International Critical Tables, is 1.14 g/cm.

Solution   First the area of an adsorbed molecule of O_2 must be calculated from Eq. (8-14):

$$\alpha = 1.09 \left[\frac{32}{(6.02 \times 10^{23}) \, 1.14} \right]^{\frac{2}{3}} = 14.2 \times 10^{-16} \text{ cm}^2$$

With this value of α the area equation [Eq. (8-13)] becomes

$$S_g = \frac{v_m(6.02 \times 10^{23})}{22,400} \, 14.2 \times 10^{-16} = 3.8 \times 10^4 v_m \quad \text{cm}^2/\text{g}$$

From curve 4 of Fig. (8-4),

$$I = 0.40 \times 10^{-3} \text{ cm}^{-3}$$

$$s = \frac{(5.4 - 0.4) \times 10^{-3}}{0.38 - 0} = 13.2 \times 10^{-3} \text{ cm}^{-3}$$

Then the monomolecular volume per gram of silica gel is, from Eq. (8-12),

$$v_m = \frac{10^3}{0.4 + 13.2} \frac{1}{0.606} = 122 \text{ cm}^3/\text{g catalyst}$$

Finally, substituting this value of v_m in the area expression gives

$$S_g = 3.8 \times 10^4 (122) = 465 \times 10^4 \text{ cm}^2/\text{g} \quad \text{or } 465 \text{ m}^2/\text{g}$$

The difference in area determined from the N_2 and O_2 data is somewhat larger than expected for these gases. The adsorption curve for N_2 at $-183°C$ gives a value in closer agreement with 550 m^2/g (see Prob. 8-1).

8-6 Void Volume and Solid Density

The void volume, or pore volume, of a catalyst particle can be estimated by boiling a weighed sample immersed in a liquid such as water. After the air in the pores has been displaced, the sample is superficially dried and weighed. The increase in weight divided by the density of the liquid gives the pore volume.

A more accurate procedure is the *helium-mercury method*. The volume of helium displaced by a sample of catalyst is measured; then the helium is removed, and the volume of mercury displaced is measured. Since mercury will not fill the pores of most catalysts at atmospheric pressure, the difference in volumes gives the pore volume of the catalyst sample. The volume of helium displaced is a measure of the volume occupied by the solid material. From this and the weight of the sample, the density of the solid phase, ρ_S, can be obtained. Then the void fraction, or porosity, of the particle, ϵ_p, may be calculated from the equation

$$\epsilon_p = \frac{\text{void (pore) volume of particle}}{\text{total volume of particle}} = \frac{m_p V_g}{m_p V_g + m_p(1/\rho_S)}$$

$$= \frac{V_g \rho_S}{V_g \rho_S + 1} \tag{8-16}$$

where m_p is the mass of the particle and V_g is the void volume per gram of particles. If the sample of particles is weighed, the mass divided by the mercury volume gives the density of the porous particles. Note that the porosity is also obtainable from the density by the expression

$$\epsilon_p = \frac{\text{void volume}}{\text{total volume}} = \frac{V_g}{1/\rho_p} = \rho_p V_g \tag{8-17}$$

From the helium-mercury measurements the pore volume, the solid density, and the porosity of the catalyst particle can be determined. Values of ϵ_p are of the order of 0.5, indicating that the particle is about half void space and half solid material. Since overall void fractions in packed beds are about 0.4, a rule of thumb for a fixed-bed catalytic reactor is that about 30% of the volume is pore space, 30% is solid catalyst and carrier, and 40% is void space between catalyst particles. Individual catalysts may show results considerably different from these average values, as indicated in Examples 8-4 and 8-5.

Example 8-4 In an experiment to determine the pore volume and catalyst-particle porosity the following data were obtained on a sample of activated silica (granular, 4 to 12 mesh size):

Mass of catalyst sample placed in chamber = 101.5 g
Volume of helium displaced by sample = 45.1 cm³
Volume of mercury displaced by sample = 82.7 cm³

Calculate the required properties.

Solution The volume of mercury displaced, minus the helium-displacement volume, is the pore volume. Hence

$$V_g = \frac{82.7 - 45.1}{101.5} = 0.371 \text{ cm}^3/\text{g}$$

The helium volume is also a measure of the density of the solid material in the catalyst; that is,

$$\rho_S = \frac{101.5}{45.1} = 2.25 \text{ g/cm}^3$$

Substituting the values of V_g and ρ_S in Eq. (8-16) gives the porosity of the silica gel particles,

$$\epsilon_p = \frac{0.371(2.25)}{0.371(2.25) + 1} = 0.455$$

To avoid excessive pressure drops and improve mechanical strength, porous particles often must be pelletted to sizes of $\frac{1}{16}$ to 1 in. Usually the pellets are cylindrical, although spherical and granular assemblies are sometimes used. Agglomeration of porous particles gives a pellet containing two void regions: small void spaces within the individual particles and larger spaces between particles. Hence such materials are said to contain *bidisperse pore systems*. Although the shape and nature of these two void regions may vary from thin cracks to a continuous region surrounding a group of particles, it has been customary to designate both regions as *pores*. The void spaces within the particles are commonly termed *micropores*, and the void regions between particles are called *macropores*. *Particle* refers only to the small individual unit from which the pellet is produced. We shall use this nomenclature in discussing solid catalysts.

Not all catalyst supports can be agglomerized. Perhaps the most widely used pellets are those of alumina. Porous alumina particles (20 to 200 microns diameter) containing micropores of 10 to 200 Å diameter are readily prepared by spray drying. These somewhat soft particles are easily made into pellets. The macroporosity and macropore diameter depend on the pelleting pressure and can be varied over a wide range. Table 8-3 shows macro and micro properties of five alumina pellets, each prepared at a different pelleting pressure. The pellet density listed in the second column is approximately proportional to the pressure used. Comparison of the least and greatest pellet density shows that the macropore volume has decreased from 0.670 to 0.120 with increased pelletting pressure, while the micropore volume has decreased only from 0.434 to 0.365.

Table 8-3   Physical properties of alumina pellets

Density, g/cm³		Pore volume, cm³/g		Void fraction		
Particle	*Pellet*	*Micro*	*Macro*	*Total*	*Macro*	*Micro*
1.292	1.121	0.365	0.120	0.543	0.134	0.409
1.264	1.010	0.383	0.198	0.587	0.200	0.387
1.238	0.896	0.400	0.308	0.634	0.275	0.359
1.212	0.785	0.416	0.451	0.680	0.353	0.327
1.188	0.672	0.434	0.670	0.725	0.450	0.275

NOTES: All properties are based on Al_2O_3. Micro refers to pore radii less than 100 Å, and macro refers to radii greater than 100 Å.

SOURCE: R. A. Mischke and J. M. Smith, *Ind. Eng. Chem., Fund. Quart.*, **1**, 288 (1962).

The external surface area of even very fine particles has been shown (Example 8-1) to be small with respect to the internal surface of the pores. Hence, in a catalyst pellet the surface resides predominantly in the small pores within the particles. The external surface of the particles, and of course the external area of the pellets, is negligible.

Macro- and micropore volumes and porosities for bidisperse catalyst pellets are calculated by the same methods as used for monodisperse pore systems. Example 8-5 illustrates the procedure.

Example 8-5   A hydrogenation catalyst is prepared by soaking alumina particles (100 to 150 mesh size) in aqueous $NiNO_3$ solution. After drying and reduction, the particles contain about 7 wt % NiO. This catalyst is then made into large cylindrical pellets for rate studies. The gross measurements for one pellet are

Mass = 3.15 g
Diameter = 1.00 in.
Thickness = $\frac{1}{4}$ in.
Volume = 3.22 cm^3

The Al_2O_3 particles contain micropores, and the pelletting process introduces macropores surrounding the particles. From the experimental methods already described, the macropore volume of the pellet is 0.645 cm^3 and the micropore volume is 0.40 cm^3/g of particles. From this information calculate:

(a)   The density of the pellet
(b)   The macropore volume in cubic centimeters per gram
(c)   The macropore void fraction in the pellet
(d)   The micropore void fraction in the pellet
(e)   The solid fraction
(f)   The density of the particles
(g)   The density of the solid phase
(h)   The void fraction of the particles

Solution

(a)   The density of the pellet is

$$\rho_P = \frac{3.15}{3.22} = 0.978 \text{ g/cm}^3$$

(b)   The macropore volume per gram is

$$(V_g)_M = \frac{0.645}{3.15} = 0.205 \text{ cm}^3/\text{g}$$

(c)   The macropore void fraction ϵ_M is obtained by applying Eq. (8-17) to the pellet. Thus

$$\epsilon_M = \frac{\text{macropore volume}}{\text{total volume}} = \frac{(V_g)_M}{1/\rho_P} = \frac{0.205}{1/0.978} = 0.200$$

(d) Since

$$(V_g)_\mu = 0.40 \text{ cm}^3/\text{g}$$

the micropore void fraction ϵ_μ in the pellet is

$$\epsilon_\mu = \frac{(V_g)_\mu}{1/\rho_P} = \frac{0.40}{1/0.978} = 0.391$$

(e) The solids fraction ϵ_S is given by

$$1 = \epsilon_M + \epsilon_\mu + \epsilon_S$$

$$\epsilon_S = 1 - 0.200 - 0.391 = 0.409$$

(f) The density ρ_p of the particles can be calculated by correcting the total volume of the pellet for the macropore volume. Thus

$$\rho_p = \frac{3.15}{3.22 - 0.645} = 1.22 \text{ g/cm}^3$$

or, in terms of 1 g of pellet,

$$\rho_p = \frac{1}{1/\rho_P - (V_g)_M} = \frac{\rho_P}{1 - (V_g)_M \rho_P}$$

$$\rho_p = \frac{0.978}{1 - 0.205(0.978)} = 1.22 \text{ g/cm}^3$$

(g) The density of the solid phase is

$$\rho_S = \frac{\text{mass of pellet}}{(\text{volume of pellet}) \, \epsilon_S}$$

$$= \frac{\rho_P}{\epsilon_S} = \frac{0.978}{0.409} = 2.39 \text{ g/cm}^3$$

(h) The void fraction of the particles is given by

$$\epsilon_p = \frac{(V_g)_\mu}{1/\rho_p} = \rho_p(V_g)_\mu$$

$$= 1.22(0.40) = 0.49$$

For this pellet a fraction equal to $\epsilon_M + \epsilon_\mu = 0.591$ is void and 0.409 is solid. Of the individual particles, a fraction 0.49 is void. Note that all these results were calculated from the mass and volume of the pellet and the measurements of macro- and micropore volumes.

8-7 Pore-volume Distribution

We shall see in Chap. 11 that the effectiveness of the internal surface for catalytic reactions can depend not only on the extent of the void spaces

(V_g), but also on the size of the openings. Therefore it is desirable to know the distribution of void volume in a catalyst according to size of the opening. This is a difficult problem because the void spaces in a given particle are non uniform in size, shape, and length, and normally are interconnected. Further, these characteristics can change radically from one type of catalyst particle to another. Figure 8-5 shows electron-microscope (scanning type) photographs of porous silver particles ($S_g = 19.7$ m^2/g). The material was prepared by reducing a precipitate of silver fumarate by heating at 350°C in a stream of nitrogen. The larger darker regions probably represent void space between individual particles, and the smaller dark spaces are intraparticle voids. The light portions are solid silver. The complex and random geometry shows that it is not very realistic to describe the void spaces as pores. It is anticipated that other highly porous materials such as alumina and silica would have similar continuous and complex void phases. For a material such as Vycor, with its relatively low porosity (0.3) and continuous solid phase, the concept of void spaces as pores is more reasonable.

In view of evidence such as that in Fig. 8-5, it is unlikely that detailed quantitative descriptions of the void structure of solid catalysts will become available. Therefore, to account quantitatively for the variations in rate of reaction with location within a porous catalyst particle, a simplified model of the pore structure is necessary. The model must be such that diffusion rates of reactants through the void spaces into the interior surface can be evaluated. More is said about these models in Chap. 11. It is sufficient here to note that in all the widely used models the void spaces are simulated as cylindrical pores. Hence the size of the void space is interpreted as a radius a of a cylindrical pore, and the distribution of void volume is defined in terms of this variable. However, as the example of the silver catalyst indicates, this does not mean that the void spaces are well-defined cylindrical pores.

There are two established methods for measuring the distribution of pore volumes. The mercury-penetration method depends on the fact that mercury has a significant surface tension and does not wet most catalytic surfaces. This means that the pressure required to force mercury into the pores depends on the pore radius. The pressure varies inversely with a; 100 psi (approximately) is required to fill pores for which $a = 10,000$ Å, and 10,000 psi is needed for $a = 100$ Å. Simple techniques and equipment are satisfactory for evaluating the pore-volume distribution down to 100 to 200 Å, but special high-pressure apparatus is necessary to go below $a = 100$ Å, where much of the surface resides. In the second method, the nitrogen-adsorption experiment (described in Sec. 8-5 for surface area measurement) is continued until the nitrogen pressure approaches the

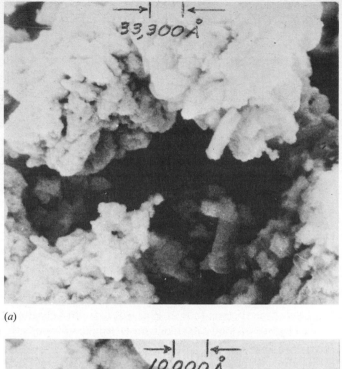

(a)

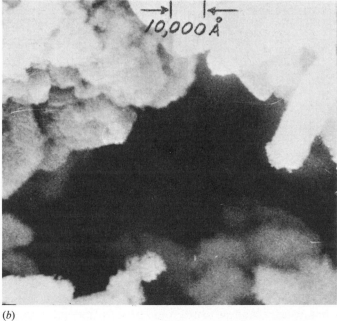

(b)

Fig. 8-5 *Electron micrographs of porous silver particles (approximate surface area = 19.7 m^2/g): (a) magnification = 3,000 (1 cm = 33,300 Å), (b) magnification = 10,000 (1 cm = 10,000 Å)*

saturation value (1 atm at the normal boiling point). At $p/p_0 \to 1.0$, where p_0 is the saturation pressure, all the void volume is filled with adsorbed and condensed nitrogen. Then a desorption isotherm is established by lowering the pressure in increments and measuring the amount of nitrogen evaporated and desorbed for each increment. Since the vapor pressure of a liquid evaporating from a capillary depends on the radius of the capillary, these data can be plotted as volume desorbed vs pore radius. Thus this procedure also gives the distribution of pore volumes. Since the vapor pressure is not affected significantly by radii of curvature greater than about 200 Å, this method is not suitable for pores larger than 200 Å.

A combination of the two methods is normally necessary to cover the entire range of pore radii (10 to 10,000 Å) which may exist in a bidisperse catalyst or support, such as alumina pellets. For a monodisperse pore distribution, such as that in silica gel, the nitrogen-desorption experiment is sufficient, since there are few pores of radius greater than 200 Å. In a bidisperse pore system the predominant part of the catalytic reaction occurs in pores less than about 200 Å (the micropore region), since that is where the bulk of the surface resides. However, the transport of reactants to these small pores occurs primarily in pores of 200 to 10,000 Å (the macropore region). Hence the complete distribution of pore volume is necessary to establish the effectiveness of the interior surface, that is, the global rate of reaction. Calculation procedures and typical results are discussed briefly in the following paragraphs.

Mercury-penetration Method   By equating the force due to surface tension (which tends to keep mercury out of a pore) to the applied force, Ritter and Drake[1] obtained

$$\pi a^2 p = -2\pi a \sigma \cos \theta$$

or

$$a = \frac{-2\sigma \cos \theta}{p} \tag{8-18}$$

where θ is the contact angle between the mercury and pore wall (Fig. 8-6). While θ probably varies somewhat with the nature of the solid surface, 140° appears to be a good average value. Then the working equation for evaluating the radius corresponding to a given pressure is

$$a\,(\text{Å}) = \frac{8.75 \times 10^5}{p\,(\text{psi})} \tag{8-19}$$

[1] H. L. Ritter and L. C. Drake, *Ind. Eng. Chem., Anal. Ed.*, **17**, 787 (1945).

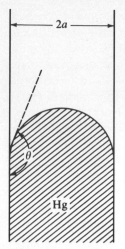

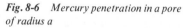

Fig. 8-6   *Mercury penetration in a pore of radius a*

Calculation of the pore-size distribution by this method is illustrated in Example 8-6.

Example 8-6   The mercury-penetration data given in Table 8-4 were obtained on a 0.624-g sample of a uranium dioxide pellet formed by sintering particles at 1000°C for 2 hr. Since the particles were nonporous, the void space was entirely between the

Table 8-4   Mercury porosimeter data for uranium dioxide pellet (*mass of sample 0.624 g*)

Pressure, psi	Mercury penetration, cm^3	Penetration, cm^3/g	
116	0.002	0.003	0.196
310	0.006	0.010	0.189
344	0.010	0.016	0.183
364	0.014	0.022	0.177
410	0.020	0.032	0.167
456	0.026	0.042	0.157
484	0.030	0.048	0.151
540	0.038	0.061	0.138
620	0.050	0.080	0.119
710	0.064	0.102	0.097
800	0.076	0.122	0.077
830	0.080	0.128	0.071
900	0.088	0.141	0.058
1,050	0.110	0.160	0.039
1,300	0.112	0.179	0.020
1,540	0.118	0.189	0.010
1,900	0.122	0.196	0.003
2,320	0.124	0.198	0.001
3,500	0.125	0.199	0

particles (macropores). At the beginning of the experiment (when the pressure was 1.77 psia) the amount of mercury displaced by the sample was found to be 0.190 cm³. Calculate the porosity and pore-volume distribution of the pellet.

Solution According to Eq. (8-19), at $p = 1.77$ psia only pores larger than about 500,000 Å (50 microns) would be filled with mercury. No pores larger than this are likely. Hence 0.190 cm³ is the total volume of the sample. At the highest pressure, 3,500 psia, only pores less than $a = 250$ Å would remain unfilled. Since the pores were entirely of the macro type, few pores smaller 250 Å are expected. If such pores are neglected, the porosity can be calculated from the porosimeter measurements alone. Thus

$$\epsilon_P = \frac{0.125}{0.190} = 0.66$$

A check on this result is available from air-pycnometer[1] data, which gave a solid volume of 0.0565 cm³. Thus the total porosity is

$$(\epsilon_P)_t = \frac{0.190 - 0.0565}{0.190} = 0.70$$

The difference between values suggests that there were a few pores smaller than 250 Å in the sample, although the comparison also includes experimental errors in the two methods.

To calculate the pore-volume distribution the penetration data are first corrected to a basis of 1 g of sample, as given in the third column of Table 8-4. If we neglect the pores smaller than 250 Å, the penetration data can be reversed, starting with $V = 0$ at 3,500 psia (250 Å). The last column shows these figures. Then, from Eq. (8-19) and the pressure, the radius corresponding to each penetration value can be established. This gives the penetration curve shown in Fig. 8-7. The penetration volume at any pore radius a is the volume of pores larger than a. The derivative of this curve, $\Delta V/\Delta a$, is the volume of pores between a and $a + \Delta a$ divided by Δa; that is, it is the distribution function for the pore volume according to pore radius. It is customary to plot the pore radius on a logarithmic coordinate as shown. Hence the distribution function is taken as the derivative of the curve so plotted, that is, $dV/d(\log a)$. The distribution function is also shown plotted against a in Fig. 8-7.

For this pellet the distribution is seen to be reasonably symmetrical with most of the volume in pores from 300 to 8,000 Å and with a most probable pore radius of 1,200 Å. Note that the flatness of the penetration curve at low pore radii justifies neglecting the pores smaller than 250 Å.

Wheeler[2] has summarized the assumptions and accuracy of the mercury-penetration method. It is important to note that erroneous results would be obtained if the porous particle contains large void spaces that are

[1] A device which uses air at two pressures to measure void volumes of porous materials. It provides the same data as the helium measurement described in Sec. 8-6.

[2] A. Wheeler, in P. H. Emmett (ed.), "Catalysis," vol. II, p. 123, Reinhold Publishing Corporation, New York, 1955.

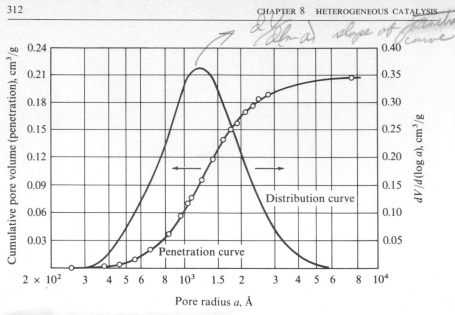

Fig. 8-7   Pore-volume distribution in a UO_2 *pellet*

connected only to smaller void spaces. Such "bottleneck" pores would fill with mercury at the higher pressure corresponding to the connecting smaller pores. For accurate results each porous region must be connected to at least one larger pore.

Nitrogen-desorption Method   As the low-temperature nitrogen-adsorption experiment (Sec. 8-5) is continued to higher pressures multilayer adsorption occurs, and ultimately the adsorbed films are thick enough to bridge the pore.[1] Then further uptake of nitrogen will result in capillary condensation. Since the vapor pressure decreases as the capillary size decreases, such condensation will occur first in the smaller pores. Condensation will be complete, as $p/p_0 \to 1.0$, when the entire void region is filled with condensed nitrogen. Now, if the pressure is reduced by a small increment, a small amount of nitrogen will evaporate from the meniscus formed at the ends of the largest pores. Pores which are emptied of condensate in this way will be those in which the vapor pressure of nitrogen is greater than the chosen pressure. The Kelvin equation gives the relationship between vapor pressure and radius of the concave surface of the meniscus of the liquid. Since some of the nitrogen is adsorbed on the surface, and therefore not present because of capillary condensation, the Kelvin relationship must be corrected for the thickness δ of the adsorbed layers. With this correction, the pore radius is

[1] L. H. Cohan, *J. Am. Chem. Soc.*, **60**, 433 (1938); A. G. Foster, *J. Phys. Colloid Chem.*, **55**, 638 (1951).

related to the saturation-pressure ratio (vapor pressure in the pore p divided by the normal vapor pressure p_0) by

$$a - \delta = \frac{-2\sigma V_l \cos \theta}{R_g T \ln (p/p_0)} \qquad \textit{Kelvin eqn} \qquad (8\text{-}20)$$

where V_l = molal volume of the condensed liquid
   σ = surface tension
   θ = contact angle between surface and condensate

Since nitrogen completely wets the surface covered with adsorbed nitrogen, $\theta = 0°$ and $\cos \theta = 1$. The thickness δ depends on p/p_0. The exact relationship has been the subject of considerable study,[1] but Halsey's form

$$\delta = A\left(\ln\frac{p_0}{p}\right)^{-1/n} \qquad (8\text{-}21)$$

where A and n depend on the nature of the catalyst surface, is generally used.

For nitrogen at $-195.8°C$ (normal boiling point) Eq. (8-20), for $a - \delta$ in Angstroms, becomes

$$a - \delta = 9.52\left(\log\frac{p_0}{p}\right)^{-1} \qquad (8\text{-}22)$$

Wheeler proposes for Eq. (8-21)

$$\delta\,(\text{Å}) = 7.34\left(\ln\frac{p_0}{p}\right)^{-\frac{1}{3}} \qquad (8\text{-}23)$$

For a chosen value of p/p_0, Eqs. (8-22) and (8-23) give the pore radius above which all pores will be empty of capillary condensate. Hence, if the amount of desorption is measured for various p/p_0, the pore volume corresponding to various radii can be evaluated. Differentiation of the curve for cumulative pore volume vs radius gives the distribution of volume as described in Example 8-6. Descriptions of the method of computation are given by several investigators.[2] As in the mercury-penetration method, errors will result unless each pore is connected to at least one larger pore.

Figure 8-8 shows the result of applying the method to a sample of Vycor (porous glass).[3] This material, which contained only micropores, had the properties

[1] A. Wheeler, in P. H. Emmett (ed.), "Catalysis," vol. II, chap. 2, Reinhold Publishing Corporation, New York, 1955; G. D. Halsey, *J. Chem. Phys.*, **16**, 931 (1948); C. G. Shull, *J. Am. Chem. Soc.*, **70**, 1405 (1948); J. O. Mingle and J. M. Smith, *Chem. Eng. Sci.*, **16**, 31 (1961).
[2] E. P. Barrett, L. G. Joyner, and P. P. Halenda, *J. Am. Chem. Soc.*, **73**, 373 (1951); C. J. Pierce, *J. Phys. Chem.*, **57**, 149 (1953); R. B. Anderson, *J. Catalysis*, **3**, 50 (1964).
[3] M. R. Rao and J. M. Smith, *AIChE J.*, **10**, 293 (1964).

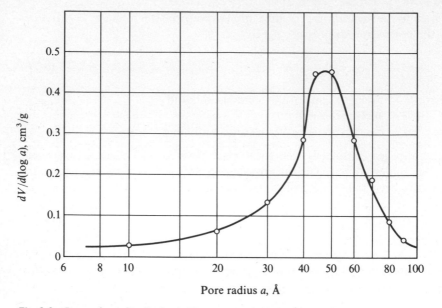

Fig. 8-8   *Pore-volume distribution in Vycor; $\rho_p = 1.46 \ g/cm^3$, $V_g = 0.208$ cm^3/g, $S_g = 90 \ m^2/g$*

Fig. 8-9   *Pore volume in alumina (boehmite) pellets*

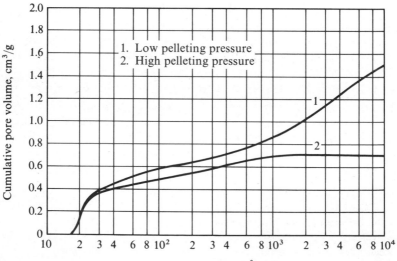

$$\rho_p = 1.46 \text{ g/cm}^3$$

$$V_g = 0.208 \text{ cm}^3/\text{g}$$

$$\epsilon_p = 0.304$$

$$S_g = 90 \text{ m}^2/\text{g}$$

The surface area was determined from nitrogen-adsorption data in the low p/p_0 range, as described in Sec. 8-5, while the distribution results in Fig. 8-8 were established from the desorption curve in the capillary-condensation (high p/p_0) region.

By combining mercury-penetration and nitrogen-desorption measurements, pore-volume information can be obtained over the complete range of radii in a pelleted catalyst containing both macro- and micropores. Figure 8-9 shows the cumulative pore volume for two alumina pellets, each prepared by compressing porous particles of boehmite ($Al_2O_3 \cdot H_2O$). The properties[1] of the two pellets are given in Table 8-5. The only difference in the two is the pelleting pressure. Increasing this pressure causes drastic reductions in the space between particles (macropore volume) but does not greatly change the void volume within the particles or the surface area. The derivative of the volume curves in Fig. 8-9 gives the pore-volume distribution, and these results are shown in Fig. 8-10. In this figure the bidisperse pore system, characteristic of alumina pellets, is clearly indicated. The micropore range within the particles is narrow, with a most probable

---

[1] The properties and pore-volume distribution were determined by M. F. L. Johnson, Sinclair Research Laboratories, Harvey, Ill., by the methods described in Secs. 8-5 to 8-7. They were orginally reported in J. L. Robertson and J. M. Smith, *AIChE J.*, **9**, 344 (1963).

Fig. 8-10   *Pore-volume distribution in alumina pellets*

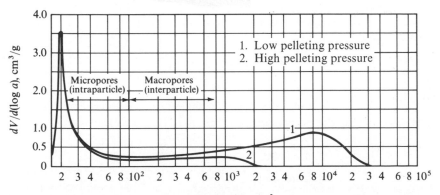

Table 8-5   Properties of boehmite ($Al_2O_3 \cdot H_2O$) pellets

Pelleting pressure	Macropore volume, cm^3/g	Micropore volume, cm^3/g	Surface area, m^2/g
Low	1.08	0.56	389
High	0.265	0.49	381

NOTES: Volume and surface area refer to mass of Al_2O_3 obtained by ignition of boehmite. The pore-volume distribution is given in Fig. 8-10. Average particle size from which pellets were made was 85 microns.

radius of 20 Å. The macropores cover a much wider range of radii and show the effect of pelleting pressure. For the high-pressure pellet all volume with pores greater than 2,000 Å has been squeezed out, while the most probable radius for the low-pressure pellet is 8,000 Å. Pelleting pressure seems to have little effect on the micropores, which suggests that the particles themselves are not crushed significantly during the pelleting process.

Some models (see Chap. 11) for quantitative treatment of the effectiveness of the internal catalyst surface require only the average pore radius $\bar{a}$, rather than the distribution of pore volumes. Wheeler[1] has developed a simple equation for $\bar{a}$ which requires only surface-area and pore-volume measurements. Suppose all the pores in a hypothetical particle are straight, cylindrical, not interconnected, and have the same radius $\bar{r}$ and length $\bar{L}$. The average pore radius may be found by writing equations for the total surface and volume in the hypothetical particle and equating these quantities to the surface $m_p S_g$ and volume $m_p V_g$ in the actual particle; i.e.

$$m_p S_g = (2\pi \bar{a} \bar{L})n \tag{8-24}$$

$$m_p V_g = (\pi \bar{a}^2 \bar{L})n \tag{8-25}$$

where m_p and n are the mass and number of pores per particle. Dividing the two equations gives the Wheeler average pore radius,

$$\bar{a} = \frac{2V_g}{S_g} \tag{8-26}$$

This expression agrees well with volume-average values obtained from the distribution curve for monodisperse pore systems. For example, from the data for the Vycor sample (Fig. 8-7) Eq. (8-26) gives

$$\bar{a} = \frac{2(0.208)}{90 \times 10^4} = 46 \times 10^{-8} \text{ cm} \qquad \text{or 46 Å}$$

[1] A. Wheeler, in P. H. Emmett (ed.), "Catalysis," vol. II, chap. 2, Reinhold Publishing Corporation, New York, 1955.

The volume-average value is calculated from the pore volume data used to obtain the distribution curve (Fig. 8-8) and the expression

$$\bar{a} = \frac{\int_0^{V_g} a \, dV}{V_g}$$

By this method $\bar{a} = 45$ Å.

Accurate values of the small areas existing in macropore systems make it difficult to use Eq. (8-26) to calculate $\bar{a}$ for interparticle pores. Hence the average radius for systems such as the UO_2 pellets discussed in Example 8-6 should be obtained by integrating under the cumulative-volume-vs-$\bar{a}$ curve shown in Fig. 8-7. Also a single value of $\bar{a}$ has no meaning for a bidisperse pore system such as that in an alumina pellet. Thus using the total pore volume in Table 8-5 for the low-pressure pellet gives

$$\bar{a} = \frac{2(1.64)}{389 \times 10^4} = 84 \times 10^{-8} \text{ cm} \qquad \text{or } 84 \text{ Å}$$

As Fig. 8-10 shows, there is very little pore volume (very few pores) in this radius region. If Eq. (8-26) is applied to the micropore region, arbitrarily taken as pores smaller than $a = 100$ Å,

$$\bar{a}_\mu = \frac{2(0.56)}{389 \times 10^4} = 29 \times 10^{-8} \text{ cm} \qquad \text{or } 29 \text{ Å}$$

Figure 8-10 shows that this is an approximate value for the *average* pore radius in the micropore region. Note that the micropore distribution is asymmetrical in such a way that the average radius is greater than the most probable value (20 Å).

Summary In concluding the treatment of physical properties of catalysts, let us review the purpose for studying properties and structure of porous solids. Heterogeneous reactions with solid catalysts occur on parts of the surface active for chemisorption. The number of these active sites and the rate of reaction is, in general, proportional to the extent of the surface. Hence it is necessary to know the surface area. This is evaluated by low-temperature-adsorption experiments in the pressure range where a mono-molecular layer of gas (usually nitrogen) is physically adsorbed on the catalyst surface. The effectiveness of the interior surface of a particle (and essentially all of the surface is in the interior) depends on the volume and size of the void spaces. The pore volume (and porosity) can be obtained by simple pycnometer-type measurements (see Examples 8-4 and 8-5). The average size (pore radius) can be estimated by Eq. (8-26) from the surface area and pore volume in some *monodisperse* systems. Determination of the complete distribution of pore volume according to pore radius requires either mercury-penetration measurements or nitrogen-adsorption data at pressures where capillary condensation occurs, or both. Accurate values of the mean pore radius can be evaluated from such pore-volume-vs-radius data. Note also that a measurement of the complete nitrogen-adsorption-desorption isotherm is sufficient to calculate surface area and pore volume, and distribution of pore sizes, in the range $10 \text{ Å} < a < 200 \text{ Å}$.

CLASSIFICATION AND PREPARATION OF CATALYSTS

8-8 Classification of Catalysts

In discussing catalysis it is well to recall proposals concerning what factors are important in making solids active as catalysts. In chronological order, Sabatier[1] suggested that a mechanism for the activity of nickel as a hydrogenation catalyst might involve formation of a chemical compound, nickel hydride. Since then, the *chemical factor* continues to be recognized as important. Later Taylor,[2] Balandin,[3] and Beeck[4] provided evidence for the significance of *geometric properties*. According to this concept the catalytic activity of a solid surface depends on the spacing of atoms so as to facilitate adsorption of reactant molecules. Over the years most of the evidence for the geometric theory has proved suspect, except that for metallic films. The work of Boudart[5] and Beeck changed the emphasis from geometric considerations to *electronic properties*. In 1948 Dowden and coworkers[6] proposed that catalysts might be classified, on the basis of electron mobility, as conductors, semiconductors, and insulators.

The conductor catalysts are the metals (silver, platinum, vanadium, iron, etc.) and have the property of chemisorption by electron transfer. The semiconductor catalysts are the oxides, such as NiO, Cu_2O, and ZnO. These materials have the capability of interchanging electrons from the filled valence bands in a compound when sufficient energy is provided,

[1] P. Sabatier, "Catalysis in Organic Chemistry," trans. by E. E. Reid, in "Catalysis, Then and Now," Franklin Publishing Company, Englewood, N.J., 1965.

[2] H. S. Taylor, *Proc. Roy. Soc. (London)*, **A108**, 105 (1925).

[3] A. A. Balandin, "Advances in Catalysis," vol. X, p. 96, Academic Press, Inc., New York, 1958.

[4] O. Beeck, *Disc. Faraday Soc.*, **8**, 118 (1950).

[5] M. Boudart, *J. Am. Chem. Soc.*, **72**, 1040 (1950).

[6] D. A. Dowden, *Research*, **1**, 239 (1948); D. A. Dowden and P. W. Reynolds, *Disc. Faraday Soc.*, **8**, 187 (1950).

for example, by heating. Upon this electron transfer the semiconductor becomes a conductor.[1] The insulator catalysts include such widely used substances as silica gel, alumina, and their combinations. Even at high temperatures, electrons are not supposed to be able to move through these two solids freely enough to justify their being called conductors. These substances are also known to be strong acids. Their activity in the many hydrocarbon reactions which they catalyze is presumably due to the formation of carbonium ions at the acid sites on the surface. While much has been written about carbonium-ion mechanisms, the basic concepts are well described in the original work of Whitmore[2] and the later work of Greensfelder.[3]

It should be emphasized that the electronic theory is not without uncertainities and at present should be considered as a concept in transition.[4] However, it does provide a convenient, and probably helpful, method of classifying solid catalysts.

Metals chemisorb oxygen and hydrogen and therefore are usually effective catalysts for oxidation-reduction and hydrogenation-dehydrogenation reactions. Thus platinum is a successful catalyst for the oxidation of SO_2, and Ni is used effectively for hydrogenation of hydrocarbons. The metal oxides, as semiconductors, catalyze the same kinds of reactions, but often higher temperatures are required. Because of the relative strength of the chemisorption bond with which such gases as O_2 and CO are attached to metals, these gases are poisons when metals are employed as hydrogenation catalysts. The semiconductor oxides are less susceptible to such poisoning. Oxides of the transition metals, such as MoO_3 and Cr_2O_3, are good catalysts for polymerization of olefins.

In connection with miscellaneous catalysts the work on polymerizing ethylene should be mentioned. It has been found[5] that aluminum alkyl–

---

[1] Semiconductors are classified as p type if they tend to attract electrons from the chemisorbed species, or as n type if they donate electrons to this species. The p type are normally the compounds, such as NiO. The n type are substances which contain small amounts of impurities, or the oxide is present in nonstoichiometric amounts (as, for example, when some of the zinc in ZnO has been reduced). Reviews of semiconductors as catalysts are given by P. H. Emmett ("New Approaches to the Study of Catalysis," 36th Annual Priestly Lectures, Pennsylvania State University, April 9–13, 1962) and by P. G. Ashmore ("Catalysis and Inhibition of Chemical Reactions," Butterworths & Co. (Publishers), London, 1963.

[2] F. C. Whitmore, *J. Am. Chem. Soc.*, **54**, 3274 (1932).

[3] B. S. Greensfelder, "Chemistry of Petroleum Hydrocarbons," vol. II, chap. 27, Reinhold Publishing Corporation, New York, 1955.

[4] An analysis of the electronic theory is given by Th. Volkenstein, in "Advances in Catalysis," vol. XII, p. 189, Academic Press Inc., New York, 1960.

[5] K. Ziegler, *Angew. Chem.*, **64**, 323 (1952); G. Natta and I. Pasquon, in "Advances in Catalysis," vol. XI, p. 1, Academic Press Inc., New York, 1959.

titanium chloride [for example, $Al(C_2H_5)_3 + TiCl_4$] constitutes an excellent catalyst for producing isotactic polymers from olefins. Alumina and silica catalysts are widely used for alkylation, isomerization, polymerization, and particularly for cracking of hydrocarbons. In each case the mechanism presumably involves carbonium ions formed at the acid sites on the catalyst. While the emphasis here has been on solid catalysts, liquid and gaseous acids, particularly H_2SO_4 and HF, are well-known alkylation and isomerization catalysts.

Often catalysts are specific. An important example is the effectiveness of iron for producing hydrocarbons from hydrogen and carbon monoxide (the Fischer-Tropsch synthesis). Dual-function catalysts for isomerization and reforming reactions consist of two active substances in close proximity to each other. For example, Ciapetta and Hunter[1] found that a silica-alumina catalyst upon which nickel was dispersed was much more effective in isomerizing n-hexane than silica alumina alone. The explanation depends on the fact that olefins are more readily isomerized than paraffin hydrocarbons. The nickel presumably acts as a hydrogenating agent, producing hexene, after which the silica alumina isomerizes the hexene to isohexene. Finally, the nickel is effective in hydrogenating hexene back to hexanes.

This discussion of catalysts for reaction types is general and superficial. Much has and is being written on the subject, and helpful sources of summary information are available.[2]

8-9 Catalyst Preparation

Experimental methods and techniques for catalyst manufacture are particularly important because chemical composition is not enough by itself to determine activity. The physical properties of surface area, pore size, particle size, and particle structure also have an influence. These properties are determined to a large extent by the preparation procedure. To begin with, a distinction should be drawn between preparations in which the entire material constitutes the catalyst and those in which the active ingredient is dispersed on a support or *carrier* having a large surface area. The first kind of catalyst is usually made by precipitation, gel formation, or simple mixing of the components.

Precipitation provides a method of obtaining the solid material in a porous form. It consists of adding a precipitating agent to aqueous solutions

---

[1] F. G. Ciapetta and J. B. Hunter, *Ind. Eng. Chem.*, **45**, 155 (1953).

[2] P. H. Emmett (ed.), "Catalysis," Reinhold Publishing Corporation, New York, 1954–; "Advances in Catalysis," Academic Press Inc., New York, 1949–; *J. Catalysis*, 1, 1962–; A. A. Balandin, "Scientific Selection of Catalysts," trans. by A. Aledjem, Davey Publishing Company, Hartford, Conn., 1968.

of the desired components. Washing, drying, and sometimes calcination and activation are subsequent steps in the process. For example, a magnesium oxide catalyst can be prepared by precipitating the magnesium from nitrate solution by adding sodium carbonate. The precipitate of $MgCO_3$ is washed, dried, and calcined to obtain the oxide. Such variables as concentration of the aqueous solutions, temperature, and time of the drying and calcining steps may influence the surface area and pore structure of the final product. This illustrates the difficulty in reproducing catalysts and indicates the necessity of carefully following tested recipes. Of particular importance is the washing step to remove all traces of impurities, which may act as poisons.

A special case of the precipitation method is the formation of a colloidal precipitate which gels. The steps in the process are essentially the same as for the usual precipitation procedure. Catalysts containing silica and alumina are especially suitable for preparation by gel formation, since their precipitates are of a colloidal nature. Detailed techniques for producing catalysts through gel formation or ordinary precipitation are given by Ciapetta and Plank.[1]

In some instances a porous material can be obtained by mixing the components with water, milling to the desired grain size, drying, and calcining. Such materials must be ground and sieved to obtain the proper particle size. A mixed magnesium and calcium oxide catalyst can be prepared in this fashion. The carbonates are milled wet in a ball machine, extruded, dried, and reduced by heating in an oven.

Catalyst *carriers* provide a means of obtaining a large surface area with a small amount of active material. This is important when expensive agents such as platinum, nickel, and silver are used. Berkman et al.[2] have treated the subject of carriers in some detail.

The steps in the preparation of a catalyst impregnated on a carrier may include (1) evacuating the carrier, (2) contacting the carrier with the impregnating solution, (3) removing the excess solution, (4) drying, (5) calcination and activation. For example, a nickel hydrogenation catalyst can be prepared on alumina by soaking the evacuated alumina particles with nickel nitrate solution, draining to remove the excess solution, and heating in an oven to decompose the nitrate to nickel oxide. The final step, reduction of the oxide to metallic nickel, is best carried out with the particles in place in the reactor by passing hydrogen through the equipment. Activation *in situ* prevents contamination with air and other gases which might

---

[1] F. G. Ciapetta and C. J. Plank, in P. H. Emmett (ed.), "Catalysis," vol I, chap. 7, Reinhold Publishing Corporation, New York, 1954.

[2] S. Berkman, J. C. Morrell, and G. Egloff, "Catalysis," Reinhold Publishing Corporation, New York, 1940.

poison the reactive nickel. In this case no precipitation was required. This is a desirable method of preparation, since thorough impregnation of all the interior surface of the carrier particles is relatively simple. However, if the solution used to soak the carrier contains potential poisons such as chlorides or sulfates, it may be necessary to precipitate the required constituent and wash out the possible poison.

The nature of the support can affect catalyst activity and selectivity. This effect presumably arises because the support can influence the surface structure of the atoms of dispersed catalytic agent. For example, changing from a silica to alumina carrier may change the electronic structure of deposited platinum atoms. This question is related to the optimum amount of catalyst that should be deposited on a carrier. When only a small fraction of a monomolecular layer is added, increases in amount of catalyst should increase the rate. However, it may not be helpful to add large amounts to the carrier. For example, the conversion rate of the ortho to para hydrogen with a NiO catalyst deposited on alumina was found[1] to be less for 5.0 wt % NiO than for 0.5 wt % NiO. The dispersion of the catalyst on the carrier may also be an important factor in such cases. The nickel particles were deposited from a much more concentrated $NiNO_3$ solution to make the catalyst containing 5.0 wt % NiO. This may have led to larger nickel particles. That is, many more nickel atoms were deposited on top of each other, so that the dispersion of nickel on the surface was less uniform than with the 0.5 wt % catalyst. It is interesting to note that a 5.0 wt % NiO catalyst prepared by 10 individual depositions of 0.5 wt % was much more active (by a factor of 11) than the 5.0 wt % added in a single treatment. This method gave a much larger active nickel surface, presumably because of better dispersion of the nickel atoms on the Al_2O_3 surface. Since the total amount of nickel was the same for the two preparations, one would conclude that the individual particles of nickel were smaller in the 10-application catalyst. These kinds of data indicate the importance of measuring surface areas for chemisorption of the reactants involved. A technique based on the chemisorption of H_2 and CO has been developed[2] to study the effect of dispersion of a catalyst on its activity and the effect of interaction between catalyst and support on activity.

8-10 Promoters and Inhibitors

As normally used, the term "catalyst" designates the composite product used in a reactor. Components of the catalyst must include the catalytically

[1] N. Wakao, J. M. Smith, and P. W. Selwood, *J. Catalysis*, **1**, 62 (1962).

[2] G. K. Boreskov and A. P. Karnaukov, *Zh. Fiz. Khim.*, **26**, 1814 (1952); L. Spenadel and M. Boudart, *J. Phys. Chem.*, **64**, 204 (1960).

active substance itself and may also include a carrier, promoters, and inhibitors.

Innes[1] has defined a *promoter* as a substance added during the preparation of a catalyst which improves activity or selectivity or stabilizes the catalytic agent so as to prolong its life. The promoter is present in a small amount and by itself has little activity. There are various types, depending on how they act to improve the catalyst. Perhaps the most extensive studies of promoters has been in connection with iron catalysts for the ammonia synthesis reaction.[2] It was found that adding Al_2O_3 (other promoters are CaO, K_2O) prevented reduction (by sintering) in surface area during catalyst use and gave an increased activity over a longer period of time. Some promoters are also believed to increase the number of active centers and so make the existing catalyst surface more active. The published information on promoters is largely in the patent literature. Innes has tabulated the data appearing from 1942 to 1952.

An *inhibitor* is the opposite of a promoter. When added in small amounts during catalyst manufacture, it lessens activity, stability, or selectivity. Inhibitors are useful for reducing the activity of a catalyst for an undesirable side reaction. For example, silver supported on alumina is an excellent oxidation catalyst. In particular, it is used widely in the production of ethylene oxide from ethylene. However, at the same conditions complete oxidation to carbon dioxide and water also occurs, so that selectivity to C_2H_4O is poor. It has been found that adding halogen compounds to the catalyst inhibits the complete oxidation and results in satisfactory selectivity.

8-11 Poisons (Catalyst Life)

In some systems the catalyst activity decreases so slowly that exchange for new material or regeneration is required only at long intervals. Examples are promoted catalysts for synthetic ammonia and catalysts containing metals such as platinum and silver. Catalysts for cracking and some other hydrocarbon reactions, however, require frequent regeneration. The decrease in activity is due to *poisons*, which will be defined here as substances, either in the reactants stream or produced by the reaction, which lower the activity of the catalyst. The frequent regeneration of cracking catalysts is necessary because of the deposition of one of the products, carbon, on the surface.

Poisons can be differentiated in terms of the way in which they operate. Many summaries listing specific poisons and classifying groups of poisons

[1] W. B. Innes, in P. H. Emmett (ed.), "Catalysis," vol. 1, chap. 7, Reinhold Publishing Corporation, New York, 1954.

[2] P. H. Emmett and S. Brunauer, *J. Am. Chem. Soc.*, **62**, 1732 (1940).

are available.[1] The following arrangement has been taken in part from Innes.

DEPOSITED POISONS   Carbon deposition on catalysts used in the petroleum industry falls into this category. The carbon covers the active sites of the catalyst and may also partially plug the pore entrances. This type of poisoning is at least partially reversible, and regeneration can be accomplished by burning to CO and CO_2 with air and/or steam. The regeneration process itself is a heterogeneous reaction, a gas-solid noncatalytic one. In the design of the reactor, attention must be given to the regeneration as well as to the reaction parts of the cycle. The quantitative description of the drop in activity with time during reaction and the increase in activity during regeneration has been investigated.[2]

CHEMISORBED POISONS   Compounds of sulfur and other materials are frequently chemisorbed on nickel, copper, and platinum catalysts. The decline in activity stops when equilibrium is reached between the poison in the reactant stream and that on the catalyst surface. If the strength of the adsorption compound is low, the activity will be regained when the poison is removed from the reactants. If the adsorbed material is tightly held, the poisoning is more permanent. The mechanism appears to be one of covering the active sites, which could otherwise adsorb reactant molecules.

SELECTIVITY POISONS   The selectivity of a solid surface for catalyzing one reaction with respect to another is not well understood. However, it is known that some materials in the reactant stream will adsorb on the surface and then catalyze other undesirable reactions, thus lowering the selectivity. The very small quantities of nickel, copper, iron, etc., in petroleum stocks may act as poisons in this way. When such stocks are cracked, the metals deposit on the catalyst and act as dehydrogenation catalysts. This results in increased yields of hydrogen and coke and lower yields of gasoline.

STABILITY POISONS   When water vapor is present in the sulfur dioxide-air mixture supplied to a platinum-alumina catalyst, a decrease in oxidation activity occurs. This type of poisoning is due to the effect of water on the structure of the alumina carrier. Temperature has a pronounced effect on

[1] R. H. Griffith, "The Mechanism of Contact Catalysis," p. 93, Oxford University Press, New York, 1936; E. B. Maxted, *J. Soc. Chem. Ind. (London)*, **67**, 93 (1948); P. H. Emmett (ed.), "Catalysis," vol. I, chap. 6, Reinhold Publishing Corporation, New York, 1954.

[2] Typical references are M. Sagara, S. Masamune, and J. M. Smith, *AIChE J.*, **13**, 1226 (1967); G. F. Froment and K. B. Bischoff, *Chem. Eng. Sci.*, **16**, 189 (1961); P. B. Weisz and R. D. Goodwin, *J. Catalysis*, **2**, 397 (1963).

stability poisoning. Sintering and localized melting may occur as the temperature is increased, and this, of course, changes the catalyst structure.

DIFFUSION POISONS This kind of poisoning has already been mentioned in connection with carbon deposition on cracking catalysts. Blocking the pore mouths prevents the reactants from diffusing into the inner surface. Entrained solids in the reactants, or fluids which can react with the catalyst to form a solid residue, can cause this type of poisoning.

Table 8-6 lists poisons for various catalysts and reactions. The materials that are added to reactant streams to improve the performance of a catalyst are called *accelerators*. They are the counterparts of poisons. For example, steam added to the butene feed of a dehydrogenation reactor appeared to reduce the amount of coke formed and increase the yield of butadiene. The catalyst in this case was iron.[1]

Table 8-6 Poisons for various catalysts

Catalyst	Reaction	Types of poisoning	Poisons
Silica, alumina	Cracking	Chemisorption Deposition Stability Selectivity	Organic bases Carbon, hydrocarbons Water Heavy metals
Nickel, platinum, copper	Hydrogenation Dehydrogenation	Chemisorption	Compounds of S, Se, Te, P, As, Zn, halides, Hg, Pb, NH_3, C_2H_2, H_2S, Fe_2O_3, etc.
Cobalt	Hydrocracking	Chemisorption	NH_3, S, Se, Te, P
Silver	$C_2H_4 + O \rightarrow C_2H_4O$	Selectivity	CH_4, C_2H_6
Vanadium oxide	Oxidation	Chemisorption	As
Iron	Ammonia synthesis Hydrogenation Oxidation	Chemisorption Chemisorption Chemisorption	O_2, H_2O, CO, S, C_2H_2 Bi, Se, Te, P, H_2O VSO_4, Bi

SOURCE: In part from W. B. Innes in P. H. Emmett (ed), "Catalysis," vol. I, chap. 7, p. 306, Reinhold Publishing Corporation, New York, 1954.

Problems

8-1. The following data were obtained at 70°C for the equilibrium adsorption of *n*-hexane on silica gel particles ($S_g = 832$ m^2/g, $\epsilon_p = 0.486$, $\rho_p = 1.13$ g/cm^3, $V_g = 0.43$ cm^3/g).

[1] K. K. Kearly, *Ind. Eng. Chem.*, **42**, 295 (1950).

Partial pressure of C_6H_{14} in gas, atm	C_6H_{14} adsorbed, g moles/(g gel)
0.0020	10.5×10^{-5}
0.0040	16.0×10^{-5}
0.0080	27.2×10^{-5}
0.0113	34.6×10^{-5}
0.0156	43.0×10^{-5}
0.0206	47.3×10^{-5}

(a) Determine how well the Langmuir isotherm fits these data. Establish the values of the constants $\bar{C}_m$ and K_c by least-mean-squares analysis of the linearized form of Eq. (8-6):

$$\frac{C_g}{\bar{C}} = \frac{1}{K_c \bar{C}_m} + \frac{C_g}{\bar{C}_m}$$

Note that a straight line should be obtained if $C_g/\bar{C}$ is plotted against C_g.

(b) Estimate the fraction of the surface covered with an adsorbed mono-molecular layer at each partial pressure of n-hexane. The surface occupied by one molecule of hexane at 70°C is estimated to be 58.5×10^{-16} cm^2.

(c) Calculate the value of $\bar{C}_m$ from the total surface area and the area occupied by one molecule of hexane. What conclusions can be drawn from the comparison of this value of $\bar{C}_m$ with that obtained in part (a)? (d) Up to what gas concentration is the isotherm linear?

8-2. Repeat Prob. 8-1 for the following data for the equilibrium adsorption of C_6H_6 on the same silica gel at 110°C. The surface occupied by one benzene molecule is 34.8×10^{-16} cm^2.

Partial pressure of C_6H_6 in gas, atm	C_6H_6 adsorbed, g moles/(g gel)
5.0×10^{-4}	2.6×10^{-5}
1.0×10^{-3}	4.5×10^{-5}
2.0×10^{-3}	7.8×10^{-5}
5.0×10^{-3}	17.0×10^{-5}
1.0×10^{-2}	27.0×10^{-5}
2.0×10^{-2}	40.0×10^{-5}

8-3. Curve 3 of Fig. 8-4 is a Brunauer-Emmett-Teller plot for the adsorption data of N_2 at -183°C on the sample of silica gel. The density of liquid N_2 at this temperature is 0.751 g/cm^3. Estimate the area of the silica gel from these data, in square meters per gram, and compare with the results of Example 8-2.

8-4. The "point-B method" of estimating surface areas was frequently used prior to the development of the Brunauer-Emmett-Teller approach. It entailed choosing from an absorption diagram such as Fig. 8-3 the point at which the central linear section begins. This procedure worked well for some systems, but it was extremely difficult, if not impossible, to select a reliable point B on an isotherm such as that shown for n-butane in Fig. 8-3. In contrast, the Brunauer-Emmett-Teller method was found to be reasonably satisfactory for this type of isotherm.

Demonstrate this by estimating the surface area of the silica gel sample from the *n*-butane curve in Fig. 8-4 (multiply the ordinate of the *n*-butane curve by 10). The density of liquid butane at 0°C is 0.601 g/cm^3.

8-5. An 8.01-g sample of Glaucosil is studied with N_2 adsorption at -195.8°C. The following data are obtained:

Pressure, mm Hg	6	25	140	230	285	320	430	505
Volume adsorbed, cm^3 (at 0°C and 1 atm)	61	127	170	197	215	230	277	335

The vapor pressure of N_2 at -195.8°C is 1 atm. Estimate the surface area (square meters per gram) of the sample.

8-6. Low-temperature (-195.8°C) nitrogen-adsorption data were obtained for an Fe-Al$_2$O$_3$ ammonia catalyst. The results for a 50.4-g sample were:

Pressure, mm Hg	8	30	50	102	130	148	233
Volume adsorbed, cm^3 (at 0°C and 1 atm)	103	116	130	148	159	163	188

	258	330	442	480	507	550
	198	221	270	294	316	365

Estimate the surface area for this catalyst.

8-7. Ritter and Drake[1] give the true density of the solid material in an activated alumina particle as 3.675 g/cm^3. The density of the particle determined by mercury displacement is 1.547. The surface area by adsorption measurement is 175 m^2/g. From this information compute the pore volume per gram, the porosity of the particles, and the mean pore radius. The bulk density of a bed of the alumina particles in a 250-cm^3 graduate is 0.81 g/cm^3. What fraction of the total volume of the bed is void space between the particles and what fraction is void space within the particles?

8-8. Two samples of silica-alumina cracking catalysts have particle densities of 1.126 and 0.962 g/cm^3, respectively, as determined by mercury displacement. The true density of the solid material in each case is 2.37 g/cm^3. The surface area of the first sample is 467 m^2/g and that of the second is 372 m^2/g. Which sample has the larger mean pore radius?

8-9. Mercury porosimeter data are tabulated below for a 0.400-g sample of UO_2 pellet. At the beginning of the measurements ($p = 1.77$ psia) the mercury displaced by the sample was 0.125 cm^3. At this low pressure no pores were penetrated. Data obtained with a pycnometer gave a true density of the solid phase of $\rho_S = 7.57$ g/cm^3.

Calculate the total porosity of the pellet and the porosity due to pores of

[1] H. L. Ritter and L. C. Drake, *Ind. Eng. Chem., Anal. Ed.*, **17**, 787 (1945).

larger than 250 Å radius. Also plot the pore-volume distribution for the pores larger than 250 Å radius, using the coordinates of Fig. 8-7.

Pressure, psi	196	296	396	500	600	700	800	900
Mercury penetration, cm³	0.002	0.004	0.008	0.014	0.020	0.026	0.032	0.038
	1,000	1,200	1,400	1,800	2,400	2,800	3,400	5,000
	0.044	0.052	0.057	0.062	0.066	0.066	0.067	0.068

9

KINETICS OF FLUID-SOLID CATALYTIC REACTIONS

In Chap. 8 heterogeneous catalysis was explained by postulating a three-step process: (1) chemisorption of at least one reactant on the solid, (2) surface reaction of the chemisorbed substance, and (3) desorption of the product from the catalytic surface. Now our objective is to formulate rate and equilibrium equations for these steps. We shall consider the kinetics and equilibrium of adsorption and then examine rate equations for the overall reaction.

CHEMISORPTION RATES AND EQUILIBRIUM

9-1 Rates of Chemisorption → *based on molecular collision models*

The rate at which molecules of a gas strike a surface, in molecules/(sec) (cm² surface), is $p/(2\pi m k_B T)^{\frac{1}{2}}$. If s is the fraction of the collisions which result in chemisorption, that is, the *sticking probability*, the rate of adsorption is

329

$$\mathbf{r}_a = \frac{sp}{(2\pi m k_B T)^{\frac{1}{2}}} \tag{9-1}$$

where p is the gas pressure and m is the mass of the molecule. As Hayward and Trapnell point out,[1] $s < 1$ for several reasons. Two are particularly important: only those molecules possessing the required activation energy can be chemisorbed, and molecules possessing the necessary energy may not be chemisorbed because the configuration of molecule and surface site may not allow the activated complex to be traversed. The fraction of the molecules possessing the required energy will be $e^{-E/R_g T}$, where E is the activation energy for chemisorption. The configuration probability for a molecule occupying a single site[2] will be proportional to the fraction of the surface that is unoccupied, that is, $1 - \theta$. Thus

$$s = \alpha(1 - \theta) e^{-E/R_g T} \tag{9-2}$$

where α is the proportionality constant, often called the *condensation coefficient*. Combining Eqs. (9-1) and (9-2) yields

$$\mathbf{r}_a = \frac{\alpha e^{-E/R_g T}}{(2\pi m k_B T)^{\frac{1}{2}}} p(1 - \theta) \tag{9-3}$$

This ideal expression for the rate of adsorption does not usually agree with experimental data. Observed rates decrease so rapidly with increasing coverage θ that they can be explained only if the activation energy increases with θ. Also, the condensation coefficient may vary with θ. These variations may be caused by surface heterogeneity; that is, the activity of the sites varies, so that different sites possess different values of α and E. The most active sites would have the lowest activation energy and would be occupied first. Alternately, interaction forces between occupied and unoccupied sites could explain the deviations. In any event, it is necessary to rewrite Eq. (9-3) as

$$\mathbf{r}_a = \frac{\alpha(\theta) e^{-E(\theta)/R_g T}}{(2\pi m k_B T)^{\frac{1}{2}}} p(1 - \theta) \tag{9-4}$$

It is instructive to compare the Langmuir equation, Eq. (8-1), with Eq. (9-4). If the latter is correct, then k in Eq. (8-1) is a function of surface coverage. However, the reason α and E are functions of θ is that the first two postulates of the Langmuir treatment (see Sec. 8-4) are not satisfied experimentally; that is, in real surfaces all sites do not have the same activity, and interactions do exist.

[1] D. O. Hayward and B. M. W. Trapnell, "Chemisorption," 2d ed., Butterworths & Co. (Publishers), London, 1964.

[2] If a molecule (for example, O_2) dissociates upon adsorption and occupies two adjacent sites, the configuration probability will be $(1 - \theta)^2$, at least for small values of θ.

For many cases of chemisorption the variation in rate with surface coverage can be accounted for entirely in the exponential term of Eq. (9-4), because $\alpha(\theta)$ and $1 - \theta$ are so much weaker functions. This leads to the result (for constant temperature)

$$\mathbf{r}_a = \beta p e^{-\gamma\theta} \qquad \Rightarrow \text{can be integrated to give} \qquad (9\text{-}5)$$
$$\text{the amount that is absorbed}$$
$$\text{see } HW \ 9\text{-}3$$

commonly known as the *Elovich equation*.[1] Equation (9-5) can be obtained[2] from Eq. (9-4) by assuming that α is constant and that E is a linear function of θ. It can also be derived by supposing that E is constant but that the number of sites is a function of the coverage and of the temperature.[3]

Transition-state theory (presented in Chap. 2 for homogeneous gas-phase reactions) has been applied to the prediction of the condensation coefficient. Lack of precise knowledge about the structure of the activated complex on the solid surface hinders these calculations and the evaluation of the results. Much experimental data exists for chemisorption, some for rate measurements,[4] and a great deal more for equilibrium systems. Conclusions from the equilibrium data are summarized in the next section, but before we leave the topic of rates it is important to mention the desorption step. Since desorption is an endothermic process,[5] it will normally have a significant activation energy. Thus desorption will be an activated process even when adsorption is not. If ΔH_a is the heat of adsorption on the site, the activation energies are related by

$$E_d = E_a - \Delta H_a \tag{9-6}$$

Rates of chemisorption of hydrogen on a given nickel catalyst were found to be nonactivated,[6] but the desorption activation energy was 13.8 kcal/g mole, about the same magnitude as the heat of adsorption (-13.5 kcal/g mole).

[1] S. Yu. Elovich and G. M. Zhabrova, *Zh. Fiz. Khim.*, **13**, 1761 (1939); S. Roginsky and Ya. Zeldovich, *Acta Physicochim.*, **1**, 554, 595 (1934).

[2] D. O. Hayward and B. M. W. Trapnell, "Chemisorption," 2d ed., Butterworths & Co. (Publishers), London, 1964.

[3] H. A. Taylor and N. J. Thon, *J. Am. Chem. Soc.*, **74**, 4169 (1952); M. J. D. Low, *Chem. Rev.*, **60**, 267 (1960).

[4] For example, see Taylor and Thon, Low, and Hayward and Trapnell for summaries of chemisorption rate measurements. Chromatographic techniques have been used to measure rates of adsorption by analyzing the moments of the peak leaving the column of adsorbent particles. For example, see G. Padberg and J. M. Smith, *J. Catalysis*, **12**, 111 (1968).

[5] A few chemisorption processes appear to be endothermic, so that desorption in these instances would be exothermic. See J. M. Thomas and W. J. Thomas, "Introduction to the Principles of Heterogeneous Catalysis," p. 30, Academic Press Inc., New York, 1967.

[6] G. Padberg and J. M. Smith, *J. Catalysis*, **12**, 111 (1968).

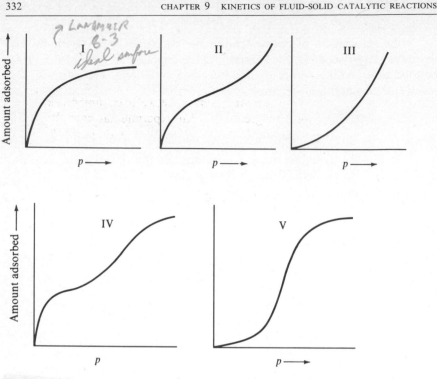

Fig. 9-1   *Types of adsorption isotherms*

9-2 Adsorption Isotherms

The adsorption isotherm is the relation between C_g, the concentration of adsorbable component in the gas (or the partial pressure p), and $\overline{C}$, the concentration on the solid surface (or θ) at equilibrium conditions and constant temperature. The Langmuir form, Eq. (8-6) or Eq. (8-3), follows curve I in Fig. 9-1. From Eq. (8-3), at low coverages (or low pressure)

$$\theta = Kp \qquad\qquad (9\text{-}7)$$

which corresponds to a linear relationship. Moreover, at high pressures θ approaches unity, and rearranging Eq. (8-3) gives

$$Kp(1 - \theta) = \theta \qquad\qquad (9\text{-}8)$$

or, as $\theta \to 1$,

$$1 - \theta \to \frac{1}{Kp} \qquad\qquad (9\text{-}9)$$

Five types of commonly recognized isotherms[1] are shown by the curves in Fig. 9-1. The linear form, Eq. (9-7), would fit several types at low pressures, but only curve I agrees with Eq. (8-3) over the whole range of pressure.

In the previous section it was observed that the Langmuir postulates of sites of equal activity and no interaction between occupied and bare sites were responsible for nonagreement with experimental data. It might be surmised that these assumptions correspond to a constant heat of adsorption. Indeed, it is possible[2] to derive the Langmuir isotherm by assuming that ΔH_a is independent of θ. The heat of adsorption can be evaluated from adsorption-equilibrium data. First the Clausius-Clapeyron equation is applied to the two-phase system of gas and adsorbed component on the surface:

$$\left(\frac{dp}{dT}\right)_\theta = \frac{-\Delta H_a}{T(V - V_a)} = \frac{H - H_a}{T(V - V_a)}$$

where V and V_a are the volumes per mole of adsorbed component in the gas and on the surface, respectively. Neglecting the latter and assuming the ideal-gas law for V gives

$$\left[\frac{d(\ln p)}{dT}\right]_\theta = \frac{-\Delta H_a}{R_g T^2} \qquad (9\text{-}10)$$

For the Clausius-Clapeyron equation to be valid the process must be univariant. This means that Eq. (9-10) can be applied only for constant concentration of adsorbate on the surface, that is, constant θ. If adsorption-equilibrium data are available at different temperatures, the slopes of p-vs-T curves at constant θ may be used with Eq. (9-10) to calculate ΔH_a. Figure 9-2, taken from Beeck,[3] shows such isosteric heats of adsorption as a function of θ for hydrogen on several metal films. These results are typical of almost all heats of adsorption in showing a decrease in $-\Delta H_a$ with increasing surface coverage.

Two other well-known isotherms may be classified in terms of the $\Delta H_a - \theta$ dependency. The *Temkin isotherm* may be derived from the Langmuir isotherm by assuming that the heat of adsorption drops linearly with increasing θ.[4] The result is

[1] D. O. Hayward and B. M. W. Trapnell, "Chemisorption," 2d ed., Butterworths & Co. (Publishers), London, 1964.

[2] K. J. Laidler, *J. Phys. Chem.*, **53**, 712 (1949).

[3] O. Beeck, *Disc. Faraday Soc.*, **8**, 118 (1950).

[4] A. Frumkin and A. Slygin, *Acta Physico chim.*, **3**, 791 (1935); S. Brunauer, K. S. Love, and R. G. Keenan, *J. Am. Chem. Soc.*, **64**, 751 (1942).

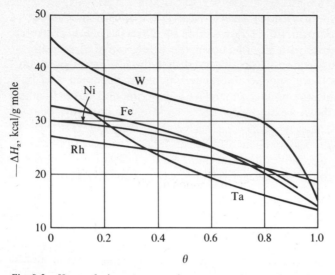

Fig. 9-2   *Heats of adsorption vs surface coverage for hydrogen on metal films*

$$\theta = k_1 \ln k_2 p \tag{9-11}$$

where k_1 and k_2 are constants at a given temperature. Figure 9-2 shows that a linear decrease in ΔH_a with θ fits some data for H_2 on metal films. Similar agreement has been obtained for N_2 chemisorbed on a promoted iron catalyst.[1]

The *Freundlich isotherm* can be derived[2] by assuming a logarithmic decrease in ΔH_a with θ; that is,

$$\Delta H_a = \Delta H_0 \ln \theta \tag{9-12}$$

The isotherm itself has the form

$$\theta = c(p)^{1/n} \tag{9-13}$$

where n has a value greater than unity. Owing to its flexibility, many investigators have found agreement between their measurements and the Freundlich isotherm.[3]

In conclusion, note two points. First, the Freundlich isotherm can be reduced to either the Langmuir or the Temkin form by proper simplication.

[1] P. H. Emmett and S. Brunauer, *J. Am. Chem. Soc.*, **56**, 35 (1934).

[2] G. Halsey and H. S. Taylor, *J. Chem. Phys.*, **15**, 624 (1947).

[3] For example, H_2 on tungsten films; B. M. W. Trapnell, *Proc. Roy. Soc. (London)*, **A206**, 39 (1951).

It may be considered a general empirical form encompassing the other more specific types. Second, a single isotherm of the Langmuir or Temkin type cannot be expected to fit data over the entire range of θ. Similarly, seldom does the Freundlich isotherm, with a single set of values of c and n, fit extensive data.

RATES OF FLUID-SOLID CATALYTIC REACTIONS

In view of the uncertainty about how the surface reaction of a chemisorbed reactant actually occurs, it is debatable how much detail should be postulated in formulating rates of reaction. In the simplest approach, we might follow the treatment of homogeneous kinetics and use only first- and second-order expressions. At the other extreme, we might separate the adsorption and surface reaction steps, using rate equations such as Eq. (9-4) for adsorption, and postulate about surface complexes to formulate the rate of the surface reaction. We shall adopt an intermediate approach here. The necessity for the occurrence of chemisorption as an explanation of catalytic activity provides strong support for separating the adsorption-desorption and surface processes. However, the functions of θ in Eq. (9-4) complicate the rate equation beyond present justification. The fundamental concepts on which the Langmuir rate and isotherm expressions are based appear sound, although, as we have seen, some of the assumptions are incorrect. The first-order relationship between gas concentration C_g, adsorbed concentration $\bar{C}$, and total concentration on the surface $\bar{C}_m$ in the Langmuir approach [Eqs. (8-4) to (8-6)] permit great simplification in formulating rate equations. Therefore Eqs. (8-4) to (8-6) for adsorption and desorption, along with simple first- and second-order expressions for the surface reaction, will be used in the sections that follow.[1]

The assumptions inherent in this procedure should not be forgotten. They have been mentioned, but it is worthwhile to emphasize one, the constancy of $\bar{C}_m$. This corresponds to an assumption that the maximum value of θ is unity, or that the total number of active sites is constant. Also, note that we are considering a general approach to the formulation of rate equations for fluid-solid catalytic reactions. In several specific cases sufficient data have been obtained to permit a more detailed analysis of the mechanism and rates of the adsorption and surface-reaction steps. An example is the

[1]This approach to kinetics of fluid-solid catalytic reactions was proposed by C. N. Hinshelwood ("Kinetics of Chemical Change," Oxford University Press, London, 1940) and developed in detail by O. A. Hougen and K. M. Watson ("Chemical Process Principles," part 3, "Kinetics and Catalysis," John Wiley & Sons, Inc., New York, 1947).

catalytic reaction $H_2 \rightarrow 2H$ on a hot tungsten filament, for which both rate of reaction and ΔH_a-vs-θ data are available.[1]

9-3 Rates of Adsorption and Desorption

In this section equations are presented for the rates of adsorption, surface reaction, and desorption. In Sec. 9-4 these equations will be combined to give expressions for the rate in terms of fluid concentrations; that is, concentrations on the surface (denoted by $\overline{C}$) will be eliminated.

Adsorption   The net rate of adsorption of a component A is given by the difference between Eqs. (8-4) and (8-5), written as[2]

$$\mathbf{r}_a = k_a C_A(\overline{C}_m - \overline{C}) - k'_a \overline{C}_A = k_a \left[C_A(\overline{C}_m - \overline{C}) - \frac{1}{K_A} \overline{C}_A \right]$$

where $\overline{C}_m - \overline{C}$ represents the concentration of vacant sites, $\overline{C}_v$. If only A were absorbed, then $\overline{C}_m - \overline{C}$ would equal $\overline{C}_m - \overline{C}_A$. However, other components of the reaction may be adsorbed, so that it is necessary to write $\overline{C}_m - \overline{C} = \overline{C}_v$, where $\overline{C}_v$ is expressed as the concentration of vacant sites per unit mass of catalyst, or

$$\mathbf{r}_a = k_a \left(C_A \overline{C}_v - \frac{1}{K_A} \overline{C}_A \right) \qquad K_A = \frac{k'_a}{k_a} = \frac{\text{desorption}}{\text{adsorption}} \tag{9-14}$$

In this equation C_A is the concentration of A in the gas phase at the catalyst surface. If the resistance to adsorption is negligible with respect to other steps in the overall conversion process, the concentration of A on the catalyst surface is in equilibrium with the concentration of A in the gas phase. The net rate of adsorption, from Eq. (9-14), approaches zero, and the equilibrium concentration of A is given by the expression

$$(\overline{C}_A)_{eq} = K_A C_A \overline{C}_v \tag{9-15}$$

where K_A designates the adsorption equilibrium constant for A.

Note that this result would reduce to the Langmuir isotherm, Eq. (8-6), if only A were adsorbed. Equations (9-14) and (9-15) are applicable when A occupies one site. Often in chemisorption a diatomic molecule, such as oxygen, will dissociate upon adsorption with each atom occupying one site. Formally, dissociative adsorption may be written

$$A_2 + 2X \rightarrow 2A \cdot X$$

[1] M. Boudart, *Ind. Chim. Belge*, **23**, 383 (1958).

[2] The gas phase concentration C_g is now written as C_A to denote component A. Also, the rate constant k_c is written as k_a to denote adsorption.

For this case the net rate of adsorption is

$$\mathbf{r}_a = k_a \left(C_{A_2} \overline{C}_v{}^2 - \frac{1}{K_{A_2}} \overline{C}_A{}^2 \right) \tag{9-14'}$$

At equilibrium the concentration of atomically adsorbed A is[1]

$$(\overline{C}_A)_{eq} = K_{A_2}{}^{\frac{1}{2}} C_{A_2}{}^{\frac{1}{2}} \overline{C}_v \tag{9-15'}$$

Surface Reaction The mechanism assumed for the surface process will depend on the nature of the reaction. Suppose that the overall reaction is of the type

$$A + B \rightleftharpoons C \tag{9-16}$$

An immediate question concerning the surface process is whether the reaction is between an adsorbed molecule of A and a gaseous molecule of B at the surface or between adsorbed molecules of both A and B on adjacent active centers. In the former case the process might be represented by the expression

$$A \cdot X + B \rightleftharpoons C \cdot X \tag{9-17}$$

If the concentration of adsorbed product C on the surface is $\overline{C}_C$, in moles per unit mass of catalyst (analogous to $\overline{C}_A$), the net rate of this step would be

Absorption - Desorption

$$\mathbf{r}_s = k_s \overline{C}_A C_B - k_s' \overline{C}_C = k_s \left(\overline{C}_A C_B - \frac{1}{K_s} \overline{C}_C \right) \tag{9-18}$$

In this equation we suppose that the rate of the forward reaction is first order in A on the solid surface and first order in B in the gas phase. Similarly, the rate of the reverse process is first order in C on the surface.

If the mechanism is a reaction between adsorbed A and adsorbed B, the process may be represented by the expression

$$A \cdot X + B \cdot X \rightarrow C \cdot X + X \tag{9-19}$$

Here only those A molecules will react which are adsorbed on sites immediately adjacent to adsorbed B molecules. Hence the rate of the forward reaction should be proportional to the concentration of the pairs of adjacent sites occupied by A and B. The concentration of these pairs will be equal to $\overline{C}_A$ multiplied by the fraction of the adjacent sites occupied by B molecules.

---

[1] If A_2 is the only adsorbable gas, then $\overline{C}_v = \overline{C}_m - \overline{C}_A$, and Eq. (9-15') becomes

$$(\overline{C}_A)_{eq} = \frac{K_{A_2}{}^{\frac{1}{2}} C_{A_2}{}^{\frac{1}{2}} \overline{C}_m}{1 + K_{A_2}{}^{\frac{1}{2}} C_{A_2}{}^{\frac{1}{2}}} \qquad \text{or} \qquad \overline{\theta}_A = \frac{K_A{}^{\frac{1}{2}} C_{A_2}{}^{\frac{1}{2}}}{1 + K_{A_2}{}^{\frac{1}{2}} C_{A_2}{}^{\frac{1}{2}}}$$

in comparison with Eqs. (8-6) and (8-7) for molecularly adsorbed A_2.

This fraction is proportional to the fraction of the total surface occupied by B molecules, i.e., to θ_B.[1] If $\overline{C}_m$ is defined as the molal concentration of total sites, then $\theta_B = \overline{C}_B/\overline{C}_m$. The rate of the forward reaction, according to the mechanism of Eq. (9-19), will be

$$\mathbf{r} = k_s \overline{C}_A \frac{\overline{C}_B}{\overline{C}_m}$$

The reverse rate is proportional to the pairs of centers formed by adsorbed product C molecules and adjacent vacant centers,

$$\mathbf{r}' = k_s' \overline{C}_C \frac{\overline{C}_v}{\overline{C}_m} \quad \rightarrow \quad \textit{vacant sites}$$

Combining these two expressions gives the net surface rate by the mechanism of Eq. (9-19),

$$\mathbf{r}_s = \frac{1}{\overline{C}_m}(k_s \overline{C}_A \overline{C}_B - k_s' \overline{C}_C \overline{C}_v) = \frac{k_s}{\overline{C}_m}\left(\overline{C}_A \overline{C}_B - \frac{1}{K_s}\overline{C}_C \overline{C}_v\right) \tag{9-20}$$

If the surface step has a negligible resistance with respect to the others, the process would occur at equilibrium, and Eq. (9-18) or Eq. (9-20) could be used to relate the concentrations of the reactants and products on the catalyst surface. For example, if the chosen mechanism is Eq. (9-19), the concentration of product C is given by Eq. (9-20) with $\mathbf{r}_s = 0$; that is,

$$K_s = \left(\frac{\overline{C}_v \overline{C}_C}{\overline{C}_A \overline{C}_B}\right)_{eq} \tag{9-21}$$

where K_s is the equilibrium constant for the surface reaction.

Desorption   The mechanism for the desorption of the product C may be represented by the expression

$$C \cdot X \rightarrow C + X$$

The rate of desorption will be analogous to Eq. (9-14) for the adsorption of A,

$$\mathbf{r}_d = k_d' \overline{C}_C - k_d C_C \overline{C}_v = -k_d\left(C_C \overline{C}_v - \frac{1}{K_C}\overline{C}_C\right) \tag{9-22}$$

Desorption - absorption

---

[1] This is strictly correct only if the fraction of the surface occupied by A molecules is small. It would be more accurate to postulate that the fraction of the adjacent centers occupied by B is equal to $\theta_B/(1 - \theta_A)$. For small values of θ_A the two results are nearly the same.

9-4 Rate Equations in Terms of Concentrations in the Fluid Phase at the Catalyst Surface

At steady state the rates of adsorption r_a, surface reaction r_s, and desorption r_d are equal. To express the rate solely in terms of fluid concentrations, the adsorbed concentrations $\overline{C}_A$, $\overline{C}_B$, $\overline{C}_C$, and $\overline{C}_v$ must be eliminated from Eqs. (9-15) to (9-22). In principle, this can be done for any reaction, but the resultant rate equation involves all the rate constants k_i and the equilibrium constants K_i. Normally neither type of constant can be evaluated independently.[1] Both must be determined from measurements of the rate of conversion from fluid reactants to fluid products. However, there are far too many constants, even for simple reactions, to obtain meaningful values from such overall rate data. The problem can be eased, with some confidence from experimental data, by supposing that one step in the overall reaction controls the rate. Then the other two steps occur at near-equilibrium conditions. This greatly simplifies the rate expression and reduces the number of rate and equilibrium constants that must be determined from experiment. To illustrate the procedure equations for the rate will be developed, for various controlling steps, for the reaction system

$$A + X \rightleftharpoons A \cdot X$$

$$B + X \rightleftharpoons B \cdot X \qquad \text{adsorption}$$

$$A \cdot X + B \cdot X \rightleftharpoons C \cdot X + X \qquad \text{surface reaction}$$

$$C \cdot X \rightleftharpoons C + X \qquad \text{desorption}$$

- -

$$A + B \rightleftharpoons C \qquad \text{overall reaction}$$

Surface Reaction Controlling   The concentrations $\overline{C}_A$, $\overline{C}_B$, and $\overline{C}_C$ will be those corresponding to equilibrium for the adsorption and desorption steps. Equation (9-15) gives the equilibrium value for $\overline{C}_A$. Similar results for $\overline{C}_B$ and $\overline{C}_C$ are

$$(\overline{C}_B)_{eq} = K_B C_B \overline{C}_v \qquad (9\text{-}23)$$

$$(\overline{C}_C)_{eq} = K_C C_C \overline{C}_v \qquad (9\text{-}24)$$

Substituting these results in Eq. (9-20) for the surface rate gives

$$\mathbf{r} = \frac{k_s}{\overline{C}_m}\left(K_A K_B C_A C_B \overline{C}_v{}^2 - \frac{K_C}{K_s} C_C \overline{C}_v{}^2\right) \qquad (9\text{-}25)$$

[1] It might be supposed that rate and equilibrium constants for adsorption or desorption could be established from pure-component adsorption data on the components involved. However, such results rarely agree with the values of the constants determined from rate data for the reaction. Interaction effects between components, other inadequacies of the Langmuir theory, and the assumption of a single controlling step, explain the deviation.

The concentration of vacant sites can be expressed in terms of the total concentration of sites $\overline{C}_m$,

$$\overline{C}_m = \overline{C}_v + \overline{C}_A + \overline{C}_B + \overline{C}_C \tag{9-26}$$

Since $\overline{C}_A$ (or $\overline{C}_B$ and $\overline{C}_C$) corresponds to the equilibrium value for adsorption, Eqs. (9-15), (9-23), and (9-24) can be combined with Eq. (9-26) to yield

$$\overline{C}_v = \frac{\overline{C}_m}{1 + K_A C_A + K_B C_B + K_C C_C} \tag{9-27}$$

Now Eqs. (9-25) and (9-27) can be combined to give a relatively simple expression for the rate in terms of fluid concentrations,

$$\mathbf{r} = k_s \overline{C}_m \frac{K_A K_B C_A C_B - (K_C/K_s)C_C}{(1 + K_A C_A + K_B C_B + K_C C_C)^2} \tag{9-28}$$

This result can be further reduced by using the relationship between the several equilibrium constants K_A, K_B, K_C, and K_s. If the equilibrium constant for the overall reaction is denoted by K, then

$$K = \left(\frac{C_C}{C_A C_B}\right)_{eq} \tag{9-29}$$

This is the conventional K for a homogeneous reaction computed from thermodynamic data, as outlined in Chap. 2. It may be related to the adsorption and surface-reaction equilibrium constants by the equilibrium equations for each of these processes, Eqs. (9-15), (9-23), and (9-24). Thus

$$K = \frac{\overline{C}_C/K_C \overline{C}_v}{(\overline{C}_A/K_A \overline{C}_v)(\overline{C}_B/K_B \overline{C}_v)} = \frac{K_A K_B}{K_C}\left(\frac{\overline{C}_v \overline{C}_C}{\overline{C}_A \overline{C}_B}\right)_{eq} \tag{9-30}$$

According to Eq. (9-21), the last group of surface concentrations is K_s, and so

$$K = \frac{K_A K_B}{K_C} K_s \tag{9-31}$$

Substituting this relationship in Eq. (9-28) gives the final expression for the rate in terms of fluid concentrations,

$$\mathbf{r} = k_s \overline{C}_m K_A K_B \frac{C_A C_B - (1/K)C_C}{(1 + K_A C_A + K_B C_B + K_C C_C)^2} \tag{9-32}$$

It is well at this point to review the chief premises involved in Eq. (9-32):

1. It is supposed that the surface reaction controls the rate of the three steps.
2. The equation applies to the simple reaction $A + B \to C$. Furthermore, it is assumed that the mechanism of the surface reaction involves the

combination of an adsorbed molecule of A and an adsorbed molecule of B.

3. The concepts of adsorption and kinetics rest on the Langmuir-Hinshelwood theories referred to earlier.

If the adsorption is weak for all components, the denominator of Eq. (9-32) approaches unity, and the rate expression reduces to the homogeneous form

$$\mathbf{r} = k_s \overline{C}_m K_A K_B (C_A C_B - \frac{1}{K} C_C)$$

or

$$\mathbf{r} = k(C_A C_B - \frac{1}{K} C_C) \tag{9-33}$$

The decomposition of formic acid on several catalytic surfaces follows such behavior.[1] Here there is only one reactant, so that the rate becomes first order in concentration of formic acid.

If the product of a reaction is strongly adsorbed and the reactant adsorption is weak, the term $K_C C_C$ is much larger than all others in the denominator. If the reaction is also irreversible, Eq. (9-32) becomes

$$\mathbf{r} = k_s \overline{C}_m K_A K_B \frac{C_A C_B}{(K_C C_C)^2}$$

Suppose there was only one reactant; that is, the reaction was of the form $A \rightarrow C$. The denominator term would be raised to a power of unity instead of being squared, so that

$$\mathbf{r} = k_s \overline{C}_m K_A \frac{C_A}{K_C C_C} \tag{9-34}$$

This result shows the retarding effect that a strongly adsorbed product can have on the rate. The decomposition of ammonia on a platinum wire has been found to follow this form of rate equation; i.e., hydrogen is strongly adsorbed and ammonia is only weakly adsorbed,[2] so that

$$\mathbf{r} = k \frac{C_{NH_3}}{C_{H_2}}$$

[1] C. N. Hinshelwood and B. Topley, *J. Chem. Soc.*, **123**, 1014 (1923).
[2] C. N. Hinshelwood and R. E. Burk, *J. Chem. Soc.*, **127**, 1105 (1925). Although there are two products in the decomposition, the nitrogen is only slightly adsorbed and does not appear in the rate equation.

Again consider a reaction of the form $A \rightarrow C$, but suppose that C is only weakly adsorbed, while A is very strongly adsorbed. Then the rate expression, in analogy with Eq. (9-34), would become zero order in A; that is

$$\mathbf{r} = k_s \bar{C}_m K_A \frac{C_A}{K_A C_A} = k \tag{9-35}$$

These special cases for the surface reaction controlling the rate all follow from the form of the Langmuir isotherm. Weak adsorption corresponds to small values of θ, and Eq. (9-7) shows that θ (or $\bar{C}_A$) is first order in p_A (or C_A), as predicted by Eq. (9-33). When the adsorption is very strong, the critical parameter is the concentration of vacant sites remaining for adsorption of reactant. Equation (9-9) shows that this is inversely proportional to the pressure of product (or $\bar{C}_C$), in agreement with Eq. (9-34).

Adsorption or Desorption Controlling   Still retaining the simple reaction $A + B \leftrightharpoons C$, let us now suppose that the adsorption of A is the slow step. Then the adsorption of B, the surface reaction, and the desorption of C will occur at equilibrium. The rate can be formulated from the adsorption equation (9-14). The adsorbed concentration $\bar{C}_A$ in this expression is obtained from the equilibrium equations for the surface rate [Eq. (9-21)], adsorption of B [Eq. (9-23)], and desorption of C [Eq. (9-24)]. Thus

$$\bar{C}_A = \frac{\bar{C}_v \bar{C}_C}{K_s \bar{C}_B} = \frac{\bar{C}_v (K_C C_C \bar{C}_v)}{K_s (K_B C_B \bar{C}_v)} = \frac{\bar{C}_v K_C C_C}{K_s K_B C_B}$$

From the relationship of the several equilibrium constants, Eq. (9-31), the expression for $\bar{C}_A$ may be simplified to

$$\bar{C}_A = \frac{K_A \bar{C}_v C_C}{K C_B} \tag{9-36}$$

Substituting this value of $\bar{C}_A$ in the rate-controlling equation (9-14) gives

$$\mathbf{r} = k_a \bar{C}_v \left(C_A - \frac{1}{K} \frac{C_C}{C_B} \right) \tag{9-37}$$

The expression for $\bar{C}_v$ can be formulated from Eq. (9-26), the equilibrium values of $\bar{C}_B$ and $\bar{C}_C$ from Eqs. (9-23) and (9-24), and $\bar{C}_A$ from Eq. (9-36). With this expression for $\bar{C}_v$ substituted in Eq. (9-37), the final rate equation for adsorption of A controlling the process is

$$\mathbf{r} = \frac{k_a \bar{C}_m [C_A - (1/K)(C_C/C_B)]}{1 + K_B C_B + (K_A/K)(C_C/C_B) + K_C C_C} \tag{9-38}$$

If, instead of adsorption, the rate of desorption of product C controls the whole reaction, the expression for $\mathbf{r}$ should be formulated from Eq. (9-22). The adsorption and surface steps will occur at equilibrium conditions. Substituting the equilibrium values of $\bar{C}_C$ and $\bar{C}_v$ in Eq. (9-22) leads to the result

$$\mathbf{r} = k_d \bar{C}_m K \; \frac{C_A C_B - (1/K)C_C}{1 + K_A C_A + K_B C_B + K_C K C_A C_B} \tag{9-39}$$

9-5 *Qualitative Analysis of Rate Equations*

The procedure for developing rate expressions in terms of fluid properties according to the Langmuir concepts has been illustrated for a single reaction. Yang and Hougen[1] have considered a number of kinds of reactions and mechanisms and examined the results when adsorption, desorption, or surface reaction controls the rate. By dividing the final equation into a kinetic coefficient [for example, $k_d \bar{C}_m K$ in Eq. (9-39)], a driving force $[C_A C_B - (1/K)C_C]$, and an adsorption term $(1 + K_A C_A + K_B C_B + K_C K C_A C_B)$, they were able to prepare tables from which the rate equation for a specific situation could be quickly assembled.[2]

Equations such as (9-32), (9-38), and (9-39) will have value if they can be used to predict the rate over a wide range of conditions and hence be suitable for designing reactors. To be useful as working expressions the various constants (kinetics and equilibrium) must be given numerical values.[3] It has not proved possible to obtain reliable K values from separate adsorption measurements. Hence all the constants must be determined from experimental kinetic data. This means, for example, that $k_d \bar{C}_m$, K_A, K_B, and K_C in Eq. (9-39) would be obtained from rate measurements. Since four-constant equations offer considerable flexibility, it is frequently possible to fit experimental data with equations based on several different mechanisms and assumptions regarding the step which is controlling. The advantage of this method of formulating rate reactions is the systematic way in which the pertinent parameters are introduced. It is a mechanized procedure which permits little flexibility and provides little insight into the actual mechanism of heterogeneous catalytic reactions. Progress in mecha-

[1] K. H. Yang and O. A. Hougen, *Chem. Eng. Progr.*, **46**, 146 (1950).

[2] See also J. M. Thomas and W. J. Thomas, in "Introduction to the Principles of Heterogeneous Catalysis," pp. 458–459, Academic Press Inc., New York, 1967, for rate equations assembled in tabular form for various controlling mechanisms for the two reactions $A \rightleftharpoons B$ and $A + B \rightleftharpoons C$.

[3] Equations (9-32), (9-38), and (9-39) were developed for constant-temperature conditions. Hence the various specific rates k and equilibrium constants K are termed constants. They are, in theory, constant with respect to pressure and conversion but change with temperature.

nism studies for specific reactions has been summarized by Thomas and Thomas. Interesting comparisons of concepts and correlations for homogeneous and heterogeneous catalytic reactions are presented by Boudart.[1]

The quantitative interpretation of kinetic data in terms of this type of rate equation is illustrated in Sec. 9-6. However, before we proceed to the evaluation of constants in rate equations, let us consider some of the implications of Eqs. (9-32), (9-38), and (9-39) from a qualitative standpoint.

The adsorption terms in the denominator are all different. For surface rate controlling, adsorption equilibrium groups ($K_i C_i$) are included for each component, and the entire term is squared because reaction is between adsorbed A and adsorbed B. For adsorption of A controlling, no group for the adsorption of A is present, and the entire term is to the first power. For desorption controlling, no group for product C is in the adsorption term. These differences result in separate relationships between the pressure and the rate. Hence the treatment of data as a function of pressure provides a useful method for distinguishing between equations. Temperature is not so useful a variable as pressure because all the constants are strong functions of temperature, and many of the same ones are part of each equation. The primary value of temperature is to determine the energies of activation for the adsorption and surface processes. Measurements at different conversion levels can be used to evaluate various equations, but in general it is difficult to draw valid conclusions about mechanisms. Quantitative treatment of conversion-vs-rate data is illustrated in Example 9-2. The importance of total pressure as a variable to study mechanism is shown in the following example.

Example 9-1 A solid-catalyzed gaseous reaction has the form

$$A + B \rightarrow C$$

Sketch curves of the *initial* rate (rate at zero conversion) vs the total pressure for the following cases:

(*a*) The mechanism is the reaction between adsorbed A and adsorbed B molecules on the catalyst. The controlling step is the surface reaction.

(*b*) The mechanism is the same as (*a*), but adsorption of A is controlling.

(*c*) The mechanism is the same as (*b*), but desorption of C is controlling. Assume that the overall equilibrium constant is large with respect to the adsorption equilibrium constants.

(*d*) The mechanism is a reaction between adsorbed A and B in the gas phase. The controlling step is the surface reaction.

In each instance suppose that the reactants are present in an equimolal mixture.

[1] Michel Boudart, "Kinetics of Chemical Processes," chaps. 8 and 9, Prentice-Hall, Inc., Englewood Cliffs, N.J., 1968.

Solution For a gaseous reaction the concentration C of any component i is proportional to its partial pressure; for an ideal-gas mixture $C_i = p_i/R_g T$. Hence at constant temperature partial pressures may be substituted for C_i in the rate equations, causing a change only in the value of the constants. At zero conversion the pressure of product C is zero, and for an equimolal mixture

$$p_A = p_B = \tfrac{1}{2}p_t$$

For the first three cases Eqs. (9-32), (9-38), and (9-39) are the appropriate rate equations.

(*a*) At initial conditions Eq. (9-32) simplifies to

$$\mathbf{r}_0 = k_s \bar{C}_m K_A K_B \frac{\tfrac{1}{4}p_t^2/(R_g T)^2}{[1 + \tfrac{1}{2}(K_A + K_B)p_t/R_g T]^2}$$

By combining constants we may write this expression as

$$\mathbf{r}_0 = \frac{ap_t^2}{(1 + bp_t)^2} \tag{A}$$

where a and b are the resulting overall constants.

(*b*) In a similar manner Eq. (9-38) for the adsorption of A controlling may be reduced to the form

$$\mathbf{r}_0 = k_a \bar{C}_m \frac{\tfrac{1}{2}p_t/R_g T}{1 + \tfrac{1}{2}K_B p_t/R_g T} = \frac{a'p_t}{1 + b'p_t} \tag{B}$$

(*c*) Equation (9-39), for the case where desorption of C is controlling, may be written

$$\mathbf{r}_0 = k_d \bar{C}_m K \frac{\tfrac{1}{4}p_t^2/(R_g T)^2}{1 + \tfrac{1}{2}(K_A + K_B)p_t/R_g T + \tfrac{1}{4}K_C K p_t^2/(R_g T)^2}$$

If the equilibrium constant K is large with respect to K_A, K_B, and K_C, only the last term in the denominator is important, and the result is

$$\mathbf{r}_0 = \frac{k_d \bar{C}_m}{K_C} = a'' \tag{C}$$

Equation (A) for surface reaction controlling shows that the initial rate will be proportional to the square of the pressure at low pressures and will approach a constant value at high pressures. This type of relation is shown in Fig. 9-3*a*. The case for adsorption controlling is indicated in Fig. 9-3*b*, and that for desorption controlling is shown in Fig. 9-3*c*. If the equilibrium constant were not very large for case (*c*), the initial rate equation would be as shown in Fig. 9-3*a*.

(*d*) For this case the rate equation can be obtained from Eq. (9-18). Combining this with Eqs. (9-15) and (9-24) for the equilibrium values for $\bar{C}_A$ and $\bar{C}_C$ gives

$$\mathbf{r} = k_s \bar{C}_v \left(K_A C_A C_B - \frac{K_C}{K_s} C_C \right)$$

Since there is no adsorption of B in this instance, Eq. (9-27) for $\bar{C}_v$ becomes

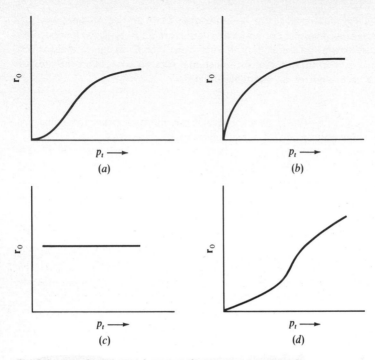

Fig. 9-3   *Initial rate vs total pressure for reaction* $A + B \rightarrow C$

$$\overline{C}_v = \frac{\overline{C}_m}{1 + K_A C_A + K_C C_C}$$

The relationship of the equilibrium constants [Eq. (9-31)] is

$$K = \frac{K_A K_s}{K_C}$$

Substituting these two expressions into the rate equation yields

$$\mathbf{r} = k_s K_A \overline{C}_m \frac{C_A C_B - (1/K)C_C}{1 + K_A C_A + K_C C_C} \tag{D}$$

Equation (D) is the appropriate expression for the surface rate controlling, with no adsorption of B. At initial conditions, and replacing concentrations with partial pressures, it becomes

$$\mathbf{r}_0 = k_s K_A \overline{C}_m \frac{\frac{1}{4}p_t^2/(R_g T)^2}{1 + \frac{1}{2}K_A p_t/R_g T} = \frac{a''' p_t^2}{1 + b''' p_t} \tag{E}$$

A schematic diagram of Eq. (E) shows the rate proportional to p_t at high pressures (Fig. 9-3d).

Suppose that experimental rate data for the reaction $A + B \to C$ were obtained over a wide range of total pressures, all at the same temperature. Comparison of a plot of the observed results with curves such as shown in Fig. 9-3 would be of value in establishing the most accurate rate equation. However, it is sometimes difficult to cover a wide enough range of pressures to observe all the changes in shape of the curves.

9-6 Quantitative Interpretation of Kinetic Data

To evaluate the rate and adsorption equilibrium constants in equations such as (9-32) rate data are needed as a function of concentrations in the fluid phase. Data are required at a series of temperatures in order to establish the temperature dependency of these constants. The proper concentrations to employ are those directly adjacent to the site. In the treatment that follows we shall suppose that these local concentrations have been established from the measurable concentrations in the bulk stream by the methods to be given in Chaps. 10 and 11. Our objective here is to find the most appropriate rate equation at a catalyst site.

Statistical methods are required to obtain the best fit of the equation to the kinetic data. Minimizing the deviations between the observed rate and that predicted from the equation is straightforward as long as the constants are linearly related in the rate equation.[1] When nonlinearities exist the analysis is more complicated.[2] There are many illustrations of the method for evaluating the constants in equations such as (9-32) from experimental rate measurements.[3] The sequence for this process is usually as follows:

1. Assume various mechanisms and controlling steps for each mechanism. Develop a rate equation for each combination of mechanism and controlling step.
2. Determine the numerical values of the constants which give the best fit of each equation to the observed rate data.
3. Choose the equation which best fits the data and agrees with available independent information about the reaction.

In the following simple example the evaluation of the constants is carried out for a single combination of mechanism and controlling step.

[1] O. A. Hougen and K. M. Watson, "Chemical Process Principles," vol. III, "Kinetics and Catalysis," John Wiley & Sons, Inc., New York, 1948.

[2] G. E. P. Box and H. L. Lucas, *Biometrika*, **46**, 77 (1959); J. R. Kittrell, W. G. Hunter, and C. C. Watson, *AIChE J.*, **11**, 105 (1965), *AIChE J.*, **12**, 5 (1966); L. Lapidus and T. I. Peterson, *AIChE J.*, **11**, 891 (1965).

[3] J. R. Kittrell, Reiji Mezaki, and C. C. Watson, *Brit. Chem. Eng.*, **11**, 15 (1966), *Ind. Eng. Chem.*, **57**, 18 (1965); W. G. Hunter and Reiji Mezaki, *AIChE J.*, **10**, 315 (1964).

Example 9-2   Olson and Schuler[1] determined reaction rates for the oxidation of sulfur dioxide, using a packed bed of platinum-on-alumina catalyst pellets. A differential reactor was employed, and the partial pressures as measured from bulk-stream compositions were corrected to fluid-phase values at the catalyst surface by the methods described in Chap. 10 (see Example 10-1). The total pressure was about 790 mm Hg.

From previous studies[2] and the qualitative nature of the rate data a likely combination appeared to be a controlling surface reaction between adsorbed atomic oxygen and unadsorbed sulfur dioxide. In order to determine all the constants in the rate equation for this mechanism, it is necessary to vary each partial pressure independently in the experimental work. Thus measuring the rate of reaction at different total pressures but at constant composition is not sufficient to determine all the adsorption equilibrium constants. Similarly, if the data are obtained at constant composition of initial reactants but varying conversions, the partial pressures of the individual components do not vary independently. However, in these cases it is possible to verify the validity of the rate equation even though values of the separate adsorption equilibrium constants cannot be ascertained. Olson and Schuler studied the effect of conversion alone and obtained the data in Table 9-1 at 480°C.

Table 9-1

r, g moles/(hr)(g catalyst)	Partial pressure (atm) at catalyst surface		
	SO_3	SO_2	O_2
0.02	0.0428	0.0255	0.186
0.04	0.0331	0.0352	0.190
0.06	0.0272	0.0409	0.193
0.08	0.0236	0.0443	0.195
0.10	0.0214	0.0464	0.196
0.12	0.0201	0.0476	0.197

Test the proposed mechanism by evaluating the constants in the rate equation and then comparing experimental rates of reaction with values computed from the equation.

Solution   To develop an expression for the rate of reaction we must postulate the method of obtaining adsorbed atomic oxygen. If it is supposed that molecular oxygen is first adsorbed on a pair of vacant centers and that this product then dissociates into two adsorbed atoms, the process may be written

$$O_2 + 2X \rightarrow \begin{matrix} O\text{---}O \\ | \quad | \\ X \quad X \end{matrix} \rightarrow 2\,O\cdot X$$

Since the surface reaction is controlling, the adsorption of oxygen must be at

[1] R. W. Olson, R. W. Schuler, and J. M. Smith, *Chem. Eng. Progr.*, **46**, 614 (1950).

[2] O. Uyehara and K. M. Watson, *Ind. Eng. Chem.*, **35**, 541 (1943).

equilibrium. Then the concentration of adsorbed atomic oxygen is given by the equilibrium equation

$$K_O = \frac{\bar{C}_O{}^2}{p_{O_2}\bar{C}_v{}^2}$$

where $\bar{C}_O$ represents the concentration of adsorbed atomic oxygen and $\bar{C}_v$ represents the concentration of vacant centers. Solving this expression for $\bar{C}_O$ yields

$$\bar{C}_O = K_O{}^{\frac{1}{2}}p_{O_2}{}^{\frac{1}{2}}\bar{C}_v \tag{A}$$

The surface-reaction step is represented by

$$SO_2 + O \cdot X \to SO_3 \cdot X$$

$$\mathbf{r}_s = k_s\left(p_{SO_2}\bar{C}_O - \frac{1}{K_s}\bar{C}_{SO_3}\right) \tag{B}$$

The concentration of SO_3 adsorbed on the catalyst is given by the conventional equilibrium expression

$$\bar{C}_{SO_3} = K_{SO_3}p_{SO_3}\bar{C}_v \tag{C}$$

Substituting the values of $\bar{C}_O$ and $\bar{C}_{SO_3}$ in Eq. (B) gives

$$\mathbf{r}_s = k_s\left(K_O{}^{\frac{1}{2}}p_{SO_2}p_{O_2}{}^{\frac{1}{2}} - \frac{K_{SO_3}}{K_s}p_{SO_3}\right)\bar{C}_v \tag{D}$$

The total concentration of centers is the summation

$$\bar{C}_m = \bar{C}_v + K_O{}^{\frac{1}{2}}p_{O_2}{}^{\frac{1}{2}}\bar{C}_v + K_{SO_3}p_{SO_3}\bar{C}_v + K_{N_2}p_{N_2}\bar{C}_v \tag{E}$$

The last term is included to take into account the possibility that N_2 may be adsorbed on the catalyst. Eliminating $\bar{C}_v$ from Eq. (D) by introducing Eq. (E) and noting that $K_O{}^{\frac{1}{2}}K_S/K_{SO_3} = K$, we may write Eq. (D) as

$$r = \frac{k_s\bar{C}_mK_O{}^{\frac{1}{2}}[p_{SO_2}p_{O_2}{}^{\frac{1}{2}} - (1/K)p_{SO_3}]}{1 + K_O{}^{\frac{1}{2}}p_{O_2}{}^{\frac{1}{2}} + K_{SO_3}p_{SO_3} + K_{N_2}p_{N_2}} \tag{F}$$

Because the partial pressures of sulfur dioxide and sulfur trioxide are both small, the value of p_{N_2} will not vary significantly with conversion. Hence $K_{N_2}p_{N_2}$ may be regarded as a constant in Eq. (F). Since conversion was the only variable causing the composition to change, p_{O_2} and p_{SO_3} and p_{SO_2} are not independent, but are related to the initial-constant composition and the conversion. Hence p_{O_2} in the denominator of Eq. (F) can be expressed in terms of conversion or, more conveniently, in terms of p_{SO_3}. Thus Eq. (F) may be simplified to

$$r = \frac{k_s\bar{C}_mK_O{}^{\frac{1}{2}}[p_{SO_2}p_{O_2}{}^{\frac{1}{2}} - (1/K)p_{SO_3}]}{A' + B'p_{SO_3}}$$

Combining the constants and rearranging leads to the form

$$A + Bp_{SO_3} = \frac{p_{SO_2}p_{O_2}{}^{\frac{1}{2}} - (1/K)p_{SO_3}}{\mathbf{r}} = R \tag{G}$$

Since conversion was the only variable, the rate equation contains only two constants, A and B. This means that individual adsorption equilibrium constants cannot be determined.

From the overall equilibrium constant K and the data for the partial pressures and rates, the right-hand side of Eq. (G), designated as R, can be evaluated for each rate. The equilibrium constant decreases with temperature, the reaction being exothermic. However, it is still about 73 at 480°C, as estimated from the expression

$$\ln K = \frac{22{,}200}{R_g T} - 10.5$$

Table 9-2

r, g moles/(hr)(g catalyst)	p_{SO_3}, atm	R	Rp_{SO_3}	$p_{SO_3}{}^2 \times 10^4$
0.02	0.0428	0.521	0.0223	18.3
0.04	0.0331	0.372	0.0123	11.0
0.06	0.0272	0.294	0.0081	7.40
0.08	0.0236	0.241	0.00569	5.57
0.10	0.0214	0.203	0.00433	4.58
0.12	0.0201	0.174	0.00349	4.03
Total	0.168	1.805	0.0562	50.8

Table 9-2 shows the values of R computed from the data, with $K = 73$. The most probable values of the constants are those corresponding to a least-mean-square fit of the data to a straight line of the general form

$$a + bx = y \tag{H}$$

Comparing Eqs. (G) and (H), we have

$$y = R \qquad x = p_{SO_3} \qquad a = A \qquad b = B$$

The values of a and b are given by the two expressions[1]

$$b = \frac{\Sigma yx - \Sigma x \Sigma y/n}{\Sigma x^2 - (\Sigma x)^2/n} \tag{I}$$

$$a = \frac{\Sigma y - b\Sigma x}{n} \tag{J}$$

where n is the number of measurements.

Equations (I) and (J) may be applied to the present case. Thus $yx = Rp_{SO_3}$, etc.

---

[1] These equations are obtained by writing an expression for the sum of the squares of the deviations of the data points from the line of Eq. (H), differentiating this equation with respect to both a and b, setting the differentiated equations equal to zero to apply the minimum-deviation restriction, and then solving the two equations for the values of a and b.

The necessary values and summations are given in the last four columns of Table 9-2. Solving for the constants, we have

$$b = B = \frac{0.0562 - 0.168(1.805)/6}{0.00508 - 0.168^2/6} = 15.3$$

$$a = A = \frac{1.805 - 15.3(0.168)}{6} = -0.127$$

With these values Eq. (G) may be written

$$R = 15.3 p_{SO_3} - 0.127 \tag{K}$$

Figure 9-4 is a plot of the experimental data and Eq. (K). Equation (K) can be used to compute values of the rate for comparison with the experimental results. For example, using Eq. (K) for the first set of data, at $p_{SO_3} = 0.0428$ atm,

$$R = 15.3(0.0428) - 0.127 = 0.525$$

Then, from Eq. (G)

$$0.525 = \frac{0.0255(0.1864)^{\frac{1}{2}} - \frac{1}{73}(0.0428)}{r}$$

Fig. 9-4 *Correlation of rate data for oxidation of* SO_2 [*Eq.*(K)]

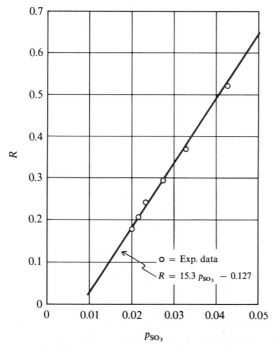

Solving for the rate, we have

$$r = 0.020 \text{ g mole}/(\text{hr})(\text{g catalyst})$$

The results for the other points are compared with the experimental rates in Table 9-3.

Table 9-3

		r	
p_{SO_3}	R	*Calculated*	*Experimental*
0.0428	0.525	0.020	0.020
0.0331	0.377	0.039	0.040
0.0272	0.287	0.061	0.060
0.0236	0.232	0.083	0.080
0.0214	0.199	0.102	0.100
0.0201	0.179	0.117	0.120

Problems

9-1.   Can the Freundlich isotherm, Eq. (9-13), represent the physical-adsorption data for *n*-hexane given in Prob. 8-1? What are the appropriate values of *c* and *n*?

9-2.   Will the Temkin isotherm fit the *n*-hexane data shown in Prob. 8-1?

9-3.   Ward[1] studied the chemisorption of hydrogen on copper powder and found the heat of adsorption to be independent of surface coverage. Equilibrium data from experiments at 25°C are as follows:

Hydrogen pressure, mm Hg	1.05	2.95	5.40	10.65	21.5	45.1	95.8	204.8
Volume adsorbed, cm^3 at 0°C and 1 atm	0.239	0.564	0.659	0.800	0.995	1.160	1.300	1.471

What kind of isotherm fits these results?

9-4.   Brunauer et al.[2] concluded that the Temkin isotherm, Eq. (9-11), explained equilibrium chemisorption data for N_2 on an ammonia-synthesis catalyst (promoted iron powder). Data taken at 396°C are as follows:

Nitrogen pressure, mm Hg	25	53	150	397	768
Volume adsorbed, cm^3 at 0°C and 1 atm	2.83	3.22	3.69	4.14	4.55

[1] A. F. H. Ward, *Proc. Roy. Soc. (London)*, **A133**, 506 (1931).

[2] S. Brunauer, K. S. Love, and R. G. Keenan, *J. Am. Chem. Soc.*, **64**, 751 (1942).

Can Brunauer's conclusion be reached by plotting the data according to linearized forms of the Langmuir, Temkin, and Freundlich isotherms? Explain your answer.

9-5. Plot the isosteric heat of adsorption vs the amount adsorbed from the following data of Shen and Smith[1] for benzene on silica gel. Using a constant value of 34.8×10^{-16} cm^2 for the surface area occupied by one benzene molecule, convert the coordinate showing the amount adsorbed to fraction of the surface covered (total surface area = 832 m^2/g).

Partial pressure of benzene, atm	Moles adsorbed/g gel $\times 10^5$			
	70°C	90°C	110°C	130°C
5.0×10^{-4}	14.0	6.7	2.6	1.13
1.0×10^{-3}	22.0	11.2	4.5	2.0
2.0×10^{-3}	34.0	18.0	7.8	3.9
5.0×10^{-3}	68.0	33.0	17.0	8.6
1.0×10^{-2}	88.0	51.0	27.0	16.0
2.0×10^{-2}		78.0	42.0	26.0

9-6. Show that the Elovich equation for the rate of chemisorption can be integrated with respect to surface coverage (at constant pressure of the adsorbing gas) to give

$$\theta = a \ln (1 + bt)$$

where t is the time from the start of the adsorption process and a and b are constants. If the activation energy for chemisorption is a linear function of surface coverage, according to $E = E_0 + c\theta$, derive the integrated form of the Elovich equation.

9-7. A gas-solid catalytic reaction of the type $A + B \rightarrow C$ is believed to occur by a molecule of B in the gas phase reacting with a molecule of A adsorbed on the catalyst surface. Preliminary experimental studies indicate that both external and internal diffusional resistances are negligible. Also, the equilibrium constant for the homogeneous reaction $K = (p_C/p_A p_B)_g$ is very large.

The rate of reaction per gram of catalyst is measured at varying total pressures, but constant composition of reactants. The results show a linear relationship between $\mathbf{r}$ and p_t all the way down to pressures approaching zero. What conclusions may be drawn from these data concerning the controlling steps in the reaction?

9-8. Two gas-solid catalytic reactions, (1) and (2), are studied in fixed-bed reactors. Rates of reaction per unit mass of catalyst, at constant composition and total pressure, indicate the variations with mass velocity and temperature shown in the figure. The interior pore surface in each case is fully effective. What do the results shown suggest about the two reactions?

[1] John Shen and J. M. Smith, *Ind. Eng. Chem. Fund.*, **7**, 100 (1968).

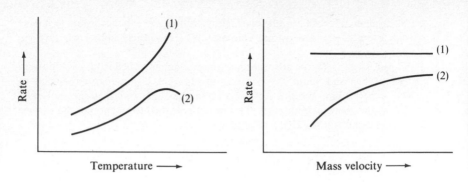

Fig. 9-5   *Rate vs temperature and mass velocity*

9-9.   In a study of the kinetics of two different gas-solid catalytic reactions it is found that all diffusional resistances are negligible. Also, both reactions are irreversible. As an aid in establishing the mechanism of the reactions the rate is measured at a constant composition over a wide range of temperature. For the first reaction, (1), the rate increases exponentially over the complete temperature range. For the second reaction, (2), the rate first increases and then decreases as the temperature continues to rise. What does this information mean with regard to the controlling step in each of the reactions?

9-10.   An isomerization reaction has the simple form $A \rightarrow B$. Assuming that operating conditions and the condition of the catalyst are such that the external- and internal-diffusion steps have negligible concentration gradients, propose rate equations for the following cases:

(*a*) The adsorption of A on the catalyst is controlling.
(*b*) The surface interaction between adsorbed A and an adjacent vacant center is controlling.
(*c*) The desorption of B from the surface is controlling.

In all cases the mechanism is adsorption of A, reaction on the surface to form adsorbed B, and desorption of B into the gas phase. Sketch the rate of reaction (per unit mass of catalyst) vs total pressure in each of the above three cases. Also, for comparison, include a sketch of the rate of the homogeneous reaction, assuming that it is first order. Sketches should be for constant composition.

9-11.   Thodos and Stutzman[1] studied the formation of ethyl chloride, using a zirconium oxide catalyst (on silica gel) in the presence of inert methane,

$$C_2H_4 + HCl \rightleftharpoons C_2H_5Cl$$

If the surface reaction between adsorbed ethylene and adsorbed HCl controls the overall kinetics, derive an expression for the rate. Neglecting external and internal transport resistances, evaluate the constants in the rate equation at 350°F from the following data:

[1]G. Thodos and L. F. Stutzman, *Ind. Eng. Chem.*, **50**, 413 (1958).

r × 10⁴, lb moles/(hr)(lb catalyst)	Partial pressures, atm			
	CH_4	C_2H_4	HCl	C_2H_5Cl
2.71	7.005	0.300	0.370	0.149
2.63	7.090	0.416	0.215	0.102
2.44	7.001	0.343	0.289	0.181
2.58	9.889	0.511	0.489	0.334
2.69	10.169	0.420	0.460	0.175

The equilibrium constant for the overall reaction at 350°F is 35.

9-12. Potter and Baron[1] studied the reaction

$$CO + Cl_2 \rightarrow COCl_2$$

at atmospheric pressure, using an activated carbon catalyst. Preliminary studies showed that the rate of reaction did not depend on the mass velocity of gases through the reactor. Analysis of the rate data indicated that the reaction occurred by adsorption of Cl_2 and CO on the catalyst and surface reaction between the adsorption complexes. It appeared that the surface reaction, rather than the adsorption or desorption steps, was controlling the overall reaction rate. Furthermore, preliminary adsorption measurements indicated that chlorine and phosgene were readily adsorbed on the catalyst, while carbon monoxide was not. Hence the adsorption equilibrium constant of carbon monoxide, although it was not zero, was considered negligible with respect to those for Cl_2 and $COCl_2$.

(a) On the basis of this information, develop an expression for the rate of reaction in terms of the bulk partial pressures in the gas phase. The reaction is irreversible. (b) Determine the best values for the adsorption equilibrium constants for Cl_2 and $COCl_2$ and the product $\bar{C}_m k_s K_{CO}$ from the following experimental data, where the temperature was 30.6°C, the catalyst size was 6 to 8 mesh, and

$\bar{C}_m$ = total concentration of active centers, in moles per gram of catalyst

k_s = specific reaction-rate constant for surface reaction

K_{CO} = adsorption equilibrium constant for CO

r, g moles/(hr)(g catalyst)	Partial pressure, atm		
	CO	Cl_2	$COCl_2$
0.00414	0.406	0.352	0.226
0.00440	0.396	0.363	0.231
0.00241	0.310	0.320	0.356
0.00245	0.287	0.333	0.376
0.00157	0.253	0.218	0.522
0.00390	0.610	0.113	0.231
0.00200	0.179	0.608	0.206

[1] C. Potter and S. Baron, *Chem. Eng. Progr.*, **47**, 473 (1951).

Assume that the 6- to 8-mesh catalyst particles are small enough that the pore surface was fully effective.

9-13. Potter and Baron also made rate measurements at other temperatures, and their results at 42.7, 52.5, and 64.0°C are as shown below. Assume that the adsorption equilibrium constants and the rate constant k_s follow an equation of the form $y = ae^{-b/R_gT}$, where a and b are constant and y is K_{Cl_2}, K_{COCl_2}, or $\bar{C}_m k_s K_{CO}$.

T, °C	r × 10³, g moles/(hr)(g catalyst)	Partial pressure, atm		
		CO	Cl₂	COCl₂
42.7	4.83	0.206	0.578	0.219
42.7	10.73	0.569	0.194	0.226
42.7	1.34	0.128	0.128	0.845
42.7	9.18	0.397	0.370	0.209
42.7	9.10	0.394	0.373	0.213
52.5	14.28	0.380	0.386	0.234
52.5	15.46	0.410	0.380	0.210
52.5	6.00	0.139	0.742	0.118
52.5	3.68	0.218	0.122	0.660
64.0	25.74	0.412	0.372	0.216
64.0	24.46	0.392	0.374	0.234
64.0	13.78	0.185	0.697	0.118
64.0	8.29	0.264	0.131	0.605

Determine values of a and b for each case.

10

EXTERNAL TRANSPORT PROCESSES IN HETEROGENEOUS REACTIONS

No matter how active a catalyst particle is, it can be effective only if the reactants can reach the catalytic surface. The transfer of reactant from the bulk fluid to the outer surface of the catalyst particle requires a driving force, the concentration difference. Whether this difference in concentration between bulk fluid and particle surface is significant or negligible depends on the velocity pattern in the fluid near the surface, on the physical properties of the fluid, and on the intrinsic rate of the chemical reaction at the catalyst; that is, it depends on the mass-transfer coefficient between fluid and surface and the rate constant for the catalytic reaction. In every case the concentration of reactant is less at the surface than in the bulk fluid. Hence the observed rate, the global rate, is less than that corresponding to the concentration of reactants in the bulk fluid.

The same reasoning suggests that there will be a temperature difference between bulk fluid and catalyst surface. Its magnitude will depend on the heat-transfer coefficient between fluid and catalyst surface, the reaction-rate constant, and the heat of reaction. If the reaction is endothermic, the temperature of the catalyst surface will be less than that in the bulk fluid, and the observed rate will be less than that corresponding to the bulk-fluid temperature; the resistances to mass and energy transfer supplement each other. If the reaction is exothermic, the temperature of the catalyst surface will be greater than that of the bulk fluid. Now the global rate may be higher or lower than that corresponding to bulk-fluid conditions; it is increased because of the temperature rise and reduced because of the drop in reactants concentration.

Our objective here is to study quantitatively how these external physical processes affect the rate. Such processes are designated as *external* to signify that they are completely *separated* from, and in series with, the chemical reaction on the catalyst surface. For porous catalysts both reaction and heat and mass transfer occur at the same internal location *within* the catalyst pellet. The quantitative analysis in this case requires *simultaneous* treatment of the physical and chemical steps. The effect of these internal physical processes will be considered in Chap. 11. It should be noted that such internal effects significantly affect the global rate only for comparatively large catalyst pellets. Hence they may be important only for fixed-bed catalytic reactors or gas-solid noncatalytic reactors (see Chap. 14), where large solid particles are employed. In contrast, external physical processes may be important for all types of fluid-solid heterogeneous reactions. In this chapter we shall consider first the gas-solid fixed-bed reactor, then the fluidized-bed case, and finally the slurry reactor.

FIXED-BED REACTORS

When a fluid flows through a bed of catalyst pellets, there will be regions near the solid surface where the velocity is very low. In this region mass and energy transfer between bulk fluid and pellet surface will be primarily by conduction. Away from the surface a convective mechanism will be dominant. The complexity of flow patterns around an individual pellet suspended in a fluid stream is considerable. When this is combined with interactions between pellets, as in a fixed-bed reactor, the problem of predicting or correlating local velocities is, at present, beyond solution. Therefore transport rates are normally defined in terms of an *average* heat- or mass-transfer coefficient. This coefficient is assumed to be the same for any particle (at the same conditions) in the bed. Even though experimental data show that variations exist, it will also be assumed that the average coefficient can be applied to all the surface of a pellet. With these assumptions it is possible to use a single value of the heat- or mass-transfer coefficient to describe the rates of transfer between bulk fluid and pellet surface in a fixed-bed reactor.

The error introduced in using an average coefficient is not as serious as might be expected, since the correlations for the mass-transfer coefficient k_m and heat-transfer coefficient h are based on experimental data for beds of particles (see Sec. 10-2). That is, the experimental results are, in general, for average values of the coefficients. Gillespie et al.,[1] however, have carried out an interesting study of the variation of heat-transfer coefficients with respect to the position of the pellet in the bed and the location on the surface of a single pellet. The results showed that h values were lower for the first two layers of pellets (measured from the entrance) than for the remainder of the bed. In addition, the coefficient was higher near the wall of the bed than in the central section of the cylindrical tube, presumably because of the higher velocity about one pellet diameter from the wall.[2] The local coefficients also varied with position; the highest values were obtained for the surface perpendicular to the direction of bulk flow in the bed. The status of analytical attempts to predict local variations in h are described by Petersen.[3]

The effect of external resistances on the rate can be important in two ways. In reactor design the rate at the surface site, as well as the temperature and concentrations in the bulk fluid, are presumed to be known (by the methods described in Chap. 9). The global rate is then calculated by simultaneous solution of the equations for the rate of reaction and mass and energy transfer, and in the process the unknown concentrations at the surface of the catalyst pellet are eliminated. If all the equations are linear in the surface concentrations and temperatures, an analytical solution is possible, as illustrated by Eqs. (10-6) and (10-7). If some equations are nonlinear, numerical solution is necessary. The other means of studying external-resistance effects on rate is through interpretations of experimental rate data measured in a fixed-bed reactor. In this case the global rate is directly obtained from the experimental measurements; for example, from differential or integral reactor data, as explained in Secs. 4-3, or from other types of laboratory reactors, as described in Chap. 12. The global rate may be used in the equations for mass and energy transfer to determine the concentrations and temperature at the surface of the catalyst pellet. Then these surface values must be correlated with the observed rate in order to establish the form of the rate equation by the procedure described in Chap. 9.

Before we proceed with quantitative illustrations, let us consider the qualitative effect of external resistances on reaction rates and summarize the available information for mass- and heat-transfer coefficients (Sec. 10-2).

[1] B. M. Gillespie, E. D. Crandall, and J. J. Carberry, *AIChE J.*, **14**, 483 (1968).

[2] C. E. Schwartz and J. M. Smith, *Ind. Eng. Chem.*, **45**, 1209 (1953).

[3] E. E. Petersen, "Chemical Reaction Analysis," pp. 129–164, Prentice-Hall, Inc., Englewood Cliffs, N.J., 1965.

10-1 The Effect of Physical Processes on Observed Rates of Reaction

Suppose an irreversible gaseous reaction on a *solid* catalyst pellet is of order n. At steady state the rate, expressed per unit mass of pellet, may be written either in terms of the diffusion rate from the bulk gas to the surface or in terms of the rate on the surface:

$$\mathbf{r}_P = k_m a_m (C_b - C_s) \tag{10-1}$$

$$\mathbf{r}_P = k a_m C_s^n \tag{10-2}$$

where C_b and C_s are the concentrations in the bulk gas and at the surface, respectively. In the first expression k_m, in centimeters per second, is the mass-transfer coefficient between bulk gas and solid surface, and a_m is the external surface area per unit mass of the pellet. Suppose that the reaction-rate constant k is very much greater than k_m. Under these conditions C_s approaches zero and Eq. (10-1) shows that the rate per pellet is

$$\mathbf{r}_P = k_m a_m C_b \tag{10-3}$$

At the other extreme, k is very much less than k_m. Then C_s approaches C_b, and the rate, according to Eq. (10-2), is

$$\mathbf{r}_P = k a_m C_b^n \tag{10-4}$$

Equation (10-3) represents the case when diffusion controls the overall process. The rate is determined by k_m; the kinetics of the chemical step at the catalyst surface are unimportant. Equation (10-4) gives the rate when the mass-transfer resistance is negligible with respect to that of the surface step; i.e., the kinetics of the surface reaction control the rate.

Consider a case where the true order of the surface reaction is 2 [according to Eq. (10-2)] but the rate is diffusion controlled, so that Eq. (10-3) is applicable. Experimental data plotted as rate vs C_b would yield a straight line. If diffusion were not considered, and Eq. (10-2) were used to interpret the data, the order would be identified as unity—a false conclusion. This simple example illustrates how erroneous conclusions can be reached about kinetics of a catalytic reaction if external mass transfer is neglected.

For a first-order reaction Eqs. (10-1) and (10-2) can be easily solved for C_s, giving

$$C_s = \frac{k_m C_b}{k + k_m} \tag{10-5}$$

Substituting this result in Eq. (10-2) gives the rate equation in terms of the bulk concentration,

$$\mathbf{r}_P = k_o a_m C_b \tag{10-6}$$

where

$$k_o = \frac{1}{1/k + 1/k_m} \tag{10-7}$$

Equations (10-6) and (10-7) show that for the intermediate case the observed rate is a function of both the rate-of-reaction constant k and the mass-transfer coefficient k_m. In a design problem k and k_m would be known, so that Eqs. (10-6) and (10-7) give the global rate in terms of C_b. Alternately, in interpreting laboratory kinetic data k_o would be measured. If k_m is known, k can be calculated from Eq. (10-7). In the event that the reaction is not first order Eqs. (10-1) and (10-2) cannot be combined easily to eliminate C_s. The preferred approach is to utilize the mass-transfer coefficient to evaluate C_s and then apply Eq. (10-2) to determine the order of the reaction n and the numerical value of k.[1] One example of this approach is described by Olson et al.[2]

Diffusion effects can also lead to a false activation energy. Suppose global reaction rates are measured for a nonporous catalyst pellet at different temperatures and that the surface rate is first order. The observed rates can be used to calculate the overall rate constant k_o from Eq. (10-6). If external diffusion is arbitrarily neglected, an apparent activation energy E' could then be calculated from the Arrhenius equation,

$$k_o = \mathbf{A}' e^{-E'/R_g T} \tag{10-8}$$

where $\mathbf{A}'$ is the apparent frequency factor. This result would give an erroneous value for E if external diffusion were a significant resistance. In fact, the data points for different temperatures would not form a straight line, but would give a curve, as indicated by the solid line in Fig. 10-1. This can be seen by rearranging Eq. (10-7) as

$$\frac{1}{k_o} = \frac{1}{k} + \frac{1}{k_m} = \frac{e^{E/R_g T}}{\mathbf{A}} + \frac{1}{k_m}$$

or

$$k_o = \frac{\mathbf{A} k_m e^{-E/R_g T}}{k_m + \mathbf{A} e^{-E/R_g T}} \tag{10-9}$$

where $\mathbf{A} e^{-E/R_g T}$ has been substituted for the true rate constant k of the

---

[1] Note that the kinetics of the surface reaction need not be expressed as a simple case of order n, as in Eq. (10-2). The procedure is equally applicable to the form of rate equation developed in Chap. 9 [e.g., Eq. (9-32)].

[2] R. W. Olson, R. W. Schuler, and J. M. Smith, *Chem. Eng. Progr.*, **46**, 614 (1950).

surface step. Here E is the true activation energy of the surface reaction. Since the mass-transfer coefficient is relatively insensitive to temperature, Eq. (10-9) shows that k_o approaches a nearly constant value equal to k_m at very high temperatures. At low temperatures $k_o = A e^{-E/R_g T}$ and a straight line is obtained on the Arrhenius plot. Figure 10-1 illustrates these results. At low temperatures a straight line would result, and its slope would give the correct activation energy of the surface reaction. At higher temperatures a curve would be obtained, and E' would vary with the temperature. When experimental rate data for fluid-solid catalytic reactions show a curved line, as in Fig. 10-1, it is possible that external diffusion resistances are important.

These illustrations of the significance of external-mass-transfer resistances have been based on isothermal conditions. Temperature differences due to external-heat-transfer resistance can also be important. Nonisothermal situations will be considered in Sec. 10-4.

10-2 Mass- and Heat-transfer Correlations

Average transport coefficients between the bulk stream and the particle surface in a fixed-bed reactor can be correlated in terms of dimensionless groups which describe the flow conditions. For mass transfer the group $k_m \rho / G$ is a function of the Reynolds number $d_P G / \mu$ and the Schmidt number $\mu / \rho \mathcal{D}$. Chilton and Colburn[1] suggested plotting j_D vs $d_P G / \mu$, where

[1] T. C. Chilton and A. P. Colburn, *Ind. Eng. Chem.*, **26**, 1183 (1934).

Fig. 10-1 *True* (E) *and apparent* (E') *activation energies*

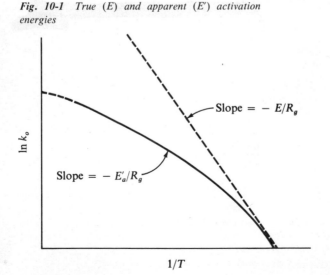

$$j_D = \frac{k_m \rho}{G} \left(\frac{\mu}{\rho \mathscr{D}} \right)^{\frac{2}{3}} = f\left(\frac{d_P G}{\mu} \right) \tag{10-10}$$

G = mass velocity based on the total (superficial) cross-sectional area of reactor

μ = viscosity of fluid

ρ = density of fluid

$\mathscr{D}$ = molecular (bulk) diffusivity of the component being transferred

This type of plot has been used widely to correlate experimental mass-transfer data. De Acetis and Thodos[1] have summarized the data available up to 1960 in a single curve of j_D vs Reynolds number, as shown in Fig. 10-2. For spherical pellets d_P is the diameter; for other shapes d_P can be taken as the diameter of a sphere with the same external area. Other investigations and correlations of mass-transfer data include those of Carberry,[2] Yeh,[3] Bradshaw and Bennett,[4] and Thoenes and Kramers.[5]

Heat transfer between a fluid and pellet in a fixed bed occurs by the same combination of molecular and convective processes as describe mass transfer. The experimental data for heat-transfer coefficients can be correlated by an expression analogous to Eq. (10-10). Thus the heat-transfer correlation is

$$j_H = \frac{h}{c_p G} \left(\frac{c_p \mu}{k} \right)^{\frac{2}{3}} = f\left(\frac{d_P G}{\mu} \right) \tag{10-11}$$

The j_H curve in Fig. 10-2 is the relationship presented by De Acetis and Thodos. The heat-transfer coefficient is defined in terms of the temperature difference between bulk fluid T_b and surface T_s as

$$Q = h a_m (T_s - T_b) \tag{10-12}$$

where Q is the heat-transfer rate from pellet to fluid per unit mass of catalyst. Heat-transfer coefficients evaluated from Eq. (10-11) and Fig. 10-2 do not include a radiation contribution. In general, radiation effects are negligible below 400°C for fixed beds of pellets not greater than 1/4 in. diameter.[6] The relationship between the temperature and concentration differences between fluid and pellet surface can be established by combining the

[1] James de Acetis and George Thodos, *Ind. Eng. Chem.*, **52**, 1003 (1960).

[2] J. J. Carberry, *AIChE J.*, **6**, 460 (1960).

[3] G. C. Yeh, *J. Chem. Eng. Data*, **6**, 526 (1961).

[4] R. D. Bradshaw and C. O. Bennett, *AIChE J.*, **7**, 48 (1961).

[5] D. Thoenes and H. Kramers, *Chem. Eng. Sci.*, **8**, 271 (1958).

[6] W. B. Argo and J. M. Smith, *Chem. Eng. Progr.*, **49**, 443 (1953).

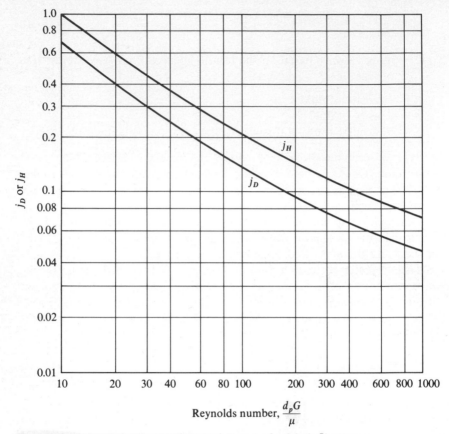

Fig. 10-2 *Heat- and mass-transfer correlations in fixed beds [by permission from J. de Acetis, and George Thodos, Ind. Eng. Chem., 52, 1003 (1960)]*

correlations for k_m and h. Thus an energy balance on the pellet requires, for steady state,

$$k_m a_m (C_b - C_s)(-\Delta H) = h a_m (T_s - T_b) \tag{10-13}$$

Using Eqs. (10-10) and (10-11) for k_m and h yields

$$T_s - T_b = (C_b - C_s) \frac{-\Delta H}{c_p \rho} \left(\frac{c_p \mu / k}{\mu / \rho \mathscr{D}} \right)^{\frac{2}{3}} \frac{j_D}{j_H} \tag{10-14}$$

This expression can be used to evaluate the temperature difference from $C_b - C_s$. For many gases the *Lewis number*, the ratio of the Prandtl and Schmidt numbers, is about 1.0. Also, from Fig. 10-2, j_D/j_H is about 0.7 for any Reynolds number. Hence for most gases Eq. (10-14) reduces approximately to

$$T_s - T_b = 0.7 \frac{-\Delta H}{c_p \rho} (C_b - C_s) \tag{10-15}$$

If the global rate is influenced by diffusion, so that $C_b - C_s$ is appreciable, Eq. (10-15) shows that significant temperature differences between fluid and pellet are possible. In fact the arrangement of properties in Eq. (10-14) is such that $T_s - T_b$ may be appreciable when $C_b - C_s$ is very small.

Equation (10-15) can be expressed in terms of the maximum, or adiabatic, temperature rise in the gas. Thus if all the heat of reaction remains in the gas, the temperature increase is given by the energy balance

$$C_b(-\Delta H) = \rho c_p (\Delta T)_{max} \quad \longrightarrow \text{ what happened to } \cdot 7$$

or

$$\frac{(-\Delta H)C_b}{c_p \rho} = (\Delta T)_{max} \tag{10-16}$$

Substituting this result in Eq. (10-15) gives

$$T_s - T_b = 0.7(\Delta T)_{max} \frac{C_b - C_s}{C_b} \tag{10-17}$$

In the extreme case where external diffusion controls the global rate, $C_s \to 0$. Then Eq. (10-17) indicates that the temperature difference for gaseous reactions would approach 0.7 of the adiabatic temperature rise. Values of 200°C for $T_s - T_b$ have been measured for the hydrogenation of oxygen in a dilute gas stream on a platinum catalyst.[1] Examples of even larger ΔT are also available.[2]

Variations of the Fixed-bed Reactor   A fixed bed of catalyst pellets with both gas and liquid flow is called a *trickle-bed reactor*. Such reactors are generally restricted to systems involving very volatile components such as hydrogen. For example, petroleum fractions may be hydrogenated or dehydrogenated (and also desulfurized) by downflow as a liquid over a bed of catalyst pellets while hydrogen gas simultaneously passes either upward or downward through the bed. The liquid wets the pellets and flows in thin layers from pellet to pellet, and the gas flows through the remaining voids. The mass-transport situation is similar in slurry reactors (see Sec. 10-7). Hydrogen must first dissolve in the fluid and then diffuse through the liquid layer in order to reach the pellet surface. Little information has been published on the performance of trickle-bed reactors. It would seem that diffusion resistances are greater than for gas-solid catalytic reactors because

[1] J. Maymo and J. M. Smith, *AIChE J.*, **12**, 845 (1966).
[2] C. N. Satterfield and H. Resnick, *Chem. Eng. Progr.*, **50**, 504 (1954); F. Yoshida, D. Rama-swami, and O. A. Hougen, *AIChE J.*, **8**, 5 (1962).

of the liquid film surrounding the pellets. Also, the advantage of using porous pellets may be considerably reduced, since the pores are likely to fill with liquid and thus offer increased diffusion resistance. Considerable convection may be induced in the liquid layers by the gas flow. This can effectively reduce the mass-transfer resistance in the liquid. Babcock et al.[1] studied the hydrogenation of α-methyl styrene by passing hydrogen upward and countercurrent to liquid styrene in beds of $\frac{1}{8} \times \frac{1}{8}$-in. alumina pellets *coated* with catalyst. Klassen[2] investigated the air oxidation of ethanol on a palladium catalyst in a trickle bed and found evidence of diffusion limitations.

The *pulsed reactor* consists of a fixed bed of catalyst pellets through which the reacting fluid moves in pulsating flow. Mass-transfer coefficients are increased because of the pulsating velocity superimposed on the steady flow. For viscous liquids, or any fluid-solid reaction system which has a high external-mass-transfer resistance, pulsation may be a practical way to increase the global reaction rate. Biskis and Smith[3] measured mass-transfer coefficients for hydrogen in α-methyl styrene in pulsed flow and found increases up to 80% over steady values. Bradford[4] found similar results based on data for the dissolution of beds of β-naphthol particles in water.

Assemblies of *screens* or grids of solids are often employed for oxidation reactions, particularly those using platinum or silver as catalysts. These reactions are frequently characterized by a high rate constant, so that mass transfer of reactants from fluid to solid surface can be a significant part of the total resistance. Data on mass transfer from single screens has been reported by Gay and Maughan.[5] Their correlation is of the form

$$j_D = \frac{\epsilon k_m \rho}{G} \left(\frac{\mu}{\rho \mathscr{D}} \right)^{\frac{2}{3}} = C \left(\frac{4G}{\beta \mu} \right)^{-m} \tag{10-18}$$

where ϵ is the porosity of the single screen and β is the external-heat-transfer area of the screen per unit volume, i.e., the reciprocal of the hydraulic radius of the screen, and the other symbols are defined as in Eq. (10-10). The coefficients C and m are given in Table 10-1 for the four screen sizes investigated. Heat-transfer results were obtained for similar screens by Coppage and London.[6]

For a matrix of 20 or more screens heat-transfer data were found[7]

[1] B. D. Babcock, G. T. Mejdell, and O. A. Hougen, *AIChE J.*, **3**, 366 (1957).

[2] J. Klassen, doctoral dissertation, University of Wisconsin, Madison, Wis., 1953.

[3] E. R. Biskis and J. M. Smith, *AIChE J.*, **9**, 677 (1963).

[4] E. Bradford, doctoral dissertation, Princeton University, Princeton, N.J., 1960.

[5] B. Gay and R. Maughan, *Intern. J. Heat Mass Transfer*, **6**, 277 (1963).

[6] J. E. Coppage and A. L. London, *Chem. Eng. Progr.*, **52**, 57 (1956).

[7] A. L. London, J. W. Mitchell, and W. A. Sutherland, *ASME J. Heat Transfer*, **82**, 199 (1960).

Table 10-1   Coefficients C and m in Eq. (10-18)

Screen mesh size	C	m	ϵ	β, ft^{-1}
10	2.62	0.73	0.817	390
16	4.26	0.85	0.795	535
24	2.80	0.81	0.763	858
60	1.46	0.77	0.690	2,030

to follow the j_H lines of Fig. 10-3. The curves given are for extremes of porosity. For intermediate porosities the curves were located between the two shown in the figure. The Reynolds number for Fig. 10-3 is defined as

$$Re' = \frac{4r_H G}{\mu} \tag{10-19}$$

where r_H, the hydraulic radius, is

$$r_H = L\,\frac{A_f}{A_h} \tag{10-20}$$

L = length of matrix of screens in the direction of flow
A_f = total cross-sectional area of screen multiplied by porosity ϵ_m
     of matrix of screens
A_h = heat-transfer area of matrix of screens
G = mass velocity based on A_f

Fig. 10-3   Heat-transfer correlation for matrix of screens

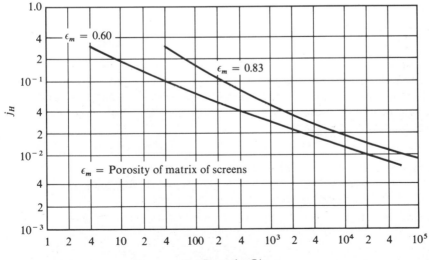

10-3 Calculation of External Concentration Differences

As mentioned earlier, in laboratory reactors the global rate is measured directly, and the question is whether these rates are influenced by external physical processes. In this section we shall consider the method of solution for an isothermal case. Combined mass- and energy-transfer limitations are discussed in Sec. 10-4.

Olson and Smith[1] measured the rate of oxidation of sulfur dioxide with air in a differential fixed-bed reactor. The platinum catalyst was deposited on the outer surface of the cylindrical pellets. The composition and the rates of the bulk gas were known. The objective was to determine the significance of external diffusion resistance by calculating the magnitude of $C_b - C_s$. If this difference is significant, then the C_s values must be used in developing a rate equation for the chemical step.

Example 10-1 Experimental, global rates are given in Table 10-2 for two levels of conversion of SO_2 to SO_3. Evaluate the concentration difference for SO_2 between bulk gas and pellet surface and comment on the significance of external diffusion. Neglect possible temperature differences. The reactor consists of a fixed bed of $\frac{1}{8} \times \frac{1}{8}$-in. cylindrical pellets through which the gases passed at a superficial mass velocity of 147 lb/(hr)(ft^2) and at a pressure of 790 mm Hg. The temperature of the catalyst pellets was 480°C, and the bulk mixture contained 6.42 mole % SO_2 and 93.58 mole % air. To simplify the calculations compute physical properties on the basis of the reaction mixture being air. The external area of the catalyst pellets is 5.12 ft^2/lb material. The platinum covers only the external surface and a very small section of the pores of the alumina carrier, so that internal diffusion need not be considered.

Table 10-2

Mean conversion of SO_2	r, g moles SO_2/(hr) (g catalyst)	p_b, atm		
		SO_2	SO_3	O_2
0.1	0.0956.	0.0603	0.0067	0.201
0.6	0.0189	0.0273	0.0409	0.187

Solution First the Reynolds number is evaluated. At 480°C the viscosity of air is about 0.09 lb/(hr)(ft). The particle diameter to employ is the diameter of the sphere with the same area as that of the cylindrical pellets. Hence πd_p^2 will equal the sum of the areas of the lateral and end surfaces of the cylinder:

$$\pi d_p^2 = \pi dL + 2\frac{\pi d^2}{4} = \pi \frac{1}{96}\left(\frac{1}{96}\right) + \frac{2\pi}{4}\left(\frac{1}{96}\right)^2$$

[1] R. W. Olson and J. M. Smith, *Chem. Eng. Progr.*, **42**, 614 (1950).

$$d_P^2 = \frac{3}{2}\left(\frac{1}{96}\right)^2$$

$$d_P = 0.0128 \text{ ft}$$

The Reynolds number is

$$\frac{d_P G}{\mu} = \frac{0.0128(147)}{0.09} = 21$$

and, from Fig. 10-2,

$$j_D = 0.37$$

Now, eliminating k_m from Eqs. (10-10) and (10-1) and solving for $C_b - C_s$, we have

$$C_b - C_s = \frac{\mathbf{r}_P}{a_m} \frac{(\mu/\rho\mathscr{D})^{\frac{2}{3}}}{j_D G/\rho} \tag{10-21}$$

In determining the Schmidt group the correct value for $\mathscr{D}$ would be the molecular diffusivity of sulfur trioxide in a mixture of nitrogen, oxygen, and sulfur dioxide, in which O_2 and SO_2 would also be diffusing. Wilke[1] has proposed procedures for evaluating diffusivities in such complex systems. However, in this instance little error will be introduced by considering $\mathscr{D}$ the binary diffusivity of SO_2 in air. This may be estimated from the Chapman-Enskog kinetic theory. The equation for $\mathscr{D}$ and illustrations of its use are given in Sec. 11-1. From Example 11-1, the molecular diffusivity of SO_2-air is 0.629 cm²/sec, or 2.44 ft²/hr.

The density of air will be

$$\frac{28.9}{359}\left(\frac{273}{480 + 273}\right)\left(\frac{790}{760}\right) = 0.0304 \text{ lb/ft}^3$$

Then the Schmidt group is

$$\frac{\mu}{\rho\mathscr{D}} = \frac{0.09}{0.0304(2.44)} = 1.21 \qquad \text{(dimensionless)}$$

For 10% conversion $\mathbf{r}_P = 0.0956$ g mole/(hr)(g), or lb mole/(hr)(lb), and $a_m = 5.12$ ft²/lb. Substituting all these results in Eq. (10-21) yields

$$C_b - C_s = \frac{0.0956}{5.12} \frac{1.21^{\frac{2}{3}}}{0.37(147/0.0304)} = 1.18 \times 10^{-5} \text{ lb mole/ft}^3$$

The numerical results are more meaningful if they are converted to partial pressures. In atmospheres, the difference between bulk and surface pressures of sulfur dioxide is

$$(p_b - p_s)_{SO_2} = R_g T(C_b - C_s) = 0.73[1.8(480 + 273)](1.18 \times 10^{-5})$$

$$= 0.0117 \text{ atm}$$

$$p_s = 0.0603 - 0.0117 = 0.0486 \text{ atm}$$

[1]C. R. Wilke, *Chem. Eng. Progr.*, **46**, 95 (1950).

If the Δp for 60% conversion is calculated in the same manner, the results at the two conversions can be summarized as in Table 10-3. The relatively large values of $p_b - p_s$ indicate that external-diffusion resistance is significant, although the effect is less at the higher conversion because the rate is less.

Table 10-3

Conversion level	Partial pressures of SO_2, atm			
	p_b	$p_b - p_s$	p_s	$(p_b - p_s)/p_b$
0.1	0.0603	1.17×10^{-2}	0.0486	0.19
0.6	0.0273	0.23×10^{-2}	0.0250	0.08

If external diffusion were neglected in Example 10-1 and C_b values were used to relate the surface rate and composition (i.e., to obtain an equation for the chemical step at the surface), serious errors would result, particularly at low conversions. The high temperature and low mass velocity lead to large external-diffusion resistances in this case. At lower temperatures the surface rate would be lower, and at higher velocities k_m would increase; both effects would reduce $C_b - C_s$. In laboratory investigations it is customary to make preliminary measurements at a series of increasing mass velocities. If all other conditions are constant during this series of runs, the importance of external diffusion will decrease with increasing mass velocity. When the experimental rate no longer increases as G is increased, external diffusion is negligible and $C_s \approx C_b$. If rate measurements are made at this mass velocity, bulk concentrations can be used to obtain an expression for the chemical step at the surface. The critical velocity must be determined at the maximum temperature if diffusion resistance is to be negligible at all other temperatures. Other methods of eliminating external resistances in laboratory reactors are described in Chap. 12. Instead of eliminating these resistances by choice of experimental conditions, we may account for them by calculating the surface values of concentration (and temperature), as illustrated in Example 10-1. However, this introduces uncertainty in the results to the extent that the correlations for k_m (and h) are in error.

10-4 Calculation of External Temperature Differences

When a fixed-bed reactor operates at steady state, an amount of energy equal to the heat released by reaction on the catalyst pellet must be transferred to the bulk fluid. In Sec. 10-2 this requirement was used to relate the concentration and temperature differences between pellet and fluid. Here we want to develop a method for predicting the magnitude of the temperature difference from the characteristics of the reaction. The heat Q_R released by the reaction (per unit mass of catalyst) is given by Eq. (10-2) multiplied by $-\Delta H$. From the Arrhenius equation,

$$Q_R = (-\Delta H)\mathbf{r}_P = (-\Delta H)a_m \mathbf{A} C_s^{\ n} e^{-E/R_g T_s} \tag{10-22}$$

where $\mathbf{A}$ is the frequency factor. The exponential term in Eq. (10-22) can be expressed in terms of $T_s - T_b$ by noting that

$$\frac{-E}{R_g T_s} = \frac{E}{R_g T_b}\left(\frac{T_s - T_b}{T_s} - 1\right) = \frac{-\alpha}{\theta + 1} \tag{10-23}$$

where $\alpha = E/R_g T_b$ and θ is a dimensionless temperature, $(T_s - T_b)/T_b$. Using Eq. (10-23) for the exponential term in Eq. (10-22) gives

$$Q_R = (-\Delta H)\,a_m \mathbf{A} C_s^{\ n} e^{(E/R_g T_b)[(T_s - T_b)/T_s - 1]} \tag{10-24}$$

or

$$Q_R = (-\Delta H)\,a_m \mathbf{A} C_s^{\ n} e^{-\alpha/(\theta + 1)}$$

The heat transferred to the bulk fluid, Q, can be simply expressed in terms of $T_s - T_b$ as

$$Q = ha_m(T_s - T_b) = ha_m T_b\theta \tag{10-25}$$

Equating Q and Q_R at steady state determines $T_s - T_b$ in terms of ΔH, the rate parameters $\mathbf{A}$ and E for the reaction, h, and the unknown surface concentration C_s. The requirement that $Q_R = Q$ introduces interesting questions about stable operating conditions. The problem is very similar to the stability situation in stirred-tank reactors, discussed in Sec. 5-4. We consider this problem and the evaluation of $T_s - T_b$ for two cases: negligible and finite external-diffusion resistance.

Negligible Diffusion Resistance With this restraint $C_s^{\ n}$ in Eq. (10-24) becomes $C_b^{\ n}$. Then equating Q_R and Q establishes $T_s - T_b$ in terms of the heat of reaction $-\Delta H$, $\mathbf{A}$, E, and h. The resultant equation cannot be solved analytically for $T_s - T_b$. However, the intersection of curves for Q and Q_R plotted against $T_s - T_b$ will give the solution. According to Eq. (10-24), Q_R will be an exponential curve in θ, as shown in Fig. 10-4 for an exothermic reaction. Equation (10-25) is linear in θ. It could intersect Q_R as indicated by curves Q_1 or Q_2, or it could be below Q_R, as shown by curve Q_3, depending on the magnitude of h and the location of Q_R.

The curves in Fig. 10-4 were originally proposed by Frank-Kamenetski[1] and are useful for describing regions where multiple values of T_s are

[1] D. A. Frank-Kamenetski, "Diffusion and Heat Exchange in Chemical Kinetics," chap. IX, Princeton University Press, Princeton, N.J., 1955.

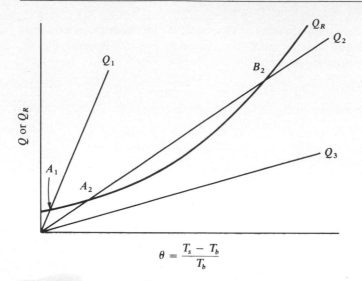

Fig. 10-4 *Temperature difference between bulk fluid and surface of catalyst pellet (negligible diffusion resistance)*

possible. If the relative position is as described by curves Q_R and Q_1, stable operation occurs at point A_1. If the rate of heat transfer from solid to fluid is less, so that the system is given by Q_R and Q_2, there can be two temperatures for stable operation, given by A_2 and B_2. The region between A_2 and B_2 is unstable in that the heat loss is greater than the heat of reaction. An initial condition in this region would stabilize at A_2. An initial condition above B_2 would be unstable, with no temperature of stabilization, because $Q_R > Q_2$. Similarly, the Q_R and Q_3 system is not usually stable, since Q_R is greater than Q_3 at any reasonable value of $T_s - T_b$.

An instructive way to quantify the effect of external resistances on the rate is by the ratio γ of the global, or actual, rate to the rate evaluated at the conditions of the bulk fluid. For the exothermic case considered in Fig. 10-4, γ is given by the ratio of rates evaluated at catalyst-pellet and bulk temperatures:

$$\gamma = \frac{\mathbf{r}_P}{\mathbf{r}_b} = \frac{a_m A C_b{}^n\, e^{-E/R_g T_s}}{a_m A C_b{}^n\, e^{-E/R_g T_b}} = e^{(E/R_g T_b)[(T_s - T_b)/T_s]} \tag{10-26}$$

or

$$\gamma = e^{\alpha\theta/(\theta + 1)}$$

This result is valid as long as the effect of temperature on the rate can be expressed by the single exponential relation (Arrhenius equation) employed in Eq. (10-22).[1]

The temperature reported in Example 10-1 was measured by inserting thermocouples in the catalyst pellets. In the usual laboratory investigation only the fluid temperature (T_b) is measured. The error introduced by assuming $T_s = T_b$ is considered in the following example.

Example 10-2 Suppose that only the fluid temperature had been measured in the problem of Example 10-1. What error would have been introduced by assuming this to be equal to the catalyst temperature? The heat of reaction for

$$SO_2 + \tfrac{1}{2}O_2 \to SO_3$$

is approximately $-23,000$ cal/g mole at 480°C, and the activation energy may be taken as 20,000 cal/g mole.

Solution $T_s - T_b$ can be obtained by equating Q_R and Q from Eqs. (10-22) and (10-25). Thus

$$T_s - T_b = \frac{\mathbf{r}_P(-\Delta H)}{ha_m} \tag{A}$$

Expressing h in terms of j_H and using Eq. (10-11) gives

$$T_s - T_b = \frac{\mathbf{r}_P}{a_m} \frac{(-\Delta H)\,Pr^{\tfrac{2}{3}}}{j_H c_p G} \tag{B}$$

For air at 480°C the *Prandtl number* is $Pr = c_p \mu / k = 0.70$ and $c_p = 0.26$ Btu/lb°F. From Example 10-1, the Reynolds number is 21. Then from Fig. 10-2, $j_H = 0.60$. Using the data given in Example 10-1, we find for the 0.1 conversion level

$$T_s - T_b = \frac{0.0956}{5.12} \frac{23,000(1.8)(0.70)^{\tfrac{2}{3}}}{0.60(0.26)(147)} = 27°F \quad \text{or } 15°C$$

$$T_b = T_s - 15° = (480 + 273) - 15 = 738°K \quad \text{or } 465°C$$

Hence

$$\theta = \frac{T_s - T_b}{T_b} = \frac{15}{738} = 0.02$$

$$\alpha = \frac{E}{R_g T_b} = \frac{20,000}{1.98(738)} = 13.5$$

Then, from Eq. (10-26),

$$\gamma = e^{13.5(0.02/1.02)} = 1.30$$

---

[1] For a Langmuir-Hinshelwood type of rate equation, such as Eq. (9-32), Eq. (10-26) is not valid because the adsorption-equilibrium-constant terms in the denominator of the rate expression are also temperature dependent.

The 15°C increase in temperature on the surface of the catalyst pellet increases the rate 30%. Evaluating the rate at T_b would introduce an error of 30%, with the assumption that the diffusion resistance is negligible. Example 10-1 showed that external diffusion is important, so that the 30% error gives only the temperature effect. The combined treatment of diffusion and heat transfer for this problem is given in Example 10-3.

The temperature difference could also have been obtained from the concentration difference (calculated in Example 10-1) by using the energy balance of Eq. (10-13). Thus, from Eq. (10-14),

$$T_s - T_b = 1.18 \times 10^{-5} \frac{23,000(1.8)}{0.26(.0304)} \left(\frac{0.70}{1.21}\right)^{\frac{2}{3}} \left(\frac{0.37}{0.60}\right) = 27°F \quad \text{or } 15°C$$

At a conversion of 0.6 the rate is reduced to 0.0189 lb mole/(hr)(lb catalyst). Equation (A) shows that Δt is directly proportional to the rate, so that

$$T_s - T_b = \frac{0.0189}{0.0956} (15) = 3°C$$

For this case $\theta = 0.004$. Hence

$$\gamma = e^{\alpha\theta/(\theta + 1)} = e^{0.054} = 1.06$$

Here the external-temperature difference is of little importance because the rate is relatively low. For both conversion levels the results apply to a stable operating point corresponding to A_2 in Fig. 10-4.

Finite Diffusion Resistance In this case $C_s < C_b$, so that C_s must be used in Eq. (10-24). Only for a first-order reaction can C_s be expressed simply in terms of C_b. To illustrate the procedure for evaluating $T_s - T_b$ we shall use first-order kinetics, with the realization that it is not necessary when the diffusion resistance is unimportant. Equation (10-5) may be written

$$C_s = \frac{k_m C_b}{k_m + k} = \frac{k_m C_b}{k_m + A e^{-E/R_g T_s}} \tag{10-27}$$

If this expression for C_s is substituted in Eq. (10-24) and the result is expressed in terms of $T_s - T_b$, from Eq. (10-23) we obtain

$$Q_R = \frac{(-\Delta H)a_m C_b}{(1/A)e^{-E/R_g T_b[(T_s - T_b)/T_s - 1]} + 1/k_m} = \frac{(-\Delta H)a_m C_b}{(1/A)e^{\alpha/(\theta + 1)} + 1/k_m} \tag{10-28}$$

The surface temperature for steady-state operation is determined by the solution of Eqs. (10-25) and (10-28) for the condition $Q_R = Q$. Again it is easiest to plot Q and Q_R, noting the intersections of the two curves. The curve for Q_R tends to flatten as $T_s - T_b$ increases to a high value, because the $1/k_m$ term in the denominator of Eq. (10-28) becomes dominant. This S-shaped curve is shown in Fig. 10-5. The relative position of the Q_R and Q curves is illustrated by drawing several lines for Q. For the set $Q_1 - Q_R$ the stable operating point is at A_1. If the initial temperature is on either side of this point, the pellet will cool or heat until A_1 is reached. For the set $Q_R - Q_2$ initial T_s values below point A_2 will increase to A_2, and initial T_s values above B_2 will fall to B_2. Intersection C_2 is, in theory, another stable condition. However, it is pseudo-stable, because small perturbations below C_2 force the temperature to A_2, while slight deviations above force the temperature to B_2. In practice, the entire region between A_2 and B_2 is unstable. For $Q_R - Q_3$ stable operation occurs only at a high value of $T_s - T_b$ corresponding to B_3. Here the Q_R curve is nearly flat and is determined by the diffusion term $1/k_m$ in Eq. (10-28). Frank-Kamenetski termed this region the *diffusion-controlled regime* and the region at A_1 and A_2 the *reaction-controlled regime*. For the set $Q_R - Q_4$ there is usually no stable region, because the heat transfer away from the pellet is normally less than the heat of reaction. Regions of large $(T_s - T_b)/T_b$ are frequently encountered for combustion, since $-\Delta H$ for these systems is high. However, hydrogenations and other reactions can exhibit both reaction and diffusion regimes at feasible operating conditions.

Fig. 10-5 *Temperature differences between bulk fluid and surface of catalyst pellet* (*general case*)

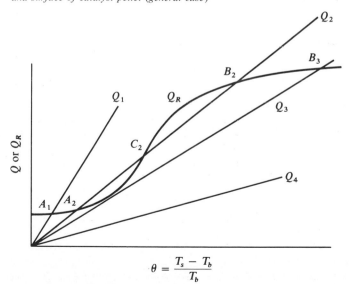

For finite diffusion resistance, the ratio γ for a first-order reaction is

$$\gamma = \frac{\mathbf{r}_P}{\mathbf{r}_b} = \frac{a_m A C_s e^{-E/R_g T_s}}{a_m A C_b e^{-E/R_g T_b}} \tag{10-29}$$

Utilizing Eq. (10-27) for C_s and Eq. (10-23), we obtain

$$\gamma = \frac{e^{\alpha\theta/(\theta+1)}}{1 + (A/k_m) e^{-\alpha/(\theta+1)}} \tag{10-30}$$

Comparison with Eq. (10-26) shows that the effect of diffusion resistance is to reduce γ.

The same type of behavior is observed for gas-solid noncatalytic reactions, such as the combustion of carbon pellets in air. This problem is a transient one, however, because the size of the pellet decreases as the carbon is consumed. This class of heterogeneous reactions is discussed in Chap. 14.

Example 10-3   For the oxidation of SO_2 calculate the ratio of the global rate to the rate if both temperature and concentration differences between gas and catalyst pellet are neglected. Use the data in Example 10-1.

Solution   The measured rates[1] were not first order, but depended on SO_2, O_2, and SO_3 concentrations. Hence Eq. (10-30) is not applicable. However, concentration differences between gas and surface were evaluated in Example (10-1), and these results may be used to establish the rates of reaction at pellet and bulk-gas conditions, to give γ directly.

For a conversion of 0.10 the results of Examples 10-1 and 10-2 give the bulk and surface concentrations and temperatures shown in Table 10-4. If both temperature and concentration differences are neglected, the rate would be evaluated at the bulk-gas conditions. From the data given by Olson and Smith, this rate is 0.333 g mole/(hr) (g catalyst), as shown in Table 10-4. The ratio of rates is therefore

$$\gamma = \frac{\mathbf{r}_P}{\mathbf{r}_b} = \frac{0.0956}{0.333} = 0.29$$

For this case the effect of diffusion more than offsets the 30% increase in rate due to the temperature rise (Example 10-2). The net result is that the global rate is only 29% of the bulk-gas rate.

Table 10-4

Location	Conc. (as partial pressure) p_{SO_2}, atm	t, °C	$\mathbf{r}$, Global rate, g moles/(hr)(g)
Catalyst surface	0.0486	480	0.0956 (exp.)
Bulk gas	0.0603	465	0.333

[1] Again, the subscript zero here is not related to initial conditions.

For the higher conversion level the results are as shown in Table 10-5. Thus

$$\gamma = \frac{0.0189}{0.0225} = 0.84$$

Table 10-5

Location	p_{SO_2}, atm	$t, °C$	r, Global rate, g moles/(hr)(g)
Catalyst surface	0.0250	480	0.0189 (exp.)
Bulk gas	0.0273	477	0.0225

It is instructive to examine how well the first-order assumption would fit this example by evaluating A and k_m for use in Eq. (10-30). If the rate is assumed to be first order, then

$$\mathbf{r}_p = a_m A C_s \, e^{-E/R_g T_s}$$

or

$$a_m A = \frac{\mathbf{r}_P}{C_s \, e^{-E/R_g T_s}} = \frac{\mathbf{r}_P}{(p_s/R_q T_s) \, e^{-E/R_g T_s}}$$

Using the actual rate at the catalyst surface (10% conversion level), we obtain

$$a_m A = \frac{0.0956}{[0.0486/0.73(753 \times 1.8)] \, e^{-20,000/1.98(753)}}$$

$$= 1.29 \times 10^9 \ \text{ft}^3/(\text{hr})(\text{lb})$$

and

$$A = \frac{1.29 \times 10^9}{5.12} = 2.5 \times 10^8 \ \text{ft/hr}$$

From Example 10-1, $j_D = 0.37$. Then in Eq. (10-10)

$$k_m = \frac{j_D G}{\rho} \left(\frac{\mu}{\rho \mathscr{D}} \right)^{-\frac{2}{3}} = \frac{0.37(147)}{0.0304} (1.21)^{-\frac{2}{3}} = 1570 \ \text{ft/hr}$$

As in Example 10-2, $\alpha = 13.5$ and $\theta = 0.02$, so that

$$\gamma = \frac{e^{0.02(13.5)/1.02}}{1 + [(2.5 \times 10^8)/1570] e^{-13.5/1.02}} = 0.97$$

This result does not agree with the actual value of 0.29, because the first-order assumption does not fit the kinetics of SO_2 oxidation. Sulfur trioxide inhibits the rate of this reversible reaction. The effect of diffusion resistance is to increase the concentration of SO_3 at the catalyst surface above that in the bulk gas. This in turn reduces the rate. For this reaction there are two effects of diffusion: reduction of SO_2 concentration at the surface and an increase of product (SO_3) concentration. Both reduce the rate. When the rate is not retarded by product concentrations but is still not first order, Eq. (10-30) is a better approximation. For example, for a second-order reaction it provides a rapid means of estimating, approximately, the combined influence of external-diffusion and heat-transfer resistances.

Hougen[1] has summarized many aspects of the problem of external-mass- and energy-transfer resistances and has included figures for predicting when such resistances need to be considered. This was done for mass transfer by eliminating k_m from Eqs. (10-10) and (10-1), thus establishing $(C_b - C_s)/C_b$ in terms of the rate of reaction, the Reynolds number, and the Schmidt number, $\mu\rho/\mathscr{D}$. If this concentration ratio is less than, say, 0.1, diffusion resistance is unimportant. In a similar way, Eqs. (10-11), (10-12), and (10-22) were used to prepare a plot of $T_s - T_b$ as a function of the rate, the Reynolds number, and the Prandtl number. The exponential effect of temperature on the rate means that small values of $T_s - T_b$ can have a large effect on the rate. Thus care must be taken when external heat-transfer resistances are neglected. The safest procedure is to evaluate the rate at T_s and at T_b and note whether the difference is significant, as was done in Example 10-2. These methods for evaluating external resistances are applicable for any form of rate equation, since they utilize an experimental rate of reaction.

As an indication of the potential magnitude of $T_s - T_b$, consider the experimental results of Maymo and Smith.[2] They studied the reaction between hydrogen and oxygen by using a single spherical pellet (1.86 cm diameter) of platinum on alumina catalyst. The pellet was suspended in the center of a spherical reactor, 4.4 cm in diameter. Measures were taken to ensure good mixing of the gas in the reactor, and a recycle system was used to provide stirred-tank operation. The external-concentration difference $(C_b - C_s)/C_b$ was calculated to be no larger than 5%. However, temperature differences were large. In one run at high rates of reaction $T_s - T_b$ was about 100°C, where $T_b = 333°K$. This example (for a large ΔH) clearly indicates the significance of Eq. (10-14); that is, external-temperature effects can be significant when the diffusion effect is negligible. Hutchings and Carberry[3] also have predicted that negligible mass-transfer and finite heat-transfer resistances would exist for some reaction conditions.

[1] O. A. Hougen, *Ind. Eng. Chem.*, **53**, 509 (1961).

[2] J. A. Maymo and J. M. Smith, *AIChE J.*, **12**, 845 (1966).

[3] John Hutchings and J. J. Carberry, *AIChE J.*, **12**, 20 (1966).

10-5 Effect of External Resistances on Selectivity

Suppose two parallel reactions

1. $A \rightarrow B$

2. $A \rightarrow C$

occur in a fixed-bed reactor through which fluid reactant A is passed. The desired product is B. For simplicity assume also that both reactions are first order and irreversible. The point selectivity S (as defined in Chap. 2) is equal to the ratio of the global rates of the two reactions, $(r_1/r_2)_p$.

If the system is isothermal, only external-diffusion resistance is involved. However, for parallel reactions involving the same reactant, diffusion has no effect on the selectivity. This is seen by using Equation (10-2) to express the rate of each reaction. Taking the global (or pellet) selectivity as S_P

$$S_P = \frac{(r_1)_P}{(r_2)_P} = \frac{k_1 a_m (C_s)_A}{k_2 a_m (C_s)_A} = \frac{k_1}{k_2} \tag{10-31}$$

Thus, S_P is equal to the ratio of the intrinsic rate constants just the same as though diffusion was not involved. For consecutive reactions, such as $A \rightarrow B \rightarrow C$, diffusion has an adverse effect on the selectivity of B with respect to C. This is because diffusion resistance reduces the surface concentration of A (from which B is produced) and increases the surface concentration of B (from which C is produced), with respect to bulk concentrations.

For nonisothermal conditions the situation is more complicated, but it can be analyzed conveniently in terms of global and bulk rates. For the parallel reaction case, the relation between bulk selectivity and γ may be written

$$S_P = \left(\frac{r_1}{r_2} \right)_P = \frac{(r_P/r_b)_1}{(r_P/r_b)_2} \left(\frac{r_1}{r_2} \right)_b = \frac{\gamma_1}{\gamma_2} S_b \tag{10-32}$$

Equation (10-30) is not applicable for the γ values, because T_s and C_s are determined by both reactions 1 and 2; that is, Eq. (10-27) is not valid. However, γ_1 and γ_2 may be formulated directly by taking ratios of the rates at surface and bulk conditions:

$$\gamma_1 = \frac{a_m (A_1 e^{-E_1/R_g T_s}) C_s}{a_m (A_1 e^{-E_1/R_g T_b}) C_b} = e^{-E_1/R_g(1/T_s - 1/T_b)} \frac{C_s}{C_b}$$

$$\gamma_2 = e^{-E_2/R_g(1/T_s - 1/T_b)} \frac{C_s}{C_b}$$

Substituting these results into Eq. (10-32) gives

$$S_P = \frac{e^{\alpha_1 \theta/(1+\theta)}}{e^{\alpha_2 \theta/(1+\theta)}} S_b \tag{10-33}$$

Consider a situation where the activation energy of the desired reaction is greater than that for the by-product reaction. Specifically, suppose that

$$T_b = 500°K$$

$$E_1 = 20 \text{ kcal/g mole}$$

$$E_2 = 15 \text{ kcal/g mole}$$

The surface temperature and, hence, θ will depend upon the rates of mass and energy transfer between bulk gas and pellet surface. Conservation of mass requires that r_p from Eq. (10-1) be equal to the *total* rate of disappearance of A by both reactions:

$$k_m a_m (C_b - C_s)_A = [a_m A_1 e^{-E_1/R_g T_s} + a_m A_2 e^{-E_2/R_g T_s}] C_s \qquad (10\text{-}34)$$

Similarly, conservation of energy requires equating Eq. (10-25) to the expression, corresponding to Eq. (10-22), for the total energy released by reaction:

$$ha_m(T_s - T_b) = a_m C_s [(-\Delta H_1) A_1 e^{-E_1/R_g T_s}$$
$$+ (-\Delta H_2) A_2 e^{-E_2/R_g T_s}] \qquad (10\text{-}35)$$

Simultaneous solution of these two expressions gives T_s and C_s, provided A_1, A_2, ΔH_1, ΔH_2, k_m, and h are known. For illustrative purposes suppose that the reactions are exothermic and that T_s turns out to be 550°K, making $\theta = 0.1$. Then

$$\alpha_1 = \frac{E_1}{R_g T_b} = \frac{20,000}{2(500)} = 20$$

$$\alpha_2 = \frac{15,000}{2(500)} = 15$$

and, from Eq. (10-33),

$$S_P = \frac{e^{20(0.1)/(1+0.1)}}{e^{15(0.1)/(1+0.1)}} S_b = 1.6 S_b$$

The selectivity is 60 percent more than that when external heat transfer is neglected.

In the foregoing illustration the temperature rise at the catalyst surface had a beneficial effect on selectivity. This is because the activation energy for the desired reaction was greater than that for the reaction producing by-product C. If E_1 were less than E_2, external heat-transfer resistance would have reduced the selectivity for exothermic reactions.

FLUIDIZED-BED REACTORS

In a fluidized bed relatively small particles of catalyst are sustained in a vertical tube by the upward motion of the reacting fluid. The range of velocities for fluidization is narrow. If the velocity is somewhat below that for fluidization, the particles remain in close contact with each other in a jiggling state. Such systems are sometimes called *expanded beds* to distinguish them from the situation at still lower velocities where the particles are motionless, i.e., a fixed bed. If the velocity is too high, the particles are carried out of the reactor by the fluid. Then the system becomes a transport reactor.

The temperature is nearly the same in all parts of a fluidized bed. The chief problems in design arise not from temperature variations, but from the complex motion of the fluid through the bed of particles. This is particularly true for gaseous systems. It has been shown[1] that the gas moves through the bed in two ways: as bubbles containing few solid particles and moving at above-average velocity, and through the interparticle spaces of a continuous region of high solids concentration. The fraction of gas flow in bubbles depends on the particle size and size distribution and increases with total gas rate. There is exchange of particles between the two types of flow. The behavior of the whole gaseous-reaction mixture is neither plug flow nor well mixed. Methods of handling this macromixing problem will be discussed in Chap. 13.

The uniformity of temperature is achieved by the high rates of heat transfer. This in turn is due to a considerable extent to the high heat-transfer rates between the gas and solid catalyst particles. Large heat-transfer coefficients are not responsible for the high rates. Rather, it is because of the small particle size (150 to 300 mesh is a normal size range) and the resultant large surface areas per unit volume. For this reason heat- and mass-transfer resistances between catalyst particle and fluid are negligible. A summary of the correlations for the transport coefficients are given here, but they are less important than those for the larger particles in fixed-bed reactors.

10-6 Mass- and Heat-transfer Correlations

The available data on heat- and mass-transfer coefficients have been correlated by Chu et al.[2] and others.[3] Most of the results agree reasonably well with the equation of Chu et al.

$$j_D \text{ or } j_H = 1.77 \left[\frac{d_p G}{\mu(1 - \epsilon)} \right]^{-0.44} \tag{10-36}$$

[1] J. J. Van Deempter, *Chem. Eng. Sci.*, **13**, 143 (1961); M. Leva, "Fluidization," McGraw-Hill Book Company, New York, 1959; F. A. Zenz and D. F. Othmer, "Fluidization and Fluid-Particle Systems," Reinhold Publishing Corporation, New York, 1960; O. Levenspiel and D. Kunii, "Fluidization Engineering," John Wiley & Sons, Inc., New York, 1969.

[2] J. C. Chu, J. Kaiil, and W. A. Wetterath, *Chem. Eng. Progr.*, **49**, 141 (1953).

[3] R. E. Riccetti and G. Thodos, *AIChE J.*, **7**, 442 (1961); A. S. Gupta and G. Thodos, *Chem. Eng. Progr.*, **58**, 58 (1962).

for the range $30 < d_p G/\mu(1 - \epsilon) < 5{,}000$. Here j_D and j_H are as defined in Eqs. (10-10) and (10-11) and

ϵ = void fraction in the bed

G = superficial mass velocity

d_p = average particle diameter

This correlation is based on data for both liquid-solid and gas-solid beds.[1]

The coefficients calculated from Eq. (10-36) are not particularly large. However, when this coefficient is multiplied by the external surface area per unit volume of bed, very high transport rates are obtained. From Eq. (10-1), the mass-transfer rate per unit volume of reactor is

$$\mathbf{r}_v = \mathbf{r}_p \rho = k_m a_m \rho (C_b - C_s)$$

where ρ is the density of the bed. The product $a_m \rho$ is the external surface of the particles per unit volume of the bed. For spherical particles of diameter d_p in a bed with a void fraction ϵ, $a_m \rho = 6(1 - \epsilon)/d_p$. Hence

$$\mathbf{r}_v = \frac{6(1 - \epsilon)}{d_p} k_m (C_b - C_s) \tag{10-37}$$

The ratio of the mass-transfer rates in fluidized and fixed beds can be obtained by using Eq. (10-36) and Fig. 10-2 to evaluate the corresponding j_D values. Then from Eqs. (10-37) and (10-10),

$$\left(\frac{\mathbf{r}_{\text{fluid}}}{\mathbf{r}_{\text{fixed}}}\right)_v = \frac{[(1 - \epsilon)/d_p]_{\text{fluid}} \; (Gj_D)_{\text{fluid}}}{[(1 - \epsilon)/d_p]_{\text{fixed}} \; (Gj_D)_{\text{fixed}}} \tag{10-38}$$

For fixed beds the curve for j_D in Fig. 10-2 is represented (at a Reynolds number greater than 100) by

$$j_D = 1.12 \left(\frac{d_p G}{\mu}\right)^{-0.46}$$

Using this result and Eq. (10-36), Eq. (10-38) becomes

$$\left(\frac{\mathbf{r}_{\text{fluid}}}{\mathbf{r}_{\text{fixed}}}\right)_v = \frac{1.77[(1 - \epsilon)^{1.44} \, G^{0.56} \, d_p^{-1.44}]_{\text{fluid}}}{1.12[(1 - \epsilon) \, G^{0.54} \, d_p^{-1.46}]_{\text{fixed}}} \tag{10-39}$$

As reasonable values for the fluidized bed, take

$d_p = 0.0063$ cm (250 mesh)

$\epsilon = 0.75$

$G = 0.02$ g/(cm^2)(sec)

[1] For an analysis of these equations and other data and correlations see O. Levenspiel and D. Kunii, "Fluidization Engineering," chap. 7, John Wiley & Sons, Inc., New York, 1969.

and for the fixed bed suppose

$$d_p = 0.635 \text{ cm } (\tfrac{1}{4} \text{ in.})$$

$$\epsilon = 0.25$$

$$G = 0.06 \text{ g/(cm}^2)(\text{sec})$$

Then

$$\left(\frac{r_{\text{fluid}}}{r_{\text{fixed}}}\right)_v = 140$$

Thus the mass-transfer rate between particle and fluid for a fluidized bed is approximately two orders of magnitude greater than for a fixed bed. With this large transport rate, it is evident that $C_b - C_s$, calculated by the methods described in Sec. 10-3, will be negligible. A similar result applies for external temperature differences.

SLURRY REACTORS

The effective hydrogenation of oils requires bringing gaseous hydrogen and liquid oil to the surface of a solid catalyst. An efficient system for accomplishing this is the *slurry reactor*. Hydrogen is bubbled into and dissolved in the oil in which catalyst particles are suspended. Unlike the fluidized bed, there is little *relative* movement between particles and liquid, even though the slurry is agitated. Both the small particle size and low relative velocity reduce the transport coefficients between particle and fluid. Furthermore, the intrinsic rate of the catalytic reaction at the particle is usually high for hydrogenations. These factors suggest that external resistances may have a significant effect on the global rate of reaction, and this is indeed the case.

Slurry reactors are also used in other situations, such as the polymerization of ethylene or propylene. Here the slurry consists of catalyst particles and a solvent, such as cyclohexane, into which the ethylene or propylene is bubbled and dissolved. Another illustration is the Fischer-Tropsch reaction between hydrogen and carbon monoxide, where these gases are dissolved in a slurry of hydrocarbon oil and catalyst (iron) particles.[1] Catalysis by colloidal metal particles and colloidal enzyme particles are other examples, although not always is one reactant a gas.

The design of slurry reactors is, as usual for heterogeneous reactions, a two-step process: formulation of the global rate of reaction, followed by design of an integral reactor by the procedures presented for homo-

[1] P. H. Calderbank, F. Evans, R. Farley, G. Jepson, and A. Poll, Proceedings of a Symposium on Catalysis in Practice, *Trans. Inst. Chem. Engrs. (London)*, 66 (1963).

geneous reactions in Chaps. 4 to 6. The second step is considered in Chap. 13. Here we want to account for the effect of external-transport resistances on the global rate.

In contrast to fixed- and fluidized-bed reactors, several physical processes must occur in series before a reactant gas can reach the catalytic surface. Some of these are likely to affect the global rate appreciably, and others are not. Various investigators[1] have considered the relative importance of the several steps. With hydrogenation as an illustration, the process may be visualized as occurring in the following sequence (see Fig. 10-6):

1. Mass transfer from the bulk concentration in the gas bubble to the bubble-liquid interface
2. Mass transfer from the bubble interface to the bulk-liquid phase
3. Mixing and diffusion in the bulk liquid
4. Mass transfer to the external surface of the catalyst particles
5. Reaction at the catalyst surface[2]

The rise of the bubbles through the liquid, along with mechanical agitation, is normally sufficient to achieve uniform conditions in the bulk liquid. Hence the resistance of step 3 can be neglected.[3] The global rate can be expressed in terms of the known bulk concentrations by writing rate equations for each of the four remaining steps. The rates of all the steps will be the same at steady state, and this equality permits elimination of the unknown interfacial concentrations. The procedure is the same as that employed in developing Eqs. (10-6) and (10-7) for a single mass-transfer step. As in that situation, a simple explicit equation for the global rate can be written only for a first-order reaction at the catalyst surface.[4] If we make the assumption of a first-order irreversible catalytic reaction, the rate per unit volume of bubble-free slurry may be written

$$\mathbf{r}_v = k a_c C_s \qquad \text{reaction at surface} \qquad (10\text{-}40)$$

[1] D. A. Frank-Kamenetski, "Diffusion and Heat Exchange in Chemical Kinetics," trans. by N. Thon, pp. 117–124, Princeton University Press, Princeton, N.J., 1955. Here slurry reactors are examined as a subdivision of "microheterogeneous processes"; this general term includes all small-particle reactions, such as fluidized-bed processes. See also C. N. Satterfield and T. K. Sherwood, "The Role of Diffusion in Catalysis," pp. 43–55, Addison-Wesley Publishing Company, Inc., Reading, Mass., 1963;and C. N. Satterfield, "Mass Transfer in Heterogeneous Catalysis," pp. 107–122, Massachusetts Institute of Technology Press, Cambridge, Mass., 1970.

[2] Since our concern in this chapter is only with external (to the catalyst particle) resistances, the possibility of resistances within a porous catalyst particle is not included. The particles are inherently small, so that intraparticle mass-transfer resistance may be unimportant. The conditions for establishing the significance of such internal resistances are given in Chap. 11.

[3] This conclusion has been verified experimentally by H. Kolbel and W. Siemes [Umschau, **24**, 746 (1957)] and by W. Siemes and W. Weiss [Dechema Monographien, **32**, 451 (1959)].

[4] For hydrogenation reactions the catalytic rate often follows first-order behavior.

Gaseous reactant

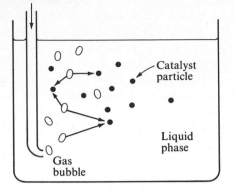

Fig. 10-6   Catalyst-slurry reactor

where a_c = external area of catalyst particles per unit volume of slurry (bubble free)

   k = first-order rate constant

   C_s = concentration of reactant (hydrogen) at the outer surface of the catalyst particle

The rates of the three mass-transfer processes may be expressed as

$$\mathbf{r}_v = k_g a_g (C_g - C_{i_g}) \qquad \text{gas to bubble interface} \qquad (10\text{-}41)$$

$$\mathbf{r}_v = k_l a_g (C_{i_l} - C_l) \qquad \text{bubble interface to bulk liquid} \qquad (10\text{-}42)$$

$$\mathbf{r}_v = k_c a_c (C_l - C_s) \qquad \text{bulk liquid to catalyst surface} \qquad (10\text{-}43)$$

where a_g is the gas bubble-liquid interfacial area per unit volume of bubble-free slurry and k_g, k_l, and k_c are the appropriate mass-transfer coefficients. If equilibrium exists at the bubble-liquid interface, C_{i_g} and C_{i_l} are related by Henry's law,

$$C_{i_g} = H C_{i_l} \qquad (10\text{-}44)$$

These five equations can be combined to eliminate C_{i_g}, C_{i_l}, C_l, and C_s and give the rate solely in terms of the concentration of reactant in the gas,

$$\mathbf{r}_v = k_o a_c C_g \qquad (10\text{-}45)$$

and

$$\frac{1}{k_o} = \frac{a_c}{a_g} \frac{1}{k_g} + \frac{a_c}{a_g} \frac{H}{k_l} + H\left(\frac{1}{k_c} + \frac{1}{k}\right) \qquad (10\text{-}46)$$

Equation (10-46) shows that the global rate is a function of the three mass-transfer coefficients, the specific reaction rate k, and the area ratio a_c/a_g. The rate will increase as this ratio falls, which corresponds to increasing the concentration of gas bubbles in the slurry. The rate is enhanced by increasing the concentration of catalyst particles in the slurry, because this increases a_c in Eq. (10-45). For low concentration of catalyst there is a sufficient supply of dissolved reactant for mass transfer to each particle to be independent of the number of particles. Under these circumstances the global rate is linear in catalyst concentration (particles per unit volume of reactor), since a_c in Eq. (10-45) is directly proportional to this concentration. As the catalyst concentration increases to higher values, there is competition between particles for reactant, and the rate ultimately approaches a constant upper limit.

The rate constant k is sensitive to temperature and, in principle, should be associated with the temperature of the catalyst particle. However, the high thermal conductivity of liquids (in comparison with that for gases) and the small particle size reduce the temperature difference between liquid and particle. Hence external temperature differences are not normally important in slurry reactors.

Not all of the four resistances indicated in Eq. (10-46) are significant in every instance. For example, in hydrogenations pure hydrogen is normally used as reactant. Then there is no resistance to diffusion from bulk gas (in the bubble) to bubble-liquid interface. Hence $C_g = C_{i_g}$ and Eq. (10-46) reduces to

$$\frac{1}{k_o H} = \frac{a_c}{a_g} \frac{1}{k_l} + \frac{1}{k_c} + \frac{1}{k} \tag{10-47}$$

Even when the gaseous reactant is in a mixture with other components in the bubbles, k_g appears to be much larger than k_l/H, so that Eq. (10-47) is applicable.[1] In this case it is sometimes desirable to define the global rate in terms of the liquid-phase concentration in equilibrium with C_g, that is, $(C_l)_{eq}$. Since $C_{i_g} = C_g = H(C_l)_{eq}$, Eq. (10-45) may be written

$$\mathbf{r}_v = k_o H a_c (C_l)_{eq} \tag{10-48}$$

where $k_o H$ is given by Eq. (10-47).

If the catalyst is very active,[2] k will be much greater than k_c or k_l/H. Then the global rate is determined by the mass-transfer coefficients k_l and k_c. In any event, k_l and k_c are the significant transport parameters. Available data for these coefficients are summarized in the following section.

---

[1] P. H. Calderbank, *Trans. Inst. Chem. Engrs. (London)*, **37**, 173 (1959).
[2] For example, nickel is an extremely active catalyst, when used in finely divided form, for many hydrogenation reactions.

10-7 Mass-transfer Correlations

Gas Bubble to Bulk Liquid(k_l)   A persistent problem in measuring transport coefficients from bubbles is the evaluation of the interfacial area. Calderbank[1] solved this problem by using a light-transmission method. His data and other results in the literature were reasonably well correlated[2] by the equation

$$k_l \left(\frac{\mu_l}{\rho_l \mathscr{D}} \right)^{\frac{2}{3}} = 0.31 \left(\frac{\Delta\rho \, \mu_l g}{\rho_l{}^2} \right)^{\frac{1}{3}} \tag{10-49}$$

where $\Delta\rho$ = difference in density between liquid phase and gas bubbles

μ_l = viscosity of liquid phase

g = acceleration of gravity

ρ_l = density of liquid phase

k_l = mass-transfer coefficient, cm/sec

This correlation is for bubbles rising through the liquid phase because of the gravitational force. Mechanical stirring is absent. It is applicable for small (less than 2.5 mm diameter) rigid bubbles of the size likely to be encountered in slurry reactors. Equation (10-49) has been applied with apparent success to transport from bubbles to liquid in systems containing catalyst particles in slurry form.[3]

In the absence of gravitational force, mass transfer from a single bubble would be by molecular diffusion through the surrounding stagnant liquid. Then the *Sherwood number* is $\mathrm{Sh} = d_b k_l / \mathscr{D} = 2$, where d_b is the diameter of the bubble. Other correlations have been developed by modifying this relation to include the effect of turbulence in the liquid phase induced by mechanical stirring or gravitational force. Hughmark[4] proposed this type of correlation for the Sherwood number when the turbulence was due to gravitational force. Both Eq. (10-49) and Hughmark's correlation are for column-type reactors, where gas bubbles rise through an appreciable height. Less information is available for the effect of agitation on k_l in a stirred-tank form of reactor. If the resistances due to k_c and k are negligible, data on rates of reaction can be employed to evaluate k_l by means of Eqs. (10-45) and (10-47). By this indirect method some data have been obtained for k_l as a function of agitation. Figure 10-7 shows the results for the adsorption

[1] P. H. Calderbank, *Trans. Inst. Chem. Engrs. (London)*, **37**, 173 (1959).

[2] P. H. Calderbank and M. B. Moo-Young, *Chem. Eng. Sci.*, **16**, 39 (1961).

[3] T. Matsuura, doctoral dissertation, Technische Universitat Berlin, Berlin, 1965; P. H. Calderbank and M. B. Moo-Young [*Chem. Eng. Sci.*, **16**, 39 (1961)] also proposed correlations for larger, nonrigid bubbles.

[4] G. A. Hughmark, *Ind. Eng. Chem., Process Design Develop. Quart.*, **6**, 218 (1967).

of oxygen from air in a 1.2 normal aqueous solution of sodium sulfite,[1] as a function of power input to the agitator. Copper ions served as a catalyst for oxidation of the sulfite. The concentration of O_2 in the liquid was taken as zero, and O_2 partial pressures in the exit gas bubbles were measured. The data are plotted as

$$\frac{k_l a_g / H R_g T}{u_g^{0.67}} \qquad \text{vs} \qquad \frac{P}{V_l}$$

where u_g = superficial velocity of gas bubbles based on inlet gas volume and cross section of vessel, ft/hr

P = power to agitator shaft, ft-lb/min

V_l = liquid volume in vessel, ft^3

k_l = bubble-interface to bulk-liquid mass-transfer coefficient, ft/hr

It is expected that k_l is influenced by reactor shape, type of agitator, method of introducing the gas, and similar quantities which are a function of the

[1]C. M. Cooper, G. A. Fernstrom, and S. A. Miller, *Ind. Eng. Chem.*, **36**, 504 (1944).

Fig. 10-7 *Gas-absorption characteristics of mechanical agitator-dispersers (by permission from T. K. Sherwood and R. L. Pigford, "Absorption and Extraction," 2d ed., p. 279, McGraw-Hill Book Company, New York, 1952)*

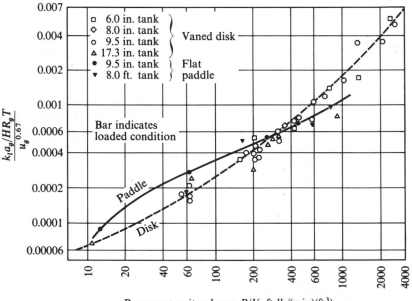

system and not accounted for by the parameters in Fig. 10-7. The data are specific for the physical system shown in Fig. 10-8. The gas was introduced through a single open-end pipe directly beneath the center of the agitator. The data are limited to the diffusion of oxygen and to a liquid whose viscosity is about that of water at room temperature.

Bulk Liquid to Catalyst Particle (k_c)   The relative velocity between the particles and the liquid determines the extent to which convection increases the Sherwood number above that for stagnant conditions, i.e., above 2. As mentioned, this relative velocity is low in slurries, because the particles are so small that they tend to move with the liquid. For these low velocities Friedlander[1] has shown that $k_c d_p/\mathscr{D}$ is a function of the *Peclet number*, Pe $= d_p u/\mathscr{D}$, where u is the relative velocity. The relation is such that the Sherwood number approaches 2 as the Peclet number falls to unity or less. In slurries the relative velocity can be due to the shearing action induced by agitation in the liquid or induced by the effect of gravitational force caused by the difference in densities of the liquid and catalyst particles. The effect of agitation is a function of particle size which is not yet well established.

[1]S. K. Friedlander, *AIChE J.*, **1**, 347 (1961).

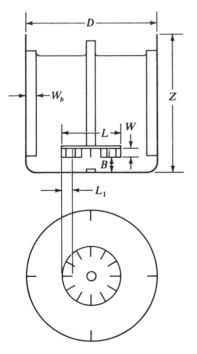

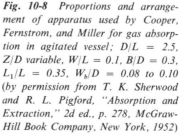

Fig. 10-8 *Proportions and arrangement of apparatus used by Cooper, Fernstrom, and Miller for gas absorption in agitated vessel; $D/L = 2.5$, Z/D variable, $W/L = 0.1$, $B/D = 0.3$, $L_1/L = 0.35$, $W_b/D = 0.08$ to 0.10 (by permission from T. K. Sherwood and R. L. Pigford, "Absorption and Extraction," 2d ed., p. 278, McGraw-Hill Book Company, New York, 1952)*

There are data to indicate that for particles 70 mesh ($d_p = 210$ microns) or larger the effect of agitation is pronounced.[1] As the particle size decreases this effect disappears. If there is little mechanical agitation of the slurry, the velocity will be related to the settling of the particles. Calderbank and Jones have equated the settling velocity due to gravitational force to the relative velocity in Friedlander's theory. The resultant correlation for k_c is

$$(k_c)_{grav}\left(\frac{\mu_l}{\rho_l \mathscr{D}}\right)^{\frac{2}{3}} = 0.34\left(\frac{\Delta\rho \, \mu_l g}{\rho_l^{\,2}}\right)^{\frac{1}{3}} \qquad (10\text{-}50)$$

This equation probably provides a minimum value for k_c in agitated slurries. For larger particles results two to four times $(k_c)_{grav}$ might be expected.[2]

10-8 The Effect of Mass Transfer on Observed Rates

Equation (10-47) shows that the relative importance of diffusion resistances from bubbles to bulk liquid and from liquid to catalyst particle depends on the area ratio. If a large concentration of very small particles is used, a_c/a_g is so large that only the first term is important. Then Eq. (10-47) reduces to

$$\frac{1}{k_o H} = \frac{a_c}{a_g}\frac{1}{k_l}$$

and the rate is, according to Eqs. (10-45) and (10-48),

$$\mathbf{r}_v = k_l a_g \frac{C_g}{H} = k_l a_g (C_l)_{eq} \qquad (10\text{-}51)$$

Calderbank et al.[3] studied the hydrogenation of ethylene using a large concentration of Raney nickel particles in a slurry reactor in order to approach these conditions. Analysis of the data[4] indicated that the controlling step was the mass transfer of hydrogen from gas bubble to bulk liquid.

Equation (10-51) indicates that the important parameters for this case are k_l and the gas-liquid interfacial area a_g. The latter depends on the size and concentration of bubbles in the slurry. Alternately, it may be expressed in terms of the bubble diameter and gas holdup. The volume of the gas bubbles is the difference between the volume of the bubbling slurry V_b, and that of the bubble-free slurry V_0. Both V_b and V_0 can be measured directly. Then for spherical particles of diameter d_b

[1] P. H. Calderbank and S. J. R. Jones, *Trans. Inst. Chem. Engrs. (London)*, **39**, 363 (1961).

[2] P. Harriott, *AIChE J.*, **8**, 93 (1962).

[3] P. H. Calderbank, F. Evans, R. Farley, G. Jepson, and A. Poll, "Proceedings of a Symposium on Catalysis in Practice," p. 66, Institution of Chemical Engineers (London), 1963.

[4] D. MacRae, doctoral dissertation, University of Edinburgh, Edinburgh, 1956.

$$a_g = \frac{6}{d_b} \frac{V_b - V_0}{V_0} \tag{10-52}$$

Koide et al.[1] give the following equation for bubbles formed over a porous plate:

$$d_b = 1.35 \frac{u/g\,\delta}{(\delta u^2 \rho_l/\sigma_l)^{\frac{1}{2}}} \left(\frac{\sigma_l \delta}{g\rho_l}\right)^{\frac{1}{3}} \tag{10-53}$$

where d_b = bubble diameter, cm

$\quad u$ = gas velocity through the porous plate

$\quad \delta$ = pore diameter of porous plate

$\quad \sigma_l$ = surface tension of liquid

$\quad \rho_l$ = density of liquid

Massimilla, Calderbank, and others also have reported experimental data on bubble sizes.[2]

At the other extreme, a large concentration of small gas bubbles (large a_g) might be combined with a low concentration of relatively large active catalyst particles (low a_c) and poor agitation (low k_c). Then Eq. (10-47) reduces to

$$\frac{1}{k_o H} = \frac{1}{k_c} + \frac{1}{k} \approx \frac{1}{k_c} \qquad \text{for very active catalyst} \tag{10-54}$$

The rate is, from Eq. (10-48),

$$\mathbf{r}_v = k_c a_c (C_l)_{\text{eq}} \tag{10-55}$$

Here the surface area of the catalyst particles is of primary importance, along with k_c.

Under many conditions a_c/a_g would be greater than unity. This follows from the fact that bubble diameters will be of the order of 1 mm, while catalyst particles will be in the region of 0.01 mm (100 microns or 140 mesh). Then for spherical shapes a_c/a_g would be $d_b/d_p = 100$ for equal volumes of bubbles and particles in the slurry. If the gas holdup were 100 times larger than the catalyst volume, a_c/a_g would be reduced to unity, but this ratio of volumes is infrequent. For many operating conditions it seems that the effect of k_l on the global rate will be greater than that of k_c or k (see Eq. 10-47). These approximations are illustrated in the following examples.

[1] K. T. Koide, T. Hirahara, and H. Kubata, *Chem. Eng. (Japan)*, **30**, 712 (1966).

[2] L. Massimilla, A. Solimando, and E. Squillace, *Brit. Chem. Eng.*, **6**, 232 (1961); P. H. Calderbank, F. Evans, and J. Rennie, *Proc. Intern. Symp. Distn., Suppl. Trans. Inst. Chem. Engrs. (London)*, p. 51 (1960).

Example 10-4   Coenan[1] has reported rates of hydrogenation of sesame seed oil with a nickel-on-silica catalyst in a slurry reactor. Hydrogen was added at the bottom of a small cylindrical vessel equipped with stator and stirrer blades. Initial rates of reaction were measured as function of catalyst concentration at 180°C, a stirrer speed of 750 rpm, atmospheric pressure, and a hydrogen rate of 60 liters/hr. The data, converted to global rates in terms of g moles/(min)(cm³ oil), are given in Table 10-6 (based on an oil density of 0.9 g/cm³). Estimate $k_l a_g / H$ from these data. Comment on the importance of the resistance of hydrogen to solution in the oil and estimate what the reaction rate would be if this resistance could be eliminated for a catalyst concentration of 0.07% Ni in oil.

Table 10-6

Catalyst particle conc., % Ni in oil	$r_v \times 10^5$	C_g/r_v, min
0.018	5.2	0.52
0.038	8.5	0.32
0.07	10.0	0.27
0.14	12.0	0.22
0.28	13.6	0.20
1.0	14.6	0.18

Solution   Equation (10-47) is applicable. Combining it with Eq. (10-45) gives

$$r_v = \frac{a_c C_g}{H[(a_c/a_g)(1/k_l) + 1/k_c + 1/k]} = \frac{C_g}{H[(1/a_g k_l) + (1/a_c)(1/k_c + 1/k)]}$$

(A)

Since the gas rate and stirrer speed are constant, k_c, k_l, and a_g will not vary from run to run. Because the temperature is constant, k will also be invariant. The catalyst consists of small particles of silica containing nickel, and its concentration is changed by adding more such particles. The activity per particle is constant. If we assume that the particles do not agglomerate, a_c will be directly proportional to the catalyst concentration, and Eq. (A) may be written

$$\frac{C_g}{r_v} = \frac{H}{a_g k_l} + \frac{AH}{C_{cat}}\left(\frac{1}{k_c} + \frac{1}{k}\right)$$

(B)

where A is the proportionality constant. This expression suggests that the data plotted as C_g/r_v vs $1/C_{cat}$ would yield a straight line. Assuming the ideal-gas law, we have

$$C_g = \frac{p}{R_g T} = \frac{1}{82(273 + 180)} = 2.7 \times 10^{-5} \text{ g mole/cm}^3$$

[1]J. W. E. Coenan, The Mechanism of the Selective Hydrogenation of Fatty Oils, in J. H. deBoer (ed.), "The Mechanism of Heterogeneous Catalysis," p. 126, Elsevier Publishing Company, New York, 1960.

The data in Table 10-6, when plotted in this way (Fig. 10-9), do yield a linear result. Extrapolation to $1/C_{cat} = 0$ gives an intercept value of 0.18 min. According to Eq. (B),

$$0.18 = \frac{H}{a_g k_l}$$

or

$$\frac{k_l a_g}{H} = 5.7 \text{ min}^{-1}$$

In Fig. 10-9 the resistance to solution of hydrogen in the slurry is represented by the horizontal dashed line. The total resistance is the ordinate of the solid line. At low catalyst concentrations the combined resistance of diffusion to the particle and chemical reaction on the catalytic surface is large, although it does not determine the rate by itself. At high concentrations the resistance to transfer of hydrogen from bubble to the bulk liquid dominates the rate. In fact at $C_{cat} = 0.28\%$, 0.18/0.20, or 90%, of the total resistance is for this step. The results show that a_c/a_g increases enough, as the catalyst concentration increases to 0.28%, that the first term in Eq. (10-47) dominates the whole quantity. The results also indicate that increases in C_{cat} beyond 0.28% would do little to speed up the reaction.

If the resistance to dissolving hydrogen could be eliminated at $C_{cat} = 0.07\%$ then $C_g/\mathbf{r}_v$ would be, from Fig. 10-9,

$$\frac{C_g}{\mathbf{r}_v} = \frac{AH}{C_{cat}} \left(\frac{1}{k_o} + \frac{1}{k} \right) = 0.27 - 0.18 = 0.09 \text{ min}$$

$$\mathbf{r}_v = \frac{2.7 \times 10^{-5}}{0.09} = 30 \times 10^{-5} \text{ g moles/(min)(cm}^3 \text{ oil)}$$

Fig. 10-9   *Effect of catalyst concentration on rate in a slurry reactor*

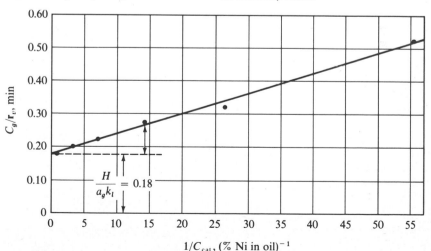

This value is threefold that of the observed (global) rate. Decreasing the bubble size would be a step in this direction. Increasing the hydrogen-flow rate at constant bubble size would have the same effect, since this would increase gas volume and a_g. Better agitation would also reduce the bubble-to-liquid resistance, as demonstrated in Example 10-5.

Example 10-5 Estimate the effect on hydrogenation rate of changing stirrer speed from 750 to 520, 910, and 1,910 rpm. The catalyst concentration is 0.07% Ni. All other conditions are the same as in Example 10-4.

Solution As an approximation assume that the effect of agitator speed on $k_l a_g$ is the same as shown in Fig. 10-7 for solution of oxygen in aqueous solutions. The power input P will be approximately proportional[1] to the 2.5 power of the stirrer speed N. Then the ratio of $(P/V_l)_N$ to $(P/V_l)_{750}$ will be as follows:

N, rpm	520	750	910	1,910
$(P/V_l)_N/(P/V_l)_{750}$	0.41	1.00	1.47	6.50

The slope of the lines in the log-log plot of Fig. 10-7 is approximately unity. Hence $k_l a_g/HR_g T$ is proportional to P/V_l. Since the temperature is constant, $k_l a_g$ is directly proportional to P/V_l. Using the results of Example 10-4 at $N = 750$, we have

N, rpm	520	750	910	1,910
$k_l a_g$, min^{-1}	2.7	5.7	8.4	37.0

If the catalyst particles are small enough that k_c is independent of N, the results of Example 10-4 can be used to evaluate $\mathbf{r}_v$ as a function of N. Equation (B) of Example 10-4 becomes, for a catalyst concentration of 0.07%,

$$\frac{C_g}{\mathbf{r}_v} = \frac{H}{k_l a_g} + \frac{AH}{C_{\text{cat}}}\left(\frac{1}{k_c} + \frac{1}{k}\right) = \frac{1}{k_l a_g} + 0.09$$

Using the $k_l a_g$ values in the preceding table, this equation gives the rate results shown in Table 10-7. The experimental values are from Coenan.[2] The agreement is perhaps better than expected, in view of the assumptions necessary to use Fig. 10-7. Note that at the highest stirrer speed the calculated results show that 25% of the total resistance is still associated with the solution of hydrogen. In contrast, the experimental rate gives $C_g/\mathbf{r}_v = 0.085$, which is almost equal to the combined resistance $(HA/C_{\text{cat}})(1/k_c + 1/k)$, indicating a negligible resistance for hydrogen solution. Thus the data suggest that k_l increases more rapidly with stirring speed than is indicated by Fig. 10-7. Differences in stirrer-stator design, vessel size, bubble size, etc., could account for the discrepancy.

[1] According to J. H. Rushton, E. W. Costich, and H. J. Everett [*Chem. Eng. Progr.*, **46**, 395 (1950)], power is proportional to N^2 for Re < 10 and N^3 for Re $> 10,000$. It is estimated for the conditions of this example that the exponent is about 2.5.

[2] J. W. E. Coenan, The Mechanism of the Selective Hydrogenation of Fatty Oils, in J. H. deBoer (ed.), "The Mechanism of Heterogeneous Catalysis," p. 126, Elsevier Publishing Company, New York, 1960.

Table 10-7

	C_g/r_v, min	$r_v \times 10^5$, g moles/(cm³)(min)	
N, rpm	Calculated	Calculated	Experimental
520	0.52	5.2	6.8
750	0.27	10.0	10.0
910	0.20	13.0	13.0
1,910	0.12	24.0	32.0

Problems

10-1. To illustrate the effect of mass velocity on external diffusion in the oxidation of sulfur dioxide with a platinum catalyst, consider the following data, all at 480°C.

Mass velocity, lb/(hr)(ft²)	Bulk partial pressures, p_b, atm			Global rate, r_P, g moles SO_2/(hr)(g catalyst)
	SO_2	SO_3	O_2	
514	0.0601	0.00668	0.201	0.1346
350	0.0599	0.00666	0.201	0.1278
245	0.0603	0.00668	0.201	0.1215
147	0.0603	0.00670	0.201	0.0956

The reactor consisted of a fixed bed of $\frac{1}{8} \times \frac{1}{8}$-in. cylindrical pellets. The pressure was 790 mm Hg. The external area of catalyst particles was 5.12 ft²/lb, and the platinum did not penetrate into the interior of the alumina particles. Calculate the partial-pressure difference between the bulk-gas phase and the surface of the catalyst for SO_2 at each mass velocity. What conclusions may be stated with regard to the importance of external diffusion? Neglect temperature differences.

10-2. The global rates of SO_2 oxidation have been measured with a platinum catalyst impregnated on the outer surface of $\frac{1}{8} \times \frac{1}{8}$-in. cylindrical pellets of Al_2O_3. The data were obtained in a differential reactor consisting of a 2-in.-ID tube packed with the catalyst pellets. The superficial mass velocity of the reaction mixture was 350 lb/(hr)(ft²). At constant conversion of 20% of a feed consisting of 6.5 mole % SO_2 and 93.5 mole % air, the rates are as shown below. The total pressure was 790 mm Hg. Consider the properties of the mixture to be those of air, except for the Schmidt number for SO_2, which is 2.9.

(a) If only the data at 460 to 500°C are used and external diffusion is neglected, what is the *apparent* activation energy? (b) By calculating $(C_b - C_s)_{SO_2}$ at various temperatures and using all the data above, estimate the true activation energy for the combined adsorption and reaction processes

t, °C	350	360	380	400	420
r_P, g moles SO_2/ (hr)(g catalyst)	0.0049	0.00788	0.01433	0.02397	0.0344

	440	460	480	500
	0.0514	0.0674	0.0898	0.122

at the catalyst surface. Neglect temperature differences between bulk gas and catalyst surface.

10-3.   Predict the global rate of reaction for the oxidation of SO_2 at bulk-gas conditions of 20% conversion at 480°C. Other conditions are as given in Prob. 10-2. The rate at the catalyst surface is to be calculated from Eq. (G) of Example 9-2. Assume isothermal conditions. The constants in this equation at 480°C are

$A = -0.127$ (atm)$^{\frac{3}{2}}$(hr)(g catalyst)/g mole

$B = 15.3$ (atm)$^{\frac{1}{2}}$ (hr)(g catalyst)/g mole

$K = K_p = 73$ (atm)$^{-\frac{1}{2}}$

Also calculate the ratio γ of the global rate and the rate evaluated at bulk conditions.

10-4.   For the conditions of Prob. 10-2 predict the temperature difference between bulk gas and pellet surface at 350 and at 500°C. Comment on the validity of the isothermal assumptions made in Probs. 10-1 to 10-3.

10-5.   (a) Estimate the *maximum* temperature difference $T_s - T_b$ for a gas-solid catalytic reaction for which

$\Delta H = -20,000$ cal/g mole

$c_p = 8.0$ cal/g mole (°K)

$p_t = 2$ atm

$T_b = 473°K (200°C)$

The mole fraction of reactant in the bulk gas is 0.25. (b) If external diffusion resistance is not controlling, but $C_b - C_s = C_b/2$, what will be the value of $T_s - T_b$?

10-6.   The reaction $H_2 + \frac{1}{2}O_2 \rightarrow H_2O$ occurs with a platinum-on-alumina catalyst at low temperatures. Following the work of Maymo and Smith,[1] suppose that the rate, in g moles O_2/(sec)(g catalyst), at the catalyst surface is given by

$r = 0.327 (p_{O_2})_s^{0.804} e^{-5,230/R_gT_s}$

where p_{O_2} is in atmospheres and T is in degrees Kelvin.

(a) Calculate the global rate of reaction at a location in a packed-bed reactor where the bulk conditions are

[1] J. A. Maymo and J. M. Smith, *AIChE J.*, **12**, 845 (1966).

$T_b = 373°K (100°C)$

$p_t = 1$ atm

$(p_{O_2})_b = 0.060$ atm

$(p_{H_2})_b = 0.94$ atm

Are external mass- and heat-transfer resistances negligible?

The diameter and mass of the nonporous catalyst pellet are 1.86 cm and 2.0 g, and the superficial mass velocity of the flowing gas is 250 lb/(hr)(ft^2). At these conditions the molecular diffusivity of oxygen in hydrogen is about 1.15 cm^2/sec. The properties of the gas mixture are essentially the same as those of hydrogen.

(b) Determine the ratio γ of the global rate to the rate evaluated at bulk-gas conditions.

10-7. The catalytic isomerization of n-butane,

$$n\text{-}C_4H_{10}(g) \rightarrow i\text{-}C_4H_{10}(g),$$

is studied in a plug-flow laboratory reactor in which the catalyst is deposited on the walls of the cylindrical tube. Pure n-butane (as a gas) is fed to the reactor at a steady rate of F (lb moles/hr), and the whole system operates isothermally. Assume that the reaction rate at the catalyst surface is first order and irreversible, according to the expression

$$\mathbf{r} = kC_s$$

where $\mathbf{r}$ is the rate of formation of iso-butane, lb moles/(hr) (ft^2 tube surface) and C_s is the surface concentration of n-butane, lb moles/ft^3

(a) Suppose that the rate of mass transfer radially from the bulk gas, where C_g is the concentration of n-C$_4$H$_{10}$, to the catalyst wall, where C_s is the concentration, is proportional to the mass-transfer coefficient k_m, in feet per hour. Derive an expression for the conversion of n-butane in terms of the reactor length L and other necessary quantities. The constant temperature is $T°R$ and the pressure is p atmospheres.

(b) Suppose the resistance to mass transfer radially is negligible. What form would the equation for the conversion take?

(c) Suppose the plug-flow model is not valid because axial diffusion is important. Let the effective diffusivity in the axial direction be D_L. Other conditions are as in part (a). Derive an equation whose solution will give the conversion as a function of reactor length and other necessary quantities.

(d) Calculate the conversion of n-butane leaving the reactor from the equation derived in part (a) and the conditions

$k = 446$ ft/hr

$d = 6$ in.

$t = 500°F$

$p = 1$ atm

$F = 20$ lb moles/hr

Reactor length $L = 10$ ft

Also calculate the conversion from the equation derived in part (c), using the above data, and $D_L/uL = 0.05$. This value of D_L/uL corresponds to an intermediate amount of dispersion.

10-8. Reconsider parts (a) to (c) of Prob. 10-7 for a reversible first-order reaction. It is suggested that the rate equation be written in terms of the equilibrium concentration of n-butane. Also show how this concentration is related to the equilibrium constant K.

10-9. Consecutive, first-order, irreversible reactions

$$A \xrightarrow{k_1} B \xrightarrow{k_2} C$$

occur on the outer surface of a solid catalyst pellet. Components A, B, and C are all gases. For isothermal conditions derive an equation for the selectivity of B with respect to C which shows the effect of external-diffusion resistance between the bulk gas and the catalyst surface. Consider a location in the reactor where the total conversion of A is x_t. The feed to the reactor contains no B or C.

10-10. Suppose that a temperature difference exists between bulk gas and catalyst surface. How would the equation derived in Prob. 10-9 be modified? What are the additional quantities that would have to be known to evaluate the effect of nonisothermal behavior?

10-11. Estimate the mass-transfer coefficient k_l for bubbles ($d_b = 1.0$ mm) of pure hydrogen rising through a nonagitated liquid ($\mu_l = 2.0$ centipoises, $\rho_l = 0.9$ g/cm^3). The system is at 150°C and 5 atm pressure.

10-12. Estimate the value of k_c if the liquid described in Prob. 10-11 contains catalyst particles of 100 mesh size. Suppose that the agitation caused by the gas bubbles increases k_c only slightly above the minimum value $(k_c)_{grav}$. The density of the particles is 1.8 g/cm^3 and the diffusivity of hydrogen in the liquid is 6×10^{-5} cm^2/sec.

10-13. If the intrinsic rate of hydrogenation at the catalyst site is very large for the conditions of Probs. 10-11 and 10-12, calculate the global rate of hydrogenation $\mathbf{r}_v$ for the following cases, corresponding to a constant concentration of gas bubbles ($a_g = 3.0$ cm^{-1}) and the following catalyst-particle concentrations. The solubility of hydrogen in the liquid is such that $C_g/(C_l)_{eq} = 10$ at 150°C.

(a) $a_c/a_g = 100$ (high particle concentration)

(b) $a_c/a_g = 10$

(c) $a_c/a_g = 2.0$ (low particle concentration)

10-14. If the rate of hydrogenation at the catalyst can be represented by Eq. (10-40) with $k = 0.0020$ cm/sec, what are the global rates of reaction for the three cases of Prob. 10-13?

11

REACTION AND DIFFUSION WITHIN POROUS CATALYSTS: INTERNAL TRANSPORT PROCESSES

Essentially all of the surface of porous catalyst pellets is internal (see page 295). Reaction and mass and heat transfer occur simultaneously at any position within the pellet. The resulting intrapellet concentration and temperature gradients cause the rate to vary with position. At steady state the average rate for a whole pellet will be equal to the global rate at the location of the pellet in the reactor. The concentration and temperature of the bulk fluid at this location may not be equal to those properties at the outer surface of the pellet. The effect of such external resistances can be accounted for by the procedures outlined in Chap. 10. The objective in the present chapter is to account for internal resistances, that is, to evaluate average rates in terms of the temperature and concentration at the outer surface. Because reaction and transport occur simultaneously, differential equations must be written to describe the process at a point, and then these equations are integrated to obtain the average rate for the whole pellet.

The differential equations involve the effective[1] diffusivity and thermal conductivity of the porous pellet. Data and theories for these quantities are given in Secs. 11-1 to 11-5, and in Secs. 11-6 to 11-11 the results are used to establish the rate for the whole pellet. The method of combining the effects of internal and external transport resistance to give the global rate in terms of bulk fluid properties is discussed in Chap. 12.

The effect of intrapellet mass transfer is to reduce the rate below what it would be if there were no internal-concentration gradient. The effect of the temperature gradient is to increase the rate for an exothermic reaction. This is because intrapellet temperatures will be greater than surface values. For endothermic reactions temperature and concentration gradients both reduce the rate below that evaluated at outer-surface conditions.

For a gaseous reaction accompanied by a change in number of moles, an intrapellet gradient in total pressure will also develop (at steady state). If the moles decrease, there will be a flow of reactant toward the center of the pellet due to this total-pressure gradient. This augments the diffusion of reactant inward toward the center of the pellet and retards the diffusion of product outward. Unless the decrease in number of moles is very large, which is unlikely for reactions, the effect on the average rate is small.[2] Also, it should be remembered that all intrapellet-transport effects will become less important as the pellet size decreases. For fluidized-bed and slurry reactors intraparticle transport processes can usually be neglected. Hence in this chapter we shall discuss such catalyst pellets as might be used in fixed-bed reactors.

INTRAPELLET MASS TRANSFER

It is seldom possible to predict diffusion rates in porous materials by simply correcting bulk-phase diffusivities for the reduction in cross-sectional area due to the solid phase. There are several reasons for this:

[1] In the remainder of the text the term "effective" is used to indicate that a transport coefficient applies to a porous material, as distinguished from a homogeneous region. Effective diffusivities D_e and thermal conductivities k_e are based on a unit of *total area* (void plus nonvoid) perpendicular to the direction of transport. For example, for diffusion in a spherical catalyst pellet at radius r, D_e is based on the area $4\pi r^2$.

[2] See A. Wheeler in P. H. Emmett (ed.), "Catalysis," vol. II, p. 143, Reinhold Publishing Corporation, New York, 1955; Von E. Wicke and Peter Hugo, Z. *Phys. Chem.*, **28**, 26 (1961); S. Otani, N. Wakao, and J. M. Smith, *AIChE J.*, **10**, 135 (1965); N. Wakao, S. Otani, and J. M. Smith, *AIChE J.*, **11**, 435 (1965).

1. The tortuous, random, and interconnected arrangement of the porous regions makes the length of the diffusion path unknown.
2. A catalyst is characterized as a substance which adsorbs reactant. When it is adsorbed it may be transported either by desorption into the pore space or by migration to an adjacent site on the surface. The contribution of surface diffusion must be added to the diffusion in the pore volume to obtain the total mass transport.
3. The diffusion in the pore volume itself will be influenced by the pore walls, provided the diffusing molecule is likely to encounter the wall rather than another molecule. This often is the case for gaseous systems. Then the pore-volume contribution to total mass transport is not dependent solely on the *bulk diffusivity*, but may be affected (or determined) by the *Knudsen diffusivity*.

Surface diffusion is not well understood, and rates of diffusion by this mechanism cannot yet be predicted. We must be content with summarizing in Sec. 11-3 the available data for surface migration. Volume diffusion in a straight cylindrical pore is amenable to analysis. Hence the procedure used (see Sec. 11-2) to predict an effective diffusivity for a porous catalyst is to combine the established equations for diffusion in a single cylindrical pore with a geometric model of the pore structure. If the surface diffusivity is known for the surface of a cylindrical pore, it can be used with the same model to obtain the total mass-transport rate (in the absence of flow due to a total-pressure gradient). The nature of gaseous diffusion in cylindrical pores is discussed in Sec. 11-1. When the reaction mixture in the pore is liquid Knudsen diffusion does not occur, but uncertainty as to how the bulk diffusivity varies with concentration is a complicating feature. The effective length of the diffusion path is determined by the pore structure of the pellet. Hence the path length is intimately connected with the model chosen to represent the porous catalyst. Such models are considered in Sec. 11-2.

It is evident from the foregoing discussion that the effective diffusivity cannot be predicted accurately for use under reaction conditions unless surface diffusion is negligible and a valid model for the pore structure is available. The prediction of an effective thermal conductivity is even more difficult. Hence sizable errors are frequent in predicting the global rate from the rate equation for the chemical step on the interior catalyst surface. This is not to imply that for certain special cases accuracy is not possible (see Sec. 11-10). It does mean that heavy reliance must be placed on experimental measurements for effective diffusivities and thermal conductivities. Note also from some of the examples and data mentioned later that intrapellet resistances can greatly affect the rate. Hence the problem is significant.

Because of the complexity of the combined problem, only a brief account will be given of the considerable theory on special aspects of isothermal mass transfer. More emphasis will be placed on interpretation of laboratory-reactor data with respect to the importance of intrapellet resistances.

11-1 Gaseous Diffusion in Single Cylindrical Pores

Basic Equations   For many catalysts and reaction conditions (especially pressure) both bulk and Knudsen diffusion contribute to the mass-transport rate within the pore volume. For some years the proper combination of the two mechanisms was in doubt. About 1961 three independent investigations[1] proposed identical equations for the rate of diffusion (in a binary gaseous mixture of A and B) in terms of the bulk diffusivity $\mathscr{D}_{AB}$ and Knudsen diffusivity $\mathscr{D}_K$. If N_A is the molal flux of A, it is convenient to represent the result as

$$N_A = -\frac{p_t}{R_g T} D \frac{dy_A}{dx} \qquad \Longrightarrow \quad BOTH \tag{11-1}$$

where y_A is the mole fraction of A, x is the coordinate in the direction of diffusion, and D is a combined diffusivity given by

$$D = \frac{1}{(1 - \alpha y_A)/\mathscr{D}_{AB} + 1/(\mathscr{D}_K)_A} \tag{11-2}$$

The quantity α is related to the ratio of the diffusion rates of A and B by

$$\alpha = 1 + \frac{N_B}{N_A} \tag{11-3}$$

For the reaction $A \rightarrow B$, reaction and diffusion (at steady state) in a pore would require equimolal counterdiffusion; that is, $N_B = -N_A$. Then $\alpha = 0$, and the effective diffusivity is

$$D = \frac{1}{1/\mathscr{D}_{AB} + 1/(\mathscr{D}_K)_A} \tag{11-4}$$

When the pore radius is large, Eqs. (11-1) and (11-2) reduce to the conventional constant-pressure form for bulk diffusion. For this condition $(\mathscr{D}_K)_A \rightarrow \infty$. Then combining Eqs. (11-1) to (11-3) gives

$$N_A = -\frac{p_t}{R_g T} \mathscr{D}_{AB} \frac{dy_A}{dx} + y_A(N_A + N_B) \qquad PURE\ BULK \tag{11-5}$$

[1] R. B. Evans, III, G. M. Watson, and E. A. Mason, *J. Chem. Phys.*, **35**, 2076 (1961); L. B. Rothfeld, *AIChE J.*, **9**, 19 (1963); D. S. Scott and F. A. L. Dullien, *AIChE J.*, **8**, 29 (1962).

If, in addition, the diffusion is equimolal, Eq. (11-5) may be written

$$N_A = \frac{-p_t}{R_g T} \mathscr{D}_{AB} \frac{dy_A}{dx} \tag{11-6}$$

If the pore radius is very small, collisions will occur primarily between gas molecules and pore wall, rather than between molecules. Then the Knudsen diffusivity becomes very low, and Eqs. (11-1) and (11-2) reduce to

$$N_A = -\frac{p_t}{R_g T} (\mathscr{D}_K)_A \frac{dy_A}{dx} \tag{11-7}$$

KNUDSEN ONLY

This equation is the usual one[1] expressing Knudsen diffusion in a long capillary.

Although Eq. (11-2) is the proper one to use for regions where both Knudsen and bulk diffusion are important, it has a serious disadvantage: the combined diffusivity D is a function of gas composition y_A in the pore. This dependency on composition carries over to the effective diffusivity in porous catalysts (see Sec. 11-2) and makes it difficult later to integrate the combined diffusion and transport equations. The variation of D with y_A is not usually strong (see Example 11-3). Therefore it has been almost universal,[2] in assessing the importance of intrapellet resistances, to use a composition-independent form for D, for example, Eq. (11-4). In fact, the concept of a single effective diffusivity loses its value if the composition dependency must be retained.

Effective diffusivities in porous catalysts are usually measured under conditions where the pressure is maintained constant by external means. The experimental method is discussed in Sec. 11-2; it is mentioned here because under this condition, and for a binary counterdiffusing system, the ratio N_B/N_A is the same regardless of the extent of Knudsen and bulk diffusion. Evans et al.[3] have shown this constant ratio to be

$$\frac{N_B}{N_A} = -\sqrt{\frac{M_A}{M_B}} \tag{11-8}$$

or *or*

$$\alpha = 1 - \sqrt{\frac{M_A}{M_B}}$$

[1] M. Knudsen, *Ann. Physik*, **28**, 75 (1909); N. Wakao, S. Otani, and J. M. Smith, *AIChE J.*, **11**, 435 (1965).

[2] N. Wakao and J. M. Smith [*Ind. Eng. Chem., Fund. Quart.*, **3**, 123 (1964)] used the general equations (11-1) and (11-2) in developing expressions for intrapellet effects for isothermal first-order reactions.

[3] R. B. Evans, III, G. M. Watson, and E. A. Mason, *J. Chem. Phys.*, **35**, 2076 (1961).

where M represents the molecular weight. Equation (11-8) applies for nonreacting conditions. When reaction occurs, stoichiometry determines α (see Example 11-3).

Calculation of Diffusivities   In analyzing Knudsen and bulk diffusivities the important parameter is the size of the pore with respect to the mean free path. The bulk diffusivity is a function of the molecular velocity and the mean free path; that is, it is a function of temperature and pressure. The Knudsen diffusivity depends on the molecular velocity v and the pore radius a. In terms of simple kinetic theory, these two diffusivities may be described by the equations

$$\mathscr{D}_{AB} = \tfrac{1}{3}\,\bar{v}\lambda \tag{11-9}$$

$$(\mathscr{D}_K)_A = \tfrac{2}{3}\,a\bar{v} \tag{11-10}$$

where λ is the mean free path. Since λ is of the order of 1,000 Å for gases at atmospheric pressure, diffusion in micropores of a catalyst pellet will be predominantly by the Knudsen mechanism. This would be the case for a material such as silica gel, where the mean pore radius is from 15 to 100 Å (see Table 8-1). For a pelleted catalyst of alumina with the pore-volume distribution shown in Fig. 8-10, the mean macropore radius is about 8,000 Å. At atmospheric pressure bulk diffusion would prevail in these pores. In the micropores of the same pellet, where $\bar{a} = 20$ Å, diffusion would be by the Knudsen process. Since the mean free path is inversely proportional to pressure, bulk diffusivity becomes more important as the pressure increases.

For accurate calculations the *Chapman-Enskog formula*[1] has been found suitable for evaluating the bulk diffusivity at moderate temperatures and pressures. The equation is

$$\mathscr{D}_{AB} = 0.0018583\,\frac{T^{\frac{3}{2}}(1/M_A + 1/M_B)^{\frac{1}{2}}}{p_t\sigma_{AB}^2\,\Omega_{AB}} \tag{11-11}$$

where $\mathscr{D}_{AB}$ = bulk diffusivity, cm²/sec

T = temperature, °K

M_A, M_B = molecular weights of gases A and B

p_t = total pressure of the gas mixture, atm

σ_{AB}, ε_{AB} = constants in the Lennard-Jones potential-energy function for the molecular pair AB; σ_{AB} is in Å.

[1] See J. O. Hirschfelder, C. F. Curtiss, and R. B. Bird, "Molecular Theory of Gases and Liquids," pp. 539, 578, John Wiley & Sons, Inc., New York, 1954.

Ω_{AB} = collision integral, which would be unity if the molecules were rigid spheres and is a function of $k_B T / \varepsilon_{AB}$ for real gases (k_B = Boltzmann's constant)

Since the Lennard-Jones potential-energy function is used, the equation is strictly valid only for nonpolar gases. The Lennard-Jones constants for the unlike molecular pair AB can be estimated from the constants for like pairs AA and BB:

$$\sigma_{AB} = \tfrac{1}{2}(\sigma_A + \sigma_B) \qquad (11\text{-}12)$$

$$\varepsilon_{AB} = (\varepsilon_A \varepsilon_B)^{\frac{1}{2}} \qquad (11\text{-}13)$$

The force constants for many gases are given in the literature and are summarized in Table 11-1. Those that are not available otherwise may be approximated by the expressions

$$\sigma = 1.18 \, V_b^{\frac{1}{3}} \qquad (11\text{-}14)$$

$$\frac{k_B T}{\varepsilon} = 1.30 \frac{T}{T_c} \qquad (11\text{-}15)$$

where k_B = Boltzmann's constant

T_c = critical temperature

V_b = volume per mole (cm³/g mole) at normal boiling point

If necessary, V_b may be estimated by adding the increments of volume for the atoms making up the molecule (Kopp's law). Such increments are given in Table 11-2. The collision integral Ω_{AB} is given as a function of $k_B T / \varepsilon_{AB}$ in Table 11-3. From these data and the equations, binary diffusivities may be estimated for any gas. For polar gases, or for pressures above 0.5 critical pressure, the errors may be greater than 10%. The effects of composition on $\mathscr{D}$ are small for gases at moderate conditions, so that the same procedure may be used as an approximation for multicomponent mixtures. An improved result for mixtures may be obtained by combining binary diffusivities by the method outlined by Wilke.[1]

For evaluating the Knudsen diffusivity we may use the following equation for the average molecular velocity $\bar{v}$ for a component of gas in a mixture:

$$\bar{v}_A = \left(\frac{8 R_g T}{\pi M_A}\right)^{1/2} \qquad (11\text{-}16)$$

[1]C. R. Wilke, *Chem. Eng. Progr.*, **46**, 95 (1950).

Table 11-1   Lennard-Jones constants and critical properties

Substance	Molecular weight	Lennard-Jones parameters*		Critical constants†		
		σ, Å	ε/k_B, °K	T_c, °K	p_c, atm	V_c, cm³/g mole
Light elements						
H₂	2.016	2.915	38.0	33.3	12.80	65.0
He	4.003	2.576	10.2	5.26	2.26	57.8
Noble gases						
Ne	20.183	2.789	35.7	44.5	26.9	41.7
Ar	39.944	3.418	124.0	151.0	48.0	75.2
Kr	83.80	3.61	190.0	209.4	54.3	92.2
Xe	131.3	4.055	229.0	289.8	58.0	118.8
Simple polyatomic substances						
Air	28.97	3.617	97.0	132.0	36.4	86.6
N₂	28.02	3.681	91.5	126.2	33.5	90.1
O₂	32.00	3.433	113.0	154.4	49.7	74.4
O₃	48.00	. . .	. . .	268.0	67.0	89.4
CO	28.01	3.590	110.0	133.0	34.5	93.1
CO₂	44.01	3.996	190.0	304.2	72.9	94.0
NO	30.01	3.470	119.0	180.0	64.0	57.0
N₂O	44.02	3.879	220.0	309.7	71.7	96.3
SO₂	64.07	4.290	252.0	430.7	77.8	122.0
F₂	38.00	3.653	112.0	. . .	. . .	. . .
Cl₂	70.91	4.115	357.0	417.0	76.1	124.0
Br₂	159.83	4.268	520.0	584.0	102.0	144.0
I₂	253.82	4.982	550.0	800.0	. . .	. . .
Hydrocarbons						
CH₄	16.04	3.822	137.0	190.7	45.8	99.3
C₂H₂	26.04	4.221	185.0	309.5	61.6	113.0

Combining this with Eq. (11-10) gives a working expression for $(\mathscr{D}_K)_A$ in a circular pore of radius a,

$$(\mathscr{D}_K)_A = 9.70 \times 10^3 a \left(\frac{T}{M_A}\right)^{\frac{1}{2}} \tag{11-17}$$

where $(\mathscr{D}_K)_A$ is in square centimeters per second, a is in centimeters, and T is in degrees Kelvin.

Example 11-1   Estimate the diffusivity of SO₂ for the conditions of Example 10-1.

Solution   The gas composition is about 94% air, with the remainder SO₂ and SO₃. Hence a satisfactory simplification is to consider the system as a binary mixture of air and SO₂.

Table 11-1 (*Continued*)

C_2H_4	28.05	4.232	205.0	282.4	50.0	124.0
C_2H_6	30.07	4.418	230.0	305.4	48.2	148.0
C_3H_6	42.08	. . .	. . .	365.0	45.5	181.0
C_3H_8	44.09	5.061	254.0	370.0	42.0	200.0
n-C_4H_{10}	58.12	. . .	. . .	425.2	37.5	255.0
i-C_4H_{10}	58.12	5.341	313.0	408.1	36.0	263.0
n-C_5H_{12}	72.15	5.769	345.0	469.8	33.3	311.0
n-C_6H_{14}	86.17	5.909	413.0	507.9	29.9	368.0
n-C_7H_{16}	100.20	. . .	. . .	540.2	27.0	426.0
n-C_8H_{18}	114.22	7.451	320.0	569.4	24.6	485.0
n-C_9H_{20}	128.25	. . .	. . .	595.0	22.5	543.0
Cyclohexane	84.16	6.093	324.0	553.0	40.0	308.0
C_6H_6	78.11	5.270	440.0	562.6	48.6	260.0
Other organic compounds						
CH_4	16.04	3.822	137.0	190.7	45.8	99.3
CH_3Cl	50.49	3.375	855.0	416.3	65.9	143.0
CH_2Cl_2	84.94	4.759	406.0	510.0	60.0	. . .
$CHCl_3$	119.39	5.430	327.0	536.6	54.0	240.0
CCl_4	153.84	5.881	327.0	556.4	45.0	276.0
C_2N_2	52.04	4.38	339.0	400.0	59.0	. . .
COS	60.08	4.13	335.0	378.0	61.0	. . .
CS_2	76.14	4.438	488.0	552.0	78.0	170.0

*Values of σ and ε/k_B are from J. O. Hirschfelder, C. F. Curtiss, and R. B. Bird, "Molecular Theory of Gases and Liquids," pp. 1110–1112, John Wiley & Sons, Inc., New York, 1954 (also Addenda and Corrigenda, p. 11). The above values are computed from viscosity data and are applicable for temperatures above 100°K.

†Values of T_c, p_c, and V_c are from K. A. Kobe and R. E. Lynn, Jr., *Chem. Rev.*, **52**, 117–236 (1952); F. D. Rosinni (ed.), *Am. Petrol. Inst. Res. Proj.*, Carnegie Institute of Technology, **44** (1952).

SOURCE: By permission from R. B. Bird, W. E. Stewart, and E. N. Lightfoot, "Transport Phenomena," John Wiley & Sons, Inc., New York, 1960.

From Table 11-1, for air

$$\frac{\varepsilon}{k_B} = 97°K \qquad \sigma = 3.617 \text{ Å}$$

and for SO_2

$$\frac{\varepsilon}{k_B} = 252°K \qquad \sigma = 4.290 \text{ Å}$$

From Eqs. (11-12) and (11-13),

$$\sigma_{AB} = \tfrac{1}{2}(3.617 + 4.290) = 3.953 \text{ Å}$$

$$\varepsilon_{AB} = k_B[97(252)]^{\frac{1}{2}}$$

Table 11-2   Volume increments for estimating molecular volume at normal boiling point

Kind of atom in molecule	Volume increment, cm³/g mole
Carbon	14.8
Chlorine, terminal as R—Cl	21.6
Chlorine, medial as —CHCl—	24.6
Fluorine	8.7
Helium	1.0
Hydrogen	3.7
Mercury	15.7
Nitrogen in primary amines	10.5
Nitrogen in secondary amines	12.0
Oxygen in ketones and aldehydes	7.4
Oxygen in methyl esters and ethers	9.1
Oxygen in ethyl esters and ethers	9.9
Oxygen in higher esters and ethers	11.0
Oxygen in acids	12.0
Oxygen bonded to S, P, or N	8.3
Phosphorus	27.0
Sulfur	25.6
For organic cyclic compounds	
3-membered ring	−6.0
4-membered ring	−8.5
5-membered ring	−11.5
6-membered ring	−15.0
Naphthalene	−30.0
Anthracene	−47.5

SOURCE: In part from C. N. Satterfield and T. K. Sherwood, "The Role of Diffusion in Catalysis," p. 9, Addison-Wesley Publishing Company, Reading, Mass., 1963.

At the temperature of 480°C,

$$\frac{k_B T}{\varepsilon_{AB}} = \frac{k_B(753)}{k_B[97(252)]^{\frac{1}{2}}} = 4.8$$

and so, from Table 11-3,

$$\Omega_{AB} = 0.85$$

Substituting all these values in Eq. (11-11) gives

$$D_{SO_2\text{-air}} = 0.0018583 \frac{753^{\frac{3}{2}}(1/64.1 + 1/28.9)^{\frac{1}{2}}}{(790/760)(3.953)^2(0.85)} = 0.629 \text{ cm}^2/\text{sec}$$

Example 11-2   A nickel catalyst for the hydrogenation of ethylene has a mean pore radius of 50 Å. Calculate the bulk and Knudsen diffusivities of hydrogen for this catalyst at 100°C, and 1 and 10 atm pressures, in a hydrogen-ethane mixture.

Table 11-3   Values of Ω_{AB} for diffusivity calculations (Lennard-Jones model)

$k_B T/\varepsilon_{AB}$	Ω_{AB}	$k_B T/\varepsilon_{AB}$	Ω_{AB}
0.30	2.662	2.0	1.075
0.35	2.476	2.5	1.000
0.40	2.318	3.0	0.949
0.45	2.184	3.5	0.912
0.50	2.066	4.0	0.884
0.55	1.966	5.0	0.842
0.60	1.877	7.0	0.790
0.65	1.798	10.0	0.742
0.70	1.729	20.0	0.664
0.75	1.667	30.0	0.623
0.80	1.612	40.0	0.596
0.85	1.562	50.0	0.576
0.90	1.517	60.0	0.560
0.95	1.476	70.0	0.546
1.00	1.439	80.0	0.535
1.10	1.375	90.0	0.526
1.20	1.320	100.0	0.513
1.30	1.273	200.0	0.464
1.40	1.233	300.0	0.436
1.50	1.198	400.0	0.417
1.75	1.128		

SOURCE: By permission from J. O. Hirshfelder, C. F. Curtiss, and R. B. Bird, "Molecular Theory of Gases and Liquids," pp. 1126, 1127, John Wiley & Sons, Inc., New York, 1954.

Solution   From Table 11-1, for H_2

$$\frac{\varepsilon}{k_B} = 38°\text{K} \qquad \sigma = 2.915 \text{ Å}$$

and for C_2H_6

$$\frac{\varepsilon}{k_B} = 230°\text{K} \qquad \sigma = 4.418 \text{ Å}$$

Then for the mixture, from Eqs. (11-12) and (11-13),

$$\sigma_{AB} = \tfrac{1}{2}(2.915 + 4.418) = 3.67 \text{ Å}$$

$$\varepsilon_{AB} = k_B[38(230)]^{\frac{1}{2}}$$

and

$$\frac{k_B T}{\varepsilon_{AB}} = \frac{273 + 100}{[38(230)]^{\frac{1}{2}}} = 4.00$$

From Table 11-3,

$$\Omega_{AB} = 0.884$$

Substituting these values in the Chapman-Enskog equation (11-11) gives the bulk diffusivity,

$$D_{H_2-C_2H_6} = 0.001858 \frac{373^{\frac{3}{2}}(1/2.016 + 1/30.05)^{\frac{1}{2}}}{p_t(3.67)^2(0.884)} = \frac{0.86}{p_t}$$

This gives 0.86 cm^2/sec at 1 atm, or 0.086 cm^2/sec at 10 atm.

The Knudsen diffusivity, which is independent of pressure, is obtained from Eq. (11-17) as

$$(\mathscr{D}_K)_{H_2} = 9.70 \times 10^3(50 \times 10^{-8})\left(\frac{373}{2.016}\right)^{\frac{1}{2}} = 0.065 \text{ cm}^2/\text{sec}$$

These results show that 1 atm pressure the Knudsen diffusivity is much less than the bulk value. Hence Knudsen diffusion controls the diffusion rate. At 10 atm both bulk and Knudsen diffusivities are important.

Example 11-3 (a) Calculate the combined diffusivity of hydrogen in a mixture of ethane, ethylene, and hydrogen in a pore of radius 50 Å at two total pressures, 1 and 10 atm. Suppose that the pore is closed at one end, and that the open end is exposed to a mixture of ethylene and hydrogen. The pore wall is a catalyst for the reaction

$$C_2H_4 + H_2 \rightarrow C_2H_6$$

The temperature is 100°C.

(b) For comparison calculate the combined diffusivity of hydrogen for diffusion through a noncatalytic capillary of radius 50Å. Hydrogen is supplied at one end and ethane at the other. The pressure is maintained the same at both ends of the capillary. Make the calculations for two compositions, $y_{H_2} = 0.5$ and 0.8.

Solution (a) Assume that the binary $H_2-C_2H_6$ will be satisfactory for representing the diffusion of hydrogen in the three-component system. From the reaction stoichiometry, the molal diffusion rates of H_2 and C_2H_6 will be equal and in opposite directions. Hence $\alpha = 0$, and Eq. (11-4) is applicable. From the results of Example 11-2,

$$D = \begin{cases} \dfrac{1}{1/0.86 + 1/0.065} = 0.060 \text{ cm}^2/\text{sec} & \text{at 1 atm} \\[3mm] \dfrac{1}{1/0.086 + 1/0.065} = 0.037 \text{ cm}^2/\text{sec} & \text{at 10 atm} \end{cases}$$

(b) For constant-pressure diffusion Eq. (11-2) should be used. From Eq. (11-3)

$$\alpha = 1 + \frac{N_{C_2H_6}}{N_{H_2}} = 1 - \sqrt{\frac{M_{H_2}}{M_{C_2H_6}}} = 1 - \sqrt{\frac{2.016}{30.05}} = 0.741$$

Then at $y_{H_2} = 0.5$, using Eq. (11-2),

$$D = \frac{1}{[1 - 0.741(0.5)]/\mathscr{D}_{\mathrm{H_2-C_2H_6}} + 1/(\mathscr{D}_K)_{\mathrm{H_2}}} = \frac{1}{0.630/\mathscr{D}_{\mathrm{H_2-C_2H_6}} + 1/(\mathscr{D}_K)_{\mathrm{H_2}}}$$

For the two pressures this expression gives

$$D = \begin{cases} \dfrac{1}{0.630/0.86 + 1/0.065} = 0.062 \text{ cm}^2/\text{sec} & \text{at 1 atm} \\[3mm] \dfrac{1}{0.630/0.086 + 1/0.065} = 0.044 \text{ cm}^2/\text{sec} & \text{at 10 atm} \end{cases}$$

At $y_{\mathrm{H_2}} = 0.8$ the two results are

$$D = \begin{cases} 0.063 \text{ cm}^2/\text{sec} & \text{at 1 atm} \\ 0.050 \text{ cm}^2/\text{sec} & \text{at 10 atm} \end{cases}$$

This example illustrates the following point. The variation of D with y_A depends on the importance of bulk diffusion. At the extreme where the Knudsen mechanism controls, the composition has no effect on D. When bulk diffusion is significant, the effect is a function of α. For equimolal counterdiffusion, $\alpha = 0$ and y_A has no influence on D. In our example, where $\alpha = 0.741$, and at 10 atm pressure, D increased only from 0.044 to 0.050 cm^2/sec as $y_{\mathrm{H_2}}$ increased from 0.5 to 0.8.

Also note that under reaction conditions in a pore, as in part (a), the ratio of the diffusion rates of the species is determined by stoichiometry. In contrast, for non-reacting systems at constant pressure, Eq. (11-8) is applicable.

11-2 Diffusion in Porous Catalysts

A considerable amount of experimental data has been accumulated for effective diffusivities.[1] Since reactors normally are operated at steady state and nearly constant pressure, diffusivities have also been measured under these restraints. The usual apparatus[2] is of the steady-flow type, illustrated in Fig. 11-1 for studying diffusion rates of H_2 and N_2. The effective diffusivity is defined in terms of such rates (per unit of total cross-sectional area) by the equation

$$(N_A)_e = -D_e \frac{dC_A}{dr} = -\frac{p}{R_g T} D_e \frac{dy_A}{dr} \tag{11-18}$$

where the subscript e on N_A emphasizes that this is a diffusion flux in a porous catalyst rather than for a single pore, as given by Eq. (11-1). Since α

[1] A summary up to 1969 is available in C. N. Satterfield, "Mass Transfer in Heterogeneous Catalysis," pp. 56–77, Massachusetts Institute of Technology Press, Cambridge, Mass., 1970. See also C. N. Satterfield and P. J. Cadle, *Ind. Eng. Chem., Fund. Quart.*, **7**, 202 (1968); *Process Design Develop.*, **7**, 256 (1968); L. F. Brown, H. W. Haynes, and W. H. Manogue, *J. Catalysis*, **14**, 220 (1969).

[2] Originally proposed by E. Wicke and R. Kallenbach [*Kolloid-Z.*, **17**, 135 (1941)].

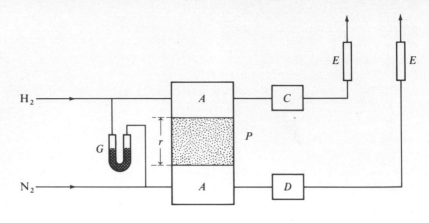

A — Mixing chambers
C — Detector for determining composition of N_2 in H_2 stream
D — Detector for determining composition of H_2 in N_2 stream
E — Flow meters
P — Catalyst pellet
G — Pressure equalization gauge

Fig. 11-1 *Constant-pressure apparatus for measuring diffusion rates in porous catalysts*

is known for the constant-pressure process via Eq. (11-8), it is possible to integrate Eq. (11-18) taking into account the variation of D_e (or D) with y_A. Thus effective values of $\mathscr{D}_{AB}$ and $(\mathscr{D}_K)_A$ can be introduced in Eq. (11-2) to give

$$D_e = \frac{1}{(1 - \alpha y_A)/(\mathscr{D}_{AB})_e + [1/(\mathscr{D}_K)_A]_e} \tag{11-19}$$

This can be substituted in Eq. (11-18) and the result integrated to give the flux in terms of $(\mathscr{D}_{AB})_e$ and $[(\mathscr{D}_K)_A]_e$. Several investigators have calculated fluxes in this way[1] and compared them with results obtained experimentally in the apparatus of Fig. 11-1. The purpose was to evaluate a geometrical model for the porous catalyst. As explained later, such a model is necessary to convert $\mathscr{D}_{AB}$ to $(\mathscr{D}_{AB})_e$ and $(\mathscr{D}_K)_A$ to $[(\mathscr{D}_K)_A]_e$ for use in Eq. (11-19). For this objective the use of a composition-dependent diffusivity is perhaps warranted. Other uncertainties make this degree of detail unjustified for evaluating the effect of intraparticle transport on reaction rate. Hence we shall employ a composition-independent D_e given by Eq. (11-4). Then Eq. (11-18) can be integrated to give

[1] N. Wakao and J. M. Smith, *Chem. Eng. Sci.*, **17**, 825 (1962); J. P. Henry, Jr., R. S. Cunningham, and C. J. Geankoplis, *Chem. Eng. Sci.*, **22**, 11 (1967); R. S. Cunningham and C. J. Geankoplis, *Ind. Eng. Chem.*, **7**, 535 (1968).

$$(N_A)_e = -\frac{p}{R_g T} D_e \frac{(y_A)_2 - (y_A)_1}{\Delta r} \tag{11-20}$$

where Δr is the thickness of the pellet. From the experiment depicted in Fig. 11-1, $(N_A)_e$ is calculated from the measured flow rates and compositions. Then an experimental effective diffusivity is easily evaluated from Eq. (11-20).

Barrer[1] has developed a transient method of measuring diffusivities in porous catalysts. Comparison with results from steady-state experiments may permit a better understanding of some aspects of the pore structure, such as the extent of dead-end pores. However, steady-state diffusivities are appropriate for assessing effects of intrapellet transport on steady-state reaction rates. Another method of evaluating effective diffusivities, provided the rate equation for the chemical step is known, is comparison of the observed and predicted global rates by combining the equation for the chemical step with the intraparticle mass-transport resistance.

In the absence of experimental data it is necessary to estimate D_e from the physical properties of the catalyst. In this case the first step is to evaluate the diffusivity for a single cylindrical pore, that is, to evaluate D from Eq. (11-4). Then a geometric *model* of the pore system is used to convert D to D_e for the porous pellet. A model is necessary because of the complexity of the geometry of the void spaces. The optimum model is a realistic representation of the geometry of the voids, with tractable mathematics, that can be described in terms of easily measurable physical properties of the catalyst pellet. As noted in Chap. 8, these properties are the surface area and pore volume per gram, the density of the solid phase, and the distribution of void volume according to pore size.

The Parallel-pore Model   Wheeler[2] proposed a model, based on the first three of these properties, to represent the monodisperse pore-size distribution in a catalyst pellet. From ρ and V_g the porosity ε is obtained from Eq. (8-16). Then a mean pore radius $\bar{a}$ is evaluated[3] by writing equations for the total pore volume and total pore surface in a pellet. The result, developed as Eq. (8-26), is

$$\bar{a} = \frac{2V_g}{S_g} \tag{11-21}$$

[1] M. Barrer, *J. Phys. Chem.*, **57**, 35 (1953).

[2] A. Wheeler, in P. H. Emmett (ed.), "Catalysis," vol II, chap. 2, Reinhold Publishing Corporation, New York, 1955; A. Wheeler, "Advances in Catalysis," vol. III, p. 250, Academic Press, Inc., New York, 1951.

[3] Recall from Sec. 8-7 that regardless of the complexity of the void structure, it is assumed that the void spaces may be regarded as cylindrical pores.

By using V_g, S_g, and ρ_s, Wheeler replaced the complex porous pellet with an assembly (having a porosity ε) of cylindrical pores of radius $\bar{a}$. To predict D_e from the model the only other property necessary is the length x_L of the diffusion path. If we assume that, on the average, the pore makes an angle of 45° with the coordinate r in the resultant direction of diffusion (for example, the radial direction in a spherical pellet), $x_L = \sqrt{2}\ r$. Owing to pore interconnections and noncylindrical shape, this value of x_L is not very satisfactory. Hence, it is customary to define x_L in terms of an adjustable parameter, the *tortuosity factor* δ, as follows:

$$x_L = \delta r \tag{11-22}$$

An effective diffusivity can now be predicted by combining Eq. (11-1) for a single pore with this parallel-pore model. To convert D, which is based on the cross-sectional area of the pore, to a diffusivity based upon the total area perpendicular to the direction of diffusion, D should be multiplied by the porosity. In Eq. (11-1), x is the length of a single, straight cylindrical pore. To convert this length to the diffusion path in a porous pellet, x_L from Eq. (11-22) should be substituted for x. With these modifications the diffusive flux in the porous pellet will be

$$(N_A)_e = -\frac{p}{R_g T}\ \frac{\varepsilon D}{\delta}\ \frac{dy_A}{dr} \tag{11-23}$$

Comparison with Eq. (11-18) shows that the effective diffusivity is

$$D_e = \frac{\varepsilon D}{\delta} \tag{11-24}$$

where D is given (in the absence of surface diffusion) by Eq. (11-4). The use of Eq. (11-24) to predict D_e is limited because of the uncertainty about δ. Comparison of D_e from Eq. (11-24) with values obtained from experimental data using Eq. (11-20) shows that δ varies from less than unity to more than 6.[1]

The Random-pore Model   This model was originally developed for pellets containing a bidisperse pore system, such as the alumina described in Chap. 8 (Table 8-5 and Fig. 8-10). It is supposed that the pellet consists of an assembly of small particles. When the particles themselves contain pores (micropores), there exists both a macro and a micro void-volume distribu-

---

[1] Tortuosity factors less than unity can occur when the surface diffusion is significant. This is because D_e is increased, while D as defined in Eq. (11-4) does not include this contribution. See Sec. 11-3 for a corrected D to include surface diffusion. C. N. Satterfield ("Mass Transfer in Heterogeneous Catalysis," Massachusetts Instiute of Technology Press, Cambridge, Mass., 1970) has summarized data from the literature and recommended the use of $\delta = 4$ when surface diffusion is insignificant. See also P. Carman, *Trans. Inst. Chem. Engrs.*, **15**, 150 (1937).

tion. The voids are not imagined as capillaries, but more as an assembly of short void regions surrounding and between individual particles, as indicated in Fig. 11-2. The nature of the interconnection of macro and micro void regions is the essence of the model. Transport in the pellet is assumed to occur by a combination of diffusion through the macro regions (of void fraction ε_M), the micro regions (of void fraction ε_μ), and a series contribution involving both regions. It is supposed that both micro and macro regions can be represented as straight, short cylindrical pores of average radii $\bar{a}_M$ and $\bar{a}_\mu$. The magnitude of the individual contributions is dependent on their effective cross-sectional areas (perpendicular to the direction of diffusion). The details of the development are given elsewhere,[1] but in general these areas are evaluated from the probability of pore interconnections. The resultant expression for D_e may be written

$$D_e = \bar{D}_M \varepsilon_M{}^2 + \frac{\varepsilon_\mu{}^2 (1 + 3\varepsilon_M)}{1 - \varepsilon_M} \bar{D}_\mu \tag{11-25}$$

Here $\bar{D}_M$ and $\bar{D}_\mu$ are obtained by applying Eq. (11-4) to macro and micro regions. Thus

$$\frac{1}{\bar{D}_M} = \frac{1}{\mathscr{D}_{AB}} + \frac{1}{(\bar{\mathscr{D}}_K)_M} \tag{11-26}$$

$$\frac{1}{\bar{D}_\mu} = \frac{1}{\mathscr{D}_{AB}} + \frac{1}{(\bar{\mathscr{D}}_K)_\mu} \tag{11-27}$$

[1] N. Wakao and J. M. Smith, *Chem. Eng. Sci.*, **17**, 825 (1962); *Ind. Eng. Chem.*, *Fund. Quart.*, **3**, 123 (1964).

Fig. 11-2 *Random-pore model*

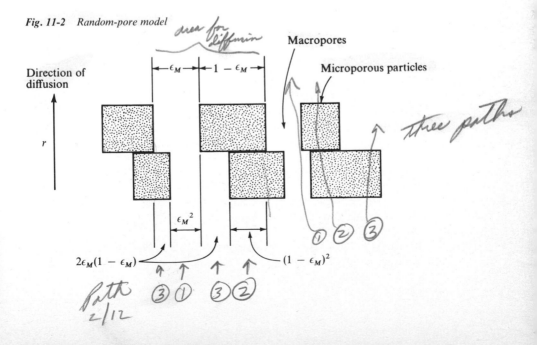

Macropores

Microporous particles

Direction of diffusion

r

ε_M $1 - \varepsilon_M$

$\varepsilon_M{}^2$

$2\varepsilon_M(1 - \varepsilon_M)$ $(1 - \varepsilon_M)^2$

It is evident from the concepts presented that no tortuosity factor is involved in this model.[1] The actual path length is equal to the distance coordinate in the direction of diffusion. To apply Eq. (11-25) requires void fractions and mean pore radii for both macro and micro regions. The mean pore radii can be evaluated for the micro region by applying Eq. (11-21) to this region. However, $\bar{a}_M$ must be obtained from the pore-volume distribution, as described in Sec. 8-7. The mean pore radii are necessary in order to calculate $(\bar{\mathscr{D}}_K)_M$ and $(\bar{\mathscr{D}}_K)_\mu$ from Eq. (11-27).

The random-pore model can also be applied to monodisperse systems. For a pellet containing only macropores, $\varepsilon_\mu = 0$ and Eq. (11-25) becomes

$$D_e = \bar{D}_M \varepsilon_M^2 \tag{11-28}$$

Similarly, for a material such as silica gel, where $\varepsilon_M = 0$, the effective diffusivity is

$$D_e = \bar{D}_\mu \varepsilon_\mu^2 \tag{11-29}$$

Comparison of these last two equations with Eq. (11-24) indicates that $\delta = 1/\varepsilon$. The significance of the random-pore model is that the effective diffusivity is proportional to the square of the porosity. This has also been proposed by Weisz and Schwartz.[2] Johnson and Stewart[3] have developed another method for predicting D_e that utilizes the pore-volume distribution. Evaluation of their model and the random-pore model with extensive experimental data has been carried out by Satterfield and Cadle[4] and Brown et al.[5]

The examples that follow illustrate the methods that have been outlined for evaluating D_e from experimental data and predicting it from a pore model.

Example 11-4   Rothfeld[6] has measured diffusion rates for isobutane, in the isobutane-helium system, through a $\frac{1}{8}$-in.-long pelleted cylinder of alumina (diameter $\frac{1}{8}$ in.). The measurements were at 750 mm Hg total pressure and 25°C, and the diffusion direction was through the pellet parallel to the central axis. The following data are available for Harshaw alumina, type Al-0104-T:

$$S_g = 76 \text{ m}^2/\text{g}$$

$$\varepsilon_M = 0.18 \qquad \varepsilon_\mu = 0.34$$

$$\bar{a}_M = 4,800 \text{ Å} \qquad \bar{a}_\mu = 84 \text{ Å}$$

[1] Thus the random-pore model does not involve an adjustable parameter.
[2] P. B. Weisz and A. B. Schwartz, *J. Catalysis*, **1**, 399 (1962).
[3] M. F. L. Johnson and W. E. Stewart, *J. Catalysis*, **4**, 248 (1965).
[4] C. N. Satterfield and P. J. Cadle, *Ind. Eng. Chem., Fund. Quart.*, **7**, 202 (1968); *Process Design Develop.*, **7**, 256 (1968).
[5] L. F. Brown, H. W. Haynes, and W. H. Manogue, *J. Catalysis*, **14**, 220 (1969).
[6] L. B. Rothfeld, *AIChE J.*, **9**, 19 (1963).

The mole fraction of isobutane is 1.0 on one face of the pellet and zero on the other face. The experimental results gave

$$\frac{(N_A)_e R_g T \, \Delta r}{\mathscr{D}_{AB} p(y_2 - y_1)} = -0.023$$

where N_A is the diffusion flux of isobutane and $\mathscr{D}_{AB}$ is the bulk diffusivity in the iso-butane-helium system.

(a) Calculate the experimental value of D_e. (b) What macropore tortuosity factor is indicated by the data? What is δ_M predicted by the parallel-pore and by the random-pore models?

Solution (a) According to the procedure of Example 11-1 (the Chapman-Enskog equation), the bulk diffusivity in the isobutane-helium system at 750 mm Hg pressure and 25°C is

$$\mathscr{D}_{AB} = 0.313 \text{ cm}^2/\text{sec}$$

Thus the measured diffusion flux of isobutane is

$$(N_A)_e = \frac{-0.023(0.313)(\frac{750}{760})(0 - 1.0)}{82(25 + 273)(\frac{1}{8})(2.54)} = 9.1 \times 10^{-7} \text{ g mole}/(\text{sec})(\text{cm}^2)$$

This value of $(N_A)_e$ can be used in Eq. (11-20) to calculate the experimental D_e,

$$D_e = -\frac{(N_A)_e R_g T \, \Delta r}{p(y_2 - y_1)} = \frac{-9.1 \times 10^{-7}(82)(298)(\frac{1}{8})(2.54)}{\frac{750}{760}(0 - 1.0)} = 0.0072 \text{ cm}^2/\text{sec}$$

(b) The *parallel-pore* model is designed for a *monodisperse* pore system and hence is not directly applicable to this catalyst. Since the macropores are much larger than the micropores, an approximate approach is to neglect the contribution of the micropores to the mass transport. Using the macro properties in Eq. (11-24) and the experimental D_e, we have

$$D_e = 0.0072 = \frac{\varepsilon_M \bar{D}_M}{\delta_M} = \frac{0.18 \bar{D}_M}{\delta_M}$$

or

$$\delta_M = \frac{0.18 \, \bar{D}_M}{0.0072}$$

The combined diffusivity in the macropores is given by Eq. (11-26)

$$\frac{1}{\bar{D}_M} = \frac{1}{\mathscr{D}_{AB}} + \frac{1}{(\mathscr{D}_K)_M}$$

From Eq. (11-17),

$$(\mathscr{D}_K)_M = 9.70 \times 10^3 (4{,}800 \times 10^{-8}) \left(\frac{298}{58.1} \right)^{\frac{1}{2}} = 1.10 \text{ cm}^2/\text{sec}$$

$$\frac{1}{\bar{D}_M} = \frac{1}{0.313} + \frac{1}{1.10}$$

$$\bar{D}_M = 0.243 \text{ cm}^2/\text{sec}$$

Then the tortuosity factor suggested by the data is

$$\delta_M = \frac{0.18(0.243)}{0.0072} = 6.1$$

If we assume that all the diffusion is in the macropores, Eq. (11-28) gives D_e for the *random-pore model*. Combining Eq. (11-28) with Eq. (11-24) yields

$$\frac{\varepsilon_M \overline{D}_M}{\delta_M} = D_e = \overline{D}_M \varepsilon_M{}^2$$

or

$$\delta_M = \frac{1}{\varepsilon_M} = \frac{1}{0.18} = 5.5$$

in reasonable agreement with the experimental result. Actually, this pellet is relatively dense (low ε_M) and probably was prepared with a high pelleting pressure. Hence the diffusion likely is affected by the micropores.[1]

Example 11-5   Vycor (porous silica) appears to have a pore system with fewer interconnections than alumina. The pore system is monodisperse, with the somewhat unusual combination of a small mean pore radius (45 Å) and a low porosity 0.31[2]. Vycor may be much closer to an assembly of individual voids than to an assembly of particles surrounded by void spaces. Since the random-pore model is based on the assembly-of-particles concept, it is instructive to see how it applies to Vycor. Rao and Smith[2] measured an effective diffusivity for hydrogen of 0.0029 cm^2/sec in Vycor. The apparatus was similar to that shown in Fig. 11-1, and data were obtained using an H_2–N_2 system at 25°C and 1 atm. Predict the effective diffusivity by the random-pore model.

Solution   Only micropores are present, and so D_e should be predicted by means of Eq. (11-29). Furthermore, mass transport in the small pores will be predominately by Knudsen diffusion. Hence $\overline{D}_\mu = (\mathscr{D}_K)_\mu$, from Eq. (11-27), and

$$D_e = (\mathscr{D}_K)_\mu \varepsilon_\mu^2$$

$$(\mathscr{D}_K)_\mu = 9.7 \times 10^3 (45 \times 10^{-8}) \left(\frac{298}{2.02}\right)^{\frac{1}{2}} = 0.053 \text{ cm}^2/\text{sec}$$

Then the predicted D_e is

$$D_e = 0.053(0.31)^2 = 0.0050 \text{ cm}^2/\text{sec}$$

[1] W. Wakao and J. M. Smith [*Chem. Eng. Sci.*, **17**, 825 (1962)] thoroughly analyzed the diffusion data of Rothfeld. These data were obtained in an apparatus of the type shown in Fig. 11-1. For the butane-helium system this means that N_{He}/N_{C_4} is 3.80. Diffusion is far from equimolal, suggesting that Eqs. (11-26) and (11-27) for D values are not exact. For this particular case Eq. (11-2) should be used. In most reaction systems the counterdiffusion of reactants and products is much closer to equimolal, so that Eqs. (11-26) and (11-27) are better approximations.

[2] M. J. Rao and J. M. Smith, *AIChE J.*, **10**, 243 (1964).

This value is 70% greater than the experimental result—evidence that the random-pore model is not very suitable for Vycor.

The experimental tortuosity, from Eq. (11-24), is

$$\delta = \frac{0.31\overline{D}}{D_e} = \frac{0.31(\overline{\mathscr{D}}_K)_\mu}{D_e} = \frac{0.31(0.053)}{0.0029} = 5.6$$

In contrast, the tortuosity predicted by the random-pore model would be $1/\varepsilon = 3.2$.

Silica gel of large surface area has even smaller pores than Vycor, but a larger void fraction; for example, for one grade of silica gel $\varepsilon = 0.486$ and $\bar{a} = 11$ Å. Schneider[1] studied the diffusion of ethane in this material at 200°C (at this temperature surface diffusion was negligible) and found an effective diffusivity such that δ from Eq. (11-24) was 3.34. The value predicted from the random-pore model is $1/\varepsilon = 2.1$. Note that the Wheeler model would predict $\delta = \sqrt{2}$ for Vycor, silica gel, or other porous material.

A final word should be said about the variety of porous materials. Porous catalysts cover a rather narrow range of possibilities. Perhaps the largest variation is between monodisperse and bidisperse pellets, but even these differences are small in comparison with materials such as freeze-dried beef, which is like an assembly of solid fibers, and freeze-dried fruit, which appears to have a structure like an assembly of ping-pong balls with holes in the surface to permit a continuous void phase.[2]

11-3 Surface Diffusion

Surface migration is pertinent to a study of intrapellet mass transfer if its contribution is significant with respect to diffusion in the pore space. When multimolecular-layer adsorption occurs, surface diffusion has been explained as a flow of the outer layers as a condensed phase.[3] However, surface transport of interest in relation to reaction occurs in the monomolecular layer. It is more appropriate to consider, as proposed by deBoer,[4] that such transport is an activated process, dependent on surface characteristics as well as those of the adsorbed molecules. Imagine that a molecule in the gas phase strikes the pore wall and is adsorbed. Then two alternatives are possible: desorption into the gas (Knudsen diffusion) or movement to an adjacent active site on the pore wall (surface diffusion). If desorption occurs,

[1] Petr Schneider and J. M. Smith, *AIChE J.*, **14**, 886 (1968).

[2] J. C. Harper, *AIChE J.*, **8**, 298 (1962).

[3] D. H. Everett in F. S. Stone (ed.), Structure and Properties of Porous Materials, *Proc. Tenth Symp. Colston Res. Soc.*, Butterworth, London, p. 178, 1958.

[4] J. H. deBoer, in "Advances in Catalysis," vol. VIII, p. 18, Academic Press, Inc., New York, 1959.

the molecule can continue its journey in the void space of the pore or be readsorbed by again striking the wall. In moving along the pore, the same molecule would be transported sometimes on the surface and sometimes in the gas phase. If this view is correct, the relative contribution of surface migration would increase as the surface area increases (or the pore size decreases). There is evidence to indicate that such is the case.[1]

Experimental verification of surface diffusion is usually indirect, since concentrations of adsorbed molecules on a surface are difficult to measure. When gas concentrations are obtained, the problem arises of separating the surface and pore-volume transport rates. One solution is to measure both N_A and N_B in the apparatus shown in Fig. 11-1, using a non-adsorbable gas for A. If the diffusion rate of B is greater than that calculated from N_A by Eq. (11-8), the excess is attributable to surface migration. Barrer and Barrie[2] used this procedure with Vycor at room temperature and found surface migration significant for such gases as CO_2, CH_4, and C_2H_4 and negligible for helium and hydrogen. Rivarola found that for CO_2 the surface contribution on an alumina pellet increased from 3.5 to 54% of the total mass-transfer rate as the macropore properties changed from $\bar{a}_M = 1,710$ Å and $\varepsilon_M = 0.33$ to $\bar{a}_M = 348$ Å and $\varepsilon_M = 0.12$.

The extent to which surface transport affects global rates of reaction has not been established. For it to be important, adsorption must occur, but this is also a requirement for catalytic activity. Indirect evidence suggests that in some cases the effect is considerable. For example, Miller and Kirk[3] found higher rates of dehydration of alcohols on silica-alumina than could be explained with only pore-volume diffusion to account for intraparticle resistances. They attributed the discrepancy to surface diffusion. Masamune and Smith[4] found that surface transport of ethanol on silica gel at temperatures as high as 175°C predominated over gas-phase diffusion in the pore. In view of the data available, it seems wise at least to consider the possibility of surface migration in any evaluation of intraparticle effects. This can be done by adding a surface-diffusion contribution to the effective diffusivity considered in the previous section. The method of doing this is presented below, but its usefulness is still limited because of inadequate experimental and theoretical aspects of surface transport.

By analogy to Fick's law, a surface diffusivity $\mathscr{D}_s$ may be defined in terms of a surface concentration C_s in moles per square centimeters of surface of adsorbate,

[1] J. B. Rivarola and J. M. Smith, *Ind. Eng. Chem., Fund. Quart.*, **3**, 308 (1964).

[2] R. M. Barrer and J. M. Barrie, *Proc. Roy. Soc. (London)*, **A213**, 250 (1952).

[3] D. N. Miller and R. S. Kirk, *AIChE J.*, **8**, 183 (1962).

[4] S. Masamune and J. M. Smith, *AIChE J.*, **11**, 41 (1965).

$$N_s = -\mathscr{D}_s \frac{dC_s}{dx} \tag{11-30}$$

where N_s is the molal rate per unit perimeter of pore surface. In order to combine surface and pore-volume contributions in a *catalyst pellet*, the surface flux should be based on the total area of the catalyst perpendicular to diffusion and on the coordinate r. If this flux is $(N_s)_e$, then

$$(N_s)_e = -D_s\rho_P \frac{d\overline{C}}{dr} \tag{11-31}$$

where ρ_P is the density of the catalyst and $\overline{C}$ now represents the moles adsorbed per gram of catalyst.[1] To be useful, Eq. (11-31) must be expressed in terms of the concentration in the gas phase. If the adsorption step is fast with respect to surface transfer from site to site on the catalyst, we may safely assume equilibrium between gas and surface concentration. Otherwise the relation between the two concentrations depends on the intrinsic rates of the two processes. Adequate theory has not been developed to treat the latter case. If equilibrium is assumed, and if the isotherm is linear, then

$$\overline{C}_A = (K_A\overline{C}_m)C_g = K_A' \frac{py_A}{R_gT} \tag{11-32}$$

where the subscript A refers to the component and K' is the linear form of the equilibrium constant, in cubic centimeters per gram. The latter is obtained from the Langmuir isotherm, Eq. (8-6), when adsorption is small enough that the linear form is valid. Now, applying Eq. (11-31) to component A,

$$(N_s)_e = -\frac{p}{R_gT} \rho_P K_A' D_s \frac{dy_A}{dr} \tag{11-33}$$

Equation (11-33) gives the surface diffusion of A in the same form as Eq. (11-18) applied earlier to transport in the gas phase of the pores. Then the total flux and total effective diffusivity are given by

$$(N_A)_t = -\frac{p}{R_gT} (D_e + \rho_P K_A' D_s) \frac{dy_A}{dr} \tag{11-34}$$

$$(D_e)_t = D_e + \rho_P K_A' D_s \tag{11-35}$$

If the density of the catalyst and the adsorption equilibrium constant

---

[1] To relate $\mathscr{D}_s$ and D_s requires a model for the porous structure. The parallel-pore and random-pore models have been applied to surface diffusion by J. H. Krasuk and J. M. Smith [*Ind. Eng. Chem., Fund. Quart.*, **4**, 102 (1965)] and J. B. Rivarola and J. M. Smith [*Ind. Eng. Chem., Fund. Quart.*, **3**, 308 (1964)].

are known, Eq. (11-35) permits the evaluation of a total effective diffusivity from D_s. Data for D_s have been reported in the literature in a variety of ways, depending on the definitions of adsorbed concentrations and equilibrium constants. Schneider[1] has summarized much of the information for light hydrocarbons on various catalysts in the form defined by Eq. (11-33). Such D_s values ranged from 10^{-3} to 10^{-6}, depending on the nature of the adsorbent and the amount adsorbed. Most values were in the interval 10^{-4} to 10^{-5} cm^2/sec. Data for other adsorbates[2] have similar magnitudes and show that the variation with adsorbed concentration can be large.

The effect of temperature on D_s, given an activated process, is described by an Arrhenius-type expression,

$$D_s = A\, e^{-E_s/R_g T} \tag{11-36}$$

where E_s is the activation energy for surface diffusion. The variation of K'_A with temperature is given by the van't Hoff equation

$$\frac{d \ln K'_A}{dT} = \frac{\Delta H}{R_g T^2}$$

or

$$K'_A = A'\, e^{-\Delta H/R_g T} \tag{11-37}$$

The observed rate of surface diffusion, according to Eq. (11-33), will be proportional to the product $K'_A D_s$. Combining Eqs. (11-36), (11-37), and (11-33), we have

$$(N_s)_e = -\frac{p}{R_g T}\, \rho_P (AA')\, e^{-(1/R_g T)(\Delta H + E_s)}\, \frac{dy_A}{dr}$$

The exponential term expresses a much stronger temperature dependency than the coefficient $1/T$. If we neglect the latter, we may express the *temperature* effect on the rate as

$$(N_s)_e = A''\, e^{-(1/R_g T)(\Delta H + E_s)}$$

where $A'' = AA' \dfrac{dy_A}{dr} \dfrac{p}{R_g T} \rho_P$

This equation shows that the observed or apparent activation energy for surface diffusion is related to E_s by

[1] Petr Schneider and J. M. Smith, *AIChE J.*, **14**, 886 (1968).

[2] R. A. W. Haul, *Angew. Chem.*, **62**, 10 (1950); P. S. Carman and F. A. Raal, *Proc. Roy. Soc. (London)*, **209A**, 38 (1951), *Trans. Faraday Soc.*, **50**, 842 (1954); D. H. Everett in F. S. Stone (ed.), Structure and Properties of Porous Materials, *Proc. Tenth Symp. Colston Res. Soc.*, Butterworth, London, p. 178, 1958.

$$E' = \Delta H + E_s \tag{11-38}$$

From available data[1] it appears that E_s is only a few kilocalories per mole. The heat of adsorption ΔH is generally greater than this, particularly for chemisorption, and is always negative. Therefore the *observed* effect is a decrease in rate of surface diffusion with increase in temperature.

From the assumptions and approximations presented, it is clear that surface diffusion is not well understood. It is hoped that improved interpretations of surface migration will permit a more accurate assessment of its effect on global rates of reaction. When we consider the effect of intraparticle resistances in Secs. 11-6 to 11-11 we shall suppose that the D_e used is the most appropriate value and includes, if necessary, a surface contribution.

INTRAPELLET HEAT TRANSFER

11-4 Concept of Effective Thermal Conductivity

The effective thermal conductivities of catalyst pellets are surprisingly low. Therefore significant intrapellet temperature gradients can exist, and the global rate may be influenced by thermal effects. The effective conductivity is the energy transferred per unit of total area of pellet (perpendicular to the direction of heat transfer). The defining equation, analogous to Eq. (11-18) for mass transfer, may be written

$$Q_e = -k_e \frac{dT}{dr} \tag{11-39}$$

where Q_e is the rate of energy transfer per unit of total area.

A major factor contributing to small values of k_e is the numerous void spaces that hinder the transport of energy. In addition, the path through the solid phase offers considerable thermal resistance for many porous materials, particularly pellets made by compressing microporous particles. Such behavior is readily understood if these materials are viewed as an assembly of particles which contact each other only through adjacent points. There is strong experimental evidence that such point contacts are regions of high thermal resistance. For example, the thermal conductivity of the bulk solid (zero porosity) from which the particles are prepared does not have a large effect on k_e. Masamune[2] found that the effective thermal conductivity of pellets of microporous particles of silver was only

[1] Petr Schneider and J. M. Smith, *AIChE J.*, **14**, 886 (1968).

[2] S. Masamune and J. M. Smith, *J. Chem. Eng. Data*, **8**, 54 (1963).

two to four times that of alumina pellets, at the same macropore void fraction, pressure, and temperature. In contrast, the thermal conductivity of solid silver is about 200 times as large as that of solid alumina. Furthermore, k_e is a strong function of the void fraction, increasing as ε decreases. Materials such as alumina pellets may be regarded as a porous assembly within a second porous system. Each particle from which the pellet is made consists of a microporous region. These particles are in point contact with like particles and are surrounded by macroporous regions. When viewed in this way, the thermal conductivity of the bulk solid should have little influence on k_e.

The pressure and nature of the fluid in the pores has an effect on the effective thermal conductivity. With liquids the effect of pressure is negligible and k_e is of the same magnitude as the true conductivity of the liquid. For gases at low pressures, where the mean free path is the same or larger than the pore size, free-molecule conduction controls the energy transfer. In this region k_e increases with pressure. At higher pressures k_e is about independent of pressure. The transition pressure depends on the gas as well as on the pore size. For air the transition pressure is about 470 mm in a silver pellet with mean pore diameter of 1,500 Å. For helium the value would be above 760 mm.[1] For alumina pellets, at 120°F and a macropore void fraction of $\varepsilon_M = 0.40$, k_e was 0.050 in vacuum, 0.082 with pores filled with air, and 0.104 Btu/(hr)(ft)(°F) with helium, at atmospheric pressure.[2] Temperature does not have a strong influence. The effect is about what would be expected for the combination of variations of thermal conductivity with temperature for the solid and fluid phases.

11-5 Effective-thermal-conductivity Data

Most of the experimental information on k_e for catalyst pellets is described by Masamune and Smith,[1] Mischke and Smith,[2] and Sehr.[3] Sehr gives single values for commonly used catalysts. The other two works present k_e as a function of pressure, temperature, and void fraction for silver and alumina pellets. Both transient and steady-state methods have been employed. Figure 11-3 shows the variation of k_e with pellet density and temperature for alumina (boehmite, $Al_2O_3 \cdot H_2O$) pellets. Different densities were obtained by increasing the pressure used to pellet the microporous particles. These data are for vacuum conditions and therefore represent the conduction of the solid matrix of the pellet. Note how low

[1] S. Masamune and J. M. Smith, *J. Chem. Eng. Data*, **8**, 54 (1963).

[2] R. A. Mischke and J. M. Smith, *Ind. Eng. Chem.*, *Fund. Quart.*, **1**, 288 (1962).

[3] R. A. Sehr, *Chem. Eng. Sci.*, **2**, 145 (1958).

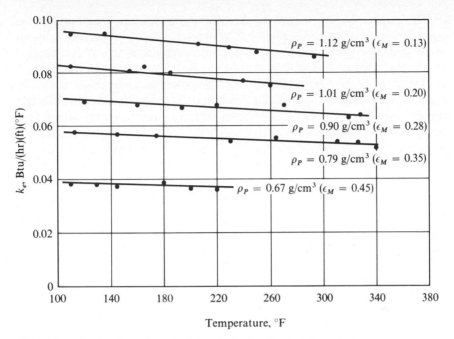

Fig. 11-3 *Effective thermal conductivity of alumina (boehmite) catalyst pellets at 10 to 25 microns Hg pressure*

k_e is in comparison with the thermal conductivity of solid alumina [about 1.0 Btu/(hr)(ft)(°F)]. The low value is due to the small heat-transfer areas at the point-to-point contacts between particles. As the pelleting pressure increases (lower macropore void fraction), these contact areas increase, and so does k_e. Figure 11-4 shows the effect of macropore void fraction on k_e for pellets of microporous silver and microporous alumina (boehmite) particles. Data are given for two conditions: vacuum and pores filled with helium at 1 atm, both at 34°C. Since helium has a high conductivity, the two curves would encompass the values expected for any reaction mixture. As 1 atm is at or beyond the transition pressure, increasing the pressure would have little effect (until the thermodynamic critical point is approached). Similarly, silver and boehmite represent close to the extremes of conductivity for the solids that might be expected in porous catalysts. Figure 11-5 is a similar plot of data for three cases: vacuum, air-filled pores at 1 atm pressure, and helium-filled pores at 1 atm pressure. These results are for boehmite pellets.[1]

The theory of heat transfer in porous materials has not been developed

[1] R. A. Mischke and J. M. Smith, *Ind. Eng. Chem., Fund. Quart.*, **1**, 288 (1962).

to the level of that for mass transfer. The contribution of the solid phase makes the problem more complex. It is not yet possible to predict k_e accurately from the properties of the fluid and solid phases. Butt[1] has applied and extended· the random-pore model to develop a valuable method of predicting the effects of void fraction, pressure, and temperature on k_e. A different and more approximate approach[2] proposes that the effective thermal conductivity is a function only of the volume fraction of the void phase and the thermal conductivities of the bulk-fluid and solid phases, k_b and k_s. The relationship is

$$k_e = k_s \left(\frac{k_b}{k_s} \right)^{1-\varepsilon} \tag{11-40}$$

In spite of the difficulties in predicting k_e, it is still possible to choose a value which will be reasonably correct, because the possible range of values (excluding vacuum conditions) is only from about 0.1 to 0.4 Btu/(hr)(ft)(°F). Furthermore, the nature of the variations within this range that are due to

[1] J. B. Butt, *AIChE J.*, **11**, 106 (1965).
[2] W. Woodside and J. H. Messmer, *J. Appl. Phys.*, **32**, 1688 (1961).

Fig. 11-4   *Effect of macropore void fraction on k_e at 34°C*

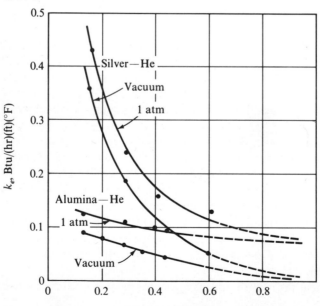

void fraction, temperature, and pressure are known approximately from Figs. 11-3 to 11-5.

MASS TRANSFER WITH REACTION

Having discussed the effective transport coefficients D_e and k_e, we now can turn to the main objective of the chapter, an expression of the rate of reaction for the whole catalyst pellet,[1] r_P, in terms of the temperature and concentrations existing at the *outer* surface. We start in a formal way by defining an effectiveness factor η as follows:

[1] r_P is the rate for the whole pellet, but based on a unit mass of catalyst. The description "whole pellet" is used to denote that intrapellet effects are accounted for in r_P. If the rate measured for one or more pellets is r and the mass of the pellets is m, then $r_P = r/m$.

Fig. 11-5 *Effective thermal conductivity of alumina (boehmite) pellets vs void fraction at 120°F*

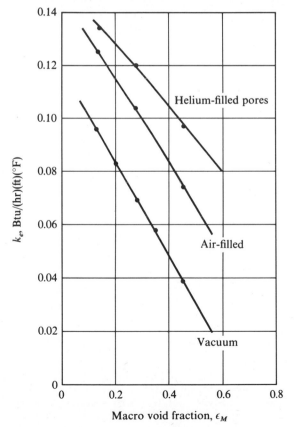

$$\eta = \frac{\text{actual rate for the whole pellet}}{\text{rate evaluated at outer surface conditions}} = \frac{\mathbf{r}_P}{\mathbf{r}_s} \qquad (11\text{-}41)$$

The equation for the local rate (per unit mass of catalyst) as developed in Chap. 9 may be expressed functionally as $\mathbf{r} = f(C,T)$, where C represents, symbolically, the concentrations of all the involved components.[1] Then Eq. (11-41) gives for $\mathbf{r}_P$

$$\mathbf{r}_P = \eta \mathbf{r}_s = \eta f(C_s, T_s) \qquad (11\text{-}42)$$

With the formulation of Eq. (11-42) the objective becomes the evaluation of η rather than $\mathbf{r}_P$. Once η is known, Eq. (11-42) gives the rate for the whole pellet in terms of the temperature and concentration at the outer surface. Then the rate can be expressed in terms of the temperature and concentration of the bulk fluid by accounting for external resistances (Chap. 10). The effectiveness factor is a function of k_e, D_e, and the rate constants associated with the chemical step at the site, i.e., the constants in the rate equations developed in Chap. 9. In the remainder of this chapter we shall develop the relationship between η and these rate parameters. In Secs. 11-6 to 11-9 isothermal conditions are assumed. With this restraint k_e is not involved, and Eq. (11-42) becomes

$$\mathbf{r}_P = \eta f(C_s) \qquad (11\text{-}43)$$

The nonisothermal problem is considered in Secs. 11-10 and 11-11.

11-6 Effectiveness Factors

Suppose that an irreversible reaction $A \rightarrow B$ is first order, so that for isothermal conditions $\mathbf{r} = f(C_A) = k_1 C_A$. Then Eq. (11-43) becomes

$$\mathbf{r}_P = \eta k_1 (C_A)_s \qquad (11\text{-}44)$$

We want to evaluate η in terms of D_e and k_1. The first step is to determine the concentration profile of A in the pellet. This is shown schematically in Fig. 11-6 for a spherical pellet (also shown is the profile from C_b to C_s). The differential equation expressing C_A vs $\mathbf{r}$ is obtained[2] by writing a mass

---

[1] For example, for the reaction $A + B \rightleftharpoons C$, Eq. (9-32) indicates that $f(C,T)$ would be

$$f(C,T) = k_s \bar{C}_m K_A K_B \frac{C_A C_B - (1/K)(C_C)}{(1 + K_A C_A + K_B C_B + K_C C_C)^2}$$

where all the k and K values are functions of temperature.

[2] This development was first presented by A. Wheeler in W. G. Frankenburg, V. I. Komarewsky, and E. K. Rideal (eds.), "Advances in Catalysis," vol. III, p. 297, Academic Press, Inc., New York, 1951; see also P. H. Emmett (ed.), "Catalysis," vol. II, p. 133, Reinhold Publishing Corporation, New York, 1955.

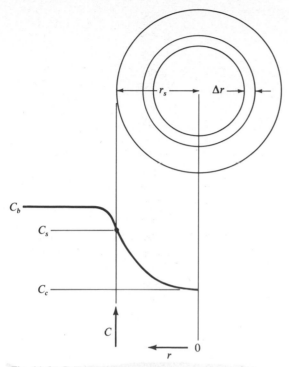

Fig. 11-6 *Reactant (A) concentration vs position for first-order reaction on a spherical catalyst pellet*

balance over the spherical-shell volume of thickness Δr (Fig. 11-6). At steady state the rate of diffusion into the element[1] less the rate of diffusion out will equal the rate of disappearance of reactant within the element. This rate will be $\rho_P k_1 C_A$ per unit volume, where ρ_P is the density of the pellet. Hence the balance may be written, omitting subscript A on C,

$$\left(-4\pi r^2 D_e \frac{dC}{dr}\right)_r - \left(-4\pi r^2 D_e \frac{dC}{dr}\right)_{r+\Delta r} = 4\pi r^2 \, \Delta r \, \rho_P k_1 C \quad (11\text{-}45)$$

If we take the limit as $\Delta r \to 0$ and assume that the effective diffusivity is independent of the concentration of reactant (see Sec. 11-2), this difference equation becomes

$$\frac{d^2 C}{dr^2} + \frac{2}{r} \frac{dC}{dr} - \frac{k_1 \rho_P}{D_e} C = 0 \quad (11\text{-}46)$$

[1] The diffusive flux into the element is given by Eq. (11-18). Note that for $A \to B$ there is equimolal counterdiffusion of A and B ($\alpha = 0$). The rate of diffusion is the product of the flux and the area, $4\pi r^2$.

At the center of the pellet symmetry requires

$$\frac{dC}{dr} = 0 \qquad \text{at } r = 0 \tag{11-47}$$

and at the outer surface

$$C = C_s \qquad \text{at } r = r_s \tag{11-48}$$

Linear differential equation (11-46) with boundary conditions (11-47) and (11-48) may be solved by conventional methods to yield

$$\frac{C}{C_s} = \frac{r_s}{r} \frac{\sinh (3\Phi_s \, r/r_s)}{\sinh 3\Phi_s} \tag{11-49}$$

where Φ_s is a dimensionless group (a Thiele-type modulus for a *spherical* pellet) defined by

$$\Phi_s = \frac{r_s}{3} \sqrt{\frac{k_1 \rho_P}{D_e}} \tag{11-50}$$

The second step is to use the concentration profile, as given by Eq. (11-49), to evaluate the rate of reaction $\mathbf{r}_P$ for the whole pellet. We have two choices for doing this: calculating the diffusion rate of reactant *into* the pellet at r_s, or integrating the local rate over the whole pellet. Choosing the first approach, we have

$$\mathbf{r}_P = \frac{1}{m_P} 4\pi r_s^2 D_e \left(\frac{dC}{dr} \right)_{r=r_s} = \frac{3}{r_s \rho_P} D_e \left(\frac{dC}{dr} \right)_{r=r_s}$$

where the mass of the pellet is $m_P = \frac{4}{3}\pi r_s^3 \rho_P$. Then, from Eq. (11-44),

$$\eta = \frac{3D_e}{r_s \rho_P k_1 C_s} \left(\frac{dC}{dr} \right)_{r=r_s} \tag{11-51}$$

Differentiating Eq. (11-49), evaluating the derivative at $r = r_s$, and substituting this into Eq. (11-51) gives

$$\eta = \frac{1}{\Phi_s} \left(\frac{1}{\tanh 3\Phi_s} - \frac{1}{3\Phi_s} \right) \tag{11-52}$$

If this equation for the effectiveness factor is used in Eq. (11-44), the desired rate for the whole pellet in terms of the *concentration at the outer surface* is

$$\mathbf{r}_P = \frac{1}{\Phi_s} \left(\frac{1}{\tanh 3\Phi_s} - \frac{1}{3\Phi_s} \right) k_1 C_s \tag{11-53}$$

Both D_e and k_1 are necessary to use Eq. (11-53). The relative importance of diffusion and chemical-reaction processes is evident from Eq.

(11-52), which gives the second lowest curve in Fig. 11-7. This curve shows that for small values of Φ_s, $\eta \rightarrow 1$. Then intraparticle mass transport has no effect on the rate per pellet; the chemical step controls the rate. From Eq. (11-50), small values of Φ_s are obtained when the pellets are small, the diffusivity is large, or the reaction is intrinsically slow (catalyst of low activity). For $\Phi_s > 5$ a good approximation for Eq. (11-52) is

$$\eta = \frac{1}{\Phi_s} \tag{11-54}$$

For such large Φ_s intraparticle diffusion has a large effect on the rate. Practically these conditions mean that diffusion into the pellet is *relatively* slow, so that reaction occurs before the reactant has diffused far into the pellet. In fact, an alternate definition for η is the fraction of the whole surface that is as active as the external surface. If $\eta \rightarrow 1$, Eq. (11-43) shows that the rate for the whole pellet is the same as the rate if all the surface were available to reactant at concentration C_s; i.e., the rate at the center is the same as the rate at the outer surface—all the surface is fully effective. In this special case the concentration profile shown in Fig. 11-6 would be horizontal, with $C = C_s$. In contrast, if $\eta \ll 1$, only the surface near the outer periphery of the pellet is effective; the concentration drops from C_s to nearly zero in a narrow region near r_s. In this case the catalyst in the central portion of the pellet is not utilized. Note that such a situation is caused by large pellets, low D_e, or high k_1. The latter factor shows that low effectiveness factors are more likely with a very active catalyst. Thus the more effective the active catalyst, the more likely it is that intrapellet diffusion resistance will reduce the rate per pellet.

Fig. 11-7 *Effectiveness factor for various pellet shapes and kinetic equations*

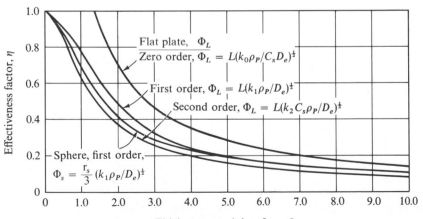

Equations (11-52) and (11-53) provide a method for accounting for intrapellet mass transport for one case, a spherical pellet, and first-order irreversible reaction. The effect of shape on the η-vs-Φ relationship has been examined by several investigators.[1] For a flat plate of catalyst *sealed from reactants on one side* and all ends,

$$\eta = \frac{\tanh \Phi_L}{\Phi_L} \tag{11-55}$$

$$\Phi_L = L \sqrt{\frac{k_1 \rho_P}{D_e}} \tag{11-56}$$

where L is the thickness of the plate. The curve for Eq. (11-55) is also shown in Fig. 11-7. The first-order curves for spherical and flat-plate geometry show little deviation—less than the error involved in evaluating k_1 and D_e. Hence the shape of the catalyst pellet is not very significant,[2] provided the different definitions of Φ_L and Φ_s must be taken into account. For example, for a spherical pellet whose radius is equal to the plate thickness, Φ_s will be equivalent to Φ_L only if $r_s = 3L$. The curve for the flat plate can also be used when both sides of the plate are exposed to reactants, provided L is one-half the plate thickness.

First-order kinetics was chosen in writing Eq. (11-46), so that an analytical solution could be obtained. Numerical solutions for η vs Φ have been developed for many other forms of rate equations.[3] Solutions include those for Langmuir-Hinshelwood equations[4] with denominator terms, as derived in Chap. 9 [e.g., Eq. (9-32)]. To illustrate the extreme effects of reaction, Wheeler[5] obtained solutions for zero- and second-order kinetics for a flat plate of catalyst, and these results are also shown in Fig. 11-7. For many catalytic reactions the rate equation is approximately represented

[1] R. Aris, *Chem. Eng. Sci.*, **6**, 262 (1957); A. Wheeler in W. G. Frankenburg, V. I. Komarewsky, and E. K. Rideal (eds.), "Advances in Catalysis," vol. III, Academic Press, Inc., New York, 1951.

[2] The curve for η vs Φ for a cylinder lies between the curves for the sphere and flat plate in Fig. 11-7. For first-order kinetics, Φ for the cylinder is defined

$$\Phi_c = (r_s/2) \sqrt{k_1 \rho_P/D_e}$$

[3] O. A. Hougen and Chu, *Chem. Eng. Sci.*, **17**, 167 (1962); P. Schneider and R. A. Mitschka, *Collection Czecho. Chem. Commun.*, **30**, 146 (1965), **31**, 1205, 3677 (1966); P. Schneider and R. A. Mitschka, *Chem. Eng. Sci.*, **21**, 455 (1965).

[4] G. W. Roberts and C. N. Satterfield, *Ind. Eng. Chem.*, *Fund. Quart.*, **4**, 288 (1965), **5**, 317, 325 (1966).

[5] A. Wheeler in W. G. Frankenburg, V. I. Komarewsky, and E. K. Rideal (eds.), "Advances in Catalysis," vol. III, Academic Press, Inc., New York, 1951. The definition of Φ_L for the second-order curve is $L(k_2 C_s \rho_P/D_e)^{\frac{1}{2}}$, where k_2 is the conventional second-order rate constant. For the zero-order case $\Phi_L = L(k_0 \rho_P/C_s D_e)^{\frac{1}{2}}$.

by a first-, second-, or intermediate-order kinetics. In these cases the curves for first- and second-order kinetics in Fig. 11-7 define a region within which the effectiveness factor will lie. For unusual situations—for example, when the desorption of a reaction product limits the rate—η values outside of this region may exist. For unusual cases the references of Hougen, Schneider, Satterfield, and their colleagues should be consulted. Krasuk and Smith[1] have presented charts of η vs Φ for flat-plate and spherical catalyst pellets for zero order, first order, and kinetics following the rate equation $kC/(1 + KC)$. The effective diffusivity is divided into pore-volume and surface contributions, so that the effect of surface migration is readily apparent, provided Eq. (11-35) is valid.

The earliest studies of diffusion and reaction in catalysts were by Thiele,[2] Damkoehler,[3] and Zeldowitsch.[4] Thiele considered the problem from the standpoint of a single cylindrical pore (see Prob. 11-9). Since the catalytic area per unit length of diffusion path does not change in a straight cylindrical pore whose walls are catalytic, the results are of the form of those for a flat plate [Eqs. (11-55) and (11-56)].

Nothing has been said as yet about reversible reactions. For the first-order case the irreversible result can be used with some modification for reversible reactions. This problem will be considered in the example which follows.

Example 11-6 Derive equations for the effectiveness factor for a first-order reversible reaction $A \rightleftharpoons B$ at isothermal conditions, for a spherical catalyst pellet.

Solution It was shown by Eq. (2-74) that the rate equation for a reversible first-order reaction could be written

$$\mathbf{r} = k_R(C - C_{eq}) \tag{A}$$

where C_{eq} is the equilibrium concentration of reactant at the temperature involved, and k_R is related to the forward-rate constant k_1 and the equilibrium constant K by

$$k_R = \frac{k_1(K + 1)}{K} \tag{B}$$

Since Eq. (A) is to be applied to a catalytic reaction, the rate is expressed as g moles/(sec)(g catalyst) and k_R has the dimensions $cm^3/(sec)(g\ catalyst)$.

The mass balance of reactant on a spherical shell (see Fig. 11-6) will be the same as Eq. (11-45) except for the reaction term; that is,

$$\left(-4\pi r^2 D_e \frac{dC}{dr}\right)_r - \left(-4\pi r^2 D_e \frac{dC}{dr}\right)_{r+\Delta r} = 4\pi r^2\,\Delta r\,\rho_P k_R(C - C_{eq})$$

[1] J. H. Krasuk and J. M. Smith, *Ind. Eng. Chem., Fund. Quart.*, **4**, 102 (1965).

[2] E. W. Thiele, *Ind. Eng. Chem.*, **31**, 916 (1939).

[3] G. Damkoehler, *Chem. Eng.*, **3**, 430 (1937).

[4] J. B. Zeldowitsch, *Acta Physicochim. U.R.S.S.*, **10**, 583 (1939).

Taking the limit as $\Delta r \rightarrow 0$ we have

$$\frac{d^2C}{dr^2} + \frac{2}{r}\frac{dC}{dr} - \frac{k_R \rho_P}{D_e}(C - C_{eq}) = 0 \tag{C}$$

If C is replaced with the variable $\mathsf{C} = C - C_{eq}$, Eq. (C) becomes

$$\frac{d^2\mathsf{C}}{dr^2} + \frac{2}{r}\frac{d\mathsf{C}}{dr} - \frac{k_R \rho_P}{D_e}\mathsf{C} = 0 \tag{D}$$

Similarly, the boundary conditions [see Eqs. (11-47) and (11-48)] are

$$\frac{d\mathsf{C}}{dr} = 0 \qquad \text{at } r = 0 \tag{E}$$

$$\mathsf{C} = C_s - C_{eq} = \mathsf{C}_s \qquad \text{at } r = r_s \tag{F}$$

Equations (B) to (D) are the same as Eqs. (11-46) to (11-48), with C replacing C and k_R replacing k_1. Hence the solution for the effectiveness factor will be identical with Eq. (11-52), but Eq. (11-50) for Φ_s becomes

$$\Phi'_s = \frac{r_s}{3}\sqrt{\frac{k_1(K+1)\rho_P}{KD_e}} \tag{11-57}$$

These results show that the first-order curves in Fig. 11-7 can be used for reversible as well as irreversible reactions, provided k_1 in the definition of Φ_s or Φ_L is replaced by $k_1(K+1)/K$. Since $(K+1)/K$ is greater than unity, Φ will be greater, and η less, for reversible reactions than for irreversible ones, other conditions being the same.

11-7 The Significance of Intrapellet Diffusion: Evaluation of the Effectiveness Factor

Isothermal effectiveness factors for practical reactions cover a wide range, from as low as 0.01 to unity. With normal pellet sizes ($\frac{1}{8}$ to $\frac{1}{2}$ in.) η is 0.7 to 1.0 for intrinsically slow reactions, such as the ammonia synthesis, and of the order of $\eta = 0.1$ for fast reactions, such as some hydrogenations of unsaturated hydrocarbons. Satterfield[1] and Sherwood[2] have summarized much of the experimental data for effectiveness factors for various reactions, temperatures, and pellet sizes. For reactor design it is important to be able to answer these questions:

1.   Should intrapellet diffusion resistance be considered in evaluating the global rate? That is, is η significantly less than unity?

---

[1] C. N. Satterfield, "Mass Transfer in Heterogeneous Catalysis," pp. 152–156, Massachusetts Institute of Technology Press, Cambridge, Mass., 1970.

[2] C. N. Satterfield and T. K. Sherwood, "The Role of Diffusion in Catalysis," pp. 72–75, Addison-Wesley Publishing Company, Reading, Mass., 1963.

2.   If $\eta < 1$, how can η be evaluated from a minimum of experimental data?

We now use the results of Sec. 11-6 to formulate answers to these questions.

Suppose that the rate $\mathbf{r}_P$ is measured at a given bulk concentration of reactant. Suppose also that either the external resistance is negligible, or the surface concentration C_s has been evaluated from the bulk value by the methods discussed in Chap. 10. Weisz[1] has provided a criterion for deciding, from these measurements and D_e, whether intrapellet diffusion may be disregarded. The basic premise is that if $\Phi_s \leq \frac{1}{3}$, then η is not much less than unity (Fig. 11-7 indicates that η will be greater than 0.9 for $\Phi_s \leq \frac{1}{3}$). Equation (11-50) shows that the criterion may be written

$$r_s \sqrt{\frac{k_1 \rho_P}{D_e}} \leq 1$$

or

$$r_s^2 \frac{k_1 \rho_P}{D_e} \leq 1 \tag{11-58}$$

The unknown rate constant k_1 can be eliminated in favor of the measured rate $\mathbf{r}_P$ from Eq. (11-44), noting that $\eta \rightarrow 1.0$. In terms of $\mathbf{r}_P$ Eq. (11-58) becomes

$$r_s^2 \frac{\mathbf{r}_P \rho_P}{C_s D_e} \leq 1 \tag{11-59}$$

The usefulness of Eq. (11-59) stems from the fact that the curves for η vs Φ for first- and higher-order reactions in Fig. 11-7 would be nearly coincident at $\Phi_s \leq \frac{1}{3}$. Hence Eq. (11-59) is satisfactory as an approximate criterion for most catalytic kinetics even though it was derived for a first-order case.

Example 11-7   The rate of isomerization of n-butane with a silica-alumina catalyst is measured at 5 atm and 50°C in a laboratory reactor with high turbulence in the gas phase surrounding the catalyst pellets. Turbulence ensures that external-diffusion resistances are negligible, and so $C_s = C_b$. Kinetic studies indicate that the rate is first order and reversible. At 50°C the equilibrium conversion is 85%. The effective diffusivity is 0.08 cm²/sec at reaction conditions, and the density of the catalyst pellets is 1.0 g/cm³, regardless of size. The measured, global rates when pure n-butane surrounds the pellets are as follows:

d_P, in.	1/8	1/4	3/8
$\mathbf{r}_P$, g moles/ (sec)(g catalyst)	4.85×10^{-4}	4.01×10^{-4}	3.54×10^{-4}

[1] P. B. Weisz, Z. Phys. Chem., **11**, 1 (1957).

(a) To reduce pressure drop in the proposed fixed-bed reactor it is desirable to use the maximum pellet size for which there will be no reduction in the global rate due to intrapellet resistances. The heat of isomerization is low enough that the whole pellet is at 50°C. What is the largest size pellet that may be used? (b) Calculate the effectiveness factor for each size.

Solution (a) In Example 11-6 the proper definition of Φ'_s for a reversible first-order reaction was shown to be given by Eq. (11-57). From this definition, the criterion for $\eta \to 1$ is

$$\left(\frac{r_s}{3}\right)^2 \frac{k_1(K+1)\rho_P}{KD_e} \leq \left(\frac{1}{3}\right)^2$$

or

$$r_s^2 \frac{k_1(K+1)\rho_P}{KD_e} \leq 1 \tag{A}$$

The rate for the whole pellet (per unit mass of catalyst) for the first-order reversible case is the same as Eq. (11-44), with $k_R(C_s - C_{eq})$ replacing $k_1 C_s$:

$$\mathbf{r}_P = \eta k_R(C_s - C_{eq}) = \eta \frac{k_1(K+1)}{K}(C_s - C_{eq}) \tag{11-60}$$

Combining Eqs. (A) and (11-60) to eliminate k_1, and noting that $\eta \to 1$, we obtain

$$r_s^2 \frac{\mathbf{r}_P \rho_P}{D_e(C_s - C_{eq})} \leq 1 \tag{11-61}$$

Equation (11-61) is the proper criterion to use for reversible reactions rather than Eq. (11-59).

For an equilibrium conversion of 85%,

$$\frac{C_s - C_{eq}}{C_s} = 0.85$$

At 5 atm and 50°C

$$C_s - C_{eq} = 0.85 \frac{p}{R_g T} = 0.85 \frac{5}{82(323)} = 1.60 \times 10^{-4} \text{ g mole/cm}^3$$

With numerical values, Eq. (11-61) becomes

$$r_s^2 \, \mathbf{r}_P \frac{1.0}{0.08(1.60 \times 10^{-4})} = 7.82 \times 10^4 \, r_s^2 \, \mathbf{r}_P \leq 1 \tag{B}$$

For $\frac{1}{8}$-in. pellets, $r_s = \frac{1}{8}(\frac{1}{2})(2.54) = 0.159$ cm. Hence

$$r_s^2 \frac{\mathbf{r}_P \rho_P}{D_e(C_s - C_{eq})} = 7.82 \times 10^4 (0.159)^2 (4.85 \times 10^{-4}) = 0.95$$

For the other two sizes

$$\frac{r_s^2 \mathbf{r}_P \rho_P}{D_e(C_s - C_{eq})} = 3.2 \qquad \frac{1}{4}\text{-in. pellets}$$

$$\frac{r_s^2 \mathbf{r}_P \rho_P}{D_e(C_s - C_{eq})} = 6.3 \qquad \frac{3}{8}\text{-in. pellets}$$

The $\frac{1}{8}$-in. pellets are the largest for which intrapellet diffusion has a negligible effect on the rate.

(b) To calculate η one relation between Φ'_s and η is obtained from Eq. (11-57) of Example 11-6 and a second is Eq. (11-52), or the Φ_s curve in Fig. 11-7. From Eqs. (11-57) and (11-60) k_1 can be eliminated, giving

$$\Phi'_s = \frac{r_s}{3} \sqrt{\frac{\mathbf{r}_P \rho_P}{\eta D_e(C_s - C_{eq})}} \tag{11-62}$$

The only unknowns in Eq. (11-62) are η and Φ'_s for a given pellet size. For the $\frac{3}{8}$-in. pellets

$$\Phi'_s = \frac{0.476}{3} \sqrt{\frac{3.54 \times 10^{-4}(1.0)}{0.08(1.60 \times 10^{-4})\eta}} = 0.158 \sqrt{\frac{27.6}{\eta}}$$

Simultaneous solution of this expression and Eq. (11-52) yields

$$\Phi'_s = 1.00 \qquad \eta = 0.68$$

Results for the other sizes, obtained the same way, are shown in Table 11-4.

Table 11-4

d_p, in.	Φ'_s	η
1/8	0.33	0.93
1/4	0.67	0.77
3/8	1.00	0.68

Example 11-7 illustrates one of the problems in scale-up of catalytic reactors. The results showed that for all but $\frac{1}{8}$-in. pellets intrapellet diffusion significantly reduced the global rate of reaction. If this reduction were not considered, erroneous design could result. For example, suppose the laboratory kinetic studies to determine a rate equation were made with $\frac{1}{8}$-in. pellets. Then suppose it was decided to use $\frac{3}{8}$-in. pellets in the commercial reactor to reduce the pressure drop through the bed. If the rate equation were used for the $\frac{3}{8}$-in. pellets without modification, the rate would be erroneously high. At the conditions of part (b) of Example 11-7 the correct $\mathbf{r}_P$ would be only 0.68/0.93, or 73% of the rate measured with $\frac{1}{8}$-in. pellets.

With respect to the second question of this section, part (b) of Example

11-7 illustrated one method of evaluating η when intrapellet mass transfer is important. In addition to an experimentally determined rate, it was necessary to know the effective diffusivity of the pellet. The need for D_e may be eliminated by making rate measurements for two or more sizes of pellets, provided D_e is the same for all sizes. To show this, note from Eq. (11-44) that the ratio of the rates for two sizes 1 and 2 is

$$\frac{(\mathbf{r}_P)_2}{(\mathbf{r}_P)_1} = \frac{\eta_2}{\eta_1} \qquad\qquad (11\text{-}63)$$

Also, from Eq. (11-50),

$$\frac{(\Phi_s)_2}{(\Phi_s)_1} = \frac{(r_s)_2}{(r_s)_1} \qquad\qquad (11\text{-}64)$$

Furthermore, Eq. (11-52) gives a relation between η and Φ_s which must be satisfied for both pellets. There are four relations between the four unknowns, η_1, $(\Phi_s)_1$, η_2, $(\Phi_s)_2$. Solution by trial is easiest. One method[1] is to assume a value of η_2 and calculate η_1 from Eq. (11-63). Determine $(\Phi_s)_1$ from Eq. (11-52); then $(\Phi_s)_2$ can be calculated from Eq. (11-64). Finally, applying Eq. (11-52) for pellet 2 gives a revised value of η_2. The calculations are continued until the initial and calculated values of η_2 agree. The method is not valid for large Φ_s, for then Eq. (11-52) reduces to Eq. (11-54), and combination with Eqs. (11-63) and (11-64) shows that the rate is inversely proportional to r_s *for all sizes.*

If one of the pellets for which $\mathbf{r}_P$ is measured is very small (e.g., powder form), η equals 1.0 for that size. Then the ratio of Eq. (11-43) for the small pellet, $(\mathbf{r}_P)_1$, and a larger size pellet, $(\mathbf{r}_P)_2$, gives, for any reaction order,

$$\frac{(\mathbf{r}_P)_2}{(\mathbf{r}_P)_1} = \frac{\eta_2 f(C_s)}{1.0 f(C_s)}$$

or

$$\eta_2 = \frac{(\mathbf{r}_P)_2}{(\mathbf{r}_P)_1} \qquad\qquad (11\text{-}65)$$

Hence the effectiveness factor for a given pellet can be obtained by measuring the rate for the pellet and for a small particle size of the same catalyst at the same concentration of reactant.

The rate measurement for the small particles determines the rate of the chemical reaction at the catalyst site, i.e., without intrapellet-diffusion resistance. Then the rate constant k can be calculated directly. Thus it is

[1] C. N. Satterfield and T. K. Sherwood, "The Role of Diffusion in Catalysis," Addison-Wesley Publishing Company, Reading, Mass., 1963.

not necessary that the measurements for the small particles and the pellet be at the same concentration of reactant, as required by Eq. (11-65). Consider an irreversible first-order reaction. From Eq. (11-44),

$$k_1 = \frac{(r_P)_1}{(C_s)_1} \tag{11-66}$$

since $\eta_1 = 1.0$. Using this value of k_1 in Eq. (11-44), applied to the pellet, we have

$$\eta_2 = \frac{(r_P)_2}{(C_s)_2 k_1} = \frac{(r_P/C_s)_2}{(r_P/C_s)_1} \tag{11-67}$$

11-8 Experimental and Calculated Effectiveness Factors

When the rate is measured for a catalyst pellet and for small particles, and the diffusivity is also measured or predicted, it is possible to obtain both an experimental and a calculated result for η. For example, for a first-order reaction Eq. (11-67) gives η_{exp} directly. Then the rate measured for the small particles can be used in Eq. (11-66) to obtain k_1. Provided D_e is known, Φ can be evaluated from Eq. (11-50) for a spherical pellet or from Eq. (11-56) for a flat plate of catalyst. Then η_{calc} is obtained from the proper curve in Fig. 11-7. Comparison of the experimental and calculated values is an overall measure of the accuracy of the rate data, effective diffusivity, and the assumption that the intrinsic rate of reaction (or catalyst activity) is the same for the pellet and the small particles. Example 11-8 illustrates the calculations and results for a flat-plate pellet of NiO catalyst, on an alumina carrier, used for the ortho-para-hydrogen conversion.

Example 11-8[1]   The pellet reactor is shown in Fig. 11-8. The reaction gases were exposed to one face of the cylindrical disk (1 in. in diameter and $\frac{1}{4}$ in. in depth), while the other face and the cylindrical surface were sealed. Vigorous turbulence near the exposed face ensured uniform composition in the gas region and eliminated external diffusion resistance. The reaction

$$o\text{-}H_2 \rightleftharpoons p\text{-}H_2$$

has a very low heat of reaction, and the whole reactor was enclosed in a liquid nitrogen bath. Thus isothermal conditions obtain at $-196°C$.

Rates were measured at $-196°C$ and 1 atm pressure for pellets of three densities. The rate of reaction was also measured for the catalyst in the form of 60-micron (average size) particles. With this small size $\Phi \ll 1$, so that $\eta = 1.0$. The rate data and pellet properties are given in Table 11-5.

[1] This example is taken from the data of M. R. Rao, N. Wakao, and J. M. Smith, *Ind. Eng. Chem.*, **3**, 127 (1964).

Table 11-5   Catalyst and rate data for ortho-hydrogen conversion on NiO/Al_2O_3

CATALYST PARTICLES

$d_P = 60$ microns
$S_g = 278$ m^2/g
$V_g = 0.44$ cm^3/g
$\rho_S = 2.63$ g/cm^3
$\rho_P = 2.24$ g/cm^3
$\bar{a}_\mu = 29$ Å (from integration of pore-volume distribution)
$k_R = 0.688$ cm^3/(sec)(g), calculated from the measured rate $(r_P)_1$, using an expression analogous to Eq. (11-62), but applicable to a *reversible* first-order reaction. This can be obtained by applying Eq. (11-60) to the small particles. Since $\eta_1 = 1.0$, $k_R = (r_P)_1/(C_s - C_{eq})$.

CATALYST PELLETS

ρ_P, g/cm^3	ε_M	Macropore radius $\bar{a}_M$, Å	$r_P/(C_s - C_{eq})$ cm^3/(sec)(g)
1.09	0.48	2,100	0.186
1.33	0.37	1,690	0.129
1.58	0.33	1,270	0.109

NOTE: The physical-property data were obtained from pore-size distribution and surface-area measurements, as described in Chap. 8.

Solution   Equation (11-67) is for an irreversible reaction, and so it cannot be used to calculate η_{exp} for the ortho-hydrogen reaction. The proper equation is obtained by applying Eq. (11-60) to the pellet. Solving for η gives

$$\eta_{exp} = \frac{(r_P)_2}{k_R(C_s - C_{eq})_2} = \frac{[r_P/(C_s - C_{eq})]_2}{[r_P/(C_s - C_{eq})]_1}$$

From this equation and the data given in Table 11-5, η_{exp} can be immediately evaluated. The results are:

ρ_P	η_{exp}
1.09	$\dfrac{0.186}{0.688} = 0.27$
1.33	$\dfrac{0.129}{0.688} = 0.19$
1.58	$\dfrac{0.109}{0.688} = 0.16$

To obtain η_{calc} we must estimate the effective diffusivity. Since the macropores are much larger than the micropores (see $\bar{a}_M$ and $\bar{a}_\mu$ in Table 11-5), it is safe to assume that diffusion is predominantly through the macropores. Then, according to the random-pore model [Eqs. (11-25) and (11-26)],

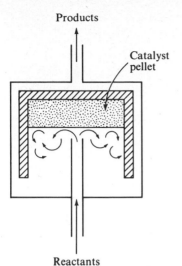

Products

Catalyst pellet

Reactants

Fig. 11-8 *Reactor for a single flat-plate catalyst pellet*

$$D_e = \varepsilon_M{}^2 \frac{1}{1/\mathscr{D}_{H_2} + 1/(\bar{\mathscr{D}}_K)_{H_2}} \tag{A}$$

For hydrogen at $-196°C$ and 1 atm, $\mathscr{D}_{H_2} = 0.14 \text{ cm}^2/\text{sec}$. Using Eq. (11-17) and the $\bar{a}_M$ values from Table 11-5, we obtain:

ρ_P	1.09	1.33	1.58
$(\bar{\mathscr{D}}_K)_{H_2}$, cm^2/sec	1.28	1.03	0.77

From Eq. (A), the effective diffusivity is, for $\rho_P = 1.09$,

$$D_e = 0.48^2 \frac{1}{1/0.14 + 1/1.28} = 0.029 \text{ cm}^2/\text{sec}$$

The results for the other pellet densities are given in the second column of Table 11-6. The Thiele-type modulus is calculated from Eq. (11-56), modified for a reversible reaction by substituting k_R for k_1:

$$\Phi'_L = L \sqrt{\frac{k_R \rho_P}{D_e}} = \frac{1}{4}(2.54) \sqrt{\frac{0.688(1.09)}{0.029}} = 3.2$$

Table 11-6

ρ_P	D_e, cm^2/sec	Φ'_L	η_{calc}	η_{exp}
1.09	0.029	3.2	0.30	0.27
1.33	0.017	4.6	0.21	0.19
1.58	0.013	5.7	0.17	0.16

From the first-order Φ_L-vs-η curve in Fig. 11-7 for a flat plate, $\eta_{calc} = 0.30$. The results for all three pellets are given in the last columns of the table.

The calculated and experimental effectiveness factors agreed well with each other in Example 11-8. Note that the calculated η required the rate for the small particles as the only experimental data. Hence the method offers an attractive procedure for predicting the rate for any size of pellet. However, there are data for other catalysts and reactions which suggest less favorable agreement. Results for the same reaction with a catalyst of NiO supported on silica gel showed good coincidence between η_{exp} and η_{calc}.[1] With Vycor as a carrier the agreement for effectiveness factors is reasonably good, even though calculated (by the random-pore model) and experimentally measured D_e were in poor accord.[2] Relatively large deviations in D_e cause small differences in η. Otani and Smith[3] applied the method to the reaction

$$CO + \tfrac{1}{2}O_2 \rightarrow CO_2$$

using spherical catalyst pellets of NiO on Al_2O_3. The calculated η were 50 to 100% higher than the experimental results, depending on the temperature. The reasons for the large deviation were obscure, but a decrease in catalyst activity may have been caused by plugging of some of the pores in the particles when the pellets were formed. Such results emphasize that the method of predicting η depends on assuming that the intrinsic rate of reaction is the same for the surface in the small particles as for the surface in the pellet. Another possible error in the method is the variation in effective diffusivity within a pellet. It has been found[4] that D_e can vary with position near the surface of a pellet, presumably because of skin effects caused by nonuniform stresses in the pelleting process.

11-9 The Effect of Intrapellet Mass Transfer on Observed Rate

In Sec. 10-1 we saw that neglecting external resistances could lead to misleading conclusions about reaction order and activation energy. Similar errors may occur when intrapellet mass transfer is neglected. Consider the situation where $\Phi_s > 5$. In this region intrapellet transport has a strong effect on the rate; Fig. 11-7 shows that η is less than about 0.2. From Eqs. (11-54) and (11-50) for a first-order reaction

---

[1] M. R. Rao and J. M. Smith, *AIChE J.*, **9**, 445 (1963).

[2] M. R. Rao and J. M. Smith, *AIChE J.*, **10**, 293 (1964).

[3] Seiya Otani and J. M. Smith, *J. Catalysis*, **5**, 332 (1966).

[4] C. N. Satterfield and S. K. Saraf, *Ind. Eng. Chem., Fund. Quart.*, **4**, 451 (1965).

$$\eta = \frac{3}{r_s} \sqrt{\frac{D_e}{k_1 \rho_P}}$$

and for second-order kinetics

$$\eta = \frac{3}{r_s} \sqrt{\frac{D_e}{k_2 \rho_P C_s}}$$

Using these expressions for η in the equations for the rates for the whole pellet [for example, Eq. (11-44)], we obtain

$$\mathbf{r}_P = \frac{3C_s}{r_s} \sqrt{\frac{D_e k_1}{\rho_P}} \qquad \text{first order}$$

$$\mathbf{r}_P = \frac{3C_s^{\frac{3}{2}}}{r_s} \sqrt{\frac{D_e k_2}{\rho_P}} \qquad \text{second order}$$

If the rate constants are expressed as Arrhenius functions of temperature, $k = \mathbf{A}\,e^{-E/R_g T}$, then

$$\mathbf{r}_P = \frac{3\mathbf{A}_1^{\frac{1}{2}} C_s}{r_s} \left(\frac{D_e}{\rho_P}\right)^{\frac{1}{2}} e^{-E/2R_g T} \qquad \text{first order} \qquad (11\text{-}68)$$

$$\mathbf{r}_P = \frac{3\mathbf{A}_2^{\frac{1}{2}} C_s^{\frac{3}{2}}}{r_s} \left(\frac{D_e}{\rho_P}\right)^{\frac{1}{2}} e^{-E/2R_g T} \qquad \text{second order} \qquad (11\text{-}69)$$

These equations give the correct influence of concentration and temperature when intrapellet diffusion is important.

Now suppose intrapellet resistance is neglected. The rate for a first-order reaction would be correlated in terms of the apparent activation energy E' by the expression

$$\mathbf{r}_P = \mathbf{A}_1\, e^{-E'/R_g T}\, C_s \qquad (11\text{-}70)$$

Comparison of Eqs. (11-70) and (11-68) indicates that the apparent activation energy determined from Eq. (11-70) would be one-half the true value, E. The measured rate, when plotted in Arrhenius coordinates, would appear as in Fig. 11-9. At low enough temperatures the data would determine a line with a slope equal to $-E/R_g$, because η would approach unity. However, at high enough temperatures intrapellet diffusion would be important; Eq. (11-68) would be applicable, and a line with a slope of $-\frac{1}{2}E/R_g$ would result. These conclusions would be the same regardless of the reaction order.

Equations (11-68) and (11-69) show that the rate $\mathbf{r}_P$ (remember that it is for the whole pellet, but per unit mass of catalyst) is inversely propor-

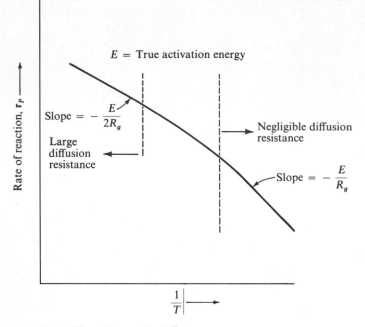

Fig. 11-9   *Effect of intrapellet diffusion on apparent activation energy*

tional to pellet size r_s. If intrapellet diffusion is neglected, the rate is independent of pellet size, as illustrated by Eq. (11-70).

If diffusion in the pores is of the Knudsen type, D_e is independent of pressure, and therefore of concentration. Then Eq. (11-68) indicates that first-order kinetics would be observed, even though intrapellet diffusion is important. However, a second-order reaction would appear to be of order $\frac{3}{2}$. If bulk diffusion were involved, $D_e \approx 1/p$. Hence when concentration is varied by changing the pressure, a first-order reaction would appear to be of order $\frac{1}{2}$. Similarly, Eq. (11-69) shows that a second-order reaction would appear to be first order.

These effects have been observed in many instances. In particular, the flattening of the line on an Arrhenius plot (as in Fig. 11-9) is frequently found when heterogeneous reactions are studied over a wide temperature range.

MASS AND HEAT TRANSFER WITH REACTION

When the heat of reaction is large, intrapellet temperature gradients may have a larger effect on the rate per pellet than concentration gradients.

Even when ΔH is low, the center and surface temperatures may differ appreciably, because catalyst pellets have low thermal conductivities (Sec. 11-5). The combined effect of mass and heat transfer on $\mathbf{r}_P$ can still be represented by the general definition of the effectiveness factor, according to Eq. (11-41). Hence Eq. (11-42) may be used to find $\mathbf{r}_P$, provided η is the nonisothermal effectiveness factor. The nonisothermal η may be evaluated in the same way as the isothermal η, except that an energy balance must be combined with the mass balance.

11-10 Nonisothermal Effectiveness Factors

Consider the same irreversible first-order reaction $A \rightarrow B$ used in Sec. 11-6 to obtain the isothermal η. If the effect of temperature on D_e is neglected, the differential mass balance and boundary conditions, Eqs. (11-46) to (11-48), are still applicable. The energy balance over the spherical shell of thickness Δr (see Fig. 11-6) is

$$\left(-4\pi r^2 k_e \frac{dT}{dr}\right)_r - \left(-4\pi r^2 k_e \frac{dT}{dr}\right)_{r+\Delta r} = 4\pi r^2 \, \Delta r \, \rho_P k_1 C \, \Delta H$$

$$\text{(11-71)}$$

Taking the limit as $\Delta r \rightarrow 0$ and assuming that k_e is independent of temperature, we find

$$\frac{d^2 T}{dr^2} + \frac{2}{r} \frac{dT}{dr} - \frac{k_1 \rho_P C}{k_e} \Delta H = 0 \qquad \text{(11-72)}$$

with boundary conditions

$$\frac{dT}{dr} = 0 \qquad \text{at } r = 0 \qquad\qquad\qquad \text{(11-73)}$$

$$T = T_s \qquad \text{at } r = r_s \qquad\qquad\qquad \text{(11-74)}$$

The solution of Eqs. (11-46) to (11-48) and (11-72) to (11-74) gives the concentration and temperature profiles within the pellet. A numerical solution is necessary because Eqs. (11-46) and (11-72) are coupled through the nonlinear dependence of k_1 on temperature; $k_1 = A e^{-E/R_g T}$. Nevertheless, the similarity of the nonreaction terms in the two differential equations does permit an analytical relation between concentration of reactant and temperature at any point in the pellet. Thus eliminating $k_1 \rho_P C$ from the two equations yields

$$D_e \left(\frac{d^2 C}{dr^2} + \frac{2}{r} \frac{dC}{dr}\right) = \frac{k_e}{\Delta H} \left(\frac{d^2 T}{dr^2} + \frac{2}{r} \frac{dT}{dr}\right)$$

or

$$D_e \frac{d}{dr}\left(r^2 \frac{dC}{dr}\right) = \frac{k_e}{\Delta H}\frac{d}{dr}\left(r^2 \frac{dT}{dr}\right) \tag{11-75}$$

If we integrate this equation once, using Eqs. (11-47) and (11-73) for the boundary conditions, and then integrate a second time, using Eqs. (11-48) and (11-74), we obtain

$$T - T_s = \frac{\Delta H\, D_e}{k_e}(C - C_s) \tag{11-76}$$

This result, originally derived by Damkoehler,[1] is not restricted to first-order kinetics, but is valid for any form of rate expression, since the rate term was eliminated in forming Eq. (11-75). The maximum temperature rise in a pellet would occur when the reactant has been consumed by the time it diffuses to the center. Applying Eq. (11-76) for $C = 0$ gives

$$(T_c - T_s)_{max} = -\frac{\Delta H\, D_e}{k_e}C_s \tag{11-77}$$

Equation (11-77) shows that the maximum temperature rise depends on the heat of reaction, transport properties of the pellet, and the surface concentration of reactant. It permits a simple method of estimating whether intrapellet temperature differences are significant (see Example 11-9).

Let us return to the nonisothermal effectiveness factor. Weisz and Hicks solved Eqs. (11-46) and (11-72) numerically[2] to obtain the concentration profile within the pellet. Then η was obtained from Eq. (11-51), which is not limited to isothermal conditions, provided k_1 is evaluated at the surface temperature. The results expressed η as a function of three dimensionless parameters:

1. The Thiele-type modulus,

$$3(\Phi_s)_s = r_s \sqrt{\frac{(k_1)_s \rho_P}{D_e}} \tag{11-78}$$

Note that Φ_s is evaluated at the surface temperature; that is, $(k_1)_s$ in Eq. (11-78) is the rate constant at T_s.

[1] G. Damkoehler, *Z. Phys. Chem.*, **A193**, 16 (1943).

[2] P. B. Weisz and J. S. Hicks, *Chem. Eng. Sci.*, **17**, 265 (1962). Actually, Eq. (11-76) permitted expression of k_1 as a function of concentration rather than of temperature ($k_1 = A\, e^{-E/R_g T}$). This uncoupled Eqs. (11-46) and (11-72), so that only Eq. (11-46) needed to be solved.

2.   The Arrhenius number,

$$\gamma = \frac{E}{R_g T_s} \tag{11-79}$$

3.   A heat-of-reaction parameter,

$$\beta = \frac{(-\Delta H)D_e C_s}{k_e T_s} \tag{11-80}$$

Figure 11-10 gives η as a function of $(\Phi_s)_s$ and β for $\gamma = 20$, which is in the middle of the practical range of γ. Weisz and Hicks give similar figures for $\gamma = 10, 30,$ and 40. The curve for $\beta = 0$ corresponds to isothermal operation $(\Delta H = 0)$ and is identical with the curve for a spherical pellet in Fig. 11-7.

For an exothermic reaction (positive β) the temperature rises going into the pellet. The increase in rate of reaction accompanying the tempera-

Fig. 11-10 *Nonisothermal effectiveness factors for first-order reactions in spherical catalyst pellets*

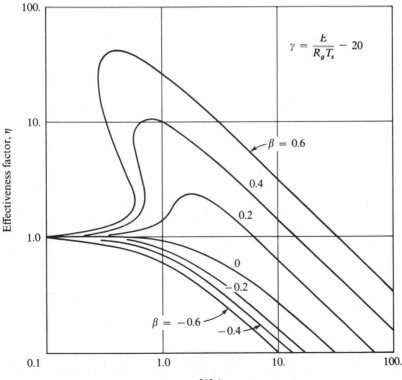

ture rise can more than offset the decrease in rate due to drop in reactant concentration. Then η values are greater than unity. While $\eta > 1$ increases the rate per pellet, and therefore the production per unit mass of catalyst, there may also be some disadvantages. With large η there will be a large increase in temperature toward the center of the pellet, resulting in sintering and catalyst deactivation. The desired product may be subject to further reaction to unwanted product, or undesirable side reactions may occur. If these reactions have higher activation energies than the desired reaction, the rise in temperature would reduce selectivity.

For an endothermic reaction there is a decrease in temperature and rate into the pellet. Hence η is always less than unity. Since the rate decreases with drop in temperature, the effect of heat-transfer resistance is diminished. Therefore the curves for various β are closer together for the endothermic case. In fact, the decrease in rate going into the pellet for endothermic reactions means that mass transfer is of little importance. It has been shown[1] that in many endothermic cases it is satisfactory to use a *thermal* effectiveness factor. Such thermal η neglects intrapellet mass transport; that is, η is obtained by solution of Eq. (11-72), taking $C = C_s$.

Figure 11-10 indicates that for $\beta > 0.3$ and $3(\Phi_s)_s < 1.0$, up to three values of η exist for a single set of γ, $3(\Phi_s)_s$, and β. Such behavior can be explained by noting that the equality of heat evolved due to reaction and heat transferred by conduction in the pellet (such equality must exist for steady-state operation) can occur with different temperature profiles. Thus the solution giving the highest η for a given β curve in Fig. 11-10 would correspond to a steep temperature profile in the pellet; physical processes dominate the rate for the whole pellet. The solution for the lowest η, which is near unity, corresponds to small temperature gradients in the pellet; the rate is controlled by the chemical-reaction step. The intermediate η is identified as a metastable state, similar to the metastable conditions described in Sec. 10-4.

Carberry[2] showed that for $(\Phi_s)_s > 2.5$ the η-vs-$(\Phi_s)_s$ relationship could be characterized approximately by the single parameter $\beta\gamma$, rather than by γ and β separately. For example, for a first-order irreversible reaction he found

$$\eta = \frac{1}{(\Phi_s)_s} e^{\beta\gamma/5} \qquad \text{for } (\Phi_s)_s > 2.5 \tag{11-81}$$

where

$$\beta\gamma = \frac{(-\Delta H)D_eC_s}{k_eT_s} \frac{E}{R_gT_s} \tag{11-82}$$

Carberry also obtained results for a second-order rate equation.

[1] J. A. Maymo, R. E. Cunningham, and J. M. Smith, *Ind. Eng. Chem., Fund. Quart.*, **5**, 280 (1966).
[2] J. J. Carberry, *AIChE J.*, **7**, 350 (1961).

In Sec. 11-7, Eq. (11-59) was developed as a criterion for deciding if intrapellet mass transport had a significant effect on the rate. Weisz and Hicks[1] have extended the analysis to the problem of combined mass and energy transport. For a first-order irreversible reaction the criterion may be expressed as

$$\Theta = r_s^2 \, \frac{\mathbf{r}_P \rho_P}{C_s D_e} \, e^{\gamma \beta / (1 + \beta)} \leq 1 \tag{11-83}$$

where γ and β are defined by Eqs. (11-79) and (11-80). When $\beta = 0$ (isothermal conditions) this criterion reduces to Eq. (11-59). As in that equation, $\mathbf{r}_P$ is the measured rate. Thus the significance of intrapellet gradients can be evaluated from measurable or known properties. If $\Theta \leq 1$, the nonisothermal effectiveness factor will be near unity, so that intrapellet gradients can be neglected.

11-11 Experimental Nonisothermal Effectiveness Factors

The magnitude of intrapellet temperature differences and nonisothermal η are illustrated by the following examples, based on experimental measurements for specific systems.

Example 11-9 Rate data for the reaction $H_2 + \frac{1}{2}O_2 \rightarrow H_2O$ have been measured[2] for single catalyst pellets (1.86-cm diameter) of platinum on Al_2O_3. Catalyst properties, k_e, D_e, and center and surface temperatures were also evaluated. The rate was obtained in a stirred-tank reactor in which the pellet was surrounded by well-mixed reaction gases. In one run the data were as follows:

Gas temperature $t_g = 90°C$
Average surface temperature $t_s = 101°C$
Pellet-center temperature $t_c = 148°C$
Density of catalyst particles in pellet[3] = 0.0602 g catalyst/cm³
Mole fraction oxygen at pellet surface = 0.0527
Effective diffusivity of pellet = 0.166 cm²/sec
Effective thermal conductivity = 6.2×10^{-4} cal/(sec)(cm)(°C)
Rate of reaction $\mathbf{r}_P = 2.49 \times 10^{-5}$ g mole O_2/(g catalyst)(sec)
Total pressure $p_t = 1$ atm

Rate data were also obtained for the small (80 to 250 mesh) particles from which the pellets were prepared. The results, expressed as the rate of oxygen consumption, g moles/(g catalyst)(sec), were correlated by

[1] P. B. Weisz and J. S. Hicks, *Chem. Eng. Sci.*, **17**, 265 (1962).

[2] J. A. Maymo and J. M. Smith, *AIChE J.*, **12**, 845 (1966).

[3] The spherical pellet was prepared by compressing a mixture of inert Al_2O_3 particles and Al_2O_3 particles containing 0.005 wt % platinum. The ratio of inert to total particles was 0.1. The density of the pellet was 0.602 g/cm³, so that the density of catalyst particles in the pellet was 0.0602.

$$\mathbf{r}_{part} = 0.327 \, p_{O_2}^{0.804} \, e^{-5.230/R_g T}$$

where p_{O_2} is in atmospheres. Do internal concentration and temperature gradients have a significant effect on the pellet rate? Estimate the maximum temperature difference, $T_c - T_s$.

Solution   Equation (11-83) will be satisfactory as a criterion, since the Φ_s-vs-η curves for a first-order reaction will be nearly the same as those for 0.8 order:

$$(C_{O_2})_s = \frac{(p_{O_2})_s}{R_g T} = \frac{1(0.0527)}{82(101 + 273)} = 1.72 \times 10^{-6} \text{ mole/cm}^3$$

$$r_s^2 \frac{\mathbf{r}_P \rho_P}{C_s D_e} = \left(\frac{1.86}{2}\right)^2 \frac{2.49 \times 10^{-5}(0.0602)}{1.72 \times 10^{-6}(0.166)} = 4.4$$

From Eqs. (11-79) and (11-80),

$$\gamma = \frac{5,230}{2(374)} = 7.0$$

$$\beta = \frac{115,400(0.166)(1.72 \times 10^{-6})}{6.2 \times 10^{-4}(374)} = 0.14$$

Note that the heat of reaction per mole of oxygen is $-57,700(2) = -115,400$ cal/g mole.

Introducing these results in Eq. (11-83) gives

$$\Theta = (4.4) \, e^{7(0.14)/1.14} = 4.4(2.36) = 10.4$$

Since the value of Θ far exceeds unity, intrapellet resistances do affect the rate.

Equation (11-77) gives the maximum value for $T_c - T_s$,

$$T_c - T_s = -\frac{(-115,400)(0.166)}{6.2 \times 10^{-4}} (1.72 \times 10^{-6}) = 52°C$$

The observed temperatures give $148 - 101 = 47°C$. Comparison between the maximum and observed $T_c - T_s$ suggests that the oxygen is nearly all consumed when it reaches the center of the pellet. Hence the rate also will vary with radial position due to the drop in reactant (O_2) concentration.

This preliminary analysis suggests that both intrapellet temperature and concentration gradients will significantly affect the rate. Since the reaction is exothermic, the two factors have opposite effects on the rate per pellet. The dominant effect can be ascertained by calculating the resultant effectiveness factor.

Example 11-10   From the experimental data given in Example 11-9 evaluate the effectiveness factor for the catalyst pellet. Also predict η, using the results of Weisz and Hicks.[1]

Solution   The particles for which rates were measured are small enough that η is unity. Hence an experimental effectiveness factor can be obtained from Eq. (11-65).

[1] P. B. Weisz and J. S. Hicks, *Chem. Eng. Sci.*, **17**, 265 (1962).

This equation was derived for isothermal operation but will be applicable for non-isothermal conditions as long as the two rates are evaluated at the same surface temperature as well as reactant concentration. Applying Eq. (11-65)

$$\eta_{\text{exp}} = \frac{\mathbf{r}_P(\text{at } T_s, C_s)}{\mathbf{r}_{\text{part}}(\text{at } T_s, C_s)}$$

The rate for the particles at T_s and C_s is obtainable from the correlation of particle rates given in Example 11-9:

$$\mathbf{r}_{\text{part}} = 0.327(0.0527^{0.804})e^{-5,230/2(374)} = 2.77 \times 10^{-5} \text{ g mole } O_2/(\text{g catalyst})$$

Then

$$\eta_{\text{exp}} = \frac{2.49 \times 10^{-5}}{2.77 \times 10^{-5}} = 0.9$$

To calculate η we first obtain $3(\Phi_s)_s$, using Eq. (11-78). The reaction is not first order, and so $(k_1)_s$ is not known. However, it may be replaced by its equivalent, $\mathbf{r}/C$, where $\mathbf{r}$ is the rate of the chemical step, that is, the rate uninfluenced by intrapellet resistances. This is the rate measured for the small catalyst particles, 2.77×10^{-5} g mole/(g catalyst)(sec). Hence

$$3(\Phi_s)_s = r_s\sqrt{\frac{\mathbf{r}_{\text{part}}}{C_s}\frac{\rho_P}{D_e}} = \frac{1.86}{2}\sqrt{\frac{2.77 \times 10^{-5}(0.0602)}{1.72 \times 10^{-6}(0.166)}} = 2.20$$

With this value for $3(\Phi_s)_s$, and $\gamma = 7.0$ and $\beta = 0.14$, a calculated η can be found from the Weisz and Hicks charts similar to Fig. (11-10). The value is difficult to estimate from the available charts, but it is approximately

$$\eta_{\text{calc}} = 0.94$$

A more accurate η can be obtained by solving equations such as (11-46) and (11-72) but avoiding the first-order restriction. This can be done by replacing $k_1 C$ in the reaction term of each equation with the more correct relationship $kp_{O_2}^{0.804}$. Maymo did this, solving the differential equations numerically by means of a high-speed digital computer. The calculated η for $3(\Phi_s)_s = 2.20$, $\gamma = 7.0$, and $\beta = 0.14$ is $\eta_{\text{calc}} = 0.96$. The calculated η for a first-order reaction agrees somewhat better with the experimental result. Probably this is not significant, because the calculated values rest on the assumption that the intrinsic rate is the same for the particles as for the pellets. There is evidence[1] to indicate that when the particles are made into pellets there is a slight reduction in intrinsic rate owing to partial blocking of the micropores in the particles. Regardless of the reasons for the small deviation (7%), the agreement between calculated and experimental η is quite adequate, particularly in view of the variety of independent data (D_e, k_e, rates, etc.) that are required to calculate the effectiveness factor.

[1] J. A. Maymo and J. M. Smith, *AIChE J.*, **12**, 845 (1966).

Since η is less than unity, the effect of the concentration gradient is more important than the effect of the temperature gradient in this instance.

The significance of thermal effects on the pellet rate is established primarily by β and secondarily by γ. For example, Otani and Smith[1] studied the reaction $CO + \frac{1}{2}O_2 \rightarrow CO_2$, making measurements similar to the $H_2 + O_2$ reaction, but at temperatures from 250 to 370°C. Here β was of the order of 0.05 and $\gamma = 9.7$. Intrapellet temperature gradients were much lower than in Example 11-9; the largest measured value for $T_c - T_s$ was 7°C. In contrast, hydrogenation reactions have large heats of reaction; D_e is relatively high for hydrogen, and often T_s is low. All these factors tend to increase β. Cunningham et al.[2] found β to be of the order of 0.5 (and γ about 25) for the hydrogenation of ethylene, and thermal effects were large. Similarly, in the work of Prater[3] for the dehydrogenation of cyclohexane, β and γ were sufficient to reduce η significantly.

Increasing reactant concentration increases β and η. When the partial pressure of oxygen was increased to 0.11 atm in the $H_2 + O_2$ reaction, with other conditions approximately the same as in Example 11-10, β became 0.34, and the experimental η was 1.10. At these conditions $T_c - T_s$ increased to $219 - 118 = 101$°C.

EFFECT OF INTERNAL RESISTANCES ON SELECTIVITY AND POISONING

In many catalytic systems multiple reactions occur, so that selectivity becomes important. In Sec. 2-10 point and overall selectivities were evaluated for homogeneous well-mixed systems of parallel and consecutive reactions. In Sec. 10-5 we saw that external diffusion and heat-transfer resistances affect the selectivity. Here we shall examine the influence of intrapellet resistances on selectivity. Systems with first-order kinetics at isothermal conditions are analyzed analytically in Sec. 11-12 for parallel and consecutive reactions. Results for other kinetics, or for nonisothermal conditions, can be developed in a similar way but require numerical solution.[4]

---

[1] S. Otani and J. M. Smith, *J. Catalysis*, **5**, 332 (1966).

[2] R. A. Cunningham, J. J. Carberry, and J. M. Smith, *AIChE J.*, **11**, 636 (1965).

[3] C. D. Prater, *Chem. Eng. Sci.*, **8**, 284 (1958), as analyzed in C. N. Satterfield and T. K. Sherwood "The Role of Diffusion in Catalysis," p. 90, Addison-Wesley Publishing Company, Reading, Mass., 1963.

[4] Selectivity for Langmuir-Hinshelwood kinetics, of the form of Eqs. (9-32) and (9-38), have been evaluated for isothermal conditions by G. Roberts and C. N. Satterfield [*Ind. Eng. Chem., Fund. Quart.*, **4**, 288 (1965)] and J. Hutchings and J. J. Carberry [*AIChE J.*, **12**, 20 (1966)].

11-12 Selectivities for Porous Catalysts

The selectivity at a position in a fluid-solid catalytic reactor is equal to the ratio of the global rates at that point. The combined effect of both external and internal diffusion resistance can be displayed easily for a set of parallel reactions. We shall do this first and then consider how internal resistance influences the selectivity for other reaction sequences.

Consider two parallel reactions of the independent form

1. $A \xrightarrow{k_1} B + C$ (desired)

2. $R \xrightarrow{k_2} S + W$

This might be the dehydrogenation of a mixed feed of propane and n-butane, where the desired catalyst is selective for the n-butane dehydrogenation. Suppose that the temperature is constant and that both external and internal diffusion resistances affect the rate. At steady state, the rate (for the pellet, expressed per unit mass of catalyst) may be written in terms of either Eq. (10-1) or Eq. (11-44),

$$\mathbf{r}_P = k_m a_m (C_b - C_s) \tag{11-84}$$

or

$$\mathbf{r}_P = \eta k_1 C_s \tag{11-85}$$

where a_m is the external area for mass transfer per unit mass of catalyst. The surface concentration can be eliminated between (11-84) and (11-85) to give

$$\mathbf{r}_P = \frac{1}{1/k_m a_m + 1/\eta k} C_b \tag{11-86}$$

Equation (11-86) expresses the combined effect of external and internal mass-transport resistance. Note that the reduction in rate due to internal diffusion (through η) is combined with the rate constant k for the chemical step, while the external effect is separate. If the external resistance is negligible, then $k_m a_m \gg \eta k_1$. If internal transport is insignificant, then $\eta \to 1$. If both conditions are satisfied, the rate is determined solely by the chemical step; that is, Eq. (11-85) reduces to $\mathbf{r}_P = k_1 C_b$.

For parallel reactions the selectivity of product B with respect to product S for a pellet in the reactor is obtained by applying Eq. (11-86) to the two reactions; thus the pellet selectivity S_P is

$$S_P = \frac{(\mathbf{r}_P)_1}{(\mathbf{r}_P)_2} = \frac{[1/(k_m)_R a_m + 1/\eta_2 k_2] (C_A)_b}{[1/(k_m)_A a_m + 1/\eta_1 k_1] (C_R)_b} \tag{11-87}$$

If there were neither external nor internal resistances, the pellet selectivity would be

$$S_P = \frac{k_1 (C_A)_b}{k_2 (C_R)_b} \tag{11-88}$$

Since $(k_m)_R$ and $(k_m)_A$ will not differ greatly, comparison of Eqs. (11-87) and (11-88) shows that external-diffusion resistance reduces the selectivity. The same conclusion was reached in Sec. 10-5. To evaluate qualitatively the effect of internal diffusion note that k_1 is presumably greater than k_2 (B is the desired product). Since η decreases as k increases (see Fig. 11-7), η_1 will be less than η_2. Then Eq. (11-87) indicates that internal diffusion also decreases the selectivity.

Equation (11-86) shows that the external effect on the rate can always be treated as a separate, additive resistance. Hence from here on we shall focus on internal effects, using Eq. (11-85) for the global rate and taking bulk and surface concentrations to be equal. The internal problem can be expressed analytically if we are satisfied with examining the extreme case of large intrapellet resistance characterized by $\Phi_s \geq 5$ ($\eta \leq 0.2$) where Eq. (11-54) is valid. Then Eq. (11-85) becomes

$$\mathbf{r}_P = \frac{1}{\Phi_s} k_1 C_b = \frac{3}{r_s} \sqrt{\frac{k_1 D_e}{\rho_P}} C_b \tag{11-89}$$

If Eq. (11-89) is applied to the stated parallel reactions, the selectivity is

$$S_P = \frac{(\mathbf{r}_P)_1}{(\mathbf{r}_P)_2} = \frac{\sqrt{k_1 (D_A)_e} (C_A)_b}{\sqrt{k_2 (D_R)_e} (C_R)_b} \tag{11-90}$$

Neglecting differences in diffusivities of A and R, we have

$$S_P = \left(\frac{k_1}{k_2}\right)^{\frac{1}{2}} \frac{(C_A)_b}{(C_R)_b} \tag{11-91}$$

Comparing Eqs. (11-88) and (11-91) shows that the effect of *strong* intrapellet diffusion resistance is to reduce the selectivity to the square root of its intrinsic value.

Wheeler[1] characterized the independent parallel reactions $A \to B$ and $R \to S$ as *type I selectivity*. *Type II selectivity* refers to parallel reactions of the form

---

[1] The treatment in this section follows in part that developed by A. Wheeler in W. G. Frankenburg, V. I. Komarewsky, and E. K. Rideal (eds.), "Advances in Catalysis," vol. III, p. 313, Academic Press, Inc., New York, 1951; and P. P. Weisz and C. D. Prater in "Advances in Catalysis," vol. VII, Academic Press, Inc., New York, 1954.

1. $A \xrightarrow{k_1} B$

2. $A \xrightarrow{k_2} C$

An example would be the dehydration of ethanol to ethylene and its dehydro-genation to acetaldehyde. If both reactions are first order, selectivity is unaffected by internal mass transport; the ratio of the rates of reactions 1 and 2 is k_1/k_2 at any position within the pellet. Equation (11-89) cannot be applied separately to the two reactions because of the common reactant A. The development of the effectiveness-factor function would require writing a differential equation analogous to Eq. (11-45) for the *total* con-sumption of A by both reactions. Hence k in Eq. (11-89) would be $k_1 + k_2$ and $\mathbf{r}_P$ would be $(\mathbf{r}_P)_1 + (\mathbf{r}_P)_2$. Such a development would shed no light on selectivity.

   If the kinetics of the two reactions are different, diffusion has an effect on selectivity. Suppose reaction 2 is second order in A and reaction 1 first order. The reduction in concentration of A due to diffusion resistance would lower the rate of reaction 2 more than that of reaction 1. For this case the selectivity of B would be improved by diffusion resistance.

   Type III selectivity applies to successive reactions of the form

$$A \xrightarrow{k_1} B \xrightarrow{k_3} D$$

where B is the desired product. Examples include successive oxidations, successive hydrogenations, and other sequences, such as dehydrogenation of butylenes to butadiene, followed by polymerization of the butadiene formed. In Chap. 2 the ratio of Eqs. (2-102) and (2-103) gave the intrinsic selectivity of B with respect to A, that is, the point selectivity determined by the kinetics of the reactions,

$$S = \frac{\text{net rate of production of } B}{\text{rate of disappearance of } A} = \frac{dC_B}{-dC_A} = 1 - \frac{k_3}{k_1} \frac{C_B}{C_A} \qquad (11\text{-}92)$$

Equation (11-92) is also applicable for the whole catalyst pellet when diffusion resistance is negligible. Furthermore, it gives the selectivity at any location within the pellet. The selectivity will vary with position in the pellet as C_B/C_A changes. Diffusion resistance causes C_A to decrease going from the outer surface toward the center of the pellet. Since B is formed within the pellet and must diffuse outward in order to enter the bulk stream, C_B increases toward the pellet center. Equation (11-92) shows qualitatively that these variations in C_A and C_B both act to reduce the global, or pellet, selectivity for B.

   A quantitative interpretation can also be developed. For first-order reactions the variation of C_A with radial position in a spherical pellet is given by Eq. (11-49). The concentration profile for B can be obtained in a similar manner by writing a differential mass balance for B. The expression will be like Eq. (11-45), except that the term on the right must represent the *net* disappearance of B; that is, it must account for the conversion of B to D and the production of B from A. Wheeler[1] solved such a differential equation to obtain $C_B = f(r)$. The solution, combined with Eq. (11-49) for C_A, can be used to obtain the selectivity of B with respect to A as a function of radial position. Integrating across the radius gives the selectivity for the whole pellet. For strong diffusion resistance ($\eta \leq 0.2$) and equal effective diffusivities the result is

$$S_P = \frac{(\mathbf{r}_P)_B}{-(\mathbf{r}_P)_A} = \frac{(k_1/k_3)^{\frac{1}{2}}}{1 + (k_1/k_3)^{\frac{1}{2}}} - \left(\frac{k_3}{k_1}\right)^{\frac{1}{2}} \frac{(C_B)_b}{(C_A)_b} \qquad (11\text{-}93)$$

Comparison of Eqs. (11-92) and (11-93) demonstrates that the selectivity is significantly reduced when diffusion resistances are large. The magnitude of the reduction depends on $(C_B/C_A)_b$, which, if external-diffusion resistance is neglected, is equal to C_B/C_A at any position in the reactor. At the entrance to a *reactor* $(C_B/C_A)_b$ is a minimum. If no B is present in the feed, $C_B = 0$ at this point. From Eq. (11-92) the selectivity is 1.0. Equation (11-93) shows that strong diffusion resistance reduces this to

$$S_P = \frac{(k_1/k_3)^{\frac{1}{2}}}{1 + (k_1/k_3)^{\frac{1}{2}}} \qquad (11\text{-}94)$$

The reduction is most severe for small values of k_1/k_3.

Suppose the pellet selectivity for B is observed to be low in the $A \to B \to D$ series of reactions, and diffusion resistance is significant. Either preparing the catalyst with larger pores or reducing the pellet size may improve the selectivity. However, these changes would be most effective if they increased η from a low value to near unity. The changes would be wholly ineffective if η remained ≤ 0.2, since Eq. (11-93) is valid for all η below 0.2. The first change would probably reduce the capacity of a fixed mass, or volume, of catalyst, since the pore surface available for reaction would decrease as pore radius increased. The second change would increase the pressure drop in a fixed-bed reactor.

Example 11-11   An ethylene stream is fed to a polymerization reactor in which the catalyst is poisoned by acetylene. The ethylene is prepared by catalytically dehydrogenating ethane. Hence it is important that the dehydrogenation catalyst be selective for dehydrogenating C_2H_6 rather than C_2H_4. The first-order reactions are

[1] A. Wheeler in W. G. Frankenburg, V. I. Komarewsky, and E. K. Rideal (eds.), "Advances in Catalysis," vol. III, Academic Press, Inc., New York, 1951.

$$\begin{array}{ccc} (A) & (B) & (D) \\ C_2H_6 \xrightarrow{k_1} C_2H_4 \xrightarrow{k_3} C_2H_2 \end{array}$$

For one catalyst $k_1/k_3 = 16$. It is suspected that intrapellet diffusion strongly retards both dehydrogenations. Estimate the potential improvement in selectivity if diffusion resistance could be eliminated. Make the estimate for a concentration ratio $(C_B/C_A)_b = 1.0$. Neglect differences in D_e between ethane and ethylene.

Solution   Equation (11-93) is applicable for the *diffusion-limited* situation. The pellet selectivity for ethylene formation with respect to ethane disappearance is

$$S_P = \frac{16^{\frac{1}{2}}}{1 + 16^{\frac{1}{2}}} - \frac{1}{16^{\frac{1}{2}}}(1) = 0.55$$

If diffusion resistance is eliminated, Eq. (11-92) gives the point as well as the pellet selectivity

$$S_P = 1 - \tfrac{1}{16}(1) = 0.94$$

These selectivities give the ratio of rates of formation of ethylene and total disappearance of ethane (ethylene plus acetylene). If we start with pure C_2H_6, a mass balance requires that the rate of formation of acetylene with respect to disappearance of ethane be $1 - S_P$. Hence eliminating diffusion resistance reduces the rate of formation of undesirable C_2H_2 from 0.45 to 0.06.

The discussion has been limited to selectivities existing at one position or conversion (set of concentrations) in a reactor. Wheeler has extended the development to show cumulative selectivities that would exist at any level of conversion in the reactor effluent.

We have seen that when intrapellet temperature gradients exist the rate per pellet (or η) cannot be expressed analytically. Hence selectivities for nonisothermal conditions must be evaluated numerically. Examples are available for several forms of rate equations.[1]

11-13 Rates for Poisoned Porous Catalysts

As mentioned in Chap. 8, the rates of fluid-solid catalytic reactions are frequently reduced by poisoning. It is important to know how the reduction in rate varies with the extent of catalytic surface which has been poisoned. Experimental data show that the rate may drop linearly with the fraction of surface poisoned, or it may fall much more rapidly. The second form of behavior may be caused by selective adsorption of the poisoning substance on those catalysts sites which are active for the main reaction. Such adsorption on a relatively small part of the total catalyst would cause a large reduction in rate. An alternate explanation is possible if the main reaction is

[1] John Beek, *AIChE J.*, 7, 337 (1961); John Hutchings and J. J. Carberry, *AIChE J.*, 12, 20 (1966).

primarily on the outer part of the catalyst pellet (low effectiveness factor) and the poison is also adsorbed primarily in this region. Then a relatively small part of the total surface will be deactivated by poison, but this small part is where the main reaction occurs. As Wheeler[1] has shown, the interaction of intrapellet diffusion on the rate of the main and poisoning reactions can lead to a variety of relations between activity and extent of poisoning. We examine some of these here as a further illustration of the effect of intrapellet mass transfer upon the rate of catalytic reactions. It is assumed that the activity of the unpoisoned catalytic surface is the same throughout the pellet and that the main reaction is first order.

Uniform Distribution of Poison Suppose the rate of the adsorption (or reaction) process which poisons the catalytic site is slow with respect to intrapellet diffusion. Then the surface will be deactivated uniformly through the pellet. If α is the fraction of the surface so poisoned, the rate constant will become $k_1(1 - \alpha)$. The rate per pellet, according to Eq. (11-44), is

$$\mathbf{r}_P = \eta k_1 (1 - \alpha) C_s \tag{11-95}$$

Consider first the case where the diffusion resistance for the main reaction also is small; that is, consider a slow main reaction. Then $\eta \to 1$, and Eq. (11-95) shows that the rate drops linearly with α. At the other extreme of large intrapellet resistance ($\Phi_s > 5$), $\eta = 1/\Phi_s$, and Eq. (11-95) becomes

$$\mathbf{r}_P = \frac{3}{r_s} \sqrt{\frac{D_e}{k_1(1 - \alpha)\rho_P}} \, k_1(1 - \alpha) C_s$$

$$= \frac{3}{r_s} \sqrt{\frac{D_e}{\rho_P}} k_1(1 - \alpha) \, C_s \tag{11-96}$$

Hence the effect of poisoning is proportional to $\sqrt{1 - \alpha}$; that is, it is less than a linear effect. This situation, corresponding to a slow poisoning reaction and a fast (large Φ_s) main reaction, was termed *antiselective poisoning* by Wheeler.

The two extreme cases are shown in Fig. 11-11 by curves A and B, where the ratio F of poisoned and unpoisoned rates is plotted against α. Note that the ratio is the quotient of Eq. (11-95) at any α and at $\alpha = 0$. Thus for curve B, from Eq. (11-96),

[1] A. Wheeler in W. G. Frankenburg, V. I. Komarewsky, and E. K. Rideal (eds.), "Advances in Catalysis," vol. III, p. 307, Academic Press, Inc., New York, 1951.

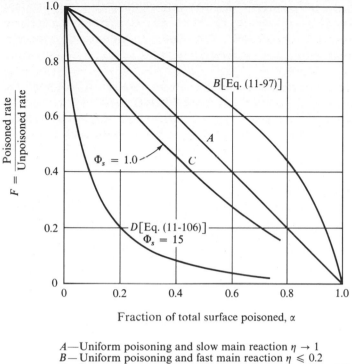

Fraction of total surface poisoned, α

A—Uniform poisoning and slow main reaction $\eta \to 1$
B— Uniform poisoning and fast main reaction $\eta \leqslant 0.2$
C—Pore-mouth poisoning and slow main reaction $\Phi_s = 1.0$
D—Pore-mouth poisoning and fast main reaction $\Phi_s = 15$

Fig. 11-11   *Intrapellet diffusion and poisoning in catalytic reactions*

$$F = \frac{(3/r_s)\sqrt{D_e k_1 (1 - \alpha)/\rho_P}\, C_s}{(3/r_s)\sqrt{D_e k_1/\rho_P}\, C_s} = \sqrt{1 - \alpha} \qquad (11\text{-}97)$$

For intermediate cases of diffusion resistance for the main reaction, the curves would fall between A and B in the figure.

Pore-mouth (Shell) Poisoning If the adsorption (or reaction) causing poisoning is very fast, the outer part of a catalyst pellet will be completely deactivated, while the central portion retains its unpoisoned activity. An example is sulfur poisoning of a platinum-on-Al_2O_3 catalyst, where the sulfur-containing molecules diffuse only a short distance into the pellet before they are adsorbed on the platinum surface. In this type of poisoning a layer of poisoned catalyst will start to grow at the outer surface and will continue to increase in thickness with time[1] until all the pellet is deactivated. During the process the boundary between the deactivated and active catalyst will remain sharply defined. This type of poisoning, described by the *progressive shell model*, will be discussed more in Chap. 14 in connection with noncatalytic reactions.

The effect of pore-mouth poisoning can be obtained by equating the rate of diffusion through the outer, deactivated layer to the rate of reaction on the inner fully active part of the pellet. Figure 11-12 depicts a spherical catalyst pellet at a time when the radius of the unpoisoned central portion is r_c, corresponding to a thickness of completely poisoned catalyst $r_s - r_c$. Consider a first-order reaction where the concentration of reactant at the outer surface is C_s. The rate of diffusion into one pellet will be

$$\mathbf{r} = D_e \, 4\pi r^2 \, \frac{dC}{dr} \tag{11-98}$$

This equation is applicable between r_s and r_c. Integrating[2] to express the rate in terms of the concentrations C_s and C_c (at $r = r_c$) gives

$$\mathbf{r} = \frac{4\pi D_e r_s r_c}{r_s - r_c} (C_s - C_c) \tag{11-99}$$

This rate is equal to the rate of reaction in the inner core, or

$$\mathbf{r} = \frac{4\pi D_e r_s r_c}{r_s - r_c} (C_s - C_c) = m_P \eta k_1 C_c \tag{11-100}$$

where m_P is the mass of the active core of the pellet. Equation (11-100) may be solved for C_c and the result inserted into Eq. (11-99). Taking $m_P = \frac{4}{3}\pi r_c^3 \rho_P$ and dividing by $\frac{4}{3}\pi r_s^3 \rho_P$ to give the rate for the whole pellet, but on a unit mass basis, gives

$$\mathbf{r}_P = \frac{\mathbf{r}}{\frac{4}{3}\pi r_s^3 \rho_P} = \frac{k_1 C_s}{r_s^3/\eta r_c^3 + (k_1 \rho_P r_s^2/3D_e)[(r_s - r_c)/r_c]} \tag{11-101}$$

The fraction of the total surface unpoisoned is

$$1 - \alpha = \frac{\frac{4}{3}\pi r_c^3 \rho_P S_g}{\frac{4}{3}\pi r_s^3 \rho_P S_g} = \frac{r_c^3}{r_s^3} \tag{11-102}$$

[1] For example, suppose that the poison is brought into contact with the catalyst pellet as a contaminant in the reactant stream flowing steadily in the reactor. Then the amount of poisoning material available is directly proportional to time.

[2] It is assumed here that the thickness of the poisoned layer is not changing in the time required for diffusion. Then **r** can be regarded as a constant during this short time interval. Such a pseudo-steady state is discussed in Chap. 14.

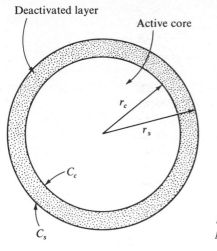

Deactivated layer

Active core

r_c

r_s

C_c

C_s

Fig. 11-12   Shell model of catalyst poisoning

In terms of α, Eq. (11-101) becomes

$$\mathbf{r}_P = \frac{k_1 C_s}{1/\eta(1 - \alpha) + 3\Phi_s{}^2[1 - (1 - \alpha)^{\frac{1}{3}}]/(1 - \alpha)^{\frac{1}{3}}} \tag{11-103}$$

where Eq. (11-50) has been used to introduce Φ_s. If there is no poisoning, then $\alpha = 0$, and Eq. (11-103) reduces to $\mathbf{r}_P = \eta k_1 C_s$ [i.e., Eq. (11-44)]. Hence the ratio of poisoned and unpoisoned rates is

$$F = \frac{1}{1/(1 - \alpha) + 3\eta(\Phi_s)^2[1 - (1 - \alpha)^{\frac{1}{3}}]/(1 - \alpha)^{\frac{1}{3}}} \tag{11-104}$$

Again consider two extremes of main reaction. If the intrinsic rate is very slow, diffusion resistance in the inner core of active catalyst will be negligible and $\eta \to 1$. At these conditions Φ_s will also be low, but not as small as would be expected. This is because η is a function of $\Phi_c = (r_c/3)$ $\sqrt{k_1 \rho_P/D_e}$, not of Φ_s. For example, the maximum value of Φ for negligible diffusion resistance was arbitrarily chosen earlier as $\frac{1}{3}$. If $\Phi_c = \frac{1}{3}$ and $r_c/r_s = \frac{1}{3}$, then $\Phi_s = (r_s/r_c)\Phi_c = 1.0$. Curve C in Fig. 11-11 is a plot of Eq. (11-104) for $\eta = 1$ and $\Phi_s = 1.0$. At these conditions poisoning has a slightly greater than linear effect on reducing the rate. Next suppose that the main reaction is fast so that $\Phi_c > 5$. Now η is given by the equation similar to (11-54), but for the active core; i.e.,

$$\eta = \frac{1}{\Phi_c} = \frac{r_s}{r_c} \frac{1}{\Phi_s} = \frac{1}{(1 - \alpha)^{\frac{1}{3}}} \frac{1}{\Phi_s} \tag{11-105}$$

Then Eq. (11-104) becomes

$$F = \frac{1}{1/(1 - \alpha) + 3\Phi_s[1 - (1 - \alpha)^{\frac{1}{3}}]/(1 - \alpha)^{\frac{2}{3}}} \tag{11-106}$$

For $\Phi_c \geq 5$, Φ_s will be even larger. For $r_c/r_s = \frac{1}{3}$, $\Phi_s \geq 15$. Curve D in Fig. 11-11 is a plot of Eq. (11-106) with $\Phi_s = 15$. In this case a sharp drop in activity is observed. Physically, the situation corresponds to a large diffusion resistance or a fast main reaction. The reactant molecules cannot penetrate far into the active part of the catalyst before reaction occurs. Thus both the poisoning and main reaction try to take place on the same (outer) surface. As the outer surface is poisoned, the rate falls dramatically.

Problems

11-1.   In a cylindrical pore of 30 Å radius at what pressure would the bulk diffusivity in a H_2–C_2H_6 mixture be equal to the Knudsen diffusivity for H_2? The temperature is 100°C.

11-2.   At 10 atm pressure and 100°C what would be the pore radius for which bulk and Knudsen diffusivities would be equal for H_2 in the hydrogen-ethane system. Locate a curve for equal values of bulk and Knudsen diffusivities on coordinates of pressure-vs-pore radius. Mark the regions on the figure where Knudsen and where bulk diffusivity would be dominant.

11-3.   Constant-pressure diffusion experiments are carried out in the apparatus shown in Fig. 11-1 with the H_2–N_2 binary system. In one experiment the nitrogen diffusion rate through the porous pellet was 0.49×10^{-5} g mole/sec. What would be the counterdiffusion rate of hydrogen?

11-4.   Diffusion rates for the H_2–N_2 system were measured by Rao and Smith[1] for a cylindrical Vycor (porous-glass) pellet 0.25 in. long and 0.56 in. in diameter, at 25°C and 1 atm pressure. A constant-pressure apparatus such as that shown in Fig. 11-1 was used. The Vycor had a mean pore radius of 45 Å, so that diffusion was by the Knudsen mechanism. The diffusion rates were small with respect to the flow rates of the pure gases on either side of the pellet. The average diffusion rate of hydrogen for a number of runs was 0.44 cm³/min (25°C, 1 atm). The porosity of the Vycor was 0.304.

(a) Calculate the effective diffusivity D_e of hydrogen in the pellet. (b) Calculate the tortuosity using the parallel-pore model. (c) Predict what the counterdiffusion rate of nitrogen would be.

11-5.   Rao and Smith also studied the first-order (at constant total pressure) reversible reaction

$$o\text{-}H_2 \rightleftharpoons p\text{-}H_2$$

at -196°C and 1 atm pressure using a NiO-on-Vycor catalyst. For NiO-on-Vycor particles, with an average diameter of 58 microns, rate measurements gave a rate constant of $k_R = 5.29 \times 10^{-5}$ g mole/(sec)(g catalyst). For NiO-

[1] M. R. Rao and J. M. Smith, *AIChE J.*, **10**, 293 (1964).

on-Vycor *pellets* $\frac{1}{8}$-in. long and $\frac{1}{2}$ in. in diameter $k_R = 2.18 \times 10^{-5}$. The catalyst pellets were encased on the cylindrical surface and on one end, so that hydrogen was available only to the other face of the cylinder, as illustrated in Fig. 11-8. The density of the pellet was 1.46 g/cm^3.

(a) From the experimental rate data evaluate the effectiveness factor for the pellet. (b) Using the random-pore model to estimate D_e, predict an effectiveness factor for comparison with the answer to part (a). Only micropores ($\bar{a}_\mu = 45$ Å) exist in Vycor, and the porosity of the pellet was $\varepsilon_\mu = 0.304$.

11-6.   Cunningham et al.[1] measured global rates of ethylene hydrogenation on copper–magnesium oxide catalysts of two types: particles of 100 mesh size and $\frac{1}{2}$-in. spherical pellets of three pellet densities. For both materials external concentration and temperature differences were negligible. The rate data at about the same surface concentrations of ethylene and hydrogen at the outer surface of the particles and pellets are as shown below.

| | | Rate **r**, g moles/(sec)(g catalyst) | | |
| | | Pellets | | |
t_s, °C	*Particles*	$\rho_p = 0.72$	$\rho_p = 0.95$	$\rho_p = 1.18$
124	1.45×10^{-5}	6.8×10^{-6}	4.3×10^{-6}	2.2×10^{-6}
112	6.8×10^{-6}	6.7×10^{-6}	4.2×10^{-6}	2.1×10^{-6}
97	2.9×10^{-6}	6.4×10^{-6}	4.0×10^{-6}	2.0×10^{-6}
84	1.2×10^{-6}	6.0×10^{-6}	3.7×10^{-6}	1.9×10^{-6}
72	$\ldots$	5.5×10^{-6}	3.4×10^{-6}	1.7×10^{-6}
50	$\ldots$	4.3×10^{-6}	2.6×10^{-6}	1.3×10^{-6}

(a) From Arrhenius plots of these data, what is the true activation energy of the chemical steps at the catalytic sites? (b) Calculate effectiveness factors for pellets of each density at all the listed temperatures. (c) Explain why some effectiveness factors are greater than unity and why some are less than unity. (d) Why do the rate and the effectiveness factor increase with decreasing pellet density? (e) Suggest reasons for the very low apparent activation energies suggested by the pellet data. Note that even for the maximum effect of intrapellet gradients the apparent activation energy would still be one-half the true value.

11-7.   Wheeler[2] has summarized the work on internal diffusion for catalytic cracking of gas-oil. At 500°C the rate data for fixed-bed operation, with relatively large ($\frac{1}{8}$-in.) catalyst particles and that for fluidized-bed reactors (very small particle size) are about the same. This suggests that the effectiveness factor for the large particles is high. Confirm this by estimating η for the $\frac{1}{8}$-in. catalyst if the

[1] R. A. Cunningham, J. J. Carberry, and J. M. Smith, *AIChE J.*, **11**, 636 (1965).

[2] A. Wheeler in W. G. Frankenburg, V. I. Komarewsky, and E. K. Rideal (eds.), "Advances in Catalysis," vol. III, pp. 250–326, Academic Press, Inc., New York, 1950.

mean pore radius is 30 Å, the particle diameter is 0.31 cm, and the pore volume is 0.35 cm^3/g catalyst. Molecular weight of oil is 120.

At atmospheric pressure with 30-Å pores the diffusion will be of the Knudsen type. The rate data, interpreted in terms of a first-order rate equation, indicate (at atmospheric pressure) that $(k_1)_{exp} = 0.25$ cm^3/(sec)(g catalyst). Assume that the parallel-pore model with a tortuosity factor of 2.0 is applicable.

11-8.   Blue et al.[1] have studied the dehydrogenation of butane at atmospheric pressure, using a chromia-alumina catalyst at 530°C. For a spherical catalyst size of $d_p = 0.32$ cm the experimental data suggest a first-order rate constant of about 0.94 cm^3/(sec)(g catalyst). The pore radius is given as 110 Å. Assuming Knudsen diffusivity at this low pressure and estimating the pore volume as 0.35 cm^3/g, predict an effectiveness factor for the catalyst. Use the parallel-pore model with a tortuosity factor of 2.0.

11-9.   (a) Develop an expression for the effectiveness factor for a straight cylindrical pore of length $2L$. Both ends of the pore are open to reactant gas of concentration $C = C_s$. A first-order irreversible reaction $A \rightarrow B$ occurs on the pore walls. Express the result as $\eta = f(\Phi_p)$, where Φ_p is the Thiele modulus for a single pore. Take the rate constant for the reaction as k_s, expressed as moles/(cm^2)(sec)(g mole/cm^3). The pore radius is a and the diffusivity of A in the pore is D.

(b) How does the relationship $\eta = f(\Phi_p)$ compare with Eq. (11-55)? What is the relationship between Φ_p and Φ_L?

(c) Comparison of the definitions of Φ_p [derived in part (a)] and Φ_L [from Eq. (11-56)] leads to what relationship between the effective diffusivity D_e in a flat plate of catalyst and the diffusivity D for a single pore? Note that L, the actual thickness of one-half the slab in Eq. (11-56), is one-half the pore length. Also note that k_1 in Eq. (11-56) is equal to $k_s S_g$.

(d) What simple model of a porous catalyst will explain the relation between D_e and D? Assume that the Wheeler equation relating the pore radius, the pore volume, and the surface area [Eq. (11-21)] is applicable.

11-10.   Consider the effect of surface diffusion on the effectiveness factor for a first-order, irreversible, gaseous reaction on a porous catalyst. Assume that the intrinsic rates of adsorption and desorption of reactant on the surface are rapid with respect to the rate of surface diffusion. Hence equilibrium is established between reactant in the gas in the pore and reactant adsorbed on the surface. Assume further that the equilibrium expression for the concentration is a linear one. Derive an equation for the effectiveness factor for each of the following two cases:

(a)   A porous slab of catalyst (thickness L) in which diffusion is in one direction only, perpendicular to the face of the slab

(b)   A spherical catalyst pellet of radius r.

Use the expression for total effective diffusivity given by Eq. (11-35) and follow the development in Sec. 11-6.

[1] R. W. Blue, V. C. F. Holm, R. B. Reiger, E. Fast, and L. Heckelsberg, *Ind. Eng. Chem.*, **44**, 2710 (1952).

11-11. A limiting case of intrapellet transport resistances is that of the *thermal effectiveness factor*.[1] In this situation of zero mass-transfer resistance, the resistance to intrapellet heat transfer alone establishes the effectiveness of the pellet. Assume that the temperature effect on the rate can be represented by the Arrhenius function, so that the rate at any location is given by

$$\mathbf{r} = A \, e^{-E/R_g T} f(C_s)$$

where $f(C_s)$ represents the concentration dependency of the rate, evaluated at the outer surface of the pellet.

(a) Derive a dimensionless form of the differential equation for the temperature profile within the pellet, using the variables

$$T^* = \frac{T}{T_s} \qquad r^* = \frac{r}{r_s}$$

where T_s is the temperature at the outer surface of the pellet (radius r_s). What are the dimensionless coefficients in the equation? One of the parameters in the coefficients should be the rate of reaction evaluated at the outer surface; that is,

$$\mathbf{r}_s = A \, e^{-E/R_g T_s} f(C_s)$$

(b) Derive an integral equation for the effectiveness factor, where the integral is a function of the dimensionless temperature profile.

(c) Are the results in parts (a) and (b) restricted to a specific form for $f(C)$, such as a first-order reaction? (d) Would you expect the thermal effectiveness factor, given by the solution of the integral equation of part (b), to be more applicable to exothermic or to endothermic reactions? (e) Determine the effectiveness factor as a function of a dimensionless coefficient involving $\mathbf{r}_s$, where the other coefficient is $E/R_g T_s = 20$.

A digital computer will facilitate the numerical solution of the equations derived in parts (a) and (b).

[1] J. A. Maymo, R. E. Cunningham, and J. M. Smith, *Ind. Eng. Chem., Fund. Quart.*, **5**, 280 (1966).

12

THE GLOBAL RATE AND LABORATORY REACTORS

It was noted in Chap. 7 that if global, or overall, rates of reaction are used, design equations for heterogeneous reactors are the same as those for homogeneous reactors. We shall take advantage of this simplification in Chap. 13, when we consider the design of heterogeneous reactors. The effects of external and internal transport processes on the global rate were analyzed in detail in Chaps. 10 and 11. In preparation for reactor design let us summarize the application of these results to relate the global and intrinsic rates at a catalyst site.

As intrinsic rate equations cannot yet be predicted, they must be evaluated from laboratory data. Such data are measurements of the global rate of reaction. The first part of the problem is to extract the equation for the intrinsic rate from the global rate data. Since laboratory reactors are small and relatively low in cost, there is great flexibility in designing them. In particular, construction and operating conditions can be chosen to reduce or eliminate the differences between the global and intrinsic rates, so that more accurate equations for the intrinsic rate can be extracted from the experimental data. Several types of laboratory reactors have been developed, and these will be discussed in Secs. 12-2 and 12-3.

After the rate equation at a catalyst site has been obtained, the relations between global and intrinsic rates must be used again, this time to obtain a global rate for a range of operating conditions. Such information is necessary for designing the commercial-scale reactor. The type of commercial reactor will be dictated by economics, and whatever the effects of physical processes, they must be included in evaluating the global rate. This aspect of the problem will be discussed in Sec. 12-4.

Because the objectives are different, operating conditions and reactor types are usually different for laboratory and large-scale units, and the global rate in the laboratory reactor cannot be used directly for designing large-scale apparatus. For example, economic considerations may require a different velocity in the commercial reactor, resulting in different external mass- and heat-transfer coefficients between catalyst pellet and reacting fluid. If external resistances are significant, there will be a different global rate in the commercial-scale reactor for the same intrinsic rate. Similarly, different sizes of catalyst pellets in the commercial and laboratory reactors will lead to different global rates if internal resistances are significant. Thus, to obtain a global rate useful for design, the global rate measured in the laboratory should first be reduced to the intrinsic rate at the catalyst site and then recombined with physical resistances corresponding to the conditions of commercial operation.

The relation between global and intrinsic rates is not always important. As a general rule, the global rate will be the same as the intrinsic rate[1] when the chemical step at the catalyst site is very slow. The difference between the two rates increases as the intrinsic rate increases, and for systems where the chemical step is extremely rapid a careful accounting of mass- and heat-transfer processes is important. In fixed-bed catalytic reactors external resistances are small for normal operating conditions. Internal resistances, however, may be important when external ones are negligible. It is usually true that external resistances are significant only when intrapellet resistances are more significant.[2] Care should be taken to distinguish between mass- and heat-transfer resistances. We have seen (Sec. 10-4) that large external temperature differences can exist when the external concentration difference is small. In fluidized-bed and slurry reactors, where the catalyst particles

---

[1] When comparisons between global and intrinsic rates are made, it is understood that the comparison is for the same temperature and concentration. Note that in a reactor operating at steady state the two rates are always the same, but the temperature and composition in the bulk fluid (i.e., the global conditions) are different from those at a site within the catalyst pellet.

[2] For porous pellets with uniform distribution of catalytic material. If the catalyst is deposited only on the outer layer of the pellet, this is not true.

are small, a situation opposite to that in fixed-bed reactors exists. Here internal resistances are negligible, but external resistances, particularly for mass transfer, may be significant (see Chap. 10).

12-1 Interpretation of Laboratory Kinetic Data

Suppose that the global rate has been measured in a laboratory reactor over a range of concentration and temperature for a single pellet size. The first step is to determine the surface concentration C_s and temperature T_s from measured bulk values by means of Eqs. (10-1) and (10-13). Next the global rate and C_s and T_s are used to establish the rate equation. In this second step the effect of the internal resistances, the *effectiveness factor*, is obtained simultaneously. The second step is a difficult one for the general case when there are significant temperature gradients within the catalyst pellet. For isothermal conditions Eq. (11-43) and the appropriate curve in Fig. 11-7 can be used to establish both the intrinsic rate [$f(C)$ in Eq. (11-43)] and the effectiveness factor. A trial procedure is usually necessary. The following steps describe a method of solution:

1. Assume a form for $f(C)$ (first order, second order, etc.).
2. For a given set of experimental data (C_s, T_s, and $\mathbf{r}_p$) assume a value of η.
3. From the appropriate curve in Fig. 11-7 obtain Φ and then evaluate k from the defining expression for Φ [for example, Eq. (11-50) for a first-order rate on spherical catalyst pellets].
4. Check the assumed value of η by employing Eq. (11-43).
5. Repeat steps 2 to 4 for all sets of data to see whether the assumed form of $f(C)$ is valid (for a first-order rate equation the evaluation would be to see if k_1 is constant for data at various concentrations but constant temperature).
6. Finally, the values of k for different temperatures would be plotted as $\ln k$ vs $1/T$ to obtain $\mathbf{A}$ and E in the Arrhenius equation, $k = \mathbf{A}e^{-E/R_g T}$.

The resultant values of $\mathbf{A}$ and E and the nature of $f(C)$ establish the desired equation for the intrinsic rate. However, for step 3 we must know or estimate the effective diffusivity. Alternately, the global rate may be measured for *small* catalyst particles for which $\eta \rightarrow 1.0$. Now internal (and usually external) transport resistances are negligible; the bulk temperature and concentrations may be taken as equal to those at a catalyst site, and the observed rate and bulk temperature and concentration data can thus be used directly to obtain $f(C)$ and $\mathbf{A}$ and E.

If intrapellet temperature gradients are significant, an analytic procedure similar to that for the isothermal case can be employed, in principle, to evaluate an equation for the intrinsic rate. However, step 1 now must include an assumption for the activation energy as well as for $f(C)$. The subsequent steps for checking the assumptions involve β and γ, defined by Eqs. (11-79) and (11-80), since η is now a function of these parameters as well as of Φ. In addition to the effective diffusivity, we need the effective thermal conductivity. The uncertainties in these quantities and the complex analysis required to obtain them severely limits the usefulness of the procedure for nonisothermal conditions. Hence it is preferable to use measured rates for small particles, or laboratory reactors in which nonisothermal conditions are eliminated.

In the laboratory either integral or differential (see Sec. 4-3) tubular units or stirred-tank reactors may be used. There are advantages in using stirred-tank reactors for kinetic studies. Steady-state operation with well-defined residence-time conditions and uniform concentrations in the fluid and on the solid catalyst are achieved. Isothermal behavior in the fluid phase is attainable. Stirred tanks have long been used for homogeneous liquid-phase reactors and slurry reactors, and recently reactors of this type have been developed for large catalyst pellets. Some of these are described in Sec. 12-3. When either a stirred-tank or a differential reactor is employed, the global rate is obtained directly, and the analysis procedure described above can be initiated immediately.

If integral data are measured, the data must first be differentiated to obtain the global rate. Alternately, the assumed rate equation may be integrated and the results compared with the observed data. For measurements made in a tubular-flow reactor packed with catalyst pellets, the mass balances of reactant, if we assume ideal flow, is

$$\mathbf{r}_P \, dW = F \, dx \tag{12-1}$$

where $\mathbf{r}_P$ = global rate of reaction per unit mass of catalyst

W = mass of catalyst

F = feed rate of reactant

x = conversion of reactant

Equation (12-1) is the same as that developed for homogeneous reactors [Eq. (3-12)], except that W has replaced the reactor volume and $\mathbf{r}_P$ is based on a unit mass of catalyst. If the volume of the reaction mixture is constant (constant temperature, pressure, and number of moles), then $dx = -(1/C_0) \, dC_b$, so that Eq. (12-1) becomes

$$\frac{W}{F} = \int \frac{dx}{\mathbf{r}_P} = -\frac{1}{C_0} \int \frac{dC_b}{\mathbf{r}_P} \tag{12-2}$$

This expression is analogous to Eq. (4-5) for homogeneous reactions. Note that C_b is the concentration of *reactant* in the *bulk* stream.

The interpretation procedure is illustrated in Example 12-1 for the reaction

$$o\text{-}H_2 \rightleftharpoons p\text{-}H_2$$

on Ni–Al$_2$O$_3$ catalyst pellets. Integral reactor data were measured under isothermal conditions. The analysis is simplified because the intrinsic rate equation is first order (at the fixed pressure existing in the reactor). Hence external and internal mass-transfer resistances can be combined in the simple way shown in Eq. (11-86). Both resistances are significant at some conditions, and the analysis permits quantitative evaluation of these resistances.

Example 12-1   Wakao et al.[1] studied the conversion of ortho hydrogen to para hydrogen in a fixed-bed tubular-flow reactor (0.50 in. ID) at isothermal conditions of $-196°C$ (liquid nitrogen temperature). The feed contained a mole fraction p-H$_2$ of $y_{b_1} = 0.250$. The equilibrium value at $-196°C$ is $y_{eq} = 0.5026$. The catalyst is Ni on Al$_2$O$_3$ and has a surface area of 155 m^2/g. The mole fraction p-H$_2$ in the exit stream from the reactor was measured for different flow rates and pressures and for three sizes of catalyst: granular particles of equivalent spherical diameter, 0.127 mm, granular particles 0.505 mm, and nominal $\frac{1}{8} \times \frac{1}{8}$-in. cylindrical pellets. The flow rate, pressure, and composition measurements are given in Table 12-1.

Table 12-1   Experimental data

Type of catalyst	Mass of catalyst in reactor, g	Pressure, psig	Flow rate, cm^3/sec (measured at 0°C, 1 atm)	Mole fraction p-H$_2$ in exit gas, $(y_b)_2$
Cylindrical pellets, $\frac{1}{8}$ in.	2.55	40	30.1	0.3790
			67.7	0.3226
			141.0	0.2886
		100	26.7	0.4031
			58.6	0.3416
		400	28.7	0.4286
			78.5	0.3480
0.505-mm particles	1.13	100	24.2	0.4261
			147.0	0.2965
		400	35.8	0.4300
			77.1	0.3485
0.127-mm particles	0.739	100	40.6	0.3378
		400	68.5	0.3381

[1]Noriaki Wakao, P. W. Selwood, and J. M. Smith, *AIChE J.*, **8**, 478 (1962).

The reaction is reversible and believed to be first order. Also, strong adsorption of hydrogen reduces the rate. The concepts of Chap. 9 suggest that a logical assumption for the form of the rate equation at a catalyst site (an intrinsic rate equation) would be

$$\mathbf{r} = f(C) = \frac{k_c(C_o - C_p/K)}{1 + K_a(C_o + C_p)} \tag{A}$$

where $\mathbf{r}$ = rate of reaction, g moles/(sec)(g catalyst)
C_o, C_p = concentrations of ortho and para hydrogen in the gas phase at a catalyst site, g moles/cm^3
k_c = specific–reaction–rate constant, cm^3/(g catalyst)(sec)
K = equilibrium constant for the reaction o-H$_2 \rightleftharpoons p$-H$_2$
K_a = adsorption constant in the rate equation

Test the suitability of the proposed rate equation and evaluate the constants k_c and K_a. Also calculate effectiveness factors for the $\frac{1}{8}$-in. pellets. The following additional data are available:

Viscosity of H$_2$ at $-196°$C = 348 × 10^{-7} poises

$$\text{Diffusivity of H}_2 \text{ at } -196°\text{C} = \begin{cases} 0.0376 \text{ cm}^2/\text{sec} & \text{at 40 psig} \\ 0.0180 \text{ cm}^2/\text{sec} & \text{at 100 psig} \\ 0.00496 \text{ cm}^2/\text{sec} & \text{at 400 psig} \end{cases}$$

Schmidt number for H$_2$ at $-196°$C = $\mu/\rho \, \mathscr{D}_{H_2}$ = 0.78
Density of catalyst (particles or pellet) = 1.91 g/cm^3

Solution   The volume of the reaction mixture is constant for any one run. If plug-flow behavior is assumed, Eq. (12-2) is applicable. The data in Table 12-1 represent integral reactor results. To use them Eq. (12-2) must be integrated, and this requires expressing the rate in terms of C_b. When external composition differences are significant and the rate equation is not first order, this is difficult. The difference $C_b - C_s$ will vary in the axial direction, so that a troublesome trial-and-error stepwise procedure is necessary to accomplish the integration. In this case the intrinsic rate is first order, so that the integration of Eq. (12-2) is simple and analytical.

For any one run the total pressure and total concentration are uniform in the reactor, so that Eq. (A) may be written

$$\mathbf{r} = k_1\left(C_o - \frac{C_p}{K}\right) = k_1 C_o - k_1' C_p \tag{B}$$

where

$$k_1 = \frac{k_c}{1 + K_a C_t} \tag{C}$$

We saw in Sec. 2-8 that for first-order reversible reactions Eqs. (2-74) and (2-75) show that Eq. (B) may be written

$$\mathbf{r} = \frac{k_1(K + 1)}{K}[C_o - (C_o)_{eq}] \tag{D}$$

where $(C_o)_{eq}$ is the equilibrium concentration of ortho hydrogen at the temperature involved. Since the total concentration of hydrogen is constant ($C_t = C_o + C_p$),

$$C_o - (C_o)_{eq} = C_t - C_p - [C_t - (C_p)_{eq}] = (C_p)_{eq} - C_p \tag{E}$$

Then Eq. (D) may be written

$$\mathbf{r} = \frac{k_1(K + 1)}{K}[(C_p)_{eq} - C_p] = \frac{k_1(K + 1)}{K}(C_{eq} - C)^1 \tag{F}$$

Equation (F) is for the intrinsic rate at a catalyst site, and so C is the p-H_2 concentration at a site within the catalyst pellet. To convert this to a rate for the pellet, $\mathbf{r}_P$, that is, to account for internal mass-transfer resistance, we use Eq. (11-43) to obtain

$$\mathbf{r}_P = \eta\frac{k_1(K + 1)}{K}(C_{eq} - C_s) \tag{G}$$

where now the concentration C_s at the outer surface of the pellet has replaced C. Next, the relation between C_s and C_b is obtained by equating the mass-transfer rate (per unit mass) from Eq. (10-1) to the rate of reaction,

$$\mathbf{r}_P = k_m a_m(C_s - C_b) = \eta\frac{k_1(K + 1)}{K}(C_{eq} - C_s) \tag{H}$$

where a_m is the external surface per unit mass, or $6/d_P\rho_P$ for a spherical pellet. Eliminating C_s from the two forms of Eq. (H) gives

$$\mathbf{r}_P = \frac{C_{eq} - C_b}{d_P\rho_P/6k_m + K/[\eta k_1(K + 1)]} = \mathbf{K}(C_{eq} - C_b) \tag{I}$$

where

$$\frac{1}{\mathbf{K}} = \frac{d_P\rho_P}{6k_m} + \frac{K}{\eta k_1(K + 1)} \tag{J}$$

Equation (I) is the required expression for the rate in terms of the bulk concentration of p-H_2. It is the form of Eq. (11-86) applicable to a reversible first-order reaction.

Parameter $\mathbf{K}$ is constant throughout the reactor for a given run. Hence when Eq. (I) is substituted in Eq. (12-2), the integration is simple. We note that $dC_p = -dC_o$, so that Eq. (12-2) becomes, for p-H_2,

$$\frac{W}{F_p} = \frac{1}{(C_b)_1}\int_1^2\frac{dC_b}{\mathbf{K}(C_{eq} - C_b)} = \frac{-1}{(C_b)_1\mathbf{K}}\ln\frac{C_{eq} - (C_b)_2}{C_{eq} - (C_b)_1} \tag{K}$$

where subscripts 1 and 2 represent entrance and exit conditions, respectively. Assuming ideal-gas behavior, we have

$$(C_b)_1 = \frac{(y_b)_1}{R_gT}p_t$$

[1] Since only p-H_2 concentrations will be used from here on, the subscript p will be omitted.

The feed rate of p-H_2 is $F_t(y_b)_1$. In terms of *total* feed rate and mole fractions, Eq. (K) may be written

$$\frac{W}{F_t} = -\frac{R_g T}{p_t K} \ln \frac{y_{eq} - (y_b)_2}{y_{eq} - (y_b)_1} \tag{L}$$

The data in Table 12-1 give $(y_b)_2$ for various values of flow rate. Also, $y_{eq} = 0.5026$ and $(y_b)_1 = 0.250$. Hence values of K can be calculated for each run. For example, for the first run at 40 psig with the $\frac{1}{8}$-in. catalyst pellets

$$\frac{W}{F_t} = \frac{2.55}{30.1/22,400} = 1,890 \text{ (g catalyst)(sec)/(g mole)}$$

Then, from Eq. (L),

$$1,890 = -\frac{R_g T}{p_t K} \ln \frac{0.5026 - 0.3790}{0.5026 - 0.250}$$

$$\frac{R_g T}{p_t K} = 2,640 \text{ (sec)(g catalyst)/(g mole)}$$

$$\frac{1}{K} = 2,640 \frac{(40 + 14.7)/14.7}{82(273 - 196)} = 1.56 \text{ (g catalyst)(sec)/cm}^3$$

The results for other runs are given in column 5 of Table 12-2.

Table 12-2   Calculated results for the reaction o-$H_2 \rightleftharpoons p$-H_2

Type of catalyst	Mass of catalyst, g	p_t, psig	W/F (g catalyst) (sec)/(g mole)	$1/K$, (g catalyst) (sec)/cm^3	$d_P \rho_P/6k_m$, (g catalyst) (sec)/cm^3	ηk_1, cm^3/(sec) (g catalyst)
Cylindrical pellets, $\frac{1}{8}$ in.	2.55	40	1,890	1.56	0.12	0.348
		40	839	1.47	0.08	0.361
		40	403	1.46	0.06	0.359
		100	2,120	2.80	0.25	0.197
		100	969	2.65	0.18	0.203
		400	1,980	7.18	0.90	0.080
		400	724	6.60	0.60	0.084
0.505-mm particles	1.13	100	1,040	1.07	0.008	0.472
		100	171	1.04	0.005	0.485
		400	701	2.50	0.020	0.202
		400	325	2.93	0.024	0.173
0.127-mm particles	0.739	100	404	1.06	0.0006	0.474
		400	240	2.51	0.0017	0.200

If the form of the intrinsic rate equation is satisfactory, K should be constant at a given pressure. The values of $1/K$ in Table 12-2 are about constant for a given catalyst and pressure. Results for different catalyst sizes may be different, because K includes the effects of external and internal mass-transfer resistances.

The next step is to evaluate and separate the external mass-transport resistance by means of Eq. (J). The mass-transfer coefficient k_m can be estimated from Fig. 10-2 and Eq. (10-10). With the first run as an example,

$$\text{Area of reactor tube} = \frac{\pi}{4} d_t^2 = \frac{\pi}{4}\left(\frac{1}{2}\right)(2.54)^2 = 1.27 \text{ cm}^2$$

$$G = \frac{30.1(2.016)}{22,400(1.27)} = 2.12 \times 10^{-3} \text{ g/(sec)(cm}^2)$$

$$d_P = \tfrac{1}{8}(2.54) = 0.318 \text{ cm}$$

$$\text{Re} = \frac{d_P G}{\mu} = \frac{0.318(2.12 \times 10^{-3})}{348 \times 10^{-7}} = 20$$

From Fig. 10-2, at this Reynolds number we have

$$j_D = 0.41$$

Using this result in Eq. (10-10) gives

$$k_m = \frac{j_D G}{\rho}\left(\frac{\mu}{\rho D}\right)^{-\frac{2}{3}} = \frac{0.41(2.12 \times 10^{-3})}{1.18 \times 10^{-3}}(0.78^{-\frac{2}{3}}) = 0.87 \text{ cm/sec}$$

where 1.18×10^{-3} g/cm^3 is the density of hydrogen at 40 psig and $-196°C$. The external resistance is given by the first term on the right side of Eq. (J),

$$\frac{d_P \rho_P}{6k_m} = \frac{0.318(1.91)}{6(0.87)} = 0.12 \text{ (g catalyst)(sec)/cm}^3$$

Similar calculations for the other runs give the results in column 6 of Table 12-2. Comparison of $1/K$ and $d_P\rho_P/6k_m$ shows that external resistance is important only for the $\frac{1}{8}$-in. pellets. Even then, the external contribution to the total resistance is less than 10% for all but the 400-psig runs. In the most severe case the external contribution is 0.90/7.18, or 13% of the total. For the two smaller particle sizes such a value would be very small. These results are typical and illustrate the previous statement about the unimportance of external resistances for small particles in fixed-bed reactors.

Since the external resistance is significant for some of the runs with $\frac{1}{8}$-in. pellets, it is expected that the internal resistance will also be important. This can be established when η is evaluated. Equation (J) can be used to separate the external resistance from K. Using the data for the first run,

$$\frac{K}{\eta k_1(K + 1)} = \frac{1}{K} - \frac{d_P \rho_P}{6k_m} = 1.56 - 0.12 = 1.44 \text{ (g catalyst)(sec)/cm}^3$$

The equilibrium constant K is

$$K = \frac{(C_p)_{eq}}{(C_o)_{eq}} = \frac{p_t(y_p)_{eq}}{p_t(y_o)_{eq}} = \frac{0.5026}{1 - 0.5026} = 1.01$$

Hence

$$\eta k_1 = \frac{K}{K + 1}\left(\frac{1}{1.44}\right) = \frac{1.01}{2.01}\left(\frac{1}{1.44}\right) = 0.348 \text{ cm}^3/(\text{sec})(\text{g catalyst})$$

Values calculated in this way for all the runs are given in the last column of Table 12-2.

If ηk_1 were available only for the $\frac{1}{8}$-in. pellets, we would have to follow the stepwise procedure outlined earlier, assuming an effective diffusivity and then using Fig. 11-7 and the ηk_1 values together to find k_1. However, the data for the small particles enable us to use the alternate procedure based on $\eta = 1$ for these particles. Since ηk_1 is known for two particle sizes, we can substantiate that η actually is unity for the particles. For a fourfold increase in particle diameter (0.505 vs 0.127 mm) the data in Table 12-2 show that ηk_1 is essentially unchanged at constant pressure. Since Φ_s, and hence η, vary with diameter [see Eq. (11-50)] for $\eta < 1$, it is clear that $\eta = 1$ for both sizes of small particles. Therefore, the values of ηk_1 for the particles in Table 12-2 are equal to k_1. We can use these results to evaluate k_c and K_a, employing Eq. (C). Averaging the three values for 100 psig, we have

$$\frac{0.472 + 0.485 + 0.474}{3} = \frac{k_c}{1 + K_a C_t}$$

$$C_t = \frac{p_t}{R_g T} = \frac{114.7/14.7}{82(77)} = 1.23 \times 10^{-3} \text{ g mole/cm}^3$$

and so

$$0.477 = \frac{k_c}{1 + 1.23 \times 10^{-3} K_a} \tag{M}$$

Similarly, at 400 psig

$$0.192 = \frac{k_c}{1 + 4.46 \times 10^{-3} K_a} \tag{N}$$

Solution of Eqs. (M) and (N), for k_c and K_a gives

$$K_a = 1.06 \times 10^3 \text{ cm}^3/\text{g mole}$$

$$k_c = 1.1 \text{ cm}^3/(\text{g catalyst})(\text{sec})$$

These two constants depend only on temperature. If data had been available at a series of temperatures, $\ln k_c$ could have been plotted against $1/T$ to obtain the activation energy E and the frequency factor $\mathbf{A}$, according to the Arrhenius equation. The adsorption equilibrium constant would also be expected to be an exponential function of temperature, according to a van't Hoff type of equation,

$$\frac{d(\ln K_a)}{dT} = \frac{\Delta H}{R_g T^2}$$

The analysis has now been completed for the assumed form for the intrinsic rate [Eq. (A)]. The results should be regarded as illustrative only. For accurate evaluation of k_c and K_a many more data points would be desired at several pressures and flow rates (W/F values). Then a statistical analysis could be used to obtain the best values of k_c and K_a.

While ηk_1 was the same for the two small particle sizes, Table 12-2 shows that ηk_1 is significantly less for the $\frac{1}{8}$-in. pellets. The comparison must be made at the same pressure because k_1 and η both vary with pressure. Therefore, considerable internal mass-transfer resistance exists. We can evaluate η from the ηk_1 data using Eq. (C). For example, at 40 psig

$$k_1 = \frac{k_c}{1 + K_a C_t} = \frac{1.1}{1 + 1.06 \times 10^3 (0.59 \times 10^{-3})} = 0.68$$

Then

$$\bar{\eta} = \frac{\bar{\eta}k_1}{k_1} = \frac{(0.348 + 0.361 + 0.359)(\frac{1}{3})}{0.68} = 0.53$$

This is an average value for the effectiveness factor at 40 psig. The results for the other pressures are shown in Table 12-3.

Table 12-3 *Effectiveness factors for $\frac{1}{8}$-in. pellets*

p, psig	C_t, moles/cm^3	k_1, cm$^3/$ (g catalyst)(sec)	$\bar{\eta}$
40	0.59×10^{-3}	0.68	0.53
100	1.23×10^{-3}	0.48	0.42
400	4.46×10^{-3}	0.19	0.42

Conclusions from this example may be summarized as follows:

1. The first-order form of rate equation with an adsorption term in the denominator fits the data.
2. For $\frac{1}{8}$-in. pellets both external and internal mass transport retard the rate. The internal mass-transfer resistance is sizable, as indicated by η values of the order of 0.5. The external resistance seldom exceeded 10% of the total.
3. For small catalyst particles both internal and external resistances were negligible, so that data for these particles could be used advantageously to calculate the constants in the intrinsic-rate equation.

12-2 Homogeneous Laboratory Reactors

Although our major interest in this chapter is heterogeneous laboratory reactors, for the sake of completeness we shall briefly review homogeneous systems. Experimental measurements in homogeneous reactors represent the influence of the intrinsic rate and the effects of mixing. The problem of extracting the intrinsic rate from the observed data is less difficult than for heterogeneous reactions. The methods of analysis for the two extremes of mixing—ideal tubular flow and ideal stirred tank—and for intermediate cases were considered in Chaps. 4 and 6. Isothermal operation is desirable for accurate results. Ideal stirred-tank performance can be achieved easily in most laboratory reactors. Ideal tubular-flow behavior is more difficult to realize. However, with differential reactors non ideal flow is not troublesome, because at low conversions the effect of a distribution of residence times is small.[1] Both stirred-tank and tubular-flow reactors are common laboratory forms. Gaseous reactions are usually studied in tubular-flow reactors, and both types are used for liquid systems. Integral tubular-flow reactors are not convenient for studying complex reactions because of the difficulty in extracting rate equations from the observed data. This difficulty was illustrated in Sec. 4-5, where we saw that analytic relations could be obtained for polymerization reactions only for a stirred-tank reactor.

The laboratory study of Fisher and Smith[2] on methane and sulfur vapor reaction illustrates typical apparatus and procedures for differential and integral tubular-flow reactors. An example of a laboratory differential reactor for a homogeneous catalytic reaction has been given by Matsuura et al.[3]

The batch recycle reactor shown in Fig. 12-1 combines some of the advantages of stirred-tank and tubular-flow operation. With high circulation rates and a small reactor volume with respect to total volume of system, the conversion per pass is very small. Differential operation is achieved in each pass through the reactor, yet over a period of time the conversion becomes significant. This is an advantage when it is difficult to make accurate measurements of the small concentration changes required in a differential tubular-flow reactor. The reaction system is charged with feed at the proper conditions, and then samples are withdrawn periodically for analysis. Suppose that the course of the reaction is followed by analyzing for a product. The data would give a curve like that in Fig. 12-2. If all the system is at the same composition, the rate at any time during the run is given by

$$\mathbf{r}V\,dt = V_t\,dC$$

$$\mathbf{r} = \frac{V_t}{V}\frac{dC}{dt} \tag{12-3}$$

[1] K. G. Denbigh, "Chemical Reactor Theory," p. 61, Cambridge University Press, Cambridge, 1965.

[2] R. A. Fisher and J. M. Smith, *Ind. Eng. Chem.*, **42**, 704 (1950).

[3] T. Matsuura, A. E. Cassano, and J. M. Smith, *AIChE J.*, **15**, 495 (1969).

where V = volume of reactor

V_t = total volume (reactor + lines + reservoir)

C = concentration of product

The slope of the curve of C vs t in Fig. 12-2 gives the rate at any time (or composition) during the run. By analyzing data for runs at different initial compositions and other runs at different temperatures, a rate equation can be obtained that includes the activation energy. A description of the apparatus and analysis procedure for a gaseous reaction (decomposition of acetone) is given by Cassano et al.[1] The validity of the assumption that the composition is uniform throughout all the system depends on the degree of mixing in the reservoir, the volume of the reservoir with respect to that of the whole system, and the conversion per pass in the reactor. If measurements at different circulation rates give the same rates of reaction, these conditions are closely approached. The apparatus can be designed so that the volumes of the reactor and connecting lines are small with respect to that of the reservoir. In the acetone study the reactor and lines had a volume of about 63 cm^3 and that of the reservoir was 5,500 cm^3. At the circulation rates employed the average residence time in the reactor was about 0.20 sec and that in the reservoir was 17 sec. Under these conditions the change in concentration of reactant per pass was only 0.0055% for a concentration change of 1% /hr in the reservoir.

As we saw from Fig. 3-1, the recycle reactor may be operated on a

[1]A. E. Cassano, T. Matsuura, and J. M. Smith, *Ind. Eng. Chem., Fund. Quart.*, **7**, 655 (1968).

Fig. 12-1   *Batch recycle reactor of tubular-flow type*

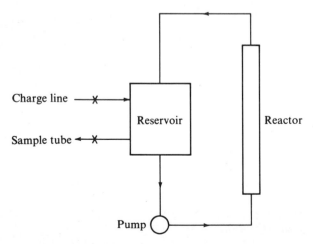

steady, continuous basis. Then ideal stirred-tank behavior is approached as the recycle rate is increased. Recycle ratios (recycle rate to feed rate) of 15 or more are usually sufficient to achieve results close to stirred-tank operation.

12-3 Heterogeneous Laboratory Reactors

Tubular-flow reactors operated in a differential, integral, or recycle manner are frequently used for investigating the kinetics of gas-solid catalytic reactions. A case of integral operation was discussed in Example 12-1. Another example is a study of hexane isomerization.[1] Illustrations of differential reactors which include experimental details are those for SO_2 oxidation,[2] C_2H_4 hydrogenation,[3] hydrogen oxidation,[4] and carbon monoxide oxidation.[5] Biskis and Smith[6] studied the hydrogenation of α-methyl styrene in a batch-type recycle reactor. The liquid stream (with

[1] A. Voorhies, Jr., and R. G. Beecher, presented at Sixty-First Annual Meeting AIChE, Los Angeles, Dec. 1–5, 1968.
[2] R. W. Olson, R. W. Schuler, and J. M. Smith, *Chem. Eng. Progr.*, **42**, 614 (1950).
[3] A. C. Pauls, E. W. Comings, and J. M. Smith, *AIChE J.*, **5**, 453 (1959).
[4] J. A. Maymo and J. M. Smith, *AIChE J.*, **12**, 845 (1966).
[5] S. Otani and J. M. Smith, *J. Catalysis*, **5**, 332 (1966).
[6] E. R. Biskis and J. M. Smith, *AIChE J.*, **9**, 677 (1963).

Fig. 12-2 *Experimental results in a batch recycle reactor*

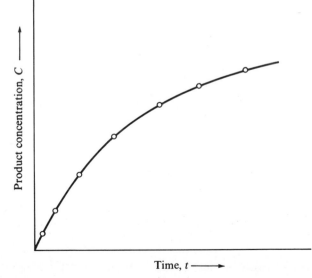

dissolved hydrogen) was pulsed as it flowed over the catalyst particles in order to reduce the external mass-transfer resistance.

Fixed-bed reactors have the disadvantage that external and, particularly, internal transport resistances may be significant. Special laboratory reactors have been designed to minimize or clarify these resistances. Carberry[1] has discussed several types. The recycle form can be used to reduce external transport resistances by increasing the recycle rate, thus increasing turbulence in the fluid around the catalyst pellets. The same result can be obtained by rotating a basket containing the catalyst pellets in a tank. Figure 12-3 shows a rotating-basket reactor developed by Tajbl et al.[2] for studying the kinetics of CO oxidation. In this apparatus external transport resistances can be independently decreased by increasing the rotation speed of the basket. Both the continuous-flow recycle and rotating-basket types can be made to perform as ideal stirred-tank reactors.

Instead of rotating the catalyst, Ford and Perlmutter[3] inserted a cylinder containing a deposited catalyst in a stirred vessel. Studies on the vapor-phase dehydrogenation of sec-butyl alcohol indicated that stirred-tank performance was achieved.

A number of single-pellet reactors have been employed in laboratory kinetic studies. Generally, large single pellets are used to facilitate measurements of intrapellet temperature and concentration profiles when the purpose is to evaluate the effects of internal transport resistances. Such reactors are not convenient for establishing the intrinsic rate equation. In a single-pellet apparatus the surrounding fluid is normally well mixed, so that ideal stirred-tank behavior can be used for interpreting experimental data. Descriptions of such systems are available.[4]

Many specialized laboratory reactors and operating conditions have been used. Sinfelt[5] has *alternately* passed reactants and inert materials through a tubular-flow reactor. This mode of operation is advantageous when the activity of the fixed bed of catalyst pellets changes with time. A system in which the reactants flow through a porous semiconductor catalyst, heated inductively, has been proposed for studying the kinetics of high-temperature (500 to 2000°C) reactions.[6] An automated microreactor

[1] J. J. Carberry, *Ind. Eng. Chem.*, **56**, 39 (1964).

[2] D. G. Tajbl, J. B. Simons, and J. J. Carberry, *Ind. Eng. Chem., Fund. Quart.*, **5**, 17 (1966).

[3] F. E. Ford and D. D. Perlmutter, presented at Fifty-fifth Annual Meeting AIChE, Chicago, Dec. 2–6, 1962.

[4] J. R. Balder and E. E. Petersen, *Chem. Eng. Sci.*, **23**, 1287 (1968); S. Otani and J. M. Smith, *J. Catalysis*, **5**, 332 (1966); J. A. Maymo and J. M. Smith, *AIChE J.*, **12**, 845 (1966); N. Wakao, M. R. Rao, and J. M. Smith, *Ind. Eng. Chem., Fund. Quart.*, **3**, 127 (1964).

[5] J. H. Sinfelt, *Chem. Eng. Sci.*, **23**, 1181 (1968).

[6] W. E. Ranz and B. A. Bydal, presented at Fifty-fifth Annual Meeting AIChE, Chicago, Dec. 2–6, 1968.

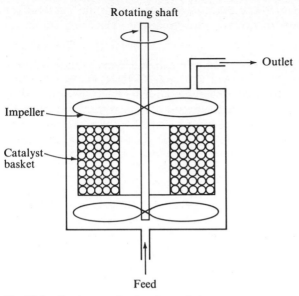

Rotating shaft

Outlet

Impeller

Catalyst
basket

Feed

Fig. 12-3   *Continuous stirred-tank catalytic reactor*
[*from D. G. Tajbl, J. B. Simons, and J. J. Carberry,*
Ind. Eng. Chem., Fund. Quart., 5, 17 (1966)]

for investigating catalytic reactions is described in detail by Harrison et
al.[1]; equipment specifications and control procedures are included.

12-4   Calculation of the Global Rate

For use in reactor design the global rate should be calculable at all locations
in the reactor. We suppose that the intrinsic rate equation is available.
The problem is to evaluate the global rate corresponding to possible bulk
concentrations C_b, bulk temperatures T_b, and flow conditions. If external
and internal temperature differences can be neglected, the problem is
straightforward and is essentially the reverse of the stepwise solution outlined in Sec. 12-1. The double-trial procedure is not necessary, because
$f(C)$ is known. The effective diffusivity of the catalyst pellet is required.
The equations we need are Eq. (10-1) for external diffusion,

$$\mathbf{r}_P = k_m a_m (C_b - C_s)$$

and Eq. (11-43) for internal diffusion,

$$\mathbf{r}_P = \eta f(C_s)$$

In addition, the relationship between η and Φ and the defining expression for

[1] D. P. Harrison, J. W. Hall, and H. F. Rase, *Ind. Eng. Chem.*, **57** (1), 18 (1965).

Φ are needed. Examples of these relationships are given in Fig. 11-7 for zero-, first-, and second-order intrinsic rate equations. The calculations are illustrated in Example 12-2.

Example 12-2   Using the intrinsic rate equation obtained in Example 12-1, calculate the global rate of the reaction o-$H_2 \rightleftharpoons p$-H_2 at 400 psig and $-196°C$, at a location where the mole fraction of ortho hydrogen in the bulk-gas stream is 0.65. The reactor is the same as described in Example 12-1; that is, it is a fixed-bed type with tube of 0.50 in. ID and with $\frac{1}{8} \times \frac{1}{8}$-in. cylindrical catalyst pellets of Ni on Al_2O_3. The superficial mass velocity of gas in the reactor is 15 lb/(hr)(ft^2). The effective diffusivity can be estimated from the random-pore model if we assume that diffusion is predominately in the macropores where Knudsen diffusion is insignificant. The macroporosity of the pellets is 0.36. Other properties and conditions are those given in Example 12-1.

Solution   The intrinsic rate equation developed in Example 12-1 is

$$r = \frac{1.1(C_o - C_p/K)}{1 + 1.06 \times 10^3\, C_t} \qquad \text{moles/(g catalyst)(sec)} \qquad \text{(A)}$$

It was shown that this expression could be written in terms of p-H_2 concentrations [Eq. (F) of Example 12-1] as

$$r = \frac{k_1(K + 1)}{K}(C_{eq} - C)_p$$

Then, the particular form for Eq. (11-43) for the rate of this reaction is [Eq. (G) of Example 12-1]

$$r_P = \eta\, \frac{k_1(K + 1)}{K}(C_{eq} - C_s)_p$$

where

$$k_1 = \frac{k_c}{1 + K_a C_t} = \frac{1.1}{1 + 1.06 \times 10^3\, C_t}$$

This equation can next be combined with Eq. (10-1), expressed in p-H_2 concentrations, to eliminate the surface concentration C_s. Then the rate can be written in terms of the bulk concentration C_b. In Example 12-1 [Eq. (I)] this resulted in

$$r_P = \frac{C_{eq} - (C_b)_p}{d_P \rho_P/6k_m + K/[\eta k_1(K + 1)]}$$

Now the problem is reduced to evaluating η, k_1, and k_m and substituting in Eq. (I) to obtain the global rate. We first calculate k_m, using properties from Example 12-1:

$$G = 15\, \frac{454}{3,600(2.54)^2(12)^2} = 2.03 \times 10^{-3}\ \text{g/(sec)(cm}^2)$$

$$\text{Re} = \frac{0.318(2.03 \times 10^{-3})}{348 \times 10^{-7}} = 19$$

From Fig. 10-2, at $Re = 19$ we have

$$j_D = 0.41$$

Then k_m is given by Eq. (10-10) as

$$k_m = \frac{j_D G}{\rho}\left(\frac{\mu}{\rho D}\right)^{-\frac{2}{3}} = \frac{0.41(2.03 \times 10^{-3})}{8.95 \times 10^{-3}}(0.78^{-\frac{2}{3}}) = 0.11 \text{ cm/sec}$$

Here 8.95×10^{-3} g/cm^3 is the density of H$_2$ at 400 psig and $-196°C$.
The total concentration of H$_2$ at 400 psig is

$$C_t = \frac{p_t}{R_g T} = \frac{414.7/(14.7)}{82(77)} = 4.46 \times 10^{-3} \text{ g mole/cm}^3$$

At this total concentration

$$k_1 = \frac{1.1}{1 + (1.06 \times 10^3)(4.46 \times 10^{-3})} = 0.19 \text{ cm}^3/(\text{g catalyst})(\text{sec})$$

According to the random-pore model, the effective diffusivity for macropore diffusion is given by Eq. (11-28). If the mass transfer is solely by bulk diffusion, Eq. (11-26) shows that $\bar{D}_M = \mathscr{D}_{AB}$, so that

$$D_e = \mathscr{D}_{AB}\varepsilon_M^2 = 0.00496(0.36)^2 = 6.4 \times 10^{-4} \text{ cm}^2/\text{sec}$$

It was proved in Example 11-6 that the curve for first-order kinetics in Fig. 11-7 could be used for a reversible first-order reaction, provided

$$\Phi_s' = \frac{r_s}{3}\sqrt{\frac{k_1(K + 1)\rho_P}{K D_e}}$$

Using the values just calculated for k_1 and D_e, we find

$$\Phi_s' = \frac{0.318}{2(3)}\sqrt{\frac{0.19(1.01 + 1)(1.91)}{1.01(6.4 \times 10^{-4})}} = 1.8$$

Then, from Fig. 11-7, $\eta = 0.43$.

The global rate can now be evaluated from Eq. (I) of Example 12-1. For $(y_o)_b = 0.65$

$$(C_{eq} - C_b)_p = C_t(y_{eq} - y_b)_p = 4.46 \times 10^{-3}\,[0.5026 - (1 - 0.65)]$$

$$= 6.8 \times 10^{-4} \text{ g mole/cm}^3$$

Hence

$$r_P = \frac{6.8 \times 10^{-4}}{0.318(1.91)/[6(0.11)] + 1.01/[0.43(0.19)(1.01 + 1)]} = \frac{6.8 \times 10^{-4}}{0.92 + 6.15}$$

$$= 0.96 \times 10^{-4} \text{ g mole/(g catalyst)(sec)}$$

The same procedure could be used to calculate the global rate for any other location in the reactor. Because the reaction was first order, the effectiveness factor is independent of concentrations, and therefore independent of location. For other intrinsic rate equations, Φ and η are dependent on the reactant concentration, as illustrated in Fig. 11-7. For these cases the effectiveness factor will vary from point to point in the reactor and cause a corresponding variation in the global rate.

Also, because the reaction is first order, C_s could be easily eliminated between Eqs. (10-1) and (11-43), and the global rate could be expressed explicitly in terms of C_b [Eq. (I) of Example 12-1]. For other kinetics a trial solution of Eqs. (10-1), (11-43), the defining equation for Φ, and the Φ-vs-η relationship would be simplest.

Finally, the first-order kinetics allowed a direct display of the relative importance of the diffusion resistances on the global rate. The quantities 0.92 and 6.15 in the denominator of the previous equation measure the resistances of external diffusion and internal diffusion plus reaction. The value of η divides the latter into internal-diffusion resistance and the resistance of the intrinsic reaction on the interior catalyst sites.

When external and internal temperature differences are also significant, trial solution is required to evaluate the global rate for a given C_b and T_b, regardless of the order of the intrinsic-rate equation. The effective thermal conductivity k_e is needed, as well as D_e. Equation (10-1) is applicable, and it is necessary to relate T_b and T_s by Eq. (10-13), written as

$$\mathbf{r}_P(-\Delta H) = ha_m(T_s - T_b)$$

To account for possible internal-temperature gradients, Eq. (11-42) is used instead of Eq. (11-43). Since the temperature effect on the intrinsic rate is expressed by the Arrhenius function, Eq. (11-42) may be written

$$\mathbf{r}_P = \eta f(C_s) \, e^{-E/R_g T_s} \tag{12-4}$$

where η is a function of Φ, β, and γ, as described in Sec. 11-10. These relations are sufficient to calculate the global rate at a given T_b and C_b. One procedure is to assume values of T_s and C_s and then calculate η from Φ, β, and γ as outlined in Sec. 11-10. Next $\mathbf{r}_P$ is obtained from Eq. (12-4), and finally, C_s and T_s are evaluated from Eqs. (10-1) and (10-13). If the calculated values of C_s and T_s do not agree with the assumed values, the process is repeated. When agreement is obtained, Eq. (12-4) gives the global rate. The procedure requires that the relation of η to Φ, β, and γ be available for the intrinsic rate of reaction. Such a relationship is available for a first-order rate (as illustrated by Fig. 11-10), but for other kinetics the information is limited.

At the beginning of this chapter it was pointed out that both external and internal resistances are frequently, but not always, important, and general criteria were discussed. Quantitative methods of evaluating $C_b - C_s$ and $T_s - T_b$ were developed in Secs. 10-3 and 10-4. Quantitative criteria for the significance of internal mass- and heat-transfer resistances were given in Chap. 11 by Eqs. (11-59) and (11-83). If preliminary calculations suggest that some of the internal or external resistances are negligible, the calculation of the global rate is simplified. For example, if external mass-transfer resistance is negligible, then $C_b \rightarrow C_s$, and Eq. (10-1) is not needed.

12-5 The Structure of Reactor Design

Before we take up quantitative aspects of heterogeneous reactor design in Chap. 13, let us survey the problem. It is appropriate to present an overview here because of the interaction between laboratory and large-scale reactors; i.e., the purpose of the laboratory study is to obtain a rate equation useful for designing the large reactor.

Suppose a new chemical reaction, or new example of a known type of reaction, has been discovered. The product has promising economic possibilities. Samples of product have been made and tested, and a preliminary economic analysis has been completed. It is decided to proceed with calculations and experimental work necessary to design a large reactor to produce the product. The overall structure of the problem is represented schematically in Fig. 12-4. The items at the right indicate the scientific dis-

Fig. 12-4 *Structure of reactor design* [*from J. M. Smith,* Chem. Eng. Progr., *64, 78 (1968)*]

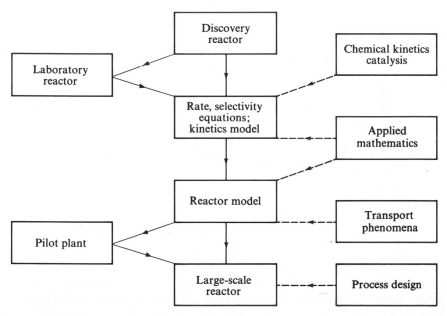

ciplines involved in the steps between the discovery of the reaction and the large-scale reactor. The vertical path from discovery to large reactor is an ideal, and usually a hypothetical, one. It implies that the intrinsic rate and selectivity equations can be obtained from analysis of data obtained in the discovery reactor. Usually this is impossible because the discovery reactor has not been designed according to the concepts discussed in Secs. 12-2 and 12-3. Therefore a laboratory (sometimes called bench-scale) reactor should be built with the objective of obtaining intrinsic rate and selectivity equations. The calculations for this step were illustrated in Example 12-1.

After the intrinsic rate equation has been established, it is used to develop a mathematical model of the large-scale reactor. The first step is to obtain the global rate at any point in the reactor, as illustrated in Example 12-2. To complete the model, equations are derived to give the conversion and selectivities in the product stream in terms of the proposed operating conditions. These equations and their solution are the subject of Chapter 13.

In principle it is possible to go directly from reactor model to large-scale reactor by the last vertical step in Fig. 12-4. However, this also is generally inadequate, and a pilot plant is necessary. The pilot plant may be omitted if the treatment of the physical processes in the model (velocity distribution, mixing, etc.) is adequate to predict their influence on conversion and selectivity, but usually this is not the case. Hence one function of the pilot plant is to evaluate the reactor model. The pilot plant is also important for investigating questions about construction materials, corrosion, instrumentation, operating procedures, and control—all important in the large-scale unit. The pilot plant does have limitations. It is not an economical or technically appropriate apparatus for obtaining intrinsic rate equations, developing a catalyst, studying the effects of contaminated feed streams, or evaluating catalyst life. It is used most efficiently for its primary purpose: helping the chemical engineer to derive an adequate model for the large-scale reactor. It serves as a testing ground for how well macro effects such as temperature and velocity variations can be combined with rate equations to predict the production rates of products.

Scale-up of a reaction process involves direct application of the results from a laboratory unit to determine the size and operating conditions for a large-scale unit. It avoids the two-step process advocated in this chapter of evaluating rate equations from laboratory data and then using them to design the large-scale unit. The simplicity of scale-up is attractive, but unfortunately it is not often useful for chemical-reactor design. For example, to maintain geometric and dynamic similarity in a fixed-bed catalytic reactor, a larger catalyst pellet would be indicated for the large-scale reactor than was used in the laboratory reactor. However, using larger pellets could introduce larger intrapellet resistances and seriously change the global rate, which would destroy the initial similarity. Beek[1] has discussed these and other scale-up problems for fixed-bed catalytic reactors.

In homogeneous reactors that are not isothermal, scale-up is dangerous because it is difficult to allow for the differences in heat-transfer conditions in laboratory and large-scale units. When precautions are taken to obtain the same global rate and the same heat-transfer conditions (for example, by requiring adiabatic or isothermal operation), scale-up concepts may be valuable. An illustration is given in Example 12-3. The similarity criteria on which scale-up methods depend are described by Johnstone and Thring.[2]

It is a simple matter to demonstrate that it is the interaction of reaction, geometric, and heat-transfer requirements that limits the value of scale-up. Consider a fixed-bed catalytic reactor. Suppose that the pressure drop has no effect on the rate and that plug flow exists. The mass balance for reactant is given by Eq. (12-2) as

$$\frac{W}{F} = \int_0^{x_e} \frac{dx}{\mathbf{r}_P}$$

where x_e is the conversion in the effluent from the reactor. Suppose plug-flow reactors of different geometries are being considered for the same chemical reaction. Equation (12-2) states that W/F will be the same for each reactor, provided that the value of the integral is the same. For the integral to be constant for a specified feed composition and conversion in the effluent, the global rate must be a function only of conversion. This is a necessary and sufficient criterion for scale-up. If this condition is met, the same conversion can be obtained in reactors of any geometry by fixing the magnitude of W/F.

The restriction that the global rate be a function only of conversion is the key to this simple scale-up procedure. Let us examine what it means. First, it rules out the possibility of intrapellet resistances, for then the catalyst-pellet size would influence the global rate. For example, the weight of catalyst W can be kept the same by reducing the size of the pellets and increasing their number. If intrapellet resistances were important, the rate would change. Similarly, external resistances are excluded, for if they were significant, the rate would become a function of velocity. Hence Eq. (12-2) is suitable for scale-up only when the global rate is determined by the

[1] John Beek, in T. B. Drew, J. W. Hoopes, Jr., and Theodore Vermeulen (eds.), "Advances in Chemical Engineering," vol. 3, p. 259, Academic Press, Inc., New York, 1962.

[2] R. E. Johnstone and M. W. Thring, "Pilot Plants, Models, and Scale-up Methods in Chemical Engineering," McGraw-Hill Book Company, New York, 1957. Chapter 15 discusses scale-up methods for chemical reactors and, in particular, describes the problems in scaling up heterogeneous types.

intrinsic rate of the chemical step at the interior site in the catalyst pellet. In addition, the whole reactor must operate isothermally or adiabatically. Otherwise the rate becomes a function of temperature in addition to conversion. The adiabatic condition is permissible because an energy balance provides a unique relation between temperature and conversion, so that the rate can be expressed solely in terms of conversion [see Eq. (5-5)].

Example 12-3 A pilot plant for a liquid-solid catalytic reaction consists of a cylindrical bed 5 cm in radius and packed to a depth of 30 cm with 0.5-cm catalyst pellets. When the liquid feed rate is 0.2 liter/sec the conversion of reactant to desirable product is 80%. To reduce pressure drop, a radial-flow reactor (Fig. 12-5) is proposed for the commercial-scale unit. The feed will be at a rate of 5 ft^3/sec and will have the same composition as that used in the pilot plant. The inside radius of the annular bed is to be 2 ft, and its length is also 2 ft. What must the outer radius of the bed be to achieve a conversion of 80%?

The same catalyst pellets will be employed in both reactors. Laboratory studies have shown that external resistances are negligible for this system. The heat of reaction is small, so that isothermal operation is achievable. Assume plug-flow behavior and that the density of the reaction liquid does not change with conversion. The bulk density of catalyst in either reactor is 1.0 g/cm^3.

Solution Under the stated conditions the global rate is a function only of conversion. Since the same conversion is desired for both reactors, the integral in Eq. (12-2) must be the same. Hence

$$\left(\frac{W}{F}\right)_{pilot} = \left(\frac{W}{F}\right)_{radial} \tag{A}$$

The molal feed rate F of reactant is equal to QC_0, where Q is the volumetric flow rate and C_0 is the concentration of reactant in the feed. Then, if V is the volume of the catalyst bed and ρ_B is its density, Eq. (A) becomes

$$\frac{\rho_B}{C_0}\left(\frac{V}{Q}\right)_{pilot} = \frac{\rho_B}{C_0}\left(\frac{V}{Q}\right)_{radial}$$

Substituting numbers, we have

$$\left(\frac{V}{Q}\right)_{radial} = \frac{\pi(5^2)(30)}{0.2(1,000)} = 3.75\pi \quad sec$$

$$V_{radial} = 3.75\pi(5) \quad ft^3$$

To obtain this volume in the radial-flow reactor the outer radius r_2 is given by

$$\pi L(r_2{}^2 - r_1{}^2) = \pi(2)(r_2{}^2 - 4) = 18.75\pi$$

$$r_2 = 3.7 \ ft$$

This result shows that the outer diameter of the radial-flow reactor should be about 7.5 ft for an inner diameter of 4 ft.

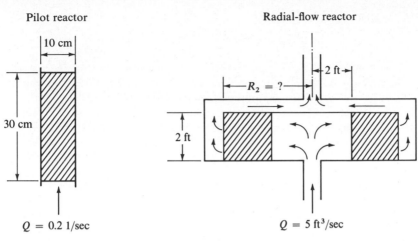

Pilot reactor   Radial-flow reactor

Fig. 12-5   *Scale-up from cylindrical to radial fixed-bed reactor*

In undergraduate courses in chemical engineering kinetics it is often advantageous to illustrate principles with laboratory experiments on suitable reactions. A recent publication[1] provides a detailed description of chemical reactions, equipment, experimental procedures, and methods of analyzing results, as developed by one educational institution. Of particular interest are actual data obtained by student groups. Experiments are described for measuring the performance of batch, stirred-tank, and tubular reactors with a homogeneous reaction (hydrolysis of acetic anhydride). Heterogeneous catalysis is studied in fixed-bed reactors, with the vapor-phase dehydrogenation and dehydration of isopropanol as the reactions. Other experiments have been conducted on fluidized-bed behavior, diffusion and reaction within catalyst pellets, diffusion in packed beds, reactor stability, biological reactions, and chemical analysis by chromatography. This report provides convenient source material for developing a laboratory course for chemical-reaction engineering.

Problems

12-1. A batch recycle reactor (Fig. 12-1), operated differentially and used for the gas-phase photolysis of acetone, has a volume of 62.8 cm^3, while the total volume of the system is 6,620 cm^3. Runs at 97°C and 870 mm Hg pressure were made at different initial concentrations of acetone in helium. Typical data for conversion to C_2H_6 are as given below for varying initial concentrations (gram moles per liter). The products of photolysis are primarily C_2H_6 and CO, according to the overall reaction

[1] J. B. Anderson, "A Chemical Reactor Laboratory for Undergraduate Instruction," Princeton University Department of Chemical Engineering, Princeton, N.J., September, 1968.

$$CH_3COCH_3 \rightarrow CO + C_2H_6$$

Therefore the rate of acetone decomposition can be followed by measuring the conversion of acetone to ethane in the reaction mixture. Preliminary measurements showed that the composition of the system was essentially uniform, indicating good mixing in the reservoir (Fig. 12-1) and a recycle rate high enough that the conversion per pass through the reactor was very small.

(a) Calculate the rate of decomposition of acetone for each run. (b) Evaluate constants k_1 and k_2, assuming that the rate expression is of the form

$$r = \frac{k_1 C}{1 + k_2 C}$$

where C is the concentration of acetone.

	Conversion to C_2H_6, %				
t, hr	4.66×10^{-3}	4.62×10^{-3}	1.99×10^{-3}	1.33×10^{-3}	1.27×10^{-3}
0	0	0	0	0	0
0.5	0.6	0.5	0.8	1.0	1.0
1.0	0.9	0.75	1.4	1.35	1.45
2.0	1.85	1.65	3.0	3.30	3.55
3.0	2.80	2.30	4.50	5.0	5.45
4.0	3.65	3.15	6.05	6.65	7.20
5.0	4.60	3.90	7.50	8.25	8.95

SOURCE: A. E. Cassano, T. Matsuura, and J. M. Smith, *Ind. Eng. Chem., Fund. Quart.*, **7**, 655 (1968).

12-2.   A tank reactor and separator (Fig. 12-6) are used to study the heterogeneous reaction between pure liquid A (phase 1) and reactant B dissolved in phase 2 (also liquid). The solvent in phase 2, reactant B, and the products of reaction are all insoluble in liquid A. No reaction occurs in the separator. The reactor operates isothermally at 25°C, and at this temperature A has a limited solubility in phase 2, the value being 2.7×10^{-5} g mole/liter. Phase 2 is dispersed as bubbles in continuous phase 1, which is recycled. There is excellent stirring in the reactor, but the fluid motion within the bubbles of phase 2 is insufficient to prevent some mass-transfer resistance. From independent measurements it is estimated that at average conditions the reaction resistance within the bubbles is 75% of the total resistance (mass-transfer plus reaction resistance).

(a) Derive a relationship between the concentration of reactant B entering the reactor in phase 2 and the concentration leaving the separator.

(b) In one experiment the following data were obtained:

Feed rate of phase 2 = 0.2 cm³/sec
Total liquid volume in reactor = 1,500 cm³
Fraction of volume in dispersed phase in reactor = 24%

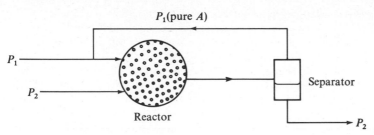

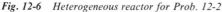

Fig. 12-6 *Heterogeneous reactor for Prob. 12-2*

Concentration of B entering reactor in phase 2 = 0.02 g mole/liter
Concentration of B (in phase 2) leaving separator = 0.0125 g mole/liter

Calculate the rate constant k for the second-order irreversible reaction of the form

$$A + B \rightarrow P$$

Assume that the average value of the reaction resistance may be used and that equilibrium exists at the interface between phase 1 and phase 2.

12-3. The liquid-phase hydrogenation of α-methyl styrene to cumene,

$$H_2(\text{dissolved}) + C_6H_5C(CH_3) = CH_2(l) \rightarrow C_6H_5CH(CH_3)_2(l),$$

has been studied in a recycle batch reactor at 80 psia and 55°C.[1] The reactor was packed with $\frac{1}{8}$-in. Al_2O_3 spheres containing 0.5 wt % palladium on the outer surface as a catalyst. The reaction was followed by analyzing micro-samples of reaction liquid at various times. The recycle rate was such that the conversion per pass was very low. The reaction liquid was always saturated with hydrogen. Data for two runs are as follows:

	Run C4	Run C5
Moles α-methyl styrene in charge	29.0	27.9
Mass of catalyst in reactor, g	30.7	30.7

t, hr	Mole fraction cumene	
0	0.0264	0.1866
4	0.0530	0.2114 (4.5 hr)
8	0.0810	0.2332
10.5	0.0995	0.2552 (11.5 hr)
17.5	0.1518	. . .
21.5	0.1866	. . .

(*a*) For the integral changes in composition (during a run) shown, a plot of cumene mole fraction vs time might be expected to be curved; yet the data

[1] E. G. Biskis and J. M. Smith, *AIChE J.*, **9**, 677 (1963).

show a linear relationship. What does this signify about the kinetics of the reaction? (b) Calculate rates of hydrogenation from the data given.

12-4. A test procedure is to be developed for measuring the activity of various catalysts. A small, fixed-bed reactor is available in the laboratory. The same gaseous reaction will be used to evaluate each catalyst. The method of analysis of the product stream is fairly accurate but not good enough to justify differential reactor operation. Hence the sample of any catalyst tested will be sufficient in amount to obtain conversions above 30%. The feed rate, temperature, pressure, feed and effluent composition (or conversion), and mass of catalyst in the reactor will be routinely measured. The activity will be defined as the ratio of a measurable variable for any catalyst to the value of that variable for a standard catalyst. The variable chosen should be a true measure of the intrinsic rate of the reaction at a catalyst site.

(a) Propose a method of operating the reactor so as to obtain the activity of any catalyst. That is, describe what variables will be held constant during a run and within the integral reactor, as well as the variables which will be held the same for the given catalyst and the standard catalyst. Also give the variable whose value will be used in formulating the activity. Illustrative of the several variables to be considered are the size of the catalyst pellet, mass velocity of the gas through the reactor, and temperature.

(b) In part (a) no consideration was given to the experimental effort needed to carry out the proposed activity tests. Suppose that the ideal procedure described in (a) requires excessive effort. Propose an alternate, perhaps less desirable, but simpler method of operation.

13

DESIGN OF HETEROGENEOUS CATALYTIC REACTORS

Our next objective is to utilize the chemical kinetics and transport information already developed to predict the performance of large heterogeneous reactors. It is not necessary to consider further the individual transport processes within and external to the catalyst pellet. Methods of combining these steps with the kinetics of the chemical reactions to obtain a global rate were summarized in Chap. 12. The goal now is to use the global rate information to evaluate the composition of the effluent from a reactor for a specified set of design conditions.[1] The design conditions that must be fixed are the temperature, pressure, and composition of the feed stream, the dimensions of the reactor and catalyst pellets, and enough about the surroundings to evaluate the heat flux through the reactor walls.

[1] An *a priori* design procedure, in contrast to a succession of larger experimental reactors, has been ably summarized by R. H. Wilhelm [*J. Pure Appl. Chem.*, **5**, 403 (1962)] for fixed beds. This review includes relations between the intrinsic and global rates and, as such, summarizes the effects of external and internal transport processes described in Chaps. 10 to 12.

One of the most common catalytic reactors is the fixed-bed type, in which the reaction mixture flows continuously through a tube filled with a stationary bed of catalyst pellets. Because of its importance, and because considerable information is available on its performance, most attention will be given to this reactor type. Fluidized-bed and slurry reactors are also considered later in the chapter. Some of the design methods given are applicable also to fluid-solid noncatalytic reactions. The global rate and integrated conversion-time relationships for noncatalytic gas-solid reactions will be considered in Chap. 14.

Only reactors operating at pseudo-steady state are discussed here; that is, the design methods presented are applicable when conditions such as catalyst activity do not change significantly in time intervals of the order of the residence time in the reactor. Brief comments about transient conditions are included in Sec. 13-8, but these refer to changes from one stable state to another. Finally, in Sec. 13-12 a brief introduction to optimization is presented. As mentioned in Chap. 1, a quantitative treatment of optimization is not discussed in this book.

FIXED-BED REACTORS

Quantitative design methods of increasing complexity are considered in Secs. 13-3, 13-4, 13-6, and 13-7. First, however, let us summarize construction and operating characteristics of fixed-bed reactors.

13-1 Construction and Operation

Fixed-bed reactors consist of one or more tubes packed with catalyst particles and operated in a vertical position. The catalyst particles may be a variety of sizes and shapes: granular, pelleted, cylinders, spheres, etc. In some instances, particularly with metallic catalysts such as platinum, instead of using single particles, wires of the metal are made into screens. Multiple layers of these screens constitute the catalyst bed. Such screen or gauze catalysts are used in commercial processes for the oxidation of ammonia and the oxidation of acetaldehyde to acetic acid.

Because of the necessity of removing or adding heat, it may not be possible to use a single large-diameter tube packed with catalyst. In this event the reactor may be built up of a number of tubes encased in a single body, as illustrated in Fig. 13-1. The energy exchange with the surroundings is obtained by circulating, or perhaps boiling, a fluid in the space between the tubes. If the heat effect is large, each catalyst tube must be small (tubes as small as 1.0-in. diameter have been used) in order to prevent excessive tem-

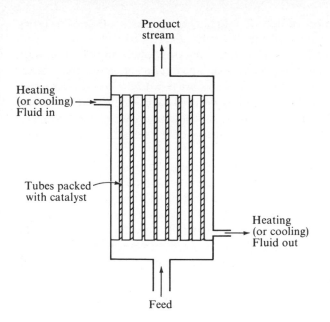

Fig. 13-1   *Multitube, fixed-bed reactor*

peratures within the reaction mixture. The problem of deciding how large
the tube diameter should be, and thus how many tubes are necessary, to
achieve a given production forms an important problem in the design of
such reactors.

A disadvantage of this method of cooling is that the rate of heat
transfer to the fluid surrounding the tubes is about the same all along
the tube length, but the major share of the reaction usually takes place
near the entrance. For example, in an exothermic reaction the rate will
be relatively large at the entrance to the reactor tube owing to the high
concentrations of reactants existing there. It will become even higher as the
reaction mixture moves a short distance into the tube, because the heat
liberated by the high rate of reaction is greater than that which can be
transferred to the cooling fluid. Hence the temperature of the reaction
mixture will rise, causing an increase in the rate of reaction. This continues
as the mixture moves up the tube, until the disappearance of reactants has a
larger effect on the rate than the increase in temperature. Farther along the
tube the rate will decrease. The smaller amount of heat can now be removed
through the wall with the result that the temperature decreases. This situation
leads to a maximum in the curve of temperature vs reactor-tube length.
An example is shown in Fig. 13-2 for a TVA ammonia-synthesis reactor.

Such a maximum temperature (hot spot) is characteristic of an exothermic reaction in a tubular reactor (Chap. 5).

As mentioned in Chap. 3, other means of cooling may be employed besides circulating a fluid around the catalyst tube. Dividing the reactor into parts with intercoolers between each part (see Fig. 13-3) is a common procedure. Another scheme which has worked satisfactorily for reactions of moderate heat of reaction, such as the dehydrogenation of butene, is to add a large quantity of an inert component (steam) to the reaction mixture.

The particular scheme employed for cooling (or heating) the fixed-bed reactor depends on a number of factors: cost of construction, cost of operation, maintenance, and special features of the reaction, such as the magnitude of ΔH. For example, the heat of reaction of naphthalene oxidation is so high that small externally cooled tubes provide about the only way to prevent excessive temperatures in fixed-bed equipment. In sulfur dioxide oxidation the much smaller heat of reaction permits the use of less expensive large-diameter adiabatic catalyst bins in series, with external intercoolers for removing the heat evolved. In dehydrogenation of butene the heat of reaction is also fairly low, so that small-diameter catalyst tubes need not be used. Here the use of external heat exchangers is possible (the reaction is endothermic, and energy must be supplied to maintain the temperature), and a satisfactory system can be designed by alternating

Fig. 13-2   *Variation in temperature profile with on-stream time in fixed-bed ammonia-synthesis reactor* [*by permission from A. V. Slack, H. Y. Allgood, and H. E. Maune,* Chem. Eng. Progr., *49, 393 (1953)*]

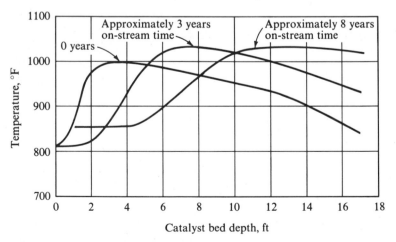

adiabatic reaction sections with heat exchangers. However, in this case there are several auxiliary advantages in adding a hot inert material (steam) to supply the energy. The blanketing effect of the steam molecules reduces the polymerization of the butadiene product. Also, the steam lowers the partial pressure of the hydrocarbons, and in so doing improves the equilibrium yield of the reaction system.

It may be observed that all the operating devices mentioned for energy exchange have the objective of preventing excessive temperatures or maintaining a required temperature level; i.e., they are attempts to achieve isothermal operation of the reactor. There are many advantages to operating at near-isothermal conditions. For example, in the naphthalene oxidation process it is necessary to control the temperature to prevent the oxidation from becoming complete, which would result in carbon dioxide and water instead of phthalic anhydride. This is a common situation in partial-oxidation reactions. The air oxidation of ethylene is a second illustration. Another reason for avoiding excessive temperatures is to prevent lowering of the catalyst activity. Changes in structure of the solid catalyst particles as the temperature is increased may reduce the activity of the

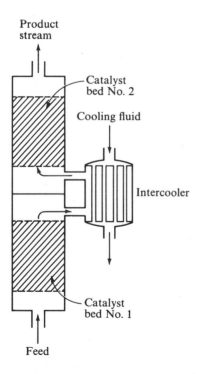

Fig. 13-3 *Divided reactor with inter-cooler between catalyst beds*

catalyst for the desired reaction. The selectivity of the catalyst may also be adversely affected. The life of the catalyst may be reduced by operating at higher than normal temperatures. For example, the iron oxide catalyst for the ammonia-synthesis reaction shows a more rapid decrease in activity with time if the synthesis unit is operated above the normal temperature range of 400 to 550°C.

The reason for limiting the temperature in sulfur dioxide oxidation is based on two factors: excessive temperatures decrease the catalyst activity, as just mentioned, and the equilibrium yield is adversely affected at high temperatures. This last point is the important one in explaining the need to maintain the temperature level in the dehydrogenation of butene. Still other factors, such as physical properties of the equipment, may require limiting the temperature level. For example, in reactors operated at very high temperatures, particularly under pressure, it may be necessary to cool the reactor-tube wall to preserve the life of the tube itself.

The problem of regenerating the catalyst to restore activity may be a serious one in the fixed-bed reactor. In a great many instances the catalyst is too valuable to discard. If the catalyst activity decreases with time, frequent regeneration may be necessary. Even when the cost is so low that regeneration is not required, shutting down the process and starting up again after new catalyst has been added is an expensive procedure. If this is necessary at frequent intervals, the entire process may become uneconomical. The exact economic limit on shutdown time depends on the particular process, but in general, if the activity cannot be maintained over a period of several months, the cost of shutdowns is likely to be prohibitive. Of course, regeneration *in situ* is one way out of this difficulty. However, *in situ* regeneration requires two or more reactors if continuous operation is to be maintained, and hence increases the initial cost of the installation. The most successful fixed-bed reactor systems are those where the catalyst activity is sustained for long periods without regeneration. The fixed-bed reactor requires a minimum of auxiliary equipment and is particularly suitable for small commercial units, where the investment of large sums for instrumentation, catalyst handling, and the like, would be uneconomical.

In order to prolong the time between regenerations and shutdowns, the reactor tube may be made longer than required for the reaction itself. For example, suppose a length of 3 ft is necessary to approach the equilibrium conversion with fresh catalyst of high activity. The reactor may be built with tubes 10 ft long. Initially, the desired conversion will be obtained in the first 3 ft. As the catalyst activity falls off, the section of the bed in which the reaction is mainly accomplished will move up the bed, until finally all 10 ft are deactivated. This technique can be used only with certain types of reactions but it has been employed successfully in the ammonia synthesis.

13-2 Outline of the Design Problem

The global rate of reaction tells how much reaction is occurring at any location in the reactor, provided the temperatures and concentrations are known. To evaluate temperatures and concentrations, energy and mass balances are formulated for a fluid flowing through a bed of catalyst pellets. These balances are normally in the form of differential equations, the solution of which gives the temperature and concentrations at any location, including the reactor exit. These concentrations are the solution to the design problem. In fixed-bed reactor design it is assumed that the variables are constant in a volume element associated with a single catalyst pellet. For example, a constant global rate is used for this element. In writing the balances as differential equations[1] it is further assumed that this volume element is small enough that such variables as temperature and concentration vary continuously with position in the reactor. In large-scale reactors the volume of a single pellet is usually so small with respect to the total volume that these assumptions are reasonable.

When temperature gradients exist in the reactor, analytical solution of the differential equations is not possible. Then the design process entails the numerical solution of a set of coupled differential equations. A stepwise procedure is employed, starting at the entrance to the reactor and moving along radially and longitudinally in increments the size of the pellet diameter. This process is carried out by machine computation. The first step is to convert the differential equations to a difference form.[2] The whole procedure is illustrated in Sec. 13-7 with an SO_2 oxidation problem. In this case, to reduce the number of calculations, the increment size is larger than the pellet diameter. With machine computation the number of calculations is not so important, and an increment the size of the pellet is practical.

The complexity of the design problem depends primarily on the amount of temperature variation in the reactor. The reactants enter the catalyst bed at uniform temperature and composition, but as they pass through the bed and reaction occurs, the accompanying heat of reaction induces both longitudinal and radial variations in temperature. To account for the effect of temperature variations we must know the heat-transfer characteristics of fixed beds. This subject is discussed in Sec. 13-5.

The difficulty in the design calculations depends directly on the type and magnitude of temperature variations that need be taken into

[1] H. A. Deans and L. Lapidus [*AIChE J.*, **6**, 656, 663 (1960)] consider the reactor to consist of an assembly of cells, with complete mixing within each cell and with a separate cell associated with each catalyst pellet. The balances then are written as *difference* equations.

[2] Numerical solution of differential equations by machine computation is described in L. Lapidus, "Digital Computation for Chemical Engineers," McGraw-Hill Book Company, chap. 4, New York, 1960.

account. In the simplest case the entire reactor operates isothermally, and the global rate is a function only of concentration. Design procedures are discussed in Sec. 13-3. For simplified flow behavior, analytical solution is possible for the conversion in the effluent leaving the reactor.

A more important case is the adiabatic reactor. As mentioned before, it is difficult in practice to achieve isothermal operation because most reactions have a significant heat effect. In adiabatic operation heat transfer to the reactor wall can be neglected, and the temperature change is only in the longitudinal direction. The global rate will vary in passing along the bed because of concentration and temperature changes. This case is considered in Sec. 13-4.

The greatest difficulty arises when heat transfer through the reactor wall must be taken into account. This type of operation occurs when it is necessary to supply or remove heat through the wall and the rate of energy transfer is not sufficient to approach isothermal operation. It is a frequent occurrence in commercial fixed-bed reactors because of their size and because fluid velocities must be low enough to allow for the required residence time. The presence of the catalyst pellets prevents sufficient turbulence and mixing to obtain uniform concentration and temperature profiles. The concentration and temperature will change in both the radial and the longitudinal direction. Then the global rate will also vary in both the longitudinal and the radial direction, and the integration of the mass balance requires the numerical-solution technique described above. A general treatment of this type of reactor entails an incremental calculation across the diameter of the reactor tube for a small longitudinal increment and the repetition of this process for each successive longitudinal increment (Sec. 13-7). If plug-flow behavior can be assumed, the stepwise calculations in the radial direction can be eliminated. A so-called "one-dimensional" simplification based on plug flow is illustrated in Sec. 13-6.[1]

13-3 Isothermal Operation

In Chap. 4 the plug-flow model was used as a basis for designing homogeneous tubular-flow reactors. The equation employed to calculate the conversion in the effluent stream was either Eq. (3-13) or Eq. (4-5). The same equations and the same calculational procedure may be used for fixed-bed catalytic reactors, provided that plug-flow behavior is a valid assumption. All that is necessary is to replace the homogeneous rate of reaction in those equations with the global rate for the catalytic reaction.

[1] One- and two-dimensional models of fixed-bed reactors are compared for a numerical case in G. F. Froment, Current Design Status, Fixed-bed Catalytic Reactors, *Ind. Eng. Chem.*, **59**, 18 (1967).

In the isothermal case deviations from plug-flow behavior arise from variations of axial velocity in the radial direction and from axial dispersion. The radial variation in velocity leads to a residence-time distribution, as does axial dispersion. The effects on the conversion of these deviations from plug-flow performance were discussed in Chap. 6. In fixed-bed reactors these effects are usually small for isothermal conditions, so that plug-flow equations are satisfactory. However, deviations are potentially large in nonisothermal reactors, and radial variations in particular must be taken into account. Hence we shall consider here the general form of the mass-balance equation, even though it is seldom needed for an isothermal reactor.

A section of a fixed-bed catalytic reactor is shown in Fig. 13-4. Consider a small volume element of radius r, width Δr, and height Δz, through which reaction mixture flows isothermally. Suppose that radial and longitudinal diffusion can be expressed by Fick's law, with D_r and D_L as effective diffusivities,[1] based on the total (void and nonvoid) area perpendicular to the direction of diffusion. We want to write a mass balance for a reactant over the volume element. With radial and longitudinal diffusion and longitudinal convection taken into account, the input term is

$$-2\pi r\,\Delta z\,D_r\left(\frac{\partial C}{\partial r}\right)_r + 2\pi r\,\Delta r\left(uC - D_L\frac{\partial C}{\partial z}\right)_z$$

The volume of the element, which is $2\pi r\,\Delta r\,\Delta z$, contains both solid catalyst pellets and surrounding fluid. The concentration in the fluid phase is constant within the element, and the global rate is known (from Chap. 12) in terms of this bulk-fluid concentration. The axial velocity of the reacting

[1] These diffusivities include both molecular and turbulent contributions. In fixed beds some convection exists even at low velocities.

Fig. 13-4   *Annular element in a fixed-bed catalytic reactor*

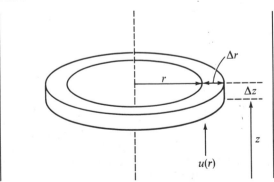

fluid can vary in the radial direction. It will be described as a local superficial velocity u based on the total (void plus nonvoid) cross-sectional area.

The output term is

$$-2\pi r \, \Delta z \, D_r \left(\frac{\partial C}{\partial r}\right)_{r+\Delta r} + 2\pi r \, \Delta r \left(uC - D_L \frac{\partial C}{\partial z}\right)_{z+\Delta z}$$

The reactant converted in the element is

$$\mathbf{r}_P \rho_B 2\pi r \, \Delta r \, \Delta z$$

and the accumulation is zero because the reactor is at steady state. In these expressions

$\mathbf{r}_P = $ *global* rate per unit mass of catalyst
$\rho_B = $ density of catalyst in the bed
$u = $ superficial velocity in the axial direction

Combining these terms according to Eq. (3-1), dividing by $2\pi \, \Delta r \, \Delta z$, and taking the limit as Δr and Δz approach zero yields

$$\frac{\partial}{\partial r}\left(rD_r \frac{\partial C}{\partial r}\right) + r \frac{\partial}{\partial z}\left(-uC + D_L \frac{\partial C}{\partial z}\right) - \mathbf{r}_P \rho_B r = 0 \qquad (13\text{-}1)$$

If the diffusivities are not very sensitive to r or to z and the velocity is not a function of z, Eq. (13-1) may be written

$$D_r \left(\frac{1}{r}\frac{\partial C}{\partial r} + \frac{\partial^2 C}{\partial r^2}\right) - u\frac{\partial C}{\partial z} + D_L \frac{\partial^2 C}{\partial z^2} - \mathbf{r}_P \rho_B = 0 \qquad (13\text{-}2)$$

If the velocity varies with z (due to changes in temperature or number of moles in a gaseous reaction), Eq. (13-1) should be used. If the concentration entering the reactor is C_0, and if there is no axial dispersion in the feed line, the boundary conditions are

$$C = C_0 \qquad \text{at } z = 0 \text{ for all } r \qquad (13\text{-}3)$$

$$uC_0 = -D_L \left(\frac{\partial C}{\partial z}\right)_{>0} + u(C)_{>0} \qquad \text{at } z = 0 \text{ for all } r \qquad (13\text{-}4)$$

$$\frac{\partial C}{\partial r} = 0 \qquad \text{at } r = r_0 \text{ for all } z \qquad (13\text{-}5)$$

$$\frac{\partial C}{\partial r} = 0 \qquad \text{at } r = 0 \text{ for all } z \qquad (13\text{-}6)$$

It is instructive to write Eq. (13-2) in dimensionless form by introducing the conversion x and dimensionless coordinates r^* and z^* based on the diameter of the catalyst pellet:

$$x = \frac{C_0 - C}{C_0} \tag{13-7}$$

$$r^* = \frac{r}{d_P} \tag{13-8}$$

$$z^* = \frac{z}{d_P} \tag{13-9}$$

In terms of these variables Eq. (13-2) becomes

$$-\frac{1}{Pe_r}\left[\frac{1}{r^*}\frac{\partial x}{\partial r^*} + \frac{\partial^2 x}{(\partial r^*)^2}\right] + \frac{\partial x}{\partial z^*} - \frac{1}{Pe_L}\frac{\partial^2 x}{(\partial z^*)^2} - \frac{\mathbf{r}_P \rho_B d_P}{C_0 u} = 0$$

$$\tag{13-10}$$

where

$$Pe_r = \frac{u d_P}{D_r} \tag{13-11}$$

$$Pe_L = \frac{u d_P}{D_L} \tag{13-12}$$

Equation (13-10) shows that the conversion depends on the dimensionless reaction-rate group $\mathbf{r}_P \rho_B d_P / C_0 u$ and the radial and axial Peclet numbers, defined by Eqs. (13-11) and (13-12).

When the velocity u varies with *radial location*, a stepwise numerical solution of Eq. (13-10) or Eq. (13-2) would be needed. Axial velocities do vary with radial position in fixed beds. The typical profile[1] is flat in the center of the tube, increases slowly until a maximum velocity is reached about one pellet diameter from the wall, and then decreases sharply to zero at the wall. The radial gradients are a function of the ratio of tube to pellet diameter. Excluding the zero value at the wall, the deviation between the actual velocity at any radius and the average value for the whole tube is small when $d/d_P > 30$. In single-tube reactors this condition is usually met; e.g., for $\frac{1}{4}$-in. pellets this corresponds to a tube diameter of about 8 in.

Radial Peclet numbers have been measured, and some of the results are shown in Fig. 13-5. Above a modified Reynolds number $d_P G/\mu$ of about 40, Pe_r is independent of flow rate and has a magnitude of about 10. The two terms involving radial gradients in Eq. (13-10) are generally small for isothermal operation. The only way[2] that concentration gradients can

---

[1] C. E. Schwartz and J. M. Smith, *Ind. Eng. Chem.*, **45**, 1209 (1953).

[2] Note that this is not so if radial temperature gradients exist. Then the rate can vary significantly with r, and large concentration gradients develop. In such a case use is made of the data in Fig. 13-5 as described in Example 13-6.

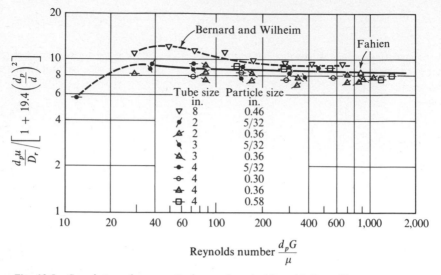

Fig. 13-5 *Correlation of average Peclet number, $d_p u / D_r$, with Reynolds number and d_p/d*

develop is through the variation of velocity with r. Also, the relatively large value of Pe_r further reduces the magnitude of these two terms. If we neglect them, Eq. (13-10) reduces to

$$\frac{\partial x}{\partial z^*} - \frac{1}{Pe_L} \frac{\partial^2 x}{(\partial z^*)^2} - \frac{\mathbf{r}_P \rho_B d_P}{C_0 u} = 0 \qquad (13\text{-}13)$$

or, in dimensional form,

$$-u\frac{\partial C}{\partial z} + D_L \frac{\partial^2 C}{\partial z^2} - \mathbf{r}_P \rho_B = 0 \qquad (13\text{-}14)$$

This expression still includes the effect of longitudinal dispersion. It is identical to Eq. (6-41), except that the rate for a homogeneous reaction has been replaced with the global rate $\mathbf{r}_P \rho_B$ per unit volume for a heterogeneous catalytic reaction. In Sec. 6-9 Eq. (6-41) was solved analytically for first-order kinetics to give Eq. (6-45). Hence that result can be adapted for fixed-bed catalytic reactors. The first-order global rate would be

$$\mathbf{r}_P = k_c C \qquad \text{g moles/(sec)(g catalyst)}$$

where k_c is the rate constant for the catalyst in the bed. The solution of Eq. (13-14) is the same as Eq. (6-45), but instead of Eq. (6-46) we have

$$\beta = \left(1 + 4k_c \rho_B \frac{D_L}{u^2}\right)^{\frac{1}{2}} \qquad (13\text{-}15)$$

If the catalytic rate equation were not first order, numerical solution of Eq. (13-14) with Eqs. (13-3) and (13-4) would be necessary.

To utilize Eqs. (6-45) and (13-15), or solutions of Eq. (13-14) for other kinetics, requires the longitudinal diffusivity D_L as well as the global rate. D_L has been measured for both gases and liquids flowing through fixed beds. The experimental data of McHenry and Wilhelm[1] (for gases) and theoretical predictions[2] both indicate that $Pe_L = 2$ for Reynolds numbers above about 10. For liquids[3] Pe_L is less, particularly at low Reynolds numbers.

The significance of the longitudinal-dispersion term in Eq. (13-14) depends on the length L of the reactor, the effective diffusivity, and the velocity. For very low velocities ($Re < 1$) in very short reactors, longitudinal dispersion can be significant when the conversion is not also low. At other conditions dispersion may be neglected. Then Eq. (13-14) takes the form

$$-u\frac{dC}{dz} = \mathbf{r}_P \rho_B \tag{13-16}$$

Multiplying the numerator and the denominator by the cross-sectional area of the reactor gives

$$-Q\frac{dC}{dV} = \mathbf{r}_P \rho_B$$

or

$$\frac{V}{Q} = -\frac{1}{\rho_B}\int_{C_0}^{C}\frac{dC}{\mathbf{r}_P} = \frac{C_0}{\rho_B}\int_{C_0}^{C}\frac{dx}{\mathbf{r}_P} \tag{13-16a}$$

Since $V\rho_B$ is the mass of catalyst in the reactor and $C_0 Q = F$, an alternate form is Eq. (12-2),

$$\frac{W}{F} = \int \frac{dx}{\mathbf{r}_P}$$

Equations (13-16a) are of the same form as the equations for homogeneous plug-flow reactors [Eq. (4-5)]; they are the constant-density version because the system is isothermal, and because for gaseous reactions no allowance has been made for a change in number of moles. Equation (12-2) is the analog of Eq. (3-13) for heterogeneous reactions and is applicable for variable-density conditions. The application of these equations to reactor design is the same as discussed, for example, in Example 4-6.

[1] K. W. McHenry, Jr., and R. H. Wilhelm, *AIChE J.*, **3**, 83 (1957).

[2] R. Aris and N. D. Amundson, *AIChE J.*, **3**, 280 (1957).

[3] J. J. Carberry and R. H. Bretton, *AIChE J.*, **4**, 367 (1958); E. J. Cairns and J. M. Prausnitz, *Chem. Eng. Sci.*, **12**, 20 (1960).

13-4 Adiabatic Operation

Large-diameter reactors approach adiabatic operation more closely than isothermal operation. With insulation around the tube the adiabatic model may be a good representation of actual operating experience. Again, radial gradients in temperature and concentration can arise only from velocity variations as in isothermal operation. Also, axial concentration gradients are insignificant, except in very shallow beds. This conclusion has been verified by Carberry and Wendel,[1] who integrated Eq. (13-13) for first-order kinetics and adiabatic operation and found little difference from plug-flow behavior except for short bed lengths. Hence Eq. (12-2) is often applicable. The solution for the temperature and composition as a function of reactor length requires, in addition, an energy balance. If plug-flow behavior is also assumed for energy transfer, the energy balance is the same as that used for adiabatic homogeneous reactors [Eq. (5-4) or Eq. (5-5)]. The method of solution for the heterogeneous case is essentially the same as for the homogeneous one (Example 5-2). It is illustrated here for the production of styrene.

Example 13-1   Wenner and Dybdal[2] studied the catalytic dehydrogenation of ethyl benzene and found that with a certain catalyst the rate could be represented by the reaction

$$C_6H_5C_2H_5 \rightleftharpoons C_6H_5CH = CH_2 + H_2$$

The global rate was given as

$$r_P = k \left(p_E - \frac{1}{K} p_S p_H \right)$$

where p_E = partial pressure of ethyl benzene

p_S = partial pressure of styrene

p_H = partial pressure of hydrogen

The specific reaction rate and equilibrium constants are

$$\log k = -\frac{4{,}770}{T} + 4.10$$

where k is the pound moles of styrene produced per hr(atm)(lb catalyst) and T is in degrees Kelvin.

t, °C	K
400	1.7×10^{-3}
500	2.5×10^{-2}
600	2.3×10^{-1}
700	1.4

[1] J. J. Carberry and M. M. Wendel, *AIChE J.*, **9**, 129 (1963).
[2] R. R. Wenner and F. C. Dybdal, *Chem. Eng. Progr.*, **44**, 275 (1948).

It is desired to estimate the volume of reactor necessary to produce 15 tons of styrene a day, using vertical tubes 4 ft in diameter, packed with catalyst pellets. Wenner and Dybdal considered this problem by taking into account the side reactions producing benzene and toluene. However, to simplify the calculations in this introductory example, suppose that the sole reaction is the dehydrogenation to styrene, and that there is no heat exchange between the reactor and the surroundings. Assume that under normal operation the exit conversion will be 45%. However, also prepare graphs of conversion and temperature vs catalyst bed depth, up to equilibrium conditions. The feed rate per reactor tube is 13.5 lb moles/hr for ethyl benzene and 270 lb moles/hr for steam. In addition,

Temperature of mixed feed entering reactor $= 625^{\circ}C$
Bulk density of catalyst as packed $= 90$ lb/cu ft
Average pressure in reactor tubes $= 1.2$ atm
Heat of reaction $\Delta H = 60,000$ Btu/lb mole
Surroundings temperature $= 70^{\circ}F$

Solution   The reaction is endothermic, so that heat must be supplied to maintain the temperature. Energy may be supplied by adding steam to the feed to provide a reservoir of energy in its heat capacity. An alternate approach of transferring heat from the surroundings is utilized for the same system in Example 13-3.

In this problem the operation is adiabatic, and the energy balance is given by Eq. (5-4) as

$$\frac{F}{M}(-\Delta H)\,dx = F_t c_p\,dT \tag{A}$$

Since x refers to the conversion of ethyl benzene, $F/M = 13.5$ lb moles/hr. As there is a large excess of steam, it will be satisfactory to take $c_p = 0.52$. Then the heat capacity of the reaction mixture will be

$$F_t c_p = (270 \times 18 + 13.5 \times 106)(0.52) = 3270 \text{ Btu/}^{\circ}\text{F(hr)}$$

Substituting numerical values in Eq. (A), we obtain

$$13.5(-60,000)\,dx = 3{,}270\,dT$$

$$-dT = 248\,dx$$

$$T - 1{,}616 = -248x \tag{B}$$

where T is in degrees Rankine, and $1616^{\circ}R$ is the entering temperature of the feed.

Equation (12-2), written in differential form, with the weight of catalyst expressed as $dW = \rho_B A_c dz$, where A_c is the cross-sectional area, is

$$F\,dx = \mathbf{r}_P \rho_B A_c dz$$

$$dz = \frac{F}{\mathbf{r}_P \rho_B A_c}\,dx = \frac{13.5\,dx}{90(0.7854)(16)\mathbf{r}_P} = \frac{0.0119}{\mathbf{r}_P}\,dx \tag{C}$$

The partial pressures can be expressed in terms of the conversion as follows. At any conversion x the moles of each component are

Steam $= 20$
Ethyl benzene $= 1 - x$
Styrene $= x$
Hydrogen $= x$
Total moles $= 21 + x$

Then

$$p_E = \frac{1 - x}{21 + x} \quad (1.2)$$

$$p_S = p_H = \frac{x}{21 + x} \quad (1.2)$$

Then the rate equation becomes

$$\mathbf{r}_P = \frac{1.2}{21 + x} k \left[(1 - x) - \frac{1.2}{K} \frac{x^2}{21 + x} \right]$$

or, with the expression for k determined by Wenner and Dybdal,[1]

$$\mathbf{r}_P = \frac{1.2}{21 + x} (12,600)\, e^{-19,800/T} \left[(1 - x) - \frac{1.2}{K} \frac{x^2}{21 + x} \right] \qquad (D)$$

Substituting this value of $\mathbf{r}_P$ in Eq. (C) gives an expression for the catalyst-bed depth in terms of the conversion and temperature,

$$dz = \frac{21 + x}{1,270,000}\, e^{19,800/T} \left[(1 - x) - \frac{1.2}{K} \frac{x^2}{21 + x} \right]^{-1} dx \qquad (E)$$

Equations (B) and (E) can be solved numerically for the bed depth for any conversion. If the coefficient of dx in Eq. (E) is designated as α, then we may write Eq. (E) as

$$\Delta z = \bar{\alpha}\, \Delta x \qquad (E')$$

At $z = 0$, $x = 0$, and $T = 1616°R$ (625°C),

$$\alpha_0 = \frac{21}{1,270,000}\, e^{12.25}\, \frac{1}{1 - 0} = 3.30$$

If an increment Δx of 0.1 is chosen, the temperature at the end of the increment is, from Eq. (B),

$$T_1 = 1,616 - 248(0.1) = 1591°R$$

Then at the end of the first increment[2]

$$\alpha_1 = \frac{21 + 0.1}{1,270,000}\, e^{12.43} \left[(1 - 0.1) - \frac{1.2}{0.28} \left(\frac{0.1^2}{21 + 0.1} \right) \right]^{-1} \qquad (F)$$

$$\alpha_1 = 4.18 \frac{1}{0.90 - 0.0020} = 4.65$$

[1] In the exponential term T has been converted to degrees Rankine.
[2] The value of K is estimated to be 0.28 at 1591°R from the tabulation of data given in the problem statement.

The bed depth required for the first increment is given by Eq. (E′) as

$$\Delta z = \frac{3.30 + 4.65}{2}(0.1) = 0.40 \text{ ft}$$

Proceeding to the second increment, we find

$$T_2 = 1{,}616 - 248(0.2) = 1566°R$$

$$\alpha_2 = \frac{21 + 0.2}{1{,}270{,}000} e^{12.66}\left[(1 - 0.2) - \frac{1.2}{0.22}\left(\frac{0.2^2}{21 + 0.2}\right)\right]^{-1} = 6.60$$

$$z_2 - z_1 = \bar{\alpha}\,\Delta x = \frac{\alpha_1 + \alpha_2}{2}(0.1) = \frac{4.65 + 6.60}{2}(0.1) = 0.56 \text{ ft}$$

$$z_2 = 0.40 + 0.56 = 0.96 \text{ ft}$$

The results of further calculations are shown in Table 13-1.

Table 13-1   Data for conversion of ethyl benzene to styrene in an adiabatic reactor

Conversion	Temperature		Catalyst-bed depth, ft
	$°R$	$°C$	
0	1616	625	0
0.10	1591	611	0.40
0.20	1566	597	0.96
0.30	1542	584	1.75
0.40	1517	570	2.93
0.50	1492	556	4.84
0.55	1480	549	6.3
0.60	1467	542	8.5
0.65	1455	536	13.2
0.69	1445	530	∞

The rate of reaction becomes zero at a conversion of about $x = 0.69$ and a temperature of 1445°R, as determined from Eqs. (B) and (D). From Fig. 13-6 it is found that a bed depth of 3.8 ft is required for a conversion of 45%. The production of styrene from each reactor tube would be

$$\text{Production/tube} = 13.5(0.45)(104)(24) = 15{,}200 \text{ lb/day} \qquad \text{or } 7.6 \text{ tons/day}$$

Hence two 4-ft-diameter reactor tubes packed with catalyst to a depth of at least 3.8 ft would be required to produce 15 tons/day of crude styrene.

13-5 Heat Transfer in Fixed-bed Reactors

Because radial temperature profiles have a large effect on reactor performance, a summary of the heat-transfer characteristics of fixed beds is given in this section. We shall return to the design problem in Sec. 13-6.

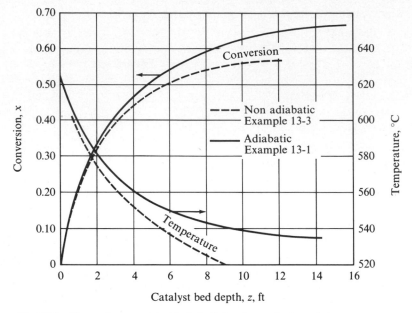

Fig. 13-6 *Conversion vs catalyst-bed depth for the production of styrene from ethyl benzene*

In homogeneous reactor design (Chap. 5) only the effect of heat exchange with the surroundings was considered. Radial mixing was supposed to be sufficiently good that all the resistance to energy transfer could be concentrated at the reactor wall. The temperature was assumed to be flat up to the wall, where a discontinuous change to the wall temperature occurred. The temperature of the reaction mixture changed only in the longitudinal direction. This one-dimensional model is normally satisfactory for homogeneous reactors, because radial mixing is sufficient to give a reasonably flat profile. However, in fixed beds this is not as good an assumption, because the catalyst pellets hinder radial mixing. As we shall see in Sec. 13-6, use of this model requires only a heat-transfer coefficient at the reactor wall, h_w. When radial temperature variations within the reactor must be taken into account, an effective thermal conductivity k_e, analogous to D_r for mass transfer, is needed. Experimental data are available for both h_w and k_e. The correlations for these parameters are presented briefly here.

Wall Heat-transfer Coefficients   If T_b is the bulk mean temperature of the reaction fluid and T_w is the wall temperature, h_w is defined by

$$dQ = h_w(T_b - T_w)\, dA_h \qquad\qquad (13\text{-}17)$$

where Q is the heat-transfer rate to the wall and A_h is the wall area. The presence of the solid particles increases the heat-transfer coefficient in a packed bed several times over that in an empty tube at the same gas flow rate. In the earliest experimental investigations of the subject,[1] the results were reported as ratios of the coefficient in the packed bed to that in the empty pipe. It was found that this ratio varied with the ratio of the pellet diameter to the tube diameter, reaching a maximum value at about $d_P/d = 0.15$. Colburn's results for the ratio of heat-transfer coefficients in packed and empty tubes, h_w/h, are:

d_P/d	0.05	0.10	0.15	0.20	0.25	0.30
h_w/h	5.5	7.0	7.8	7.5	7.0	6.6

Presumably the large increase in heat-transfer coefficient for a packed tube over that for an empty tube is due to the mixing, or turbulence, caused by the presence of the solid particles. This turbulence tends to prevent the buildup of a slow-moving layer of fluid next to the wall and also increases the radial transfer of heat within the fluid in the tube. Up to a point, decreasing the particle size increases the importance of these factors, and the heat-transfer coefficient continues to increase. However, the maximum at a certain d_P/d ratio suggests that another factor is involved. This is concerned with the size of the radial eddies in the fluid in the bed. As the particle size continues to decrease, the size of the eddies decreases, and the distance over which each mixing process occurs is decreased. Also, there is a larger number of more-or-less stagnant films between the fluid and the solid particles which the heat must cross in reaching the wall. The maximum in the heat-transfer coefficient would represent the point at which this second factor counterbalances favorable effects of the mixing process obtained with the smaller particles.

Leva and others[2] have correlated fixed-bed transfer coefficients for air over a wide range of variables. These can be used to predict the heat-transfer rate by Eq. (13-17) when the variation in temperature within the bed itself is neglected. Leva's expressions are

$$\frac{h_w d}{k} = 0.813 \left(\frac{d_P G}{\mu}\right)^{0.9} e^{-6d_P/d} \qquad \text{heating} \qquad (13\text{-}18)$$

[1] A. P. Colburn, *Ind. Eng. Chem.*, **28**, 910 (1931); *Trans. AIChE*, **26**, 166 (1931); E. Singer and R. H. Wilhelm, *Chem. Eng. Progr.*, **46**, 343 (1950).

[2] M. Leva, *Ind. Eng. Chem.*, **40**, 857 (1947); M. Leva and M. Grummer, *Ind. Eng. Chem.*, **40**, 415 (1948); D. Thoenes, Jr., and H. Kramers, *Chem. Eng. Sci.*, **8**, 271 (1958); T. J. Hanratty, *Chem. Eng. Sci.*, **3**, 209 (1954); S. Yagi and N. Wakao, *AIChE J.*, **5**, 79 (1960); G. F. Froment, *Ind. Eng. Chem.*, **59**, 18 (1967).

$$\frac{h_w d}{k} = 3.50 \left(\frac{d_P G}{\mu}\right)^{0.7} e^{-4.6 d_P/d} \qquad \text{cooling} \qquad (13\text{-}19)$$

Heat Transfer in a Packed Bed (Effective Thermal Conductivity)   In a bed of solid particles through which a reacting fluid is passing, heat can be transferred in the radial direction by a number of mechanisms. However, it is customary to consider that the bed of particles and the gas may be replaced by a hypothetical solid in which conduction is the only mechanism for heat transfer. The thermal conductivity of this solid has been termed the *effective thermal conductivity* k_e. With this scheme the temperature T of any point in the bed may be related to k_e and the position parameters r and z by the differential equation

$$-\mathbf{r}_P \rho_B \, \Delta H - G c_p \frac{\partial T}{\partial z} + k_e \left(\frac{1}{r}\frac{\partial T}{\partial r} + \frac{\partial^2 T}{\partial r^2}\right) = 0 \qquad (13\text{-}20)$$

This expression is derived in the same way as the mass-transfer expression, Eq. (13-2). However, the axial-dispersion term has been omitted, since its contribution is negligible except for very shallow beds and low velocities.[1] Also, in Eq. (13-20) k_e has been assumed to be independent of radial location. Because the velocity varies with r, k_e can also change. Data suggest that the conductivity decreases somewhat as the reactor wall is approached.[2] For simplicity we shall use an average value in our discussions.

Equation (13-20) has been used by several investigators to evalute k_e from experimental data. The data are obtained by flowing heated fluids, usually air, through a bed of pellets in a tube equipped with a cooling jacket. Then the term involving $\mathbf{r}_P$ does not exist. The measured temperatures are compared either directly with Eq. (13-20) or with an integrated form of it to extract k_e. By analyzing the mechanisms of radial heat transfer, correlations have been developed for predicting the conductivity from observable properties. Owing to the many mechanisms in the heterogeneous system, k_e is not an ordinary thermal conductivity, but a property of the bed that depends on a large number of variables, such as gas flow rate, particle diameter, porosity, true thermal conductivity of the gas and of the solid phases, and temperature level. Therefore the most logical method of correlating data is to divide k_e into separate contributions, each of which corresponds to a mechanism of heat transfer. Such methods have evolved from a very simple division,[3] through several extensions,[4] to more recent

[1] J. J. Carberry and M. M. Wendel, *AIChE J.*, **9**, 129 (1963).

[2] R. W. Schuler, V. P. Stallings, and J. M. Smith, *Chem. Eng. Prog., Symp. Ser. 4*, **48**, 19 (1952).

[3] T. E. W. Schumann and V. Voss, *Fuel*, **13**, 249 (1934).

[4] G. Damkoeler, "Der Chemie Ingenieur," Euken Jacob 3, part 1, p. 44, Akademische Verlagsgesellschaft m.b.H., Leipzig, 1937; R. H. Wilhelm, W. C. Johnson, R. Wynkoop, and

developments.[1] The approach of Argo attempts to account for all the processes that would be expected to contribute to radial energy transfer. His development is summarized in the following paragraphs and leads to Eq. (13-40) for predicting k_e.

Prediction of Effective Thermal Conductivity   At steady-state conditions the heat-transfer rate through a cylindrical plane parallel to the centerline of a packed cylindrical bed of catalyst will be the sum of the part passing through the void space and the part passing through the solid material. If q is this total rate of heat flow per unit area of the plane, the point effective thermal conductivity k_e per unit of plane area is defined by[2]

$$q = -k_e \frac{\partial T}{\partial r} = q_{\text{void}} + q_{\text{solid}} \tag{13-21}$$

This concept is illustrated in the upper half of Fig. 13-7. The temperature gradient in Eq. (13-21) applies to the bed as a whole and deserves some explanation because of the possibility of temperature differences between the solid and fluid phases in a packed bed.

Bunnell et al.[3] measured temperatures at the same radial position in the gas and in the center of the solid particle, under nonreacting conditions, and found no significant difference. The solid particles were activated alumina, a material of relatively low conductivity. On the basis of these data, we shall assume that the average temperature of the particle is the same as that of the gas at the same radial position. This does not require that the temperature gradient within a single solid particle coincide with that of the fluid phase. As an illustration, Bunnell measured effective thermal conductivities of the order of 0.10 to 0.30 Btu/(hr)(ft)(°F), while the conductivity of the alumina pellets is estimated to be about 0.5. Clearly the solid particles are not conducting heat under the gross radial gradient observed for the bed as a whole. For solid packings with thermal conductivities of 0.5 or greater, the gradients within the pellets would normally

---

D. W. Collier, *Chem. Eng. Progr.*, **44**, 105 (1948); D. G. Bunnell, H. B. Irvin, R. W. Olson, and J. M. Smith, *Ind. Eng. Chem.*, **41**, 1977 (1949); H. B. Irvin, R. W. Olson, and J. M. Smith, *Chem. Eng. Progr.*, **47**, 287 (1951); H. Vershoor and G. C. A. Schuit, *Appl. Sci. Res.*, **42** (A.2), 97 (1950).

[1] E. Singer and R. H. Wilhelm, *Chem. Eng. Progr.*, **46**, 343 (1950); W. B. Argo and J. M. Smith, *Chem. Eng. Progr.*, **49**, 443 (1953); John Beek, Design of Packed Catalytic Reactors, in "Advances in Chemical Engineering," vol. 3, p. 229, Academic Press, Inc., New York, 1962.

[2] Note that q_{void} is the heat-transfer rate through the void space per unit of void plus nonvoid area. It is not based on a unit of void area. Similarly, q_{solid} is based not on a unit of solid area, but on a unit of void plus solid area.

[3] D. G. Bunnell, H. G. Irvin, R. W. Olson, and J. M. Smith, *Ind. Eng. Chem.*, **41**, 1977 (1948).

be less than those for the whole bed. This situation is shown graphically in the lower portion of Fig. 13-7. The higher the true thermal conductivity of the solid, the less the temperature change in the solid phase, so that with metallic pellets the temperature within a single particle would be expected to approach a constant value. The gradient in Eq. (13-21) is the value for the bed as a whole and is designated as the *observed gradient* in Fig. 13-7. It is the same as that in the continuous-fluid phase.

As shown in Fig. 13-7, there may be a considerable difference in temperature between the gas and the particle. This difference results in heat transfer to the particle on one side and from the particle back to the gas stream on the other side. The temperature gradient within the particle would then be just sufficient to transfer this heat from one side to the other.

The heat passing across the plane in the void space is the sum of that due to molecular conduction, turbulent diffusion, and radiation. Since these paths are in parallel,

$$q_{\text{void}} = -\epsilon_B (k_c + k_{td} + k_r) \frac{\partial T}{\partial r} \tag{13-22}$$

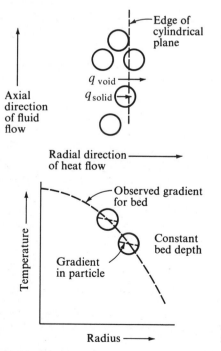

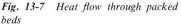

Fig. 13-7 *Heat flow through packed beds*

where ϵ_B is the void fraction and k indicates the conductivity is based on a unit of void area.[1]

In order to avoid geometrical difficulties an ideal model of the packed bed will be employed to evaluate the heat transfer through the particle. The methods by which heat can enter a particle from its inner side are radiation, convection from the gas stream, and conduction through point contacts and stagnant fillets, as indicated in Fig. 13-8. Heat is transferred through the particle and leaves the other side by the same three mechanisms. The three processes are in series, and the whole will be designated as the *series mechanism.* Hence

$$q_{solid} = -k_{series} \left(1 - \epsilon_B\right) \frac{\partial T}{\partial r} \qquad (13\text{-}23)$$

[1] Conductivities based on a unit of void plus nonvoid area are related to the void-area values by equations such as $k_r = k_r \epsilon_B$

Fig. 13-8   Heat transfer in a single spherical particle

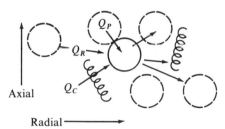

Q_R = Heat transfer by radiation

Q_C = Heat transfer by convection from gas stream

Q_P = Heat transfer by conduction through point contacts and stagnant gas

Axial

Radial

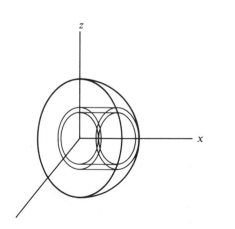

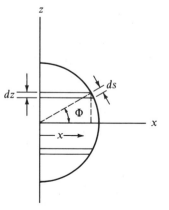

Combining Eqs. (13-21) to (13-23) gives an expression for the point effective thermal conductivity in terms of contributions for each mechanism responsible for radial heat transfer,

$$k_e = \epsilon_B(k_c + k_{td} + k_r) + (1 - \epsilon_B)k_{\text{series}}$$
$$= k_c + k_{td} + k_r + k_{\text{series}} \tag{13-24}$$

FLUID-PHASE CONDUCTION   The value of k_c in Eq. (13-24) is the molecular conductivity of the fluid. Its value will change with radial position in the bed because of temperature variations. Molecular conductivity in gases is so low that it is not an important contribution to k_e, while in liquids it may be quite important.

TURBULENT DIFFUSION   The contribution of turbulent diffusion k_{td} is a measure of heat transfer as a result of turbulent mixing of portions of the gas stream at different temperatures. As Singer and Wilhelm[1] have pointed out, its value can be estimated from measurements of mass transfer radially by the same mechanism. On this basis

$$k_{td} = \epsilon_B k_{td} = \rho_b c_p D_r$$

where ρ_b = density of the bulk fluid

c_p = specific heat

D_r = effective diffusivity in the radial direction

Figure 13-5 summarizes data for D_r in terms of Pe_r. In terms of the radial Peclet number determined from mass-transfer data, if d_P is the particle diameter and u is the superficial velocity, the turbulent-diffusion contribution is

$$k_{td} = \frac{\rho_b c_p d_P u}{\epsilon_B Pe_r} = \frac{c_p d_P G}{\epsilon_B Pe_r} \tag{13-25}$$

The value of k_{td} may vary significantly with radial position, especially near the tube wall, where the void fraction and velocity are rapidly changing. The results in Fig. 13-5 are average figures.

RADIATION CONTRIBUTION   Schuler et al.[2] used measured temperature gradients and found that k_r reached a maximum at 0.4 to 0.6 of the distance from the center to the wall, and that the temperature level was more important in establishing k_r than the temperature gradient. Thus Damkoehler's[3]

[1] E. Singer and R. H. Wilhelm, *Ind. Eng. Chem.*, **46**, 343 (1950).

[2] R. W. Schuler, V. P. Stallings, and J. M. Smith, *Chem. Eng. Progr.*, *Symp. Ser. 4*, **48**, 19 (1952).

[3] G. Damkoehler, "Der Chemie Ingenieur," Euken Jacob **3**, part 1, p. 44, Akademische Verlagsgesellschaft m.b.H., Leipzig, 1937.

simplified expression for radiation (corrected for emissivities σ less than 1.0),

$$k_r = 4\frac{\sigma}{2-\sigma}d_P(0.173)\frac{\bar{T}^3}{100^4} \tag{13-26}$$

was found to agree well with more elaborate methods, even though it included only the bulk mean temperature of the bed, $\bar{T}$ [in degrees Rankine in Eq. (13-26)].

Schuler's investigations were made at temperatures as high as 400°C and with large temperature gradients. At low mass velocities the radiation contribution to k_e was of the order of 10 to 15%. Hence an approximate equation such as Eq. (13-26) is satisfactory for estimating k_r, except for reactions at very high temperatures.

THE SERIES MECHANISM   In order to obtain a useful solution to the problem of radial heat transfer through the pellet, the following assumptions are made:

1. A constant temperature gradient exists in the fluid phase for a radial distance of one pellet diameter.
2. A constant temperature gradient exists in the solid particle.
3. Heat flow in the particle is radial in direction.
4. The mean temperature of the particle is equal to the temperature of the surrounding fluid at the centerline of the particle.
5. An average heat-transfer coefficient may be considered applicable to the entire surface of the particle.

Assumption 3 permits the heat transfer in the particle to be summed over elements normal to the direction of flow. One-half of such a spherical particle with the normal element is illustrated in the lower part of Fig. 13-8. An energy balance on the element leads to the expression

$$dQ_s = k_s(2\pi z\, dz)\frac{T-T_s}{x} = h2\pi z(T_s - T_b)\, ds \tag{13-27}$$

where dQ_s = rate of heat flow through an element of solid particle

$\quad\quad k_s$ = molecular thermal conductivity of solid particle, based on a unit area of solid

$\quad\quad z$ = coordinate normal to heat flow (parallel to axis of bed)

$\quad\quad x$ = coordinate in direction of heat flow

$\quad\quad T$ = temperature at center plane of particle = temperature of fluid phase at same radial position

T_s = temperature of surface of particle at a distance x

$T_b = T + x\, dT/dx$ = temperature of bulk fluid at a distance x

ds = element of surface through which heat is transferred to gas stream or to other particles

h = total heat-transfer coefficient from particle surface to fluid or to other particles

The element of surface area available for heat transfer to the gas stream may be related to dz by

$$ds = \frac{dz}{\cos \Phi} = \frac{d_P\, dz}{2x} \tag{13-28}$$

Combining Eqs. (13-28) and (13-27) gives the following expression for T_s:

$$T_s = \frac{2k_s T + h d_P T_b}{2k_s + h d_P}$$

With the fact that $T_b = T + x\, dT/dx$, the equation for T_s becomes

$$\frac{T - T_s}{x} = \frac{-h d_P}{2k_s + h d_P}\, \frac{dT}{dx} \tag{13-29}$$

This value of $(T - T_s)/x$ may now be substituted into the second member of Eq. (13-27) to obtain an expression for dQ_s in terms of the constant gradient dT/dx. The differential equation so obtained can be integrated from 0 to z, to obtain the total heat flow through the particle at a vertical plane a distance x from the center. The result is

$$Q_s = -\frac{h k_s d_P}{2k_s + h d_P}\, \pi z^2\, \frac{dT}{dx} \tag{13-30}$$

Inasmuch as πz^2 is the cross-sectional area of the solid at this point, and dT/dx is the radial temperature gradient in the fluid phase, Eq. (13-30) may be written

$$Q_s = -\frac{h k_s (d_P/2)}{k_s + h d_P/2}\, A\, \frac{dT}{dr} \tag{13-31}$$

Here Q_s is the heat transferred through the solid, based on the temperature gradient of the bed, and A is the surface area of one half of the spherical particle.

Since q_{solid} is based on a *unit* area of normal plane, including both void and nonvoid surface,

$$q_{\text{solid}} = \frac{Q_s}{A}(1 - \epsilon_B) = -\frac{hk_s(d_P/2)}{k_s + hd_P/2}(1 - \epsilon_B)\frac{dT}{dr} \tag{13-32}$$

Comparison of Eqs. (13-23) and (13-32) gives the desired expression,

$$k_{\text{series}} = \frac{hk_s d_P}{2k_s + hd_P} \tag{13-33}$$

Similar derivations for cylindrical pellets, oriented in the bed with the cylindrical axis either parallel or normal to the axis of the tube, yield the same result as Eq. (13-33), provided that d_P is equal to the length of the cylinder.

The evaluation of k_{series} requires a knowledge of the total heat-transfer coefficient h, which will be the sum of the convection coefficient from the fluid surrounding the particle, the radiation contribution from adjacent particles, and a conduction contribution from pellets in contact with each other. Thus h may be defined as

$$h = \frac{Q_s}{A(\Delta T)_m} = \frac{q_c + q_r + q_P}{A(\Delta T)_m} = h_c + h_r + h_P \tag{13-34}$$

The convection coefficient h_c can be predicted from the data on heat transfer between solids and fluids flowing in packed beds. Such data were given in Fig. 10-2. The radiation and conduction coefficients, h_r and h_P, depend on the value of $(\Delta T)_m$ defined by Eq. (13-34). A derivation based on the same assumptions employed in obtaining Eq. (13-33) leads to the results

$$(\Delta T)_m = -\frac{d_P k_s}{2(2k_s + hd_P)}\frac{dT}{dr} \tag{13-35}$$

Since Eq. (13-26) gives the radiation contribution in terms of k_r, h_r must be expressed in terms of k_r,

$$h_r = -k_r \frac{A'}{A}\frac{dT/dr}{(\Delta T)_m} \tag{13-36}$$

where $A' = \pi d_P^2/4$ is the projected area of one-half the spherical particle and $A = \pi d_P^2/2$ is the area of one-half the spherical particle. Combining Eqs. (13-35) and (13-36) to eliminate $(\Delta T)_m$ gives

$$h_r = \frac{k_r(2k_s + hd_P)}{d_P k_s} \tag{13-37}$$

Equation (13-37) permits the evaluation of h_r by employing Eq. (13-26) for k_r. The analogous expression for h_P is

$$h_P = \frac{k_P(2k_s + hd_P)}{d_P k_s} \tag{13-38}$$

The conductivity k_P determines the heat transfer through the pellet by solid contact with an adjacent pellet, and through stagnant gas fillets surrounding the contact points. The correlation of Wilhelm et al.,[1] summarized by Eq. (13-39), may be used with Eq. (13-38) to determine h_P.

$$\log k_P = -1.76 + 0.0129 \frac{k_s}{\epsilon_B} \tag{13-39}$$

COMBINATION OF SEPARATE CONTRIBUTIONS   Summation of the expressions for each mechanism in accordance with Eq. (13-24) results in a composite equation for the effective thermal conductivity,

$$k_e = \epsilon_B \left[k_c + \frac{d_P c_p G}{Pe_r \epsilon_B} + 4 \frac{\sigma}{2 - \sigma} d_P (0.173) \frac{\bar{T}^3}{100^4} \right]$$

$$+ (1 - \epsilon_B) \frac{h k_s d_P}{2k_s + hd_P} \tag{13-40}$$

In this expression the heat-transfer coefficient h is given by Eq. (13-34), Fig. 10-2, and Eqs. (13-37) and (13-38). Equation (13-40) predicts the effect of a number of basic variables on k_e. Since Pe_r is relatively insensitive to particle diameter, the second, third, and fourth terms in the equation require that k_e increase with particle diameter. Since h increases with mass velocity G, the equation predicts that k_e will increase with G. It is also expected that k_e will increase with the conductivity of the solid particle, although the increase will be slight because k_s occurs in both the numerator and the denominator of the last term.

When the temperature in the bed is less than 300°C, the radiation contributions to k_e may be neglected. Under these circumstances Eq. (13-40) can be simplified to the form

$$k_e = \epsilon_B(k_c + k_{td}) + (1 - \epsilon_B)k_s \frac{d_P h_c + 2k_P}{d_P h_c + 2k_s} \tag{13-41}$$

where k_P is the conductivity between solid particles in contact, as given by Eq. (13-39).

Example 13-2   Predict k_e for a fixed-bed reactor for the oxidation of SO_2. The reactor is a 2.06-in.-ID tube through which a mixture of air (93.5 mole %) and SO_2 (6.5 mole %) flowed at a superficial mass velocity of $G = 350$ lb/(hr)(ft²). The catalyst pellets contained a surface coating of platinum (0.2 wt % of pellet) and were $\frac{1}{8} \times \frac{1}{8}$-in. cylinders

[1] R. H. Wilhelm, W. C. Johnson, R. Wynkoop, and D. W. Collier, *Chem. Eng. Progr.*, **44**, 105 (1948).

of alumina. The total pressure was 790 mm. The bulk density of the catalyst bed was 64 lb/ft^3. Additional data are as follows:

$k_s = 0.5$ Btu/(hr)(ft)(°F)
Emissivity $\sigma = 0.5$
Bed porosity $\epsilon_B = 0.40$
Average bed temperature $T = 350°C$ (662°F)

Solution   First estimate Pe$_r$, using Fig. 13-5:

$$\frac{d_P}{d} = \frac{1/8}{2.06} = 0.0607$$

The reaction mixture is predominantly air, so the viscosity of air [estimated to be 0.08 lb/(hr)(ft) at 350°C] may be used in calculating the Reynolds number,

$$\text{Re} = \frac{d_P G}{\mu} = \frac{1}{8(12)} \left(\frac{350}{0.08}\right) = 46$$

Then, from Fig. 13-5,

$$\text{Pe}_r = \frac{u d_P}{D_r} = 9.0[1 + 19.4(0.0607^2)] = 9.6 \tag{A}$$

Next, the separate contributions to k_e can be estimated: The radiation contribution, from Eq. (13-26), is

$$k_r = 4 \frac{0.5}{2 - 0.5} \frac{1}{8(12)} \frac{0.173(662 + 460)^3}{100^4} = 0.033$$

The point conductivity, from Eq. (13-39), is

$$\log k_P = -1.76 + 0.0129 \left(\frac{0.5}{0.4}\right)$$

$$k_P = 0.018$$

The coefficient h_c, from Fig. 10-2, is

$$j_H = 0.34$$

and

$$h_c = 30 \text{ Btu/(hr)(ft}^2)(°F)$$

To estimate coefficients h_r and h_P from Eqs. (13-37) and (13-38) we assume

$$h = h_c + h_P + h_r = 43$$

Then, from Eq. (13-37),

$$h_r = \frac{0.033[2(0.5) + 43(0.0104)]}{0.0104(0.5)} = 9.2$$

From Eq. (13-38),

$$h_P = \frac{0.018[2(0.5) + 43(0.0104)]}{0.0104(0.5)} = 5.0$$

$$h = 30 + 9.2 + 5.0 = 44 \text{ Btu/(hr)(ft}^2)(°F)$$

This value agrees well with the assumed h.

The series contribution, from Eq. (13-33), is

$$k_{series} = \frac{43(0.5)(0.0104)}{2(0.5) + 43(0.0104)} = 0.155$$

The turbulent-diffusion contribution is estimated from Eq. (13-25). The Peclet number for mass transfer has been estimated to be 9.6. Then, from Eq. (13-25),

$$k_{td} = \frac{d_P c_p G}{Pe_r \epsilon_B} = \frac{0.0104(0.26)(350)}{9.6(0.4)} = 0.247$$

The thermal conductivity of air at 350°C is approximately 0.028, so that the molecular-conductivity contribution is

$$k_c = 0.028$$

The total effective thermal conductivity and Pe_h are found from Eq. (13-40). Summing the individual contributions leads to a value of the average effective conductivity for the bed:

$$k_e = \epsilon_B(k_c + k_{td} + k_r) + (1 - \epsilon_B)k_{series}$$

$$k_e = 0.4(0.028 + 0.247 + 0.033) + (1 - 0.4)(0.155)$$

$$= 0.123 + 0.093 = 0.216 \text{ Btu/(hr)(ft)(°F)}$$

These results will be used in Example 13-6 when the design of a fixed-bed reactor for the oxidation of SO_2 is illustrated.

13-6 The One-dimensional Model

The form of the radial temperature profile in a nonadiabatic fixed-bed reactor has been observed experimentally to have a parabolic shape. Data for the oxidation of sulfur dioxide with a platinum catalyst on $\frac{1}{8} \times \frac{1}{8}$-in. cylindrical pellets in a 2-in.-ID reactor are illustrated in Fig. 13-9. Results are shown for several catalyst-bed depths. The reactor wall was maintained at 197°C by a jacket of boiling glycol. This is an extreme case. The low wall temperature resulted in severe radial temperature gradients, more so than would exist in a commercial reactor, where the wall temperature would be higher. The longitudinal profiles are shown in Fig. 13-10 for the same experiment. These curves show the typical hot spots, or maxima, characteristic of exothermic reactions in a nonadiabatic reactor. The greatest increase above the reactants temperature entering the bed is at the center,

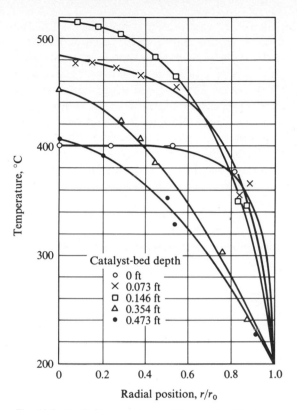

Fig. 13-9   *Radial temperature profiles in a fixed-bed reactor for the oxidation of SO₂ with air*

$r/r_0 = 0$, as would be expected. This rise decreases as the wall is approached and actually disappears at a radial position of 0.9. The temperature is so low here, even in the entering stream, that very little reaction occurs. Hence the curve in Fig. 13-10 at $r/r_0 = 0.9$ is essentially a cooling curve, approaching 197°C as the bed depth increases.

A completely satisfactory design method for nonadiabatic reactors entails predicting the radial and longitudinal variations in temperature, such as those shown in Figs. 13-9 and 13-10, analogous concentration profiles, and the bulk mean conversion. To make such predictions we must know the effective thermal conductivity and diffusivity for heat and mass transfer, as discussed in preceding sections. However, it is worthwhile first to consider a simplified approach that eliminates the need for effective conductivities and diffusivities, but still gives a reasonable prediction for the average temperature across the bed.

The parabolic shape of the radial temperature curves in Fig. 13-9 suggests that most of the resistance to heat transfer is near the wall of the reactor and only a small amount is in the central core. To carry this idea further, if we assume that all the heat-transfer resistance is in a very thin layer next to the wall, the temperature profile will be as shown by the dashed lines in Fig. 13-11. The solid line is the 0.146-ft bed-depth curve of Fig. 13-9. The horizontal dashed line represents the bulk mean temperature obtained by integration of the data on the solid line. If the actual situation is replaced by this approximate model, the only data necessary to establish the energy exchange with the surroundings is the heat-transfer coefficient at the wall, h_w. Under such conditions the design procedure would be the same as for nonadiabatic homogeneous tubular reactors, as illustrated in Example 5-2.

This one-dimensional approach is useful as a rapid procedure for

Fig. 13-10 *Longitudinal temperature profiles in SO_2 reactor*

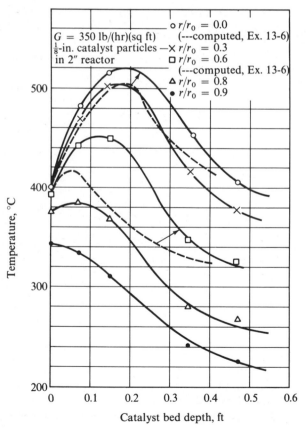

estimating the reactor size and predicting the effect of such variables as tube diameter. As the tube diameter decreases, the ratio of the heat-transfer area to the reactor volume will increase. Therefore the temperature rise of the reaction mixture as it passes through the bed will be less, and the radial temperature variation within the bed will also be less. Hence where it is necessary not to exceed a certain temperature in the catalyst bed, small-diameter tubes are indicated. The approximate size necessary for a given temperature can be determined by means of this simplified procedure. Problem 13-6 illustrates calculations for a phthalic anhydride reactor. The oxidation of naphthalene has a high heat of reaction, so that the size of the catalyst tubes is a critical point in the design.

Examples 13-3 to 13-5 illustrate the simplified design method for different cases. The first is for the endothermic styrene reaction, where the temperature decreases continually with catalyst-bed depth. Example 13-4 is for an exothermic reaction carried out under conditions where radial temperature gradients are not large. Example 13-5 is also for an exothermic case, but here the gradients are severe, and the simplified solution is not satisfactory.

Example 13-3   Under actual conditions the reactor described in Example 13-1 would not be truly adiabatic. Suppose that with reasonable insulation the heat loss U would correspond to a heat-transfer coefficient of 1.6 Btu/(hr)(ft^2 inside tube area) (°F). This value of U is based on the difference in temperature between the reaction mixture and the surroundings at 70°F. Determine revised curves of temperature and conversion vs catalyst-bed depth for this nonadiabatic operation.

Solution   The design equation (C) and the rate expression (D) of Example 13-1 are

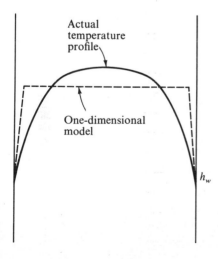

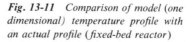

Fig. 13-11   *Comparison of model (one dimensional) temperature profile with an actual profile (fixed-bed reactor)*

applicable here, as is their combination, Eq. (E). The energy balance will be different and will follow the form of Eq. (3-18); that is,

$$\frac{F}{M}(-\Delta H)\, dx + U(T_s - T)\, dA_h = F_t c_p\, dT$$

$$13.5\, dx(-60,000) - 1.6(\pi 4)(T - 530)\, dz = 3,270\, dT$$

$$dT = -248\, dx - 0.00615(T - 530)\, dz$$

Written in difference form, this becomes

$$\Delta T = -248\, \Delta x - 0.00615(T - 530)_{av}\, \Delta z \tag{A}$$

Equation (E') of Example 13-1 and Eq. (A) determine the solution for the bed depth and temperature as a function of conversion. As in Example 13-1, it is convenient to choose an increment of conversion. However, in this nonadiabatic case it is necessary to assume a temperature at the end of the increment. Then Δz is computed from Eq. (E') and the temperature assumption is checked in Eq. (A).

Since the heat loss from the reactor will be small, assume the temperature at the end of the first increment, $\Delta x = 0.1$, to be the same as in Example 13-1, $T_1 = 1591°R$. From Example 13-1,

$$\alpha_0 = 3.30$$

$$\alpha_1 = 4.65$$

Then from Eq. (E'),

$$\Delta z = \frac{3.30 + 4.65}{2}(0.1) = 0.40 \text{ ft}$$

Now, checking the assumed temperature in Eq. (A), we find

$$\Delta T = -248(0.1) - 0.00615\left(\frac{1,616 + 1,591}{2} - 530\right)(0.40)$$

$$= -24.8 - 2.6 = -27.4°F$$

$$T_1 = 1,616 - 27 = 1589°R$$

in reasonable agreement with 1591°R.

At the end of the second increment, assume $T_2 = 1560°R$. From Eq. (E) of Example 13-1,

$$\alpha_2 = \frac{21 + 0.2}{1,270,000}e^{12.70}\frac{1}{(1 - 0.2) - 1.2/0.22[0.2^2/(21 + 0.2)]} = 5.51\left(\frac{1}{0.791}\right)$$

$$= 6.97$$

From Eq. (B),

$$z_2 - z_1 = \bar{\alpha}\, \Delta x = \frac{4.65 + 6.97}{2}(0.1) = 0.58 \text{ ft}$$

$$z_2 = 0.40 + 0.58 = 0.98 \text{ ft}$$

Checking the temperature in Eq. (A), we have

$$\Delta T = -24.8 - 0.00615 \left(\frac{1{,}589 + 1{,}560}{2} - 530 \right) 0.58$$

$$= -24.8 - 3.7 = -28.5°R$$

$$T_2 = 1{,}589 - 28.5 = 1560°R$$

This agrees with the assumption.

Proceeding further with the calculations leads to the results shown in Table 13-2. The results are also plotted in Fig. 13-6 (labeled nonadiabatic operation). The bed depth for a given conversion is greater than for the adiabatic case, because the temperature is less. For example, for a conversion of 50%, 5.8 ft of catalyst is required, in comparison with 4.8 ft in Example 13-1.

Table 13-2 *Data for conversion of ethyl benzene to styrene in a non-adiabatic reactor*

Conversion	Mean bulk temperature		Catalyst-bed depth, ft
	°R	°C	
0.0	1616	625	0
0.10	1589	610	0.40
0.20	1560	594	0.98
0.30	1530	578	1.80
0.38	1505	564	2.80
0.45	1478	549	4.2
0.50	1456	537	5.8
0.55	1426	520	9.0
0.57	1412	512	∞

Example 13-4 A bench-scale study of the hydrogenation of nitrobenzene was published by Wilson[1] in connection with reactor design studies. Nitrobenzene and hydrogen were fed at a rate of 65.9 g moles/hr to a 3.0-cm-ID reactor containing the granular catalyst. A thermocouple sheath, 0.9 cm in diameter, extended down the center of the tube. The void fraction was 0.424 and the pressure atmospheric. The feed entered the reactor at 427.5°K, and the tube was immersed in an oil bath maintained at the same temperature. The heat-transfer coefficient from the mean reaction temperature to the oil bath was determined experimentally to be 8.67 cal/(hr)(cm^2)(°C). A large excess of hydrogen was used, so that the specific heat of the reaction mixture may be taken equal to that for hydrogen and the change in total moles as a result of reaction may be neglected. The heat of reaction is approximately constant and equal to $-152{,}100$ cal/g mole.

The entering concentration of nitrobenzene was 5.0×10^{-7} g mole/cm^3. The global rate of reaction was represented by the expression

[1] K. B. Wilson, *Trans. Inst. Chem. Engrs. (London)*, **24**, 77 (1946).

$$\mathbf{r}_P = 5.79 \times 10^4 C^{0.578} e^{-2,958/T}$$

where $\mathbf{r}_P$ = g moles nitrobenzene reacting/cm^3(hr), expressed in terms of void volume in reactor

C = concentration of nitrobenzene, g moles/cm^3

T = temperature, °K

The experimental results for temperature vs reactor length are shown in Fig. 13-12. From the data given, calculate temperatures up to a reactor length of 25 cm and compare them with the observed results.

Solution   The concentration depends on temperature as well as conversion. If Q is the volumetric flow rate at a point in the reactor where the concentration is C, and Q_0 is the value at the entrance, the conversion of nitrobenzene is

$$x = \frac{C_0 Q_0 - CQ}{C_0 Q_0} = 1 - \frac{C}{C_0} \frac{Q}{Q_0}$$

Since there is no change in pressure or number of moles, Q changes only because of temperature changes. Hence, assuming perfect-gas behavior, we have

$$x = 1 - \frac{C}{C_0} \frac{T}{T_0}$$

or

$$C = (1 - x) \frac{C_0 T_0}{T} = (5 \times 10^{-7}) \frac{427.5}{T} (1 - x) \tag{A}$$

Fig. 13-12   *Longitudinal temperature profile in a reactor for hydrogenation of nitrobenzene*

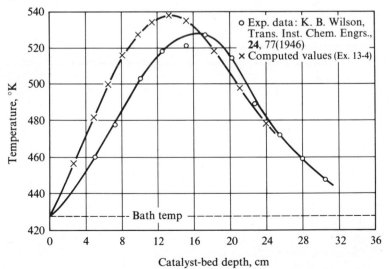

Substituting this expression into the rate equation gives the global rate in terms of the temperature and conversion,

$$\mathbf{r}_P = 439\left(\frac{1-x}{T}\right)^{0.578} e^{-2,958/T} \tag{B}$$

The mass balance, even though the reaction is catalytic, must be written in terms of the void volume because of the form in which the rate data are reported. Thus

$$F \, dx = \mathbf{r}_P \, 0.424 \, dV$$

The feed rate of nitrobenzene is

$$F = 65.9(22,400)\left(\frac{427.5}{273}\right)(5.0 \times 10^{-7}) = 1.15 \text{ g moles/hr}$$

$$1.15 \, dx = \mathbf{r}_P \, 0.424 \frac{\pi}{4}(9 - 0.81) \, dz$$

$$dz = \frac{0.423}{\mathbf{r}_P} \, dx \tag{C}$$

The additional relationship required is the energy balance. Equation (3-18) is applicable and yields

$$-(-152,100)F \, dx - 8.67\pi(3)(T - 427.5) \, dz = 65.9(6.9 \, dT)$$

where 6.9 cal/(g mole)(°K) is the molal heat capacity of hydrogen at 427.5°K. This equation may be simplified to

$$dT = 385 \, dx - 0.180(T - 427.5) \, dz \tag{D}$$

Equations (B), (C), and (D) are, respectively, the rate, design, and energy-balance relationships needed to solve the problem. A numerical approach is necessary and may be carried out as follows:[1]

1. Choose an incremental value of conversion, say, $\Delta x = 0.1$.
2. Calculate $\mathbf{r}_P$ at the beginning and end of the increment from Eq. (B) by assuming a temperature at the end of the increment.
3. Solve Eq. (C) for Δz.
4. Use Eq. (D) to check the assumed temperature.

Let us illustrate these calculations by starting with a conversion increment of $\Delta x = 0.10$ and assuming that $T_1 = 427.5 + 15 = 442.5$°K at $x = 0.10$. Then

$$\mathbf{r}_0 = 439\left(\frac{1}{427.5}\right)^{0.578} e^{-2,958/427.5} = 0.0130 \text{ g mole/(hr)(cm}^3)$$

$$\mathbf{r}_1 = 439\left(\frac{1-0.1}{442.5}\right)^{0.578} e^{-2,958/442.5} = 0.0152$$

[1] Note that the method of solution is exactly the same as in Example 5-2 for a homogeneous reaction.

From Eq. (C),

$$\Delta z = z_1 - 0 = 0.423 \left(\frac{1}{r}\right)_{av} (0.10)$$

$$z_1 = 0.423 \left(\frac{1}{0.0130} + \frac{1}{0.0152}\right)\left(\frac{1}{2}\right)(0.10) = 3.02 \text{ cm}$$

From Eq. (D),

$$T_1 - 427.5 = 385\,\Delta x - 0.180(T - 427.5)_{av}\,\Delta z$$

$$= 38.5 - 0.180(7.5)(3.02) = 34.4°C$$

This temperature rise of 34°C is much larger than the assumed value of 15°C. For the second trial assume $T_1 = 427.5 + 30 = 457.5°K$. Then

$$r_1 = 0.0186$$

$$\frac{1}{r_1} = 53.8$$

$$z_1 = z_1 - 0 = 0.423(65.4)(0.10) = 2.77 \text{ cm}$$

$$T_1 - 427.5 = 38.5 - 0.180(15)(2.77) = 38.5 - 7.5 = 31°C$$

The assumed and calculated values are now close, so that further calculations are unnecessary. At the end of the first increment

$$x_1 = 0.10$$

$$z_1 = 2.8 \text{ cm}$$

$$T_1 = 427.5 + 31 = 458.5°K$$

For the end of the second increment assume that $T_2 = 458.5 + 23 = 481.5°K$. Then, from Eq. (B),

$$r_2 = 439 \left(\frac{1 - 0.2}{481.5}\right)^{0.578} e^{-2,958/481.5} = 0.0225$$

$$\frac{1}{r_2} = 44.4$$

Equation (C) gives the second increment of Δz as

$$\Delta z = 0.423 \frac{53.8 + 44.4}{2}(0.1) = 2.08 \text{ cm}$$

Checking the assumed value of T_2 in Eq. (D), we have

$$\Delta T = 38.5 - 0.180 \left(\frac{481.5 + 458.5}{2} - 427.5\right)(2.08)$$

$$= 38.5 - 15.9 = 22.6°C \qquad \text{vs the assumed value of } 23°C$$

Thus at the end of the second increment

$x_2 = 0.20$

$z_2 = 2.77 + 2.08 = 4.85$ cm

$T_2 = 458.5 + 22.6 = 481°K$

For T_3 assume a value of $481 + 18 = 499°K$. Then

$r_3 = 0.0260$

$\dfrac{1}{r_3} = 38.4$

$\Delta z = 0.423 \dfrac{44.4 + 38.4}{2} (0.1) = 1.75$ cm

$z_3 = 6.6$ cm

$\Delta T = 38.5 - 0.180 \left(\dfrac{481 + 499}{2} - 427.5 \right)(1.75)$

$\quad\quad = 38.5 - 19.7 = 18.8°C \quad\quad$ vs 18°C assumed

It is apparent from the calculations thus far that the heat transfer to the reactor-tube wall is becoming larger as the conversion increases. This is owing to the increased temperature difference between reaction mixture and wall. The trend continues until the heat transferred to the oil bath is as large as that evolved as a result of reaction. The temperature reaches a maximum at this point, the so-called "hot spot." The results of further calculations, summarized in Table 13-3 and Fig. 13-12, show that the computed hot spot is reached at about 14 cm from the entrance to the reactor. This location is 3 cm before the experimental hot spot. Also, the temperature is 10°C less than the

Table 13-3   *Temperatures and conversions in a nitrobenzene hydrogenation reactor*

Conversion	Mean bulk temperature, °C	Catalyst-bed depth, cm
0	427.5	0
0.1	458.5	2.8
0.2	481	4.85
0.3	500	6.6
0.4	516	8.2
0.5	527	9.8
0.6	534	11.4
0.7	538	13.1
0.8	536	15.1
0.9	518	18.3
0.95	498	21.1
0.97	478	24.1

computed value at the maximum. Three points should be mentioned in making this comparison. First, the measured temperature corresponds to the center of the tube, while the computed results are for a bulk mean temperature. Second, the thermo-couples were contained in a metal sheath entering down the center of the reactor. The sheath would reduce the observed temperatures, because of longitudinal con-duction, and make them more nearly comparable with the bulk mean computed values. Third, using a specific heat of hydrogen for the whole reaction mixture results in a value that is too low, causing the calculated temperatures to be high.

Actually, the agreement shown in Fig. 13-12 is quite good, in view of these three points and that the one-dimensional model neglects radial temperature gradients. The fact that a heat-transfer coefficient was determined experimentally in the same apparatus probably improved the agreement. Unfortunately, no information is avail-able on the catalyst-particle size, so that a heat-transfer coefficient cannot be estimated from existing correlations.

The simplified approach to the design of nonadiabatic reactors led to good results in Example 13-4 partly, at least, because radial tem-perature variations were not large. This is because the wall temperature was the same as the temperature of the entering reactants. Thus radial variations in temperature at the entrance to the reactor were zero. At the hot spot the maximum temperature difference between the center and the wall was about 100°C. In Example 13-5 much larger radial temperature gradients exist, and the simplified method is not so suitable.

Example 13-5   Using the one-dimensional method, compute curves for temperature and conversion vs catalyst-bed depth for comparison with the experimental data shown in Figs. 13-10 and 13-14 for the oxidation of sulfur dioxide. The reactor consisted of a cylindrical tube, 2.06 in. ID. The superficial gas mass velocity was 350 lb/(hr)(ft^2), and its inlet composition was 6.5 mole % SO_2 and 93.5 mole % dry air. The catalyst was prepared from $\frac{1}{8}$-in. cylindrical pellets of alumina and contained a surface coating of platinum (0.2 wt % of the pellet). The measured global rates in this case were not fitted to a kinetic equation, but are shown as a function of temperature and conversion in Table 13-4 and Fig. 13-13. Since a fixed inlet gas composition was used, independent variations of the partial pressures of oxygen, sulfur dioxide, and sulfur trioxide were not possible. Instead these pressures are all related to one variable, the extent of con-version. Hence the rate data shown in Table 13-4 as a function of conversion are suf-ficient for the calculations. The total pressure was essentially constant at 790 mm Hg. The heat of reaction was nearly constant over a considerable temperature range and was equal to $-22,700$ cal/g mole of sulfur dioxide reacted. The gas mixture was pre-dominantly air, so that its specific heat may be taken equal to that of air. The bulk density of the catalyst as packed in the reactor was 64 lb/ft^3.

From Fig. 13-9 it is apparent that the entering temperature across the diameter is not constant, but varies from a maximum value of about 400°C at the center down to 197°C at the wall. Since the one-dimensional method of solution to be used in this example is based on a uniform temperature radially, we use a mean value of 364°C.

Table 13-4   *Experimental global rates* [*g moles/(hr)(g catalyst)*] *of oxidation of SO₂ with a 0.2% Pt-on-Al₂O₃ catalyst*

t, °C	% conversion* of SO₂						
	0	*10*	*20*	*30*	*40*	*50*	*60*
350	0.011	0.0080	0.0049	0.0031	. . .	. . .	. . .
360	0.0175	0.0121	0.00788	0.00471	0.00276	0.00181	. . .
380	0.0325	0.0214	0.01433	0.00942	0.00607	0.00410	. . .
400	0.0570	0.0355	0.02397	0.01631	0.0110	0.00749	0.00488
420	0.0830	0.0518	0.0344	0.02368	0.0163	0.0110	0.00745
440	0.1080	0.0752	0.0514	0.03516	0.0236	0.0159	0.0102
460	0.146	0.1000	0.0674	0.04667	0.0319	0.0215	0.0138
480	. . .	0.1278	0.0898	0.0642	0.0440	0.0279	0.0189
500	. . .	0.167	0.122	0.0895	0.0632	0.0394	0.0263

*Conversion refers to a constant feed composition of 6.5 mole % SO₂ and 93.5 mole % air.
SOURCE: R. W. Olson, R. W. Schuler, and J. M. Smith, *Chem. Eng. Progr.*, **46**, 614 (1950).

Fig. 13-13   *Rate of oxidation of SO₂ on ⅛-in. catalyst particles containing 0.2% platinum* [*mass velocity 350 lb/(hr)(ft²)*]

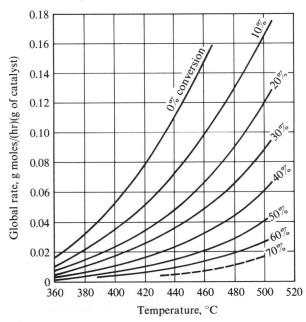

In the two-dimensional model, considered in Example 13-6, the actual entering-temperature profile can be taken into account.

Solution   Before carrying out the reactor calculations it is necessary to estimate a wall heat-transfer coefficient. Leva's correlations [Eq. (13-19)] may be used for this purpose:

$$\frac{d_p G}{\mu} = 46 \qquad \text{(from Example 13-2)}$$

$$\frac{d_p}{d} = \frac{1}{8(2.06)} = 0.0607$$

$$\frac{h_w d}{k} = 3.50(46)^{0.7}\, e^{-4.6(0.0607)} = 38.7$$

$$h_w = \frac{12(0.028)}{2.06}\,(38.7) = 6.3 \text{ Btu/(hr)(ft}^2\text{)(}^\circ\text{F)}$$

The energy balance for an element of reactor height dz is, according to Eq. (3-18),

$$\frac{F}{M}\,22{,}700(1.8)\,dx + 6.3\,\pi\,\frac{2.06}{12}\,(197 - T)1.8\,dz = F_t c_p\,dT(1.8) \tag{A}$$

In this expression the conversion factor of $1.8^\circ\text{F}/^\circ\text{C}$ has been introduced so that the temperature may be expressed in degrees centigrade:

$$\frac{F}{M} = 350\pi\!\left(\frac{1.03}{12}\right)^2 \frac{1}{31.2}\,(0.065) = 0.017 \text{ lb mole/hr}$$

The number 31.2 is the molecular weight of the feed containing 6.5 mole % SO_2. The heat capacity of the reaction mixture, assumed to be that of air at an average temperature of 350°C, is 0.26 Btu/(lb)($^\circ$F). Hence

$$F_t c_p = 350\pi\!\left(\frac{1.03}{12}\right)^2 (0.26) = 2.11 \text{ Btu/}^\circ\text{F}$$

Substituting these values in Eq. (A), we find

$$0.017(22{,}700)\,dx - 3.40\,(T - 197)\,dz = 2.11\,dT$$

$$dT = 182\,dx - 1.61\,(T - 197)\,dz \tag{B}$$

The mass balance may be written

$$\pi\!\left(\frac{1.03}{12}\right)^2 \mathbf{r}_P \rho_B\,dz = \frac{F}{M}\,dx = 0.017\,dx$$

The bulk density of the catalyst as packed is given as 64 lb/ft^3. Hence the mass balance simplifies to the form

$$dz = 0.0112\,\frac{dx}{\mathbf{r}_P} \tag{C}$$

In contrast to the previous examples, the rate of reaction is not expressed in equation form, but as a tabulation of experimental data. Actually, a simple first- or second-order expression would not correlate the data for this reaction. Instead, a Langmuir type of equation, based on adsorption of oxygen on the catalyst, was necessary. This equation was developed and tested in Example 9-2. The tabulation of rates is more convenient to use in this problem than the rate equation. The units of r_P in Eq. (C) should be lb moles reacted/(hr)(lb catalyst), but numbers in these units are numerically equivalent to g moles/(hr)(g catalyst), so that the data in Table 13-4 can be used directly.

The method of solution is the same as in Examples 13-3 and 13-4:

1. An increment of conversion is chosen.
2. A temperature at the end of the increment is assumed.
3. The rates are obtained at the beginning and end of the increment from Table 13-4 or Fig. 13-13.
4. The increment of depth Δz is computed from Eq. (C).
5. The assumed temperature is checked in Eq. (B).

For the first increment choose $\Delta x = 0.05$. From Fig. 13-13, the rate at the beginning of the increment, zero conversion and 364°C, is 0.020. If the temperature at the end of the increment is assumed to be $364 + 1 = 365$°C, the rate at this temperature and $x = 0.05$ is 0.018. Then in Eq. (C)

$$\Delta z = 0.0112\left(\frac{1}{r_P}\right)_{av}(0.05) = 0.0112\ \frac{1/0.02 + 1/0.018}{2}(0.05) = 0.0294 \text{ ft}$$

Then in Eq. (B)

$$T_1 - 364 = 182(0.05) - 1.61(364 - 197)(0.0294) = 9.1 - 8.0 = 1.1°C$$

$$T_1 = 365.1°C$$

Continuing the computations gives the results summarized in Table 13-5 and shown in Fig. 13-14. It is apparent that the mean computed temperature never rises appreciably above the entering value of 364°C. Indeed, when a bed depth of about 0.2 ft is reached, the temperature is so low that the rate is no longer high enough to

Table 13-5   *Conversion and temperature in an SO_2 reactor calculated by the one-dimensional model*

Conversion	t, °C	Catalyst-bed depth, ft
0	364	0
0.05	365	0.029
0.10	364.5	0.065
0.15	361	0.112
0.18	357	0.149
0.21	347	0.210
0.24	(300)	(0.5)

give a significant change in conversion. The last calculation, at a conversion of 24%, is estimated by extrapolating the tabulated rate data.

In the experimental reactor used by Schuler et al.[1] the low wall temperature results in severe radial temperature gradients. Hence the mean bulk temperature may be low, but the temperatures near the center of the bed may be high enough to cause a significant amount of reaction. This is a major factor in the large difference between the experimental conversion curve and the results computed in this example, both shown in Fig. 13-14. A critical quantity in the simplified design method applied to this problem is the wall heat-transfer coefficient. Small changes in h_w can cause large differences in the conversion. The use of Leva's correlation [Eq. (13-19)] at the low Reynolds number 46 is somewhat uncertain. In the next section this example is repeated, this time with radial temperature and concentration gradients taken into account.

The one-dimensional model is particularly useful for investigating complex reactor systems. For example, Lucas and Gelbin[2] have used the approach to design multistage reactors for ammonia synthesis. Design calculations which included radial gradients would be time consuming, even with large digital computers. With the one-dimensional method, approximate results and the effects of changes in operating conditions can be evaluated for a series of reactors with a modest amount of computer time.

13-7 The Two-dimensional Model

In this section the design procedure is made more accurate by taking into account radial variations in temperature and concentration. A thorough treatment would take into consideration the radial distribution of velocity, would account for radial concentration and temperature gradients by using Peclet numbers which themselves varied with radial position, and would allow for axial dispersion if that were significant. With machine computation all these refinements may be included. However, to illustrate the design calculations let us make some simplifying assumptions: the catalyst bed is deep enough that longitudinal dispersion can be neglected, and both the mass velocity G and radial Peclet numbers for mass and heat transfer are constant across the reactor tube.

The mass balance of reactant for these conditions is similar to Eq. (13-1). The term for longitudinal diffusion is omitted, and the mass velocity

[1] R. W. Schuler, V. P. Stallings, and J. M. Smith, *Chem. Eng. Progr.*, *Symp. Ser. 4*, **48**, 19 (1952).
[2] K. Lucas and D. Gelbin, *Brit. Chem. Eng.*, **7**, 336 (1962).

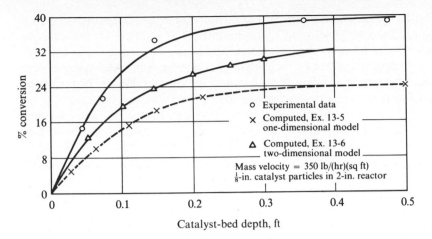

Fig. 13-14 *Comparison of calculated and experimental conversion values in an SO_2 reactor*

is assumed to be constant.[1] The result, expressed in terms of the conversion of reactant, is

$$\frac{\partial x}{\partial z} - \frac{d_P}{Pe_r}\left(\frac{1}{r}\frac{\partial x}{\partial r} + \frac{\partial^2 x}{\partial r^2}\right) - \frac{r_P \rho_B}{(G/\overline{M})y_0} = 0 \qquad (13\text{-}42)$$

where $(G/\overline{M})y_0 = u_0 C_0 =$ molal feed rate of reactant per unit area of reactor

$\overline{M} =$ average molecular weight of the feed stream

$y_0 =$ reactant mole fraction in the feed

The energy balance is Eq. (13-20). In comparison with the one-dimensional model, Eq. (13-42) has replaced Eq. (12-1) [or Eq. (C) of Example 13-1] as the mass balance, and Eq. (13-20) has replaced Eq. (3-18) as the energy balance.

The objective is to solve Eqs. (13-42) and (13-20) for the temperature and conversion at any point in the catalyst bed. As boundary conditions at the entrance we need the temperature and conversion profile across the diameter of the reactor. Further boundary conditions applicable at any axial location are that the conversion is flat $(\partial x/\partial r = 0)$ at both the centerline and the wall of the tube. The temperature gradient at the centerline is zero, but the condition at the wall is determined by the heat-transfer character-

[1] Variations in linear velocity due to variations in temperature and number of moles are accounted for.

istics. In the illustration that follows, the wall temperature is maintained constant by a boiling liquid (ethylene glycol) in a jacket surrounding the reactor. In other cases the wall temperature may vary with z. For example, if the fluid in the jacket or surroundings is at T_s and the heat-transfer coefficient between the inside wall surface and the surroundings is U, the proper boundary condition is

$$U(T_w - T_s) = -k_e \left(\frac{\partial T}{\partial r} \right)_w \tag{13-43}$$

Equations (13-42) and (13-20) are solved by a stepwise numerical procedure, starting at the entrance to the reactor.[1] The equations are first written in difference form. Let n and L represent the number of increments in the radial and axial directions, respectively, and Δr and Δz their magnitude, so that

$$r = n \, \Delta r \tag{13-44}$$

$$z = L \, \Delta z \tag{13-45}$$

The temperature at any point in the bed can be represented by $T_{n,L}$, that is, the temperature at $r = n \, \Delta r$ and $z = L \, \Delta z$. Note that r is measured from the center of the bed and z from the feed entrance.

The first difference in temperature in the r direction can be written

$$\Delta_r T = T_{n+1,L} - T_{n,L} \tag{13-46}$$

Similarly, the first difference in the z direction may be written

$$\Delta_z T = T_{n,L+1} - T_{n,L} \tag{13-47}$$

The second difference in the r direction is

$$\Delta_r^2 T = (T_{n+1,L} - T_{n,L}) - (T_{n,L} - T_{n-1,L}) \tag{13-48}$$

With these definitions, the approximate difference form of Eq. (13-20) is

$$T_{n,L+1} = T_{n,L} + \frac{\Delta z}{(\Delta r)^2} \frac{k_e}{Gc_p} \left[\frac{1}{n} (T_{n+1,L} - T_{n,L}) + T_{n+1,L} \right.$$
$$\left. - 2T_{n,L} + T_{n-1,L} \right] - \frac{\Delta H \mathbf{r}_P \rho_B \Delta z}{Gc_P} \tag{13-49}$$

Similarly, Eq. (13-42) may be written in difference form as

[1] An unsophisticated numerical method is used here for simplicity. Various implicit and explicit methods of writing difference equations and their limitations are discussed by Leon Lapidus in "Digital Computation for Chemical Engineers," McGraw-Hill Book Company, New York, 1960. The Crank-Nicholson implicit method [G. Crank and P. Nicholson, *Proc. Cambridge Phil. Soc.*, **43**, 50 (1947)] is well suited for machine computation.

$$x_{n,L+1} = x_{n,L} + \frac{\Delta z}{(\Delta r)^2} \frac{d_P}{Pe_r} \left[\frac{1}{n} (x_{n+1,L} - x_{n,L}) + x_{n+1,L} \right.$$

$$\left. - 2x_{n,L} + x_{n-1,L} \right] + \frac{\bar{r}_P \rho_B \bar{M} \Delta z}{G y_0} \quad (13\text{-}50)$$

Provided the magnitude of the reaction terms involving r_P can be estimated, Eqs. (13-49) and (13-50) can be solved step by step to obtain the conversion. The first step is to compute values of T and x across the diameter, at $z = 1\,\Delta z$, or $L = 1$, from known values at $L = 0$. Then continue to the next longitudinal increment, $L = 2$, etc. The indeterminate form of the equations at $n = 0$ can be avoided by using the special expressions

$$T_{0,L+1} = T_{0,L} + \frac{2\,\Delta z}{(\Delta r)^2} \frac{k^e}{c_p G} (2T_{1,L} - 2T_{0,L}) - \frac{\Delta H r_P \rho_B \Delta z}{G c_P} \quad (13\text{-}51)$$

$$x_{0,L+1} = x_{0,L} + \frac{2\,\Delta z}{(\Delta r)^2} \frac{d_P}{Pe_r} (2x_{1,L} - 2x_{0,L}) + \frac{\bar{r}_P \rho_B \bar{M} \Delta z}{G y_0} \quad (13\text{-}52)$$

derived from L'Hôpital's rule.

The effect of the reaction terms in Eqs. (13-49) and (13-50) is, for an exothermic reaction, to increase both the temperature and the conversion. Since the rate depends on the temperature and composition, and the average value for the increment L to $L + 1$ is not known until Eqs. (13-49) and (13-50) are solved, a trial-and-error procedure is indicated. The calculations are illustrated in Example 13-6, where the SO_2 reactor problem of Example 13-5 is recomputed with radial variations taken into account. In this example experimental data for the global rate are available. These data were obtained for the same size of catalyst pellets and for the same gas velocity as for the integral reactor. Hence external and internal transport effects on the global rate were the same for the laboratory conditions in which r_P was measured as for the integral reactor to be designed. In this way the calculations discussed in Chap. 12 for evaluating a global rate to fit the transport situation in the large-scale reactor were avoided. In most cases global rates have to be evaluated from intrinsic rates, effectiveness factors, and external transport coefficients.[1]

Example 13-6 Recompute the conversion-vs-bed-depth curve for the SO_2 reactor of Example 13-5 by the two-dimensional method, with the assumption that k_e, D_e, and G are constant. The temperature profile at the entrance to the reactor is given in the following table and also plotted in Fig. 13-15:

[1] See S. L. Liu, "The Influence of Intraparticle Diffusion in Fixed-bed Catalytic Reactors," presented at Sixty-first Annual Meeting AIChE, Los Angeles, December, 1968; J. J. Carberry and M. M. Wendel, *AIChE J.*, **9**, 129 (1963).

Feed temperature, °C	376.5	400.4	400.1	399.5	400.1	376.1	197.0
Radial position	0.797	0.534	0.023	0.233	0.474	0.819	1.000

The reactants composition across the diameter may be assumed to be uniform.

Solution   The effective thermal conductivity and Pe_r for the conditions of this example were calculated in Example 13-2. The quantities needed in Eqs. (13-49) and (13-50) are

$$\frac{k_e}{c_p G} = \frac{0.216}{0.26(350)} = 0.00238 \text{ ft}$$

$$\frac{d_p}{Pe_r} = \frac{1/8}{12(9.6)} = 0.00109 \text{ ft}$$

TEMPERATURE AND CONVERSION EQUATIONS   It is convenient to divide the radius of the bed into five increments, so that

$$\Delta r = 0.2 r_0 = 0.2 \left(\frac{1.03}{12} \right) = 0.0172 \text{ ft}$$

If Δz is chosen to be 0.05 ft,

$$\frac{\Delta z}{(\Delta r)^2} = \frac{0.05}{0.0172^2} = 170 \text{ ft}^{-1}$$

Then the coefficients in Eqs. (13-49) and (13-50) are

$$\frac{k_e}{c_p G} \frac{\Delta z}{(\Delta r)^2} = 0.00238(170) = 0.404 \qquad \text{dimensionless}$$

$$\frac{d_p}{Pe_r} \frac{\Delta z}{(\Delta r)^2} = 0.00109(170) = 0.185 \qquad \text{dimensionless}$$

$$\frac{\Delta H \mathbf{r}_p \rho_B \Delta z}{G c_p} = \frac{-22{,}700(64)(0.05)\mathbf{r}_p}{350(0.26)} = -798\mathbf{r}_p \qquad °C$$

$$\frac{\mathbf{r}_p \rho_B \overline{M} \Delta z}{G y_0} = \frac{64(31.2)(0.05)\mathbf{r}_p}{350(0.065)} = 4.38\mathbf{r}_p \qquad \text{dimensionless}$$

Substituting these values in Eqs. (13-49) and (13-50) gives working expressions for calculating the temperature and conversion at a bed depth $L + 1$, from data at the previous bed depth L. Thus

$$T_{n,L+1} = T_{n,L} + 0.404 \left[\frac{1}{n} (T_{n+1,L} - T_{n,L}) + T_{n+1,L} - 2T_{n,L} + T_{n-1,L} \right]$$

$$+ 798\bar{\mathbf{r}}_p \quad (A)$$

$$x_{n,L+1} = x_{n,L} + 0.185 \left[\frac{1}{n} (x_{n+1,L} - x_{n,L}) + x_{n+1,L} - 2x_{n,L} + x_{n-1,L} \right]$$

$$+ 4.38\bar{\mathbf{r}}_p \quad (B)$$

CALCULATIONS FOR THE FIRST BED-DEPTH INCREMENT $(L = 1)$ The entering-temperature distribution is known (see Fig. 13-15), and the entering conversion will be zero at all radial positions. Starting at $n = 1$, the temperatures $T_{0,0}$, $T_{1,0}$, and $T_{2,0}$, as read from Fig. 13-15, are all 400°C. Substituting these values into Eq. (A) gives $T_{1,1}$ in terms of the average rate $\bar{r}_P$ over the increment of bed depth from 0 to $L = 1$ ($z = 0.05$ ft):

$$T_{1,1} = T_{1,0} + 0.404\left[\frac{1}{1}(T_{2,0} - T_{1,0}) + T_{2,0} - 2T_{1,0} + T_{0,0}\right] + 798\bar{r}_P$$

$$= 400 + 0.404(0) + 798\bar{r}_P = 400 + 798\bar{r}_P \tag{C}$$

Since $x_{0,0}$, $x_{1,0}$, and $x_{2,0}$ are all 0, Eq. (B) gives, for the conversion at $n = 1$ and $L = 1$,

$$x_{1,1} = 0 + 0.185(0) + 4.38\bar{r}_P = 4.38\bar{r}_P \tag{D}$$

Equations (C) and (D) and the rate data, Table 13-4 or Fig. 13-13, constitute three relationships of the unknown quantities $T_{1,1}$, $x_{1,1}$, and $\bar{r}_P$. One method of solution is the following four-step process:

1. Assume a value of $\bar{r}_P$, after obtaining $r_{1,0}$ from Fig. 13-13.
2. Compute $T_{1,1}$ and $x_{1,1}$ from Eqs. (C) and (D).
3. Evaluate the rate $r_{1,1}$ at the end of the increment from Fig. 13-13.
4. Average $r_{1,1}$ and $r_{1,0}$, and compare the result with the assumed $\bar{r}_P$. If agreement is not obtained, repeat the sequence with a revised value of $\bar{r}_P$.

Let us carry out these steps. At 400°C and zero conversion we have

$$r_{1,0} = 0.055$$

We assume $\bar{r}_P = 0.051$. Then from Eq.(C),

$$T_{1,1} = 400 + 798(0.051) = 441°C$$

$$x_{1,1} = 4.38(0.051) = 0.223$$

From Fig. 13-13 at 441°C and 22.3% conversion, $r_{1,1} = 0.046$. Hence

$$\bar{r}_P = \frac{0.055 + 0.046}{2} = 0.0505$$

This result is close to the assumed value of 0.051, so the calculated temperature and conversion at $n = 1$ and $L = 1$ may be taken as 441°C and 22.3%.

The same result would apply at $n = 0$ and 2, because the entering temperature is 400°C up to a radial position of $n = 3(r/r_0 = 0.6)$, as noted in Fig. 13-15. At $n = 3$ the situation changes because $T_{4,0} = 376°C$. Let us make the stepwise calculations at this radial position. Starting with

$$r_{3,0} = 0.055$$

we assume $\bar{r}_P = 0.046$. Then, from Eq. (A),

$$T_{3,1} = 400 + 0.404[\tfrac{1}{3}(376 - 400) + 376 - 2(400) + 400] + 798(0.046)$$

$$= 400 - 13 + 37 = 424°C$$

$$x_{3,1} = 0 + 4.38(0.046) = 0.201$$

From Fig. 13-13 at 424°C and 20.1% conversion, $r_{3,1} = 0.037$, and so

$$\bar{r}_P = \frac{0.055 + 0.037}{2} = 0.046$$

Continuing the calculations at $n = 4(r/r_0 = 0.8)$, where $T_{4,0} = 376°$ and $x_{4,0} = 0$, we have

$$\mathbf{r}_{4,0} = 0.029$$

and we assume $\bar{r}_P = 0.015$. Then

$$T_{4,1} = 376 + 0.404[\tfrac{1}{4}(197 - 376) + 197 - 2(376) + 400] + 798(0.015)$$

$$= 376 - 83 + 12 = 305°C$$

$$x_{1,1} = 0 + 4.38(0.015) = 0.066$$

From Fig. 13-13 at 305°C and $x = 0.066$ it is evident that the rate is close to zero; thus

$$\bar{r}_P = \frac{0.029 + 0}{2} = 0.015$$

which agrees with the assumed value. Hence at $n = 4$ and $L = 1$ the calculated temperature will be 305°C and the conversion 6.6%.

Fig. 13-15 *Calculated temperatures in an SO_2 reactor*

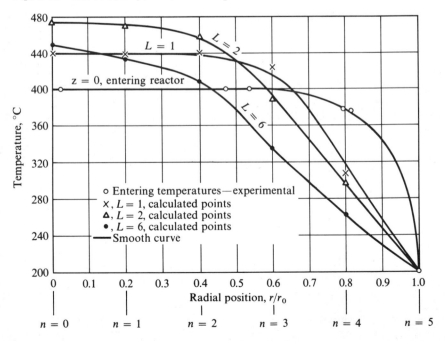

Since at $n = 5$ the wall is reached, the temperature remains 197°C. The rate is zero at this temperature, and so there will be no conversion due to reaction. Hence $T_{5,1} = 197°C$, and $x_{5,1} = 0$. At higher bed depths the conversion at the wall will not be zero, not because of reaction, but because of diffusion of the product SO_2 from the center of the tube.

The computations have now been made across the radius of the reactor at $L = 1$, and $z = 0.05$ ft. The temperature results are indicated by the points marked × in Fig. 13-15. The results computed by this stepwise procedure do not form a smooth curve at low bed depths. It is desirable before proceeding to the next increment to draw a smooth curve for this bed depth, as indicated in Fig. 13-15. The computed values and corresponding smooth conversion curve are shown in Fig. 13-16. The temperatures and conversions read from the smoothed curves, and to be used in the calculations at $L = 2$, are given in Table 13-6.

RESULTS FOR SUCCESSIVE INCREMENTS ($L = 2$ to 6)   By similar calculations temperature and conversion profiles may be obtained for successive bed lengths. The results at $L = 2$ are also plotted in Figs. 13-15 and 13-16. The computed values fall more nearly on a smooth curve than those at $L = 1$. The quantities at $L = 2$ shown in Table 13-6 were read from the smooth curves. These, rather than the computed points, are used in the calculations for the next bed depth. At higher bed depths the tabulated values are those directly computed from the equations. A conversion of 68% is reached at

Fig. 13-16   *Calculated conversions in an SO_2 reactor*

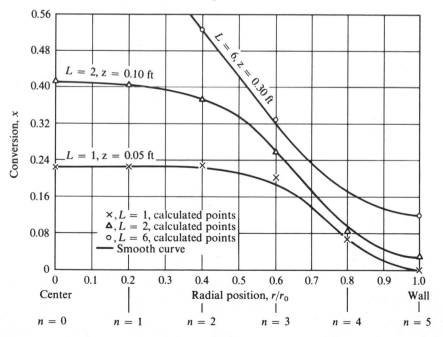

Table 13-6   Temperatures and conversions for SO_2 oxidation by the two-dimensional model

Bed depth, ft	0 (center)	0.2	0.4	0.6	0.8	1.0 (wall)
			Radial position			
			TEMPERATURE, °C			
$L = 0, z = 0$	400	400	400	400	376	197
$L = 1, z = 0.05$	441	441	437	418	315	197
$L = 2, z = 0.10$	475	471	458	390	298	197
$L = 3, z = 0.15$	496	488	443	378	285	197
$L = 4, z = 0.20$	504	476	437	360	278	197
$L = 5, z = 0.25$	470	466	415	350	269	197
$L = 6, z = 0.30$	451	435	412	334	265	197
			CONVERSION			
$L = 0, z = 0$	0	0	0	0	0	0
$L = 1, z = 0.05$	0.223	0.223	0.216	0.186	0.066	0
$L = 2, z = 0.10$	0.411	0.402	0.380	0.258	0.090	0.027
$L = 3, z = 0.15$	0.557	0.540	0.464	0.293	0.110	0.053
$L = 4, z = 0.20$	0.658	0.607	0.510	0.311	0.130	0.072
$L = 5, z = 0.25$	0.686	0.638	0.527	0.318	0.150	0.096
$L = 6, z = 0.30$	0.684	0.650	0.525	0.337	0.173	0.122

the center when $z = 0.30$ ft, and 12% is obtained at the wall. The temperature reaches a maximum value of 504°C at $z = 0.20$ ft and then decreases at higher bed depths because the radial transfer of heat toward the wall exceeds the heat evolved due to reaction. The temperature and conversion profiles at $z = 0.30$ are also plotted in Figs. 13-15 and 13-16.

MEAN CONVERSION AND TEMPERATURE   The bulk mean conversions and temperatures at any bed depth are obtained by graphical integration of the radial temperature and conversion profiles. The bulk mean temperature is the value resulting when the stream through the reactor is completely mixed in the radial direction. Hence the product of the heat capacity and the temperature at each radial position should be averaged. For an element dr the heat capacity of the flowing stream will be $G(2\pi r\,dr)c_p$. Hence the bulk mean temperature is given by the equation

$$\bar{T}_b = \frac{\displaystyle\int_0^{r_0} G(2\pi r\,dr)c_p T}{\pi G r_0{}^2 \bar{c}_p} = \frac{2\displaystyle\int_0^{r_0} T c_p r\,dr}{r_0{}^2 \bar{c}_p}$$

If the variable $n = r/r_0$ is substituted for r, this expression becomes

$$\bar{T}_b = \frac{2}{\bar{c}_p}\int_0^1 T c_p n\,dn \tag{E}$$

In Eq. (E) c_p is the specific heat at the temperature T and $\bar{c}_p$ is for the bulk mean temperature $\bar{T}_b$. Equation (E) can be integrated by plotting the product $Tc_p n$ vs n and evaluating the area under the curve.

Similarly, the bulk mean conversion $\bar{x}_b$ corresponds to complete radial mixing of the flow through the reactor. The moles of SO_2 converted in an element of thickness dr are $x(G/\bar{M})y_0 2\pi r\, dr$, where $G/\bar{M}$ represents the total moles per unit area entering the reactor and y_0 is the mole fraction SO_2 in the feed.

Integrating over all radial elements gives

$$\frac{\pi G r_0^2 y_0 \bar{x}_b}{\bar{M}} = \int_0^{r_0} x\frac{G}{\bar{M}} y_0 2\pi r\, dr$$

$$\bar{x}_b = \frac{2\int_0^{r_0} xr\, dr}{r_0^2}$$

If r/r_0 is replaced by n, then[1]

$$\bar{x}_b = 2\int_0^1 xn\, dn \qquad\qquad\qquad (F)$$

From the data in Table 13-6, graphs can be made of $Tc_p n$ and xn vs n to represent the values of the integrals in Eqs. (E) and (F). Actually, the mean conversion is the quantity of interest. Table 13-7 shows xn at a bed depth of $z = 0.30$ ft. These data are plotted in Fig. 13-17, and the area is

$$\int_0^1 xn\, dn = 0.150$$

Then, from Eq. (F),

$$\bar{x}_b = 2(0.150) - 0.300 \qquad \text{or } 30\% \text{ conversion}$$

Similar evaluations of the mean conversion have been made at the other bed depths. The results are plotted in Fig. 13-14, which also shows the measured conversions and the conversions calculated by the simplified method of Example 13-5.

Table 13-7   *Data for calculation of mean conversion at $L = 6$ and $z = 0.30$ ft*

n	x	xn
0	0.684	0
0.2	0.650	0.130
0.4	0.525	0.210
0.5	0.422	0.211
0.6	0.337	0.202
0.8	0.173	0.139
1.0	0.122	0.122

[1] The velocity does not appear in Eqs. (E) and (F) because it is assumed that G is constant across the diameter of the reactor.

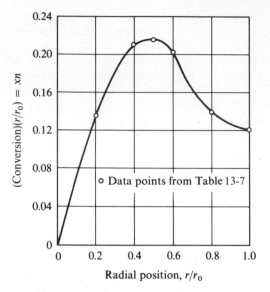

Fig. 13-17　*Graph for obtaining mean conversion*

The two-dimensional method has resulted in better agreement than the simplified approach, but the computed conversions are still less than the experimental results. In view of the problems in estimating the radial heat- and mass-transfer rates, and possible uncertainties in kinetic rate data, the comparison is reasonably good. The net effect of allowing for radial heat and mass transfer is to increase the conversion. The computed results are sensitive to rather small variations in the effective thermal conductivities and diffusivities, which emphasizes the need for the best possible information concerning these quantities.

The temperatures shown in Fig. 13-10 as solid curves represent experimental data for the conditions of this example. The computed values for r/r_0 of 0.0 and 0.6 have been taken from Table 13-6 and plotted as dashed lines on the figure for comparison. Referring to the centerline ($r/r_0 = 0$), we see that the computed results are about 10 to 20°C below the experimental values, although the location of the hot spot is predicted accurately. The comparison at $r/r_0 = 0.6$ is not as good.

When more than one reaction occurs the calculation procedures are similar to those illustrated in Example 13-6. A difference equation is written for each component, and these equations are solved simultaneously with the difference equation for the conservation of energy. Froment[1] has used one- and two-dimensional models to predict conversion and temperatures in a fixed-bed reactor for the oxidation of o-xylene to phthalic anhydride, CO, and CO_2, with a V_2O_5 catalyst. The reaction scheme is

[1] G. F. Froment, *Ind. Eng. Chem.*, **59**, 18 (1967).

o-xylene $\xrightarrow{\text{air}}$ phthalic anhydride $\xrightarrow{\text{air}}$ CO

$\quad\quad \llcorner \xrightarrow{\text{air}} CO_2$

Calculations were made for a reactor containing a bundle of 2,500 1-in. tubes packed with catalyst pellets and surrounded by molten salt to absorb the heat of reaction. The calculations showed that even with tubes of only 1-in. diameter, radial temperature gradients are severe for this extremely exothermic system.

A thorough study of a nonisothermal fixed-bed reactor for the reaction

$$C_6H_6 + 3H_2 \rightarrow C_6H_{12}$$

has been carried out by Otani.[1] An experimental differential reactor was used to obtain an equation of the Langmuir-Hinshelwood form [e.g., Eq. (9-32)] for the global rate, and then conversion and temperature measurements were made in an integral reactor. Both sets of data were obtained with the same cylindrical catalyst pellets (5 × 5 mm), for which the effectiveness factor was about 0.12. Then the global rate equation and estimated values of k_e and D_e were used to predict temperature and conversion data in the integral reactor tubes. The approach was the same as that described by Eqs. (13-42) and (13-20) and the one used in Example 13-6. The agreement between the predicted and experimental results was good.

13-8 Dynamic Behavior

Thus far we have considered only steady-state operation of fixed-bed reactors. The response to variations in feed composition, temperature, or flow rate is also of significance. The dynamic response of the reactor to these involuntary disturbances determines the control instrumentation to be used. Also, if the system is to be put on closed-loop computer control, a knowledge of the response characteristics is vital for developing a control policy. It is beyond the scope of this book to treat the dynamic behavior of reactors, but it is necessary to draw attention to the availability of information on the subject.

The solution of the steady-state form of the mass- and energy-conservation equations for fixed-bed reactors has been found to be complex. When transient conditions are considered the situation becomes rather intractable. For the special case of isothermal operation only the mass balance is involved, and it is possible for many types of kinetics to

[1] S. Otani, "Some Practices in Petrochemical Process Development," presented at Sixty-fourth National Meeting AIChE, New Orleans, March, 1969.

obtain solutions for conversion at any point in the reactor in response to fluctuations in the feed. However, the nonisothermal case is most important because of the possibility of instabilities, and it is just this problem that is so difficult to solve. For this reason effort has been directed toward establishing criteria for predicting when disturbances in the feed would grow and when they die out, instead of trying to solve the whole problem of conversion and temperature at any point in the reactor. We saw in Example 13-4 (Fig. 13-12) that sharp hot spots can develop when an exothermic reaction occurs in a cooled fixed-bed reactor. In that example conditions were such that the temperature rise was moderately large. Given other combinations of the heat of reaction, activation energy, rate equation, and heat-transfer rate to the surroundings, the temperature rise could be more pronounced. In such instances a positive fluctuation in feed temperature or composition could cause so large a temperature rise in the reactor that reaction would be complete within a short section of the bed. Such behavior could be undesirable because of catalyst deactivation and reduced selectivity, but most important, the reactor would become uncontrollable. It is clear that a means for predicting the conditions at which this instability could occur would be helpful.

Beek[1] has reviewed the stability criterion of Barkelew,[2] which is applicable for several rate expressions, including second-order, autocatalytic, and product-inhibited ones. A simpler, but, unnecessarily conservative, proposal is that of Wilson,[3] developed in connection with the study of the hydrogenation of nitrobenzene (Example 13-4). This criterion states that instability cannot occur if

$$\frac{E(T_{\max} - T_{\text{sur}})}{R_g T_{\max}^2} < 1 \tag{13-53}$$

where $T_{\max}$ and T_{sur} are the maximum and surroundings (cooling-medium) temperatures. For example, using the results of Example 13-4, $E/R_g = 2958 \, °K^{-1}$, at the hot spot we have

$$\frac{E(T_{\max} - T_{\text{sur}})}{R_g T_{\max}^2} = \frac{2,958(527 - 427)}{(527)^2} = 1.06$$

According to this criterion, the reactor operation is at the point of instability, so that a perturbation in the feed could cause an uncontrollable situation.

The calculations of Froment[4] for a reactor for the oxidation of o-xylene

[1] J. Beek, in T. B. Drew, J. W. Hoopes, Jr., and Theodore Vermeulen (eds.), "Advances in Chemical Engineering," vol. 3, Academic Press, Inc., New York, 1962.

[2] C. H. Barkelew, *Chem. Eng. Progr., Symp. Ser.*, **55**, 37 (1959).

[3] K. B. Wilson, *Trans. Inst. Chem. Eng.*, **24**, 77 (1946).

[4] G. F. Froment, *Ind. Eng. Chem.*, **59**, 18 (1967).

nicely illustrate the effect of a small fluctuation in feed temperature. The results showed that a 3°C rise in feed temperature (from 357 to 360°C) would lead to a continuous increase of the center temperature with reactor length, instead of the stable behavior of a rise and fall in temperature (for example as in Fig. 13-12). Such an unstable situation for this reaction results in a great loss in selectivity; the xylene is converted almost entirely to CO and CO_2 rather than to phthalic anhydride.

FLUIDIZED-BED REACTORS

The fluidized-bed reactor differs from its fixed-bed counterpart in that the solid catalyst particles usually are smaller (10 to 200 microns), the porosity in the reactor is larger, and the particles are in motion. In Secs. 1-6 (Fig. 1-4) and 3-8 the overall characteristics of fluidized beds were briefly described. Now, in preparation for design problems, it is necessary to describe the internal behavior. The fluid mechanics of solid particles in a gas stream is complex, and to some extent poorly understood. Summaries of present knowledge are available.[1]

Consider a cylindrical bed of small particles through which a gas is flowing. As the velocity is gradually increased, a point is reached where the upward drag force on the particles equals their weight. With a further increase in velocity, the particles jiggle and then move upward, expanding the bed. The particles are separate from one another and become suspended in the gas stream; the bed is fluidized. This ideal condition is *smooth fluidization*. Additional increase in velocity further expands the bed but does not cause a commensurate increase in pressure drop. A condition is reached eventually where gas bubbles, of low particle concentration, form and rise through the bed, causing the dense phase (higher particle concentration) to be disturbed. Visual observation shows *slugs* of dense phase being moved upward and outward by the more rapidly moving *bubbles*. The heterogeneous nature of the bed affects its performance as a reactor. To predict the conversion leaving the bed it would be necessary to know the residence time of the gas in both the bubbles and the dense phase, as well as catalyst-particle concentrations. Also, transport of both gas and particles occurs between bubbles and dense phase, and this should also be taken into account. Because our knowledge of these phenomena is not yet definitive, the analysis must be partially qualitative.[2]

[1] M. Leva, "Fluidization," McGraw-Hill Book Company, New York, 1959; J. F. Davidson and D. Harrison, "Fluidized Particles," Cambridge University Press, Cambridge, 1963; *Trans. Inst. Chem. Engrs. Symp. Fluidization*, **39**, 166 (1961); O. Levenspiel and D. Kunii, "Fluidization Engineering," John Wiley & Sons, Inc., New York, 1969.

[2] The available quantitative information is well summarized by Levenspiel and Kunii.

Let us first consider the mass- and heat-transfer characteristics of fluidized beds.

13-9 Heat Transfer and Mixing in Fluidized-bed Reactors

Under smooth fluidization the motion, heat capacity, and small size of the particles result in a remarkably uniform temperature throughout the bed. Radial gradients, so important in fixed beds, are negligible. The transfer of heat to or from the reactor can be considered by assuming that a finite heat-transfer coefficient exists at the wall, and that the temperature across the bed is uniform. This situation is depicted in Fig. 13-18, where curve c applies to the fluidized bed. For comparison, curve b represents a homogeneous tubular reactor in turbulent flow, where the temperature profile is not so flat as in the fluidized bed but is still more uniform than for the packed bed, case (a).[1]

Heat Transfer to the Wall   A number of investigations of heat-transfer coefficients at the wall in fluidized beds have been reported,[2] and in all cases the values found for h_w were considerably larger than those for an empty tube at the same fluid velocity. Presumably this is because the motion of solid particles near the wall tends to prevent the development of a slow-moving layer or film of gas, and the heat-carrying capacity of the particles themselves as they move between the center and the wall of the reactor is significant.

Dow and Jakob proposed the following empirical correlation of the heat-transfer coefficient at the inside wall:

$$\frac{h_w d_p}{k_g} = 0.55 \left(\frac{d}{d_p}\right)^{0.03} \left(\frac{d}{L}\right)^{0.65} \left(\frac{d_p G}{\mu}\right)^{0.80} \left(\frac{\rho_s c_s}{\rho_g c_g}\right)^{0.25} \left(\frac{1 - \epsilon_B}{\epsilon_B}\right)^{0.25}$$

$$(13\text{-}54)$$

where h_w = heat-transfer coefficient at inside-wall surface

d_p = diameter of particle

d = diameter of tube

---

[1] The comparison between the homogeneous tubular reactor and the packed-bed case depends on the velocity level and d_p/d. It is possible for the temperature profile in the packed bed to be more uniform than the profile shown for case (b).

[2] C. Y. Wen and M. Leva, *AIChE J.*, **2**, 482 (1956); O. Levenspiel and J. S. Walton, *Chem. Eng. Progr., Symp. Ser. 9*, **50**, 1 (1954); H. S. Mickley and C. A. Trilling, *Ind. Eng. Chem.*, **41**, 1135 (1949); W. M. Dow and Max Jakob, *Chem. Eng. Progr.*, **47**, 637 (1951); L. Wender and G. T. Cooper, *AIChE J.*, **4**, 15 (1958).

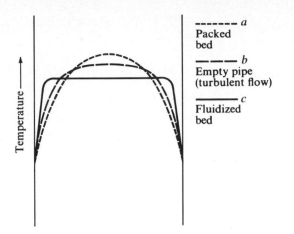

Fig. 13-18   *Temperature profiles (cooling conditions) in reactors*

L = length of tube

μ = viscosity of gas

k_g = thermal conductivity of gas

ρ_g = density of gas

ρ_s = density of solid particles

c_g = specific heat of gas

c_s = specific heat of solid particles

ϵ_B = void fraction in fluidized bed

Example 13-7   Estimate the heat-transfer coefficient to the wall in a 6-in.-ID tube through which air is passed at a mass rate of 20 lb/hr at 100°F and 1 atm pressure, under each of the following bed conditions:

(a) Solid particles [c_s = 0.5 Btu/(lb)(°F) and ρ_s = 90 lb/ft³] of average diameter of 100 microns are fluidized in the tube. The tube length is 10 ft and the void fraction is 0.95.
(b) The tube is empty.
(c) The tube is packed with ½-in. spheres.

Solution   (a) In the fluidized-bed case

$$d = \frac{6}{12} = 0.50 \text{ ft}$$

d_p = 100 microns, or $100 \times 10^{-4}/[2.54(12)] = 3.28 \times 10^{-4}$ ft

$$\mu = 0.0448 \text{ lb/(hr)(ft)}$$

$$k_g = 0.0156 \text{ Btu/(hr)(ft)(°F)}$$

$$G = \frac{40}{0.7854(\frac{6}{12})^2} = 204 \text{ lb/(hr)(ft}^2)$$

Hence

$$\frac{d}{d_p} = \frac{0.50}{0.000328} = 1{,}524$$

$$\frac{d}{L} = \frac{0.50}{10} = 0.05$$

$$\frac{d_p G}{\mu} = \frac{3.28 \times 10^{-4}(204)}{0.0448} = 1.49$$

$$\rho_g = \frac{28.9}{359(100 + 460)/492} = 0.0707 \text{ lb/ft}^3$$

$$\frac{\rho_s c_s}{\rho_g c_g} = \frac{90(0.5)}{0.0707(0.24)} = 265$$

Substituting these quantities in Eq. (13-54) yields

$$\frac{h_w d_p}{k_g} = 0.55(1{,}524)^{0.03}(0.05)^{0.65}(1.49)^{0.80}(265)^{0.25} \left(\frac{1 - 0.95}{0.95}\right)^{0.25}$$

$$\frac{h_w(3.28 \times 10^{-4})}{0.0156} = 0.259$$

$$h_w = 12.3 \text{ Btu/(hr)(ft}^2)(°F)$$

(b) In the empty tube the flow is just inside the turbulent zone, with a Reynolds number of $dG/\mu = 2{,}280$. The coefficient for an empty tube may be estimated from the expression

$$\frac{h_w d}{k_g} = 0.023(dG/\mu)^{0.8} \left(\frac{c_p \mu}{k_g}\right)^{0.4}$$

$$\frac{c_p \mu}{k_g} = \frac{0.24(0.0448)}{0.0156} = 0.69$$

$$\frac{h_w d}{k_g} = 0.023(2{,}280)^{0.8}(0.69)^{0.4} = 10.5$$

$$h_w = \frac{10.5(0.0156)}{0.50} = 0.33 \text{ Btu/(hr)(ft}^2)(°F)$$

(c) With $\frac{1}{2}$-in. spherical packing,

$$\frac{d_p}{d} = \frac{0.5}{6} = 0.083$$

The ratio of h_P/h_w is estimated (from table in Sec. 13-5) to be 6.5. Hence the heat-transfer coefficient for the packed-bed case is

$$6.5(0.33) = 2.1 \text{ Btu/(hr)(ft}^2)(°F)$$

For this particular case the coefficient in the fluidized bed is six times that in the packed tube and thirty-seven times that in an empty tube.

Mixing Characteristics Although the temperature is essentially uniform in fluidized beds, the same cannot be said of the gas composition, particularly when slugging occurs. Even for smooth fluidization some mixing occurs in the gas. For this ideal state the dispersion model (Sec. 6-5), with a suitable value of axial diffusivity, may represent mixing of gas reasonably well. Axial diffusivities are available for beds fluidized both with gases[1] and with liquid.[2] For the slugging or bubbling regime models have been proposed to account for the extent of mixing,[3] but reliable data for the mixing parameters are not yet available. As noted in Chap. 6, another approximate way of accounting for mixing in nonideal reactors is to measure the distribution of residence times. Gilliland et al.[4] used this technique in an experimental study and found that the fluidized bed performed between the extremes of ideal stirred-tank and plug-flow reactors.

13-10 Design Procedures

Fluidized beds are particularly suitable when frequent catalyst regeneration is required or for reactions with a very high heat effect. Frequently the reactors are vessels of large diameter (10 to 30 ft is not unusual for catalytic cracking units in the petroleum industry). A typical system is shown in Fig. 13-19. The movable catalyst permits continuous regeneration in place. Part of the catalyst is continuously withdrawn from the reactor in line *A* and flows into the regenerator. The regenerator shown in the figure is another fluidized bed, from which reactivated catalyst is returned to the reactor through line *B*. It is not necessary to carry out the regeneration in a fluidized bed, as catalyst could be withdrawn through line *A* continuously, and reactivated catalyst returned through line *B*, with the regeneration accomplished by any procedure. However, the process is most economical if both reaction and regeneration are carried out in fluidized beds by means of the integral setup illustrated in Fig. 13-19.

[1]E. R. Gilliland, E. A. Mason, and R. C. Oliver, *Ind. Eng. Chem.*, **45**, 1177 (1953); P. V. Danckwerts et al., *Chem. Eng. Sci.*, 3, 26 (1954); E. R. Gilliland and E. A. Mason, *Ind. Eng. Chem.*, **41**, 1191 (1949).

[2]E. J. Cairns and J. M. Prausnitz, *AIChE J.*, **6**, 400 (1960).

[3]W. G. May, *Chem. Eng. Progr.*, **55**, 49 (1959); J. J. van Deemter, *Chem. Eng. Sci.*, **13**, 143 (1961); O. Levenspiel and D. Kunii, *Ind. Eng. Chem., Fund. Quart.*, **7**, 446 (1968); *Ind. Eng. Chem., Process Design Develop.*, 7, 481 (1968).

[4]E. R. Gilliland, E. A. Mason, and R. C. Oliver, *Ind. Eng. Chem.*, **45**, 1177 (1953).

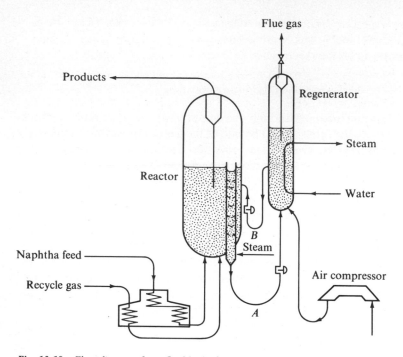

Fig. 13-19   *Flow diagram for a fluid hydroformer, a fluidized-bed reactor-regenerator combination (by permission from Esso Standard Oil Company, Baton Rouge, La.)*

An important feature of the fluidized-bed reactor is that it operates at a nearly constant temperature and hence is easy to control. There is no opportunity for hot spots to develop, as in the case of the fixed-bed unit. The fluidized bed does not possess the flexibility of the fixed bed for adding or removing heat. A diluent can be added to control the temperature level, but this may not be desirable for other reasons (requires separation after the reactor, lowers the rate of reaction, increases the size of the equipment). A heat-transfer fluid can be circulated through a jacket around the reactor, but if the reactor is large in diameter, the energy exchange by this method is limited.[1]

Another characteristic of the fluidized bed is the small size and density of catalyst particles necessary to maintain proper fluidization. The particles provide a much larger external surface per unit mass of catalyst than those in a fixed-bed unit. This results in a higher rate of reaction (per unit mass) for a nonporous catalyst. Also, internal transport effects are negligible.

---

[1] A heat-transfer fluid may also be circulated through tubes placed within the fluidized bed.

From a practical standpoint catalyst loss due to carryover with the gas stream from the reactor and regenerator may be an important problem. Attrition of particles decreases their size to a point where they are no longer fluidized, but move with the gas stream. It has been customary to recover most of these catalyst fines by cyclone separators and electrical precipitation equipment placed in the effluent lines from reactor and regenerator.

Deterioration of lines and vessels due to the abrasive action of the sharp, solid particles in the fluidized cracking process has caused concern. This problem has been particularly severe in the small-diameter transfer lines, where the particle velocity is high. These and other matters relating to the commercial operation of fluidized catalytic cracking plants have been discussed in the literature.[1]

The quantitative design of fluidized-bed reactors is similar to that for ideal homogeneous systems presented in Chaps. 4 and 5. The difference is that the flow pattern is neither that of an ideal stirred-tank reactor nor that of a plug-flow reactor. When the flow pattern can be described as plug flow modified by axial dispersion, the results in Chap. 6 are applicable. For example, Eq. (6-54) gives the conversion for a first-order reaction in terms of the diffusivity D_L. The most serious design problem arises when both dense (emulsion) and bubble phases exist in the bed. The average residence time of the gas in the bubbles is much less than that of the gas moving through the dense phase. Furthermore, the catalyst concentration in the bubbles is very low. The effect is one of channeling, which reduces the conversion in the effluent stream. Several models for the flow pattern in such cases have been used to derive equations for the conversion. For example, van Deemter[2] took into account mass transfer between bubbles and dense phase, through the use of the height of a transfer unit, and longitudinal dispersion of the gas in the dense phase. The resultant procedure for evaluating the conversion requires information about both mass-transfer-rate coefficients. May's model[3] is based on the solution of mass-conservation equations for the bubble and dense phases. After analyzing available data, Levenspiel and Kunii[4] proposed a *bubbling-bed* model and recommended correlations for the necessary mass-transfer rates. They also illustrated the application of the model to predicting the conversion in fluid-bed, catalytic reactors.

[1] E. V. Murphree, C. L. Brown, E. J. Goba, C. E. Hohnig, H. Z. Martin, and C. W. Tyson, *Trans. AIChE*, **41**, 19 (1945); A. L. Conn, W. F. Meeham, and R. V. Shankland, *Chem. Eng. Progr.*, **46**, 176 (1950).

[2] J. J. van Deemter, *Chem. Eng. Sci.*, **13**, 143 (1961).

[3] W. G. May, *Chem. Eng. Progr.*, **55**, 49 (1959).

[4] O. Levenspiel and D. Kunii, "Fluidization Engineering," John Wiley & Sons, Inc., New York, 1969. See also J. G. Yates, P. N. Rowe, and S. T. Whang, *Chem. Eng. Sci.*, **25**, 1387 (1970).

Lewis et al.,[1] measured conversions for the catalytic hydrogenation of ethylene as a function of fluidization velocity. Figure 13-20 shows the data and calculated conversions for ideal tubular-flow and stirred-tank reactors at the same conditions. The steep drop in conversion in the fluidized bed at high velocities presumably results from the increased extent of bubbling. The difficulty in representing a fluidized bed by a model is due not only to the nonhomogeneity of the gas phase, but also to the tendency of a layer of more-or-less stagnant gas to form around a particle and move with the particle in the bed. Hence the efficiency of contact between gas and catalyst is low in comparison with fixed beds. It should be noted that the bubbles rising through the dense phase increase mixing of the particles and promote uniformity of temperature in the bed. Thus mixing of particles is more efficient than gas-particle contacting. The relatively low mass-transfer coefficients between gas and particles can reduce selectivity. The effect is the same as that described in Sec. 10-5; that is, for two parallel

[1]W. K. Lewis, E. R. Gilliland, and W. Glass, *AIChE J.*, **5**, 419 (1959).

Fig. 13-20   *Hydrogenation of ethylene in a fluidized-bed reactor [from W. K. Lewis, E. R. Gilliland, and W. Glass,* AIChE J., *5, 419 (1959)]*

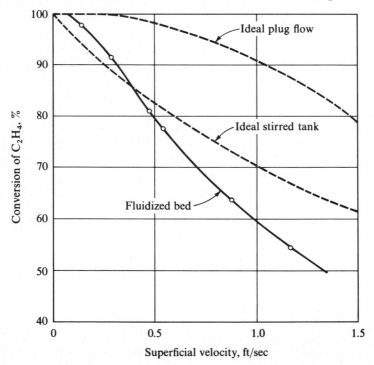

first-order reactions the intrinsic selectivity k_1/k_2 for the chemical steps is reduced in accordance with Eq. (10-31).

SLURRY REACTORS

Slurry reactors, first discussed in the last sections of Chap. 10 (see Fig. 10-6), are similar to fluidized-bed reactors in that a gas is passed through a reactor containing solid catalyst particles suspended in a fluid. In slurry reactors the catalyst is suspended in a liquid; in fluidized-bed reactors the suspending fluid is the reacting gas itself. The advantages of slurry reactors over fixed beds are similar to those of fluidized beds: uniform temperature, easy regeneration of catalyst, better temperature control for highly exothermic reactions, and the lack of intraparticle diffusion resistance. For very active catalysts this last factor means that the global rate can be much higher than for fixed-bed reactors.

In Chap. 10 rate equations were developed (for the overall transfer of reactant from gas bubble to catalyst surface) in terms of the individual mass-transfer and chemical-reaction steps [Eqs. (10-45) and (10-46)]. The purpose there was to show how the global rate $\mathbf{r}_v$ (per unit volume of bubble-free slurry) was affected by such variables as the gas-bubble–liquid interface a_g, the liquid-solid catalyst interface a_c, and the various mass-transfer coefficients, and the rate constant for the chemical step. Now the objective is to evaluate the performance of a slurry reactor in terms of the results from Chap. 10; that is, we suppose that the global rate, or overall rate coefficient k_o, is known, and the goal is to design the reactor.

The feature of slurry reactors that distinguishes them from homogeneous or fixed-bed catalytic reactors is that there are two mobile phases. This increases the possible classifications of reactor operation. Referring to the ideal extremes, both the gas and liquid phases may be in either plug flow or a completely mixed condition. Furthermore, the liquid phase may be in continuous flow or may remain in the reactor as in batch operation. The performance of two-phase reactors will depend on the specific type of operation. The procedure for evaluating the behavior is the same as that given in Chap. 3 for single-phase systems. A mass balance is written for each reaction according to the general Eq. (3-1).[1] If reactants exist in both gas and liquid, separate balances must be written for each phase. Solution of these equations, which contain the global rate, determines the relationship between volume of reactor and extent of reaction, just as Eqs. (3-3), (3-10),

[1] An energy-conservation equation is usually not necessary in two-phase reactors, because the large surface area and the turbulence caused by the gas bubbles prevent significant temperature variations.

and (3-13) do for single-phase systems. Schaftlein and Russell[1] have developed mass-balance equations for several types of operation of tank-type gas-liquid reactors:[2]

1. Continuous flow of both gas and liquid
   a. Plug flow of gas, well-mixed liquid (stirred tank)
   b. Well-mixed gas, well-mixed liquid
2. Batch liquid
   a. Plug flow of gas, well-mixed liquid
   b. Well-mixed gas, well-mixed liquid
3. Batch liquid and gas (both phases well mixed)

Note that the same mass-balance equations apply whether the reaction in the liquid phase is homogeneous or catalyzed by solid particles as in a slurry reactor. The difference between catalyzed and noncatalytic systems is accounted for in the global rate. If the reactants are introduced only in the gas phase, a mass balance is needed only for that phase. This situation exists for some slurry reactors where the liquid phase is inert and its purpose is simply to suspend the catalyst particles.

As an example of a two-phase system, consider a slurry reactor in which bubbles of gas are in plug flow upward through a well-mixed batch of inert liquid carrying catalyst particles. This corresponds to Schaftlein and Russell's classification 2a. Suppose that the reactants enter only in the gas bubbles, so that a mass balance need be written only for this phase.[3] Under these conditions the gas bubbles may be imagined to undergo a reaction at a rate r_v as they pass in plug flow through a reactor of volume equal to that of the bubble-free slurry. The mass balance for a reactant is then given by the expression for plug-flow behavior, Eq. (3-13),

$$\frac{V}{F} = \int_0^x \frac{dx}{r_v}$$

where V is the volume of bubble-free slurry in the reactor and F is the feed rate of gaseous reactant. Consider an overall reaction of the form

$$A(g) + B(g) \rightarrow C(g)$$

which occurs by solution of A and B from the bubbles into the liquid, followed by diffusion to the catalyst particles, as described in Chap. 10. Suppose that the reaction is first order with respect to A, so that the rate is given by Eq. (10-45). Substituting this rate in Eq. (3-13) gives

---

[1] R. W. Schaftlein and T. W. F. Russell, *Ind. Eng. Chem.*, **60**, 12 (1968).

[2] Gas-liquid reactor design is also considered in J. Bridgwater and J. J. Carberry, *Brit. Chem. Eng.*, **12**, 58 (1967); **12**, 217 (1967).

[3] Assuming that the holdup of reactants in the liquid is constant.

$$\frac{V}{F_A} = \int_0^{x_1} \frac{dx}{k_o a_c (C_A)_g} \tag{13-55}$$

This expression can be used to predict the conversion in the bubbles leaving the reactor from a knowledge of k_o.

Equation (13-55) is applied to a slurry reactor for the hydrogenation of ethylene in Example 13-8. This is a very simple illustration of a two-phase reactor, since the liquid carrying the catalyst particles does not take part in the reaction. For other applications, such as a reaction between one reactant in the gas phase and another in the liquid phase, see Schaftlein and Russell.[1]

Example 13-8 Ethylene is to be hydrogenated by bubbling mixtures of H_2 and C_2H_4 through a slurry of Raney nickel catalyst particles suspended in toluene. The gas bubbles are formed at the bottom of a tubular reactor and rise in plug flow through the slurry. The slurry is well mixed, so that its properties are the same throughout the tube. A large concentration of small catalyst particles will be used. The temperature and pressure are to be 50°C and 10 atm. At these conditions the overall rate has been shown to be determined by the rate of diffusion of hydrogen from the bubble interface to the bulk liquid.[2]

Estimate the volume of bubble-free slurry required to obtain a conversion of 30% for a hydrogen feed rate of 100 ft³/min (at 60°F and 1 atm). By a light-transmission technique, Calderbank measured gas-liquid interfacial areas of 0.94 to 2.09 cm²/cm³ for bubble sizes likely to be encountered in this system. Suppose for this illustration $a_g = 1.0$ cm²/cm³ of bubble-free slurry. The Henry's law constant for hydrogen in toluene at 50°C is 9.4 (g mole/cm³)/(g moles/cm³), and its diffusivity is 1.1×10^{-4} cm²/sec. The density and viscosity of toluene at 50°C are 0.85 g/cm³ and 0.45 centipoises, respectively. Equimolal feed rates of ethylene and hydrogen will be used.

Solution For the reaction

$$C_2H_4(g) + H_2(g) \rightarrow C_2H_6(g)$$

the number of moles of each component at a conversion level x is

$H_2 = F(1 - x)$

$C_2H_4 = F(1 - x)$

$C_2H_6 = Fx$

Total moles $= F(2 - x)$

where F is the molal feed rate of H_2 or C_2H_4. Then the concentration of H_2 in the gas bubbles is

[1] R. W. Schaftlein and T. W. F. Russell, *Ind. Eng. Chem.*, **60**, 12 (1968).
[2] P. H. Calderbank, F. Evans, R. Farley, G. Jepson, and A. Poll, Catalysis in Practice, *Symp. Inst. Chem. Engrs. (London)*, 1963.

$$(C_{H_2})_g = \frac{p_t}{R_g T} y_{H_2} = \frac{p_t}{R_g T} \frac{1-x}{2-x} \tag{A}$$

Equation (13-55) is applicable to this reactor. If we take k_o and a_c as constant, this expression becomes

$$\frac{V}{F} = \frac{R_g T}{k_o a_c p_t} \int_0^{x_1} \frac{2-x}{1-x} dx \tag{B}$$

Integrating from zero to the exit conversion gives

$$\frac{V}{F} = \frac{1}{k_o a_c} \frac{R_g T}{p_t} [x_1 - \ln(1-x_1)] \tag{C}$$

The overall rate constant k_o would, in general, be a function of several rate parameters. Since it is known that only the resistance to diffusion of hydrogen from the bubble interface is significant, Eq. (10-46) reduces to

$$\frac{1}{k_o} = \frac{a_c}{a_g} \frac{H}{k_l}$$

or

$$k_o a_c = \frac{a_g k_l}{H} \tag{D}$$

Hence Eq. (C) may be written

$$\frac{V}{F} = \frac{H}{a_g k_l} \frac{R_g T}{p_t} [x_1 - \ln(1-x_1)] \tag{E}$$

Everything in Eq. (E) is known except the mass-transfer coefficient for hydrogen in the liquid. This may be estimated from Eq. (10-49), since this correlation was based on data for gas bubbles rising through a liquid phase. Thus

$$k_l \left[\frac{0.45 \times 10^{-2}}{0.85(1.1 \times 10^{-4})} \right]^{\frac{2}{3}} = 0.31 \left[\frac{(0.85 - 0.8 \times 10^{-3})(0.45 \times 10^{-2})(980)}{0.85^2} \right]^{\frac{1}{3}}$$

$$k_l = \frac{0.31}{13.2}(1.74) = 0.041 \text{ cm/sec}$$

Substituting this result for k_l and the other numerical value in Eq. (E) gives

$$\frac{V}{F} = \frac{9.4}{1.0(0.041)} \frac{82(273 + 50)}{10}[0.3 - \ln(1 - 0.3)]$$

$$= 0.40 \times 10^6 \text{ cm}^3/(\text{g mole/sec})$$

For a hydrogen feed rate of 100 ft^3/min, at 60°F and 1 atm, the slurry volume required would be

$$V = \frac{100}{379}\left(\frac{454}{60}\right)(0.40 \times 10^6)(10^{-3})\left(\frac{1}{28.32}\right) = 28 \text{ ft}^3$$

In this example many important properties of the slurry have been omitted. For example, the questions of bubble diameter and volume fraction of gas in the slurry have been avoided by giving a directly measured value of a_g. Alternately, the interfacial area can be estimated from Eqs. (10-52) and (10-53).

It should not be inferred from Example 13-8 that diffusion from the bubble to the bulk liquid always controls the global rate. For example, data for different slurry reactions given by Sherwood and Farkas[1] show other results. For the hydrogenation of α-methyl styrene containing a slurry of palladium black (diameter 55 microns), the global rate was controlled by k_l and k_c, with the significance of k_l decreasing at low catalyst concentrations. For catalyst concentrations below about 0.5 g of catalyst per liter of slurry, and at temperatures less than 30°C, the rate of diffusion of dissolved hydrogen to the catalyst particles (that is, k_c) essentially controlled the rate. Results for the hydrogenation of cyclohexene in an aqueous suspension of 30-micron palladium-black particles indicated again that the predominant resistance was the diffusion to catalyst particles. The data of Kolbel and Maennig[2] for the hydrogenation of ethylene in a slurry of Raney nickel particles in a paraffin oil show still different results. Here the catalyst particles were small enough (5 microns) and their concentration high enough that $k_c a_c$ was very large. Sherwood and Farkas were able to correlate these data by assuming that the chemical step on the catalyst particles controlled the rate.

OPTIMIZATION

Although we shall not consider the quantitative aspects of optimization, some general comments are necessary in order to give proper emphasis to the material that has been covered. As indicated in Sec. 1-1, the purpose of this text has been to present the concepts necessary to design a reactor. We started with chemical kinetics (Chap. 2) and then discussed physical processes in terms of the process-design features of large-scale reactors, first for homogeneous reactions (Chaps. 3 to 6) and then for heterogeneous catalytic reactions (Chaps. 7 to 13). The general approach was that the reactor form was known, and the objective was to predict the performance for a single set of operating conditions. Nevertheless, aspects of optimum performance were often introduced. As early as Chap. 1, the interrelationship between kinetics and thermodynamics for an exothermic reversible reaction was employed. We saw that the maximum attainable conversion in such reactions decreases as the temperature increases, but the rate of

[1] T. K. Sherwood and E. J. Farkas, *Chem. Eng. Sci.*, **21**, 573 (1966).
[2] H. Kolbel and H. G. Maennig, *Elektrochem.*, **66**, 744 (1962).

reaction increases. These contrasting effects suggest that improved conversion (per unit volume of reactor) could be obtained by operating the system at different temperatures: first in a high-temperature reactor, where most of the conversion is obtained at a high rate, and then in a second reactor operated at a lower temperature to achieve the higher conversion dictated by thermodynamics. This concept was applied to the oxidation of SO_2 to SO_3. In Chaps. 4 and 5 (particularly Sec. 5-2) conclusions were reached about operating conditions and reactor types for maximum yield of the desired product in multiple-reaction systems. A great amount of work has been expended in developing quantitative methods of optimization from these simple concepts, particularly for the performance of specific types of reactors.[1]

The vague term "optimizing the performance" does not properly describe the goal, since the ultimate objective concerns economics. However, this term does indicate the dilemma of optimization studies. Rarely can the profit from a chemical reactor be described quantitatively in terms of operating conditions. First, the reactor is probably only one unit of a plant, and the most economical operation of the reactor may conflict with the economy of subsequent separation processes. Hence overall economy may in fact require operating the reactor at nonoptimum conditions. Second, market conditions for the reaction products, even when they are known, are subject to fluctuation. Hence the economics of the entire plant may be time dependent or uncertain. Third, it is difficult to establish valid figures for *all* the costs that accrue to a reactor or a plant. Because of these uncertainties, optimization studies have dealt with conversion and selectivity rather than with profits. While this approach does not take into consideration any of the cost factors, it does provide a constant solution to the technical problem. However, for commercial reactors a unit value must be applied to each product, determined on the basis of operating and initial costs and marketing uncertainties. Such studies require a knowledge of how operating conditions, such as temperatures, pressures, and feed compositions, affect the production rate of the products, and the problem is to determine the particular operating conditions that will give the maximum profit.

It is supposed that the most profitable reactor type has already been chosen, and the question is: What conditions will afford the most profitable operation of this type? Actually, complete optimization would require simultaneous solution for both considerations. Consider a highly exothermic

---

[1] Much of the literature has been summarized in R. Aris, "The Optimal Design of Chemical Reactors," Academic Press, Inc., New York, 1961; K. G. Denbigh, "Chemical Reactor Theory," chap. 5, Cambridge University Press, Cambridge, 1965; H. Kramers and K. R. Westerterp, "Elements of Chemical Reactor Design and Operation," chap. VI, Academic Press, Inc., New York, 1963.

catalytic reaction (such as air oxidation of naphthalene to phthalic anhydride) which must be carried out below an upper temperature limit to prevent undesirable side reactions (oxidation to CO_2 and H_2O). Several choices of reactor exist. A single, large-diameter, adiabatic fixed bed could be used, with an excess of inert diluent added to the feed to absorb the heat of reaction. This would reduce the initial cost of the reactor. Alternately, a large number of parallel small tubes, packed with catalyst pellets and surrounded by a cooling fluid, might be employed; in this case the diluent might be reduced, and the temperature rise would be limited by heat transfer to the cooling fluid. Operating costs might be reduced, but the initial cost would be high for a reactor consisting of hundreds of small tubes manifolded together. Another possibility would be a large-diameter fluidized bed, either with extensive diluent in the feed and no internal cooling tubes, or with little diluent but a bank of tubes in the bed through which a cooling medium flowed. To decide among these three types optimization studies and a comparison of the results for each would be necessary. It appears that rather more of this broader type of optimization study would be rewarding.

Let us return, however, to the problem of optimizing total conversion and selectivity. To begin with, it is important to note the relation between the two functions. For a single reaction there is no problem: optimum operation is the maximum production rate of the product per unit mass of catalyst. For plug-flow reactions the mass balance of reactant is given by Eq. (12-1), which may be rearranged to the form

$$\frac{F\,dx}{dW} = \mathbf{r}_P$$

In an element of catalyst dW the production rate per unit mass is $F\,dx/dW$. Hence Eq. (12-1) states that optimum performance is attained in the element when $\mathbf{r}_P$ is at a maximum. For the entire reactor to operate at optimum conditions the rate should have its maximum permissible value at every axial position. This concept has been used to predict optimum temperature profiles for single exothermic reversible reactions.[1]

When two or more reactions are involved the situation is more complicated, because both total conversion and selectivity, that is, the production rate of each product, are likely to be important. Moreover, this importance may depend on factors other than the reactor, specifically the difficulty (cost) of separation and recycling unreacted feed components and the separation of desirable and undesirable products. Sometimes separation is not feasible.

[1] K. G. Denbigh, *Trans. Faraday Soc.*, **40**, 352 (1944); P. H. Calderbank, *Chem. Eng. Progr.*, **49**, 585 (1953); F. Horn, *Chem. Eng. Sci.*, **14**, 77 (1961).

As an example, consider the catalytic reforming of naphtha in a fixed bed, with a platinum catalyst. The aromatization and cracking processes that occur with the naphtha feed (which itself contains many components) lead to a product that contains literally hundreds of individual components. An overall measure of selectivity is usually all that is possible, and this is generally the octane number of product; separation of specific products is not attempted. The profitability of the reactor depends on the total production of reformed product and its selectivity (octane number). As temperature increases, total conversion increases, but octane number decreases (many other operating conditions, of course, affect these two measures of performance). Hence a profit function giving appropriate emphasis to total conversion and octane number must be calculated, and then operating conditions that maximize this function must be determined. A complete mathematical analysis, taking into account all the reactions, is impossible—in this case because all the reactions and their kinetics are not known, and not because of the magnitude of the calculations. In simpler situations involving only a few variables, complete analysis is not difficult. Kramers and Westerterp[1] consider several cases; for example, maximizing the production rate of B in the reactions $A \rightarrow B \rightarrow C$ in an ideal stirred-tank reactor for various temperature conditions, and maximizing the profit in a tubular-flow reactor where three reactions occur, $A \rightarrow B$, $A \rightarrow C$, and $A \rightarrow D$, and B is the valuable product. For somewhat more complex cases with more variables, solutions are still feasible if machine computation is used for the extensive calculations. In these cases mathematical concepts such as those embodied in the optimality theory of Bellman[2] or the "method of steepest descent"[3] are necessary.

In summary, the requirement for economical commercial operation suggests the need for further optimization studies. Mathematical procedures are available for carrying out almost any reactor optimization problem. Thus the limitations do not reside so much in optimization methods, but in formulation of the profit function and complete knowledge of the technical aspects of the design problem.

Problems

13-1.   Suppose it were possible to operate an ethyl benzene–dehydrogenation reactor under approximately isothermal conditions. If the temperature is 650°C, prepare a curve for conversion vs catalyst-bed depth which extends to the

[1] H. Kramers and K. R. Westerterp, "Elements of Chemical Reactor Design and Operation," chap. 6, Academic Press, Inc., New York, 1963.
[2] See R. Aris, *Chem. Eng. Sci.*, **13**, 18 (1960).
[3] F. Horn and U. Trolten, *Chem. Eng. Techol.*, **32**, 382 (1960).

equilibrium conversion. The catalyst to be used is that for which rate data were presented in Example 13-1. Additional data are:

Diameter of catalyst tube = 3 ft
Feed rate of ethyl benzene per tube = 8.0 lb moles/hr
Feed rate of steam per tube = 225 lb moles/hr
Bulk density of catalyst as packed = 90 lb/ft³

Equilibrium-constant data are given in Example 13-1.

13-2. In this case assume that the reactor in Prob. 13-1 operates adiabatically and that the entrance temperature is 650°C. If the heat of reaction is $\Delta H = 60{,}000$ Btu/lb mole, compare the curve for conversion vs bed depth with that obtained in Prob. 13-1.

13-3. Begley[1] has reported temperature data taken in a bed consisting of $\frac{1}{4} \times \frac{1}{4}$-in. alumina pellets packed in a 2-in. pipe (actual 2.06 in. ID) through which heated air was passed. The tube was jacketed with boiling glycol to maintain the tube wall at about 197°C. For a superficial mass velocity (average) of air equal to 300 lb/(hr)(ft²) the experimental temperatures at various packed-bed depths are as follows:

Radial position	Experimental temperature, °C				
	0.076 ft	0.171 ft	0.255 ft	0.365 ft	0.495 ft
0.0	378.7	354.7	327.8	299.0	279.3
0.1	377.2	353.7	327.0	298.0	278.9
0.2	374.6	349.9	324.1	294.7	277.0
0.3	369.5	343.9	319.7	289.2	273.2
0.4	360.3	336.3	313.8	282.1	267.6
0.5	347.7	327.4	306.4	274.0	260.8
0.6	331.9	316.1	298.2	265.0	252.7
0.7	313.2	300.7	287.9	254.8	243.8
0.8	291.0	282.8	273.1	242.2	234.5
0.9	256.5	257.9	244.2	224.8	224.6

The mass-velocity profile at an average value of 300 lb/(hr)(ft²) for $\frac{1}{4} \times \frac{1}{4}$-in. cylindrical packing is as follows:

Radial position, r/r_0	0	0.1	0.2	0.3	0.4	0.5	0.6	0.7	0.8	0.9
Superficial mass velocity, lb/(hr)(ft²)	247	254	262	282	307	349	400	476	382	234

As the first step in calculating k_e, estimate an average value for the mass-transfer Peclet number Pe_r for the bed, and then evaluate the effective diffusivity D_r.

[1] J. W. Begley, master's thesis, Purdue University, Purdue, Ind., February, 1951.

13-4.   The temperature-profile data in Prob. 13-3 are to be represented by a constant k_e (across the tube diameter) and a wall heat-transfer coefficient h_w. Estimate the values of k_e and h_w which best fit the temperature data. Note that at the boundary layer between the wall and the central core of the bed the following relation must apply:

$$h_w(T_i - T_w) = -k_e\left(\frac{\partial T}{\partial r}\right)_i$$

where k_e = constant for central core of bed

T_i = temperature at the interface between film and central core

$\left(\dfrac{\partial T}{\partial r}\right)_i$ = gradient in central core at the interface between core and wall film

13-5.   Using Equation (13-40), predict k_e for comparison with the result obtained in Prob. 13-4. The bed porosity is 0.40 and the emissivity of the alumina pellets is 0.9.

13-6.   In the German phthalic anhydride process[1] naphthalene is passed over a vanadium pentoxide (on silica gel) catalyst at a temperature of about 350°C. Analysis of the available data indicates that the rate of reaction (pound moles of naphthalene reacted to phthalic anhydride per hour per pound of catalyst) can be represented empirically by the expression

$$\mathbf{r} = 305 \times 10^5 p^{0.38}\, e^{-28,000/R_g T}$$

where p is partial pressure of naphthalene in atmospheres and T is in degrees Kelvin. The reactants consist of 0.10 mole % naphthalene vapor and 99.9% air. Although there will be some complete oxidation to carbon dioxide and water vapor, it is satisfactory to assume that the only reaction (as the temperature is not to exceed 400°C) is

$$C_{10}H_8 + 4.5O_2 \rightarrow C_8H_4O_3 + 2H_2O + 2CO_2$$

The heat of this reaction is $\Delta H = -6300$ Btu/lb of naphthalene, but use a value of -7300 Btu/lb in order to take into account the increase in heat release owing to the small amount of complete oxidation. The properties of the reaction mixture may be taken as equivalent to those for air.

The reactor will be designed to operate at a conversion of 80% and have a production rate of 6,000 lb/day of phthalic anhydride. It will be a multitube type (illustrated in Fig. 13-1) with heat-transfer salt circulated through the jacket. The temperature of the entering reactants will be raised to 340°C by preheating, and the circulating heat-transfer salt will maintain the inside of the reactor-tube walls at 340°C.

Determine curves for temperature vs catalyst-bed depth for tubes of three different sizes: 1.0, 2.0, and 3.0 in. ID. In so doing, ascertain how large the tubes can be without exceeding the maximum permissible temperature of

[1] *FIAT Repts. 984, 649; BIOS Repts. 1597, 957, 753, 666; CIOS Rept. XXVIII 29; XXVII 80, 89.*

400°C. The catalyst will consist of 0.20 by 0.20-in. cylinders, and the bulk density of the packed bed may be taken as 50 lb/ft³ for each size of tube.[1] The superficial mass velocity of gases through each tube will be 400 lb/(hr) (ft² tube area). Use the one-dimensional design procedure.

13-7. In order to compare different batches of catalysts for fixed-bed cracking operations it is desired to develop a numerical catalyst activity by comparing each batch with a so-called "standard catalyst" for which the x-vs-W/F curve is known. If the activity of a batch is defined as the rate of reaction of that batch divided by the rate for the standard catalyst at the same conditions, which of the following two procedures would be the better measure of catalyst activity?

(a) Determine from the curves the values of W/F required to obtain the same conversion x, and call the activity the ratio

$$\frac{(W/F)_{\text{standard}}}{(W/F)_{\text{actual}}}$$

(b) Determine from the curves the values of x at the same W/F, and define the activity as

$$\frac{x_{\text{actual}}}{x_{\text{standard}}}$$

Sketch the x-vs-W/F curve for a standard catalyst and a curve for a catalyst with an activity less than unity.

13-8.[2] Design a reactor system to produce styrene by the vapor-phase catalytic dehydrogenation of ethyl benzene. The reaction is endothermic, so that elevated temperatures are necessary to obtain reasonable conversions. The plant capacity is to be 20 tons of crude styrene (styrene, benzene, and toluene) per day. Determine the bulk volume of catalyst and number of tubes in the reactor by the one-dimensional method. Assume that two reactors will be needed for continuous production of 20 tons/day, with one reactor in operation while the catalyst is being regenerated in the other. Also determine the composition of the crude styrene product.

With the catalyst proposed for the plant, three reactions may be significant:

$$C_6H_5C_2H_5 \overset{k_1}{\rightleftharpoons} C_6H_5CH = CH_2 + H_2$$

$$C_6H_5C_2H_5 \overset{k_2}{\rightleftharpoons} C_6H_6 + C_2H_4$$

$$H_2 + C_6H_5C_2H_5 \overset{k_3}{\rightleftharpoons} C_6H_5CH_3 + CH_4$$

The mechanism of each reaction follows the stoichiometry indicated by these

---

[1] This represents an approximation, since the bulk density will depend to some extent on the tube size, especially in the small-diameter cases.

[2] From an example suggested by R. R. Wenner and F. C. Dybdal, *Chem. Eng. Progr.*, **44**, 275 (1948).

reactions. The forward rate constants determined for this catalyst by Wenner and Dybdal are

$$\log k_1 = \frac{-11,370}{4.575T} + 0.883$$

$$\log k_2 = \frac{-50,800}{4.575T} + 9.13$$

$$\log k_3 = \frac{-21,800}{4.575T} + 2.78$$

where T is in degrees Kelvin, and

k_1 = lb moles styrene/(hr)(atm)(lb catalyst)
k_2 = lb moles benzene/(hr)(atm)(lb catalyst)
k_3 = lb moles toluene/(hr)(atm)^2(lb catalyst)

The overall equilibrium constants for the three reversible reactions are as follows:

t, °C	K_1	K_2	K_3
400	1.7×10^{-3}	2.7×10^{-2}	5.6×10^4
500	2.5×10^{-2}	3.1×10^{-1}	1.4×10^4
600	2.3×10^{-1}	2.0	4.4×10^3
700	1.4	8.0	1.8×10^3

The reactor will be heated by flue gas passed at a rate of 6,520 lb/(hr)(tube) countercurrent (outside the tubes) to the reaction mixture in the tubes. The flue gas leaves the reactor at a temperature of 1600°F. The reactant stream entering the reactor will be entirely ethyl benzene. The reactor tubes are 4.03 in. ID, 4.50 in. OD, and 15 ft long. The feed, 425 lb of ethyl benzene/(hr)(tube), enters the tubes at a temperature of 550°C and a pressure of 44 psia; it leaves the reactor at a pressure of 29 psia. The heat-transfer coefficient between the reaction mixture and the flue gas is 9.0 Btu/(hr)(ft²)(°F) (based on outside area) and $\rho_B = 61$ lb/ft³.

The thermodynamic data are as follows:

Average specific heat of reaction mixture = 0.63 Btu/(lb)(°F)
Average specific heat of flue gas = 0.28 Btu/(lb)(°F)
Average heat of reaction 1, $\Delta H_1 = 53,600$ Btu/lb mole
Average heat of reaction 2, $\Delta H_2 = 43,900$ Btu/lb mole
Average heat of reaction 3, $\Delta H_2 = -27,700$ Btu/lb mole

To simplify the calculations assume that the pressure drop is directly proportional to the length of catalyst tube. Point out the possibility for error in this assumption, and describe a more accurate approach.

13-9. A pilot plant for the hydrogenation of nitrobenzene is to be designed on the basis of Wilson's rate data (see Example 13-4). The reactor will consist of a 1-in.-ID tube packed with catalyst. The feed, 2.0 mole % nitrobenzene and 98% hydrogen, will enter at 150°C at a rate of 0.25 lb mole/hr. To reduce temperature variations the wall temperature of the tube will be maintained at 150°C by a constant-temperature bath. The heat-transfer coefficient between reaction mixture and wall may be taken equal to 20 Btu/(hr)(ft^2)(°F).

Determine the temperature and conversion as a function of catalyst-bed depth for the conversion range 0 to 90%. Convert the rate equation in Example 13-4 to a form in which r_P is expressed as lb moles nitrobenzene reacting/(hr)(lb catalyst) by taking the void fraction equal to 0.424 and the bulk density of catalyst as 60 lb/ft^3. The heat of reaction is $-274,000$ Btu/lb mole. The properties of the reaction mixture may be assumed to be the same as those for hydrogen.

13-10. Use the two-dimensional method to compute the conversion for bed depths up to 0.30 ft for the oxidation of sulfur dioxide under conditions similar to those described in Examples 13-5 and 13-6. The same reactor conditions apply, except that the superficial mass velocity in this case is 147 lb/(hr)(ft^2), and the temperature profile at the entrance to the reactor is as shown below:

t, °C	352.0	397.5	400.4	401.5	401.2	397.4	361.2	197.0
Radial position	0.797	0.534	0.248	0.023	0.233	0.474	0.819	1.00

The measured conversions are given for comparison:

Catalyst-bed depth, in.	0	0.531	0.875	1.76	4.23	5.68
% SO$_2$ converted	0	26.9	30.7	37.8	41.2	42.1

13-11. A fluidized catalyst bed has been suggested for the oxidation of ethylene to ethylene oxide, with a feed gas containing 8% C$_2$H$_4$, 19% O$_2$, and 73% N$_2$. A fairly uniform temperature can be maintained in the bed, and it has been decided to operate at 280°C and 1 atm pressure. The fluidized catalyst will have a density of 20 lb/ft^3, while the feed gas enters at a superficial velocity of 2 ft/sec at 280°C. Assuming plug flow and neglecting pressure drop, estimate the minimum bed length required for 40% conversion of ethylene to ethylene oxide. Two competing reactions occur,

1. $C_2H_4 + \frac{1}{2}O_2 \rightarrow C_2H_4O$

2. $C_2H_4 + 3O_2 \rightarrow 2CO_2 + 2H_2O$

The particular catalyst used is selective to the extent that the selectivity of ethylene oxide (with respect to CO$_2$) is constant at 1.5 for the whole range of conversions. The experimental rate (global) equation is

$$r_p = kp_{C_2H_4}(p_{O_2})^{0.2} \quad \text{lb moles ethylene oxide produced/(hr)(lb catalyst)}$$

The partial pressures are in atmospheres, and k is 0.12 at 280°C.

13-12. An irreversible first-order gaseous reaction $A \rightarrow B$ is carried out in a fluidized-bed reactor at conditions such that the rate constant is $k_1 = 0.076$ ft³/(sec)(lb catalyst). The superficial velocity is 1.0 ft/sec and the bulk density of catalyst in the bed is 5.25 lb/ft³. Suppose that the mixing conditions correspond to plug flow modified by axial dispersion. The extent of dispersion can be evaluated from Gilliland and Mason's[1] equation,

$$\frac{u}{D_L} = 2.6\left(\frac{1}{u}\right)^{0.61}$$

where u is superficial velocity, in feet per second, and D_L is axial diffusivity, in square feet per second.

What will be the conversion in the effluent from a reactor with a catalyst-bed depth of 5.0 ft? What would the conversion be if plug-flow behavior were assumed? If stirred-tank performance were assumed?

13-13. A first-order gaseous reaction $A \rightarrow B$ is carried out in a fluidized bed at 500°F and 2 atm pressure. At this temperature $k_1 = 0.05$ ft³/(sec)(lb catalyst). The bulk density of the catalyst bed is 3 lb/ft³ at a superficial mass velocity of 0.15 lb/(sec)(ft²). If the bed height is 10 ft, what will be the exit conversion? The molecular weight of component A is 44.

13-14. Repeat Prob. 13-13 for a reversible reaction for which the equilibrium constant is 0.6.

13-15. A reaction $2A \rightarrow B$ is being studied in a fluidized-bed reactor at atmospheric pressure and 200°F. It appears that the global rate of reaction may be approximated by a second-order irreversible equation

$$\mathbf{r}_p = k_2 p_A^2$$

where $k_2 =$ lb moles/(sec)(atm²)(lb catalyst). At the operating temperature, $k_2 = 4.0 \times 10^{-6}$. The linear velocity in the reactor is 1.0 ft/sec, and the bulk density of the fluidized catalyst 4.0 lb/ft³. (a) Calculate the conversion of A for plug-flow conditions, with bed heights of 5, 10, and 15 ft. (b) Correct for the effect of longitudinal diffusion, using the diffusivity data given in Prob. 13-12.

[1] E. R. Gilliland and E. A. Mason, *Ind. Eng. Chem.*, **41**, 1191 (1949).

14

FLUID-SOLID NONCATALYTIC REACTIONS

There is an important class of fluid-solid reactions in which the solid is a reactant rather than a catalyst. Illustrations are[1]

$$CaCO_3(s) \rightarrow CaO(s) + CO_2(g) \qquad \text{lime kiln}$$

$$Fe_2O_3(s) + 3C(s) + \tfrac{1}{2}O_2(g) \rightarrow 2Fe(s) + CO_2(g) + 2CO(g)$$

blast furnace

$$ZnS(s) + \tfrac{3}{2}O_2(g) \rightarrow ZnO(s) + SO_2(g) \qquad \text{roasting of ores}$$

$$CaCO_3(aq) + 2NaR(s) \rightarrow Na_2CO_3(aq) + CaR_2(s) \qquad \text{ion-exchange reaction}$$

$$2C(s) + \tfrac{3}{2}O_2(g) \rightarrow CO_2(g) + CO(g) \qquad \text{catalyst regeneration, combustion of coal, etc.}$$

In the first four cases solid products are formed, so that the original particles are replaced with another solid phase. In the fifth reaction the solid phase

[1] For a more complete survey see C. Y. Wen, *Ind. Eng. Chem.*, **60**, 34 (1968).

disappears (except for the small amount of ash formed). Since the amount of reactant surface and its availability (in the first four cases) change with extent of reaction, the global rate changes with time. Calculating reactor performance for complex examples of heterogeneous noncatalytic reactions can be difficult. The blast furnace is an example of such a complex system. However, there are other situations in which the chemical system and reactor operating conditions are simpler. For such cases quantitative design methods can be developed which are relatively simple and are also good representations of the actual process. The design of noncatalytic reactors will be illustrated by treating some of these tractable cases.

14-1 Design Concepts

The development of global rate equations for fluid-solid noncatalytic reactions entails the same concepts of combining physical and chemical steps as for catalytic reactions (Chap. 12). Also, the reactor design problems are similar to those considered in Chap. 13. The special features of the noncatalytic case are that the rate of reaction is often a function of time as well as of position, and the solid phase is more often (but not always) in continuous flow through the reactor. When the fluid phase is not well mixed its composition varies with position. Plug flow of fluid is the extreme case. A noncatalytic reaction carried out in a fixed-bed reactor could approach this behavior. Then the design involves the usual two steps: formulation of a global rate equation applicable for any location at any time, and use of the global rate in a differential mass balance that describes the reactor behavior.

    In other reactors the composition of the fluid phase may be nearly uniform. The stirred-tank reactor is one example. Another is a transport reactor (solid particles move with the fluid through the vessel), where there is a large excess of reactant in the fluid phase. When the composition is uniform the design is greatly simplified. The global rate equation can be integrated immediately to obtain a relationship of conversion (of solid reactant particles) vs time. If the reactant particles have different sizes, the conversion-vs-time relationship will not be the same for all particles. The average conversion is evaluated by summing the results in terms of the particle-size distribution. In a batch reactor (with respect to solid phase) all particles will have the same residence time. In a flow reactor this will be true only for plug flow of solids. When a distribution of residence times exists the average conversion for all the particles may be obtained by the methods presented in Sec. 6-8 for segregated flow.[1]

[1] Note that the movement of solid particles in a vessel follows exactly the concepts of segregated flow, i.e., the solid particles cannot mix.

In the next section various reaction models are considered. Then global rate equations are developed in Sec. 14-3 for one model (shrinking core). In Sec. 14-4 integrated conversion-vs-time relationships (for single particles) are presented. Such relationships are suitable for use in design of reactors in which the fluid phase is completely mixed. In Secs. 14-5 and 14-6 all these results will be applied to reactor design.

14-2 Reaction Models

When components in solid and fluid phases react, the sequence of steps must be similar to those for fluid-solid catalytic reactions. In Sec. 8-2 catalytic reactions were explained in terms of a three-step process: adsorption of the fluid molecule on the solid surface, reaction on the surface involving the adsorbed molecule, and desorption of product. We shall not be concerned here with the mechanism of these processes; instead we shall start out on the basis that the rate equation is known from experimental measurements. Often observed data agree with a rate equation which is first order in concentration of the reactant in the fluid phase and directly proportional to the surface of the reactant in the solid phase. For example, despite the stoichiometry of the reaction

$$UO_2(s) + 4HF(g) \rightarrow UF_4(s) + 2H_2O(g)$$

the rate has been found[1] to be first order with respect to gaseous hydrogen fluoride. The experimental procedure used to study the kinetics consists of measuring the conversion of the solid reactant, either as a large single pellet or as a sample of small particles. This is commonly carried out by observing the change in mass of the solid with time. Figure 14-1 shows an apparatus for such a study. The UO_2 pellet (original diameter 2 cm) was weighed at various times by attaching a balance to the wire holding the pellet. From the weight change the conversion of UO_2 to UF_4 could be calculated from the stoichiometry of the reaction. The stirrer was rotated about the suspended pellet in order to obtain a uniform gas composition (ideal stirred-tank performance). The resultant conversion-vs-time data represent the integration of the global rate equation from the start of the experiment ($t = 0$). To evaluate a suitable rate equation, various possibilities are integrated and compared with the data.

For the integration it is necessary to know how the surface of the solid reactant exists initially and how its surface changes with time. There are several possible models.[2]

[1] L. Tomlinson, S. A. Morrow, and S. Graves, *Trans. Faraday Soc.*, **57**, 1008 (1961).

[2] An extensive discussion of various models is given in C. Y. Wen, *Ind. Eng. Chem.*, **60**, 34 (1968).

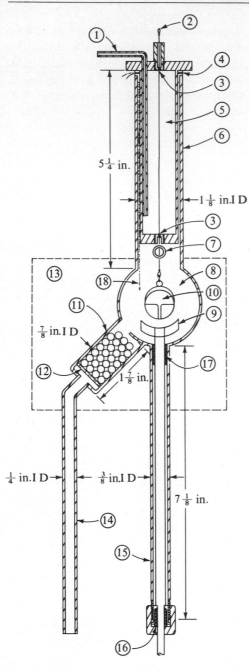

$5\frac{1}{4}$ in.

$1\frac{1}{8}$ in.I D

$\frac{7}{8}$ in.I D

$1\frac{7}{8}$ in.

$\frac{1}{4}$ in.I D

$\frac{3}{8}$ in.I D

$7\frac{1}{8}$ in.

①	N₂ inlet to pressure chamber
②	Weighing wire
③	Pinhole
④	Teflon gaskets
⑤	N₂ pressure chamber
⑥	Reactor neck
⑦	Reactor effluent line
⑧	Reaction chamber
⑨	Stirrer impeller (1-in.diam.)
⑩	Pellet of UO₂
⑪	Ball preheater
⑫	Perforated plate
⑬	Furnace
⑭	HF inlet line
⑮	Stirrer extension tube
⑯	Teflon bushing
⑰	Graphite bearing
⑱	Gas thermocouple

Fig. 14-1 *Stirred-tank single-pellet reactor for hydrofluorination of uranium dioxide*

SHRINKING CORE   If the reactant is nonporous, the reaction will occur at its outer surface. This surface recedes with extent of reaction, as shown in Fig. 14-2a. As reaction occurs, a layer of product [for example, $UF_4(s)$] builds up around the unreacted core of reactant. A porous particle might also behave in this way if the resistance to reaction is much less than the resistance to diffusion of fluid reactant in the pores of the particle. The key factor in this model is that the reaction always occurs at a surface boundary, that is, at the interface between unreacted core and surrounding solid product.

Fig. 14-2   *Models for gas-solid noncatalytic reactions of the type $A(g) + bB(s) \rightarrow E(g) + F(s)$: (a) shrinking-core model, (b) highly porous reactant, (c) porous reactant pellet composed of nonporous particles*

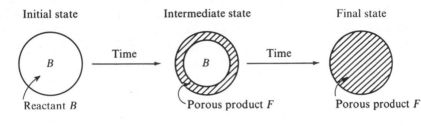

(a)

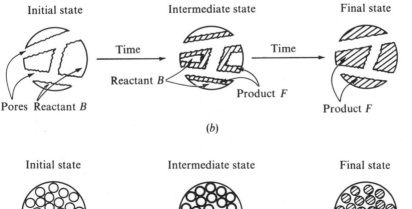

(b)

(c)

HIGHLY POROUS REACTANT (no pore-diffusion resistance) Suppose the solid reactant is so porous that the fluid reactant can reach all parts of the solid without diffusion resistance; this is depicted in Fig. 14-2b. Now the rate per particle will vary as the surface of solid reactant changes with time and the layer of solid product accumulates. The key factor here is that the concentration of reactant in the fluid phase is the same at any location within the particle.

POROUS REACTANT (intermediate pore-diffusion resistance) An example of this case would be a solid reactant formed by compressing nonporous particles into a porous pellet, as shown in Fig. 14-2c. The pores surrounding the particles are supposed to be small enough that the fluid reactant concentration decreases significantly toward the center of the pellet.

All three models have been used as bases for integrating rate equations. Choice of the most appropriate one depends on the initial form of the solid reactant and the changes that occur with reaction. However, in the remainder of this chapter the shrinking-core model will be used. It is amenable to quantitative treatment and represents many real systems rather well.

14-3 Global Rate Equations (Shrinking-core Model)

Suppose that the gas-solid reaction

$$A(g) + bB(s) \rightarrow E(g) + F(s)$$

obeys the shrinking-core model (Fig. 4-2a), where solid reactant B is initially a sphere of radius r_s. The solid sphere is in contact with gas A, whose bulk concentration is C_b. Consider the case where the temperature is uniform throughout the heterogeneous region. As reaction occurs, a layer of product F forms around the unreacted core of reactant B. It is supposed that this layer is porous, so that reaction occurs by diffusion of A through the layer of F to react at the interface between F and unreacted core. This situation is shown in Fig. 14-3, where the concentrations of A are labeled at various locations. The shape of the concentration profile from bulk gas to reacting surface is also indicated. It is assumed that the pellet retains its spherical shape during reaction. It is also assumed, for convenience, that the densities of the porous product and the reactant B are the same, so that the total radius of the pellet does not change with time and there is no gaseous region between the pellet and the product layer F.[1]

---

[1] It is interesting to note that K. G. Denbigh and G. S. G. Beveridge [*Trans. Inst. Chem. Engrs.*, **40**, 23 (1962)], in studying the oxidation of ZnS at high temperatures, observed a gaseous region between solid ZnS and solid ZnO. This was due to vaporization of some of the ZnS.

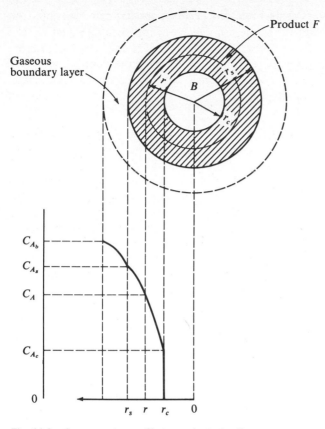

Fig. 14-3 *Concentration profile in a spherical pellet (shrinking-core model)*

One more restriction must be placed on the system before a simple mathematical analysis is feasible: the rate of movement of the reaction interface at r_c, that is, dr_c/dt, is small with respect to the velocity of diffusion of A through the product layer. The requirement for this pseudo-steady-state concept has been carefully developed,[1] but approximately stated, it is valid if the density of the gas in the pores of the product layer is small with respect to the density of solid reactant B. This is usually the case.

Granting pseudo-steady-state conditions, the three rates—diffusion of A through the boundary layer, diffusion through the layer of product, and reaction at the interface—are identical. By equating the expressions for each of these processes, the concentration $(C_A)_c$ can be expressed in terms of the known $(C_A)_b$ and the radius of the unreacted core r_c. The three

[1] K. B. Bischoff, *Chem. Eng. Sci.*, **18**, 711 (1963).

rate equations, expressed as moles of A disappearing per unit time per particle, are

$$-\frac{dN_A}{dt} = 4\pi r_s^2 k_m[(C_A)_b - (C_A)_s] \qquad \text{external diffusion} \qquad (14\text{-}1)$$

$$-\frac{dN_A}{dt} = 4\pi r_c^2 D_e\left(\frac{dC_A}{dr}\right)_{r=r_c} \qquad \text{diffusion through product} \qquad (14\text{-}2)$$

$$-\frac{dN_A}{dt} = 4\pi r_c^2 k(C_A)_c \qquad \text{reaction at } r_c \qquad (14\text{-}3)$$

In Eq. (14-1) k_m is the external mass-transfer coefficient discussed in Chap. 10 [Eq. (10-10)]. The rate through the product layer, Eq. (14-2), is evaluated at $r = r_c$; D_e is the effective diffusivity of A through this porous layer. In writing Eq. (14-3) the chemical reaction at r_c is assumed to be first order in A and irreversible. It is also taken as directly proportional to the outer surface area of unreacted core B.[1]

To evaluate the gradient in Eq. (14-2) consider the diffusion of A through the layer of F. With the pseudo-steady-state assumption this process can be evaluated independently of the change in r_c. Consider a small element of thickness Δr at location r in the product layer (Fig. 14-3). The steady-state mass balance of A around this layer is

$$-\left(\pi r^2 D_e \frac{dC_A}{dr}\right)_r - \left(\pi r^2 D_e \frac{dC_A}{dr}\right)_{r+\Delta r} = 0 \qquad (14\text{-}4)$$

Taking the limit as $\Delta r \to 0$ gives

$$\frac{d}{dr}\left(r^2 D_e \frac{dC_A}{dr}\right) = 0 \qquad (14\text{-}5)$$

If this expression is integrated twice, with the boundary conditions

$$C_A = \begin{cases} (C_A)_s & \text{at } r = r_s \\ (C_A)_c & \text{at } r = r_c \end{cases}$$

used to evaluate the integration constants, the result is

$$C_A - (C_A)_c = [(C_A)_s - (C_A)_c]\frac{1 - r_c/r}{1 - r_c/r_s} \qquad (14\text{-}6)$$

[1] The number of active sites per unit surface of B will presumably be constant, and the rate proportional to the total number of sites. Hence the rate should be proportional to the surface area of the unreacted core. A feature of the shrinking-core model is that this area is known and is equal to $4\pi r_c^2$ for a spherical core. This may not be a realistic area for reaction in real situations, but it is the characteristic of the model that permits mathematical analysis of the process.

This expression for the concentration profile through the product layer may be differentiated with respect to r and then evaluated at $r = r_c$ to give

$$\left(\frac{dC_A}{dr}\right)_{r=r_c} = \frac{(C_A)_s - (C_A)_c}{r_c(1 - r_c/r_s)} \tag{14-7}$$

Substituting Eq. (14-7) in Eq. (14-2) gives

$$-\frac{dN_A}{dt} = 4\pi r_c \, D_e \frac{(C_A)_s - (C_A)_c}{1 - r_c/r_s} \tag{14-8}$$

Now $(C_A)_s$ and dN_A/dt can be eliminated from Eqs. (14-1), (14-3), and (14-8) to give $(C_A)_c$ in terms of $(C_A)_b$ and r_c. The result is

$$(C_A)_c = \frac{(C_A)_b}{1 + (r_c^2/r_s^2)(k/k_m) + (kr_c/D_e)(1 - r_c/r_s)} \tag{14-9}$$

Using Eq. (14-9) for $(C_A)_c$ in Eq. (14-3) gives the global rate in terms of $(C_A)_b$ and r_c,

$$\begin{array}{c}\text{Rate per}\\\text{particle}\end{array} = -\frac{dN_A}{dt} = \frac{4\pi r_c^2 (C_A)_b k}{1 + (r_c^2/r_s^2)(k/k_m) + (kr_c/D_e)(1 - r_c/r_s)} \tag{14-10}$$

This rate may be expressed in terms of the rate per unit volume of reactor by multiplying $-dN_A/dt$ by the particle density (particles per unit volume).

Since r_c is a variable, Eq. (14-10) is not useful for reactor design until we express r_c as a function of time. According to the spherical geometry of the particle, the rate of reaction of B (moles per unit time per particle) may be written as

$$\frac{dN_B}{dt} = \frac{\rho_B}{M_B} \frac{d}{dt}\left(\frac{4}{3}\pi r_c^3\right) = \frac{4\pi r_c^2 \rho_B}{M_B} \frac{dr_c}{dt} \tag{14-11}$$

From the stoichiometry of the reaction,

$$\frac{dN_A}{dt} = \frac{1}{b}\frac{dN_B}{dt} = \frac{4\pi r_c^2 \rho_B}{b M_B}\frac{dr_c}{dt} \tag{14-12}$$

Combining this result with Eq. (14-3) for dN_A/dt gives

$$\frac{dr_c}{dt} = -\frac{b M_B k}{\rho_B}(C_A)_c \tag{14-13}$$

Finally, substituting Eq. (14-9) for $(C_A)_c$ in Eq. (14-13) provides a differential equation whose solution gives $r_c = f(t)$,

$$-\frac{dr_c}{dt} = \frac{bM_Bk(C_A)_b/\rho_B}{1 + (r_c^2/r_s^2)(k/k_m) + (kr_c/D_e)(1 - r_c/r_s)} \qquad (14\text{-}14)$$

Equations (14-10) and (14-14) together provide the global rate in terms of $(C_A)_b$ and time. The integration of Eq. (14-14) will depend on how C_b varies with time, that is, on the operation of the reactor. As mentioned in Sec. 14-1, for a stirred-tank type (well-mixed fluid phase) C_b is constant. The conversion-vs-time relationship for constant C_b is considered next.

14-4 Conversion vs Time for Single Particles (Constant Fluid Concentration)

If $(C_A)_b$ is constant, Eq. (14-14) is easily integrated. Starting with $r = r_s$ at $t = 0$,

$$-\frac{bM_Bk(C_A)_b}{\rho_B}\int_0^t dt = \int_{r_s}^{r_c}\left[1 + \frac{r_c^2}{r_s^2}\frac{k}{k_m} + \frac{kr_c}{D_e}\left(1 - \frac{r_c}{r_s}\right)\right]dr_c \qquad (14\text{-}15)$$

It is convenient to express the result in terms of a dimensionless time,

$$t^* = \frac{bM_Bk(C_A)_b}{\rho_Br_s}t \qquad (14\text{-}16)$$

and two groups relating the diffusion and reaction resistances,

$$Y_1 = \frac{D_e}{k_mr_s} = \frac{\text{external diffusion resistance}}{\text{diffusion resistance in product layer}} \qquad (14\text{-}17)$$

$$Y_2 = \frac{kr_s}{D_e} = \frac{\text{diffusion resistance in product layer}}{\text{reaction resistance at }r_c} \qquad (14\text{-}18)$$

In terms of these parameters Eq. (14-15) can be integrated to give

$$t^* = \left(1 - \frac{r_c}{r_s}\right)\left\{1 + \frac{Y_1Y_2}{3}\left[\left(\frac{r_c}{r_s}\right)^2 + \frac{r_c}{r_s} + 1\right] + \frac{Y_2}{6}\times\right.$$
$$\left.\left[\left(\frac{r_c}{r_s} + 1\right) - 2\left(\frac{r_c}{r_s}\right)^2\right]\right\} \qquad (14\text{-}19)$$

The conversion of B is related to the radius of the unreacted core by the expression

$$x_B = \frac{\text{initial mass} - \text{mass at }t}{\text{initial mass}} = \frac{\frac{4}{3}\pi r_s^3\rho_B - \frac{4}{3}\pi r_c^3\rho_B}{\frac{4}{3}\pi r_s^3\rho_B}$$

or

$$x_B = 1 - \left(\frac{r_c}{r_s}\right)^3 \tag{14-20}$$

Equations (14-19) and (14-20) establish the conversion as a function of time for single particles in a situation where $(C_A)_b$ is constant. Constant total radius r_s has been assumed, as well as an irreversible first-order reaction in A. Solutions when these assumptions are not made are also possible,[1] but the results are more complicated. Isothermal conditions have also been chosen. This restriction can be eliminated by writing expressions similar to Eqs. (14-1) to (14-3) in order to account for the effect of the heat of reaction and the effect of temperature on the rate. The nonisothermal case also has been solved.[1]

When all three resistances are not significant Eq. (14-19) is less complicated. Some of these simpler results are considered in the following paragraphs.

Chemical Reaction Controlling If the gas-phase velocity relative to that of the solid particle is high, as in a fixed-bed reactor, external-diffusion resistance may be negligible. Also, for a highly porous product layer and for low conversions, diffusion resistance through the product may be small. Under such conditions the chemical step at r_c will determine the rate, and $Y_2 \rightarrow 0$. Then Eq. (14-19) reduces to

$$t^* = \frac{bM_B k(C_A)_b t}{\rho_B r_s} = 1 - \frac{r_c}{r_s}$$

or

$$t = \frac{\rho_B r_s}{bM_B k(C_A)_b}\left(1 - \frac{r_c}{r_s}\right) = \frac{\rho_B r_s}{bM_B k(C_A)_b}\left[1 - (1 - x_B)^{\frac{1}{3}}\right] \tag{14-21}$$

The time for complete conversion ($x_B = 1$ and $r_c = 0$) is

$$t_{x_B=1} = \frac{\rho_B r_s}{bM_B k(C_A)_b} \tag{14-22}$$

Diffusion through Product Controlling For rapid chemical reactions at the interface and a low D_e, diffusion through the product layer may determine the rate, even at small conversions. If this is the case, $Y_1 = 0$, Y_2 is large, and Eq. (14-19) becomes

$$t^* = \left(1 - \frac{r_c}{r_s}\right)\frac{Y_2}{6}\left[\frac{r_c}{r_s} + 1 - 2\left(\frac{r_c}{r_s}\right)^2\right] \tag{14-23}$$

[1] J. Shen and J. M. Smith, *Ind. Eng. Chem., Fund. Quart.*, **4**, 293 (1965).

or, from the definitions of t^* and Y_2,

$$t = \frac{\rho_B r_s^2}{6D_e b M_B (C_A)_b} \left[1 - 3\left(\frac{r_c}{r_s}\right)^2 + 2\left(\frac{r_c}{r_s}\right)^3 \right] \qquad (14\text{-}24)$$

In this case the time for complete conversion of B depends on D_e and is given by

$$t_{x_B=1} = \frac{\rho_B r_s^2}{6D_e b M_B (C_A)_b} \qquad (14\text{-}25)$$

Weisz and Goodwin[1] have used Eq. (14-24) successfully to explain how the extent of carbon removal on a deactivated catalyst varies with time. In this case the reactions are

1. $C(s) + O_2(g) \rightarrow CO_2(g)$

2. $C(s) + \frac{1}{2}O_2(g) \rightarrow CO(g)$

and so no solid product is formed. However, the carbon has been deposited throughout the pores of a porous catalyst pellet (for example, petroleum cracking catalysts become deactivated by such carbon deposition). If the rate of the chemical reaction is rapid with respect to the rate of diffusion of oxygen into the pellet, the carbon will be burned off the pellet according to the shrinking-core model. Since such oxidation reactions are normally carried out at a high temperature, they are intrinsically fast, and the shrinking-core assumption is often a reasonable one. The shrinking-core concept of reaction at a sharp interface has been applied to several other systems, including hydrofluorination of UO_2 and poison deposition on porous catalysts.[2]

The application of Eq. (14-19) is illustrated in the following example.

Example 14-1   The reduction of FeS_2 particles,

$$FeS_2(s) + H_2(g) \rightarrow FeS(s) + H_2S(g)$$

has been studied under conditions where the concentration of hydrogen in the gas phase was essentially constant.[3] Hydrogen at a high flow rate was passed at atmospheric pressure through beds of FeS_2 particles. The results indicated a first-order (with respect to hydrogen) reversible reaction. Measurements were made at 450°, 477°, and 495°C, and an activation energy of 30,000 was proposed. The experimental data for conversion of FeS_2 vs time are shown in Fig. 14-4.

[1] P. B. Weisz and R. D. Goodwin, *J. Catalysis*, **2**, 397 (1963); *J. Catalysis*, **6**, 227 (1966).

[2] P. B. Weisz and C. D. Prater, "Advances in Catalysis," vol. VI, p. 143, Academic Press, Inc., New York, 1954; E. C. Costa and J. M. Smith, *Proc. Fourth European Symp. Chem. Reaction Eng.*, *Brussels*, Sept. 9–11, 1968; J. J. Carberry and R. L. Gorring, *J. Catalysis*, **5**, 529 (1966).

[3] G. M. Schwab and J. Philinis, *J. Am. Chem. Soc.*, **69**, 2588 (1947).

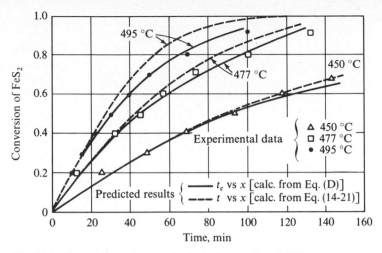

Fig. 14-4 *Conversion-vs-time results for hydrogenation of FeS$_2$*

Determine whether the shrinking-core model will fit these data and evaluate the rate constant (the frequency factor in the rate equation) and an effective diffusivity D_e. The particles were granular and varied in size from 0.01 to 0.1 mm, but assume that a spherical particle with an average radius of 0.035 mm can represent the mixture.

Solution Since the gas-flow rate is high, the external-diffusion resistance should be small, and Y_1 will approach zero. Furthermore, at low conversions the product layer of FeS will be thin, so that the chemical reaction at the interface may control the rate. This is more likely to be a good assumption at the lowest temperature, where the reaction rate is smallest. Under these conditions Eq. (14-21) is applicable, and an appropriate way to initiate the analysis is to apply this equation to the data at 450°C. Substituting numerical values ($\rho_{\text{FeS}_2} = 5.0$ g/cm^3), we have

$$t = \frac{5.0(0.0035)}{1(120)(C_A)_b k} \left[1 - (1 - x_{\text{FeS}_2})^{\frac{1}{3}}\right] \tag{A}$$

From the ideal-gas law at 450°C,

$$(C_A)_b = \frac{p}{R_g T} = \frac{1}{82(273 + 450)} = 1.69 \times 10^{-5} \text{ g mole/cm}^3$$

Hence Eq. (A) for 450°C is

$$t = \frac{8.6}{k} \left[1 - (1 - x_{\text{FeS}_2})^{\frac{1}{3}}\right] \tag{B}$$

The data points in Fig. 14-4 can be used with Eq. (B) to evaluate k. It is found that $k = 0.019$ cm/min, or 3.2×10^{-4} cm/sec, gives a curve in good agreement with the data. This is the dashed curve for 450°C in Fig. 14-4.

According to the Arrhenius equation,

$$k = A \, e^{-E/R_gT}$$

Hence

$$A = \frac{3.2 \times 10^{-4}}{e^{-30,000/(1.98)(723)}} = 3.8 \times 10^5 \text{ cm/sec}$$

Then the rate constant for any temperature is given by

$$k = 3.8 \times 10^5 \, e^{-30,000/R_gT} \tag{C}$$

By calculating k from Eq. (C) and using it in Eq. (B) we can calculate t-vs-x curves for 477 and 495°C. These are the dashed lines in Fig. 14-4. At these higher temperatures the predicted conversions are higher than the experimental values. Particularly at the higher conversions, there is considerable deviation. These results suggest that the diffusion resistance through the product layer (FeS) may not be negligible. In fact, the value of Y_2 necessary for agreement with the observed results can be evaluated by applying Eq. (14-19) to this case. If we still assume external diffusion to be negligible ($Y_1 = 0$), Eq. (14-19) reduces to the form

$$t_c = \frac{\rho_B r_s (1 - r_c/r_s)}{b M_B k (C_A)_b} \left\{ 1 + \frac{Y_2}{6} \left[\frac{r_c}{r_s} + 1 - 2 \left(\frac{r_c}{r_s} \right)^2 \right] \right\} \tag{D}$$

where the subscript on t indicates that t_c is corrected for diffusion resistance in the product layer. Comparison with Eq. (14-21) shows that the term in braces in Eq. (D) is a correction factor for this diffusion resistance; that is,

$$\frac{t_c}{t} = 1 + \frac{Y_2}{6} \left[1 + \frac{r_c}{r_s} - 2 \left(\frac{r_c}{r_s} \right)^2 \right] \tag{E}$$

or, in terms of conversion,

$$\frac{t_c}{t} = 1 + \frac{Y_2}{6} \left[1 + (1 - x)^{\frac{1}{3}} - 2(1 - x)^{\frac{2}{3}} \right] \tag{F}$$

The t values (without subscripts) in Eq. (E) refer to those calculated from Eq. (14-21) and correspond to the dashed lines in Fig. 14-4. Y_2 was evaluated at 477°C by finding what value would give the best agreement of t_c with the experimental data, making the calculations with Eq. (F). The solid line for 477°C (Fig. 14-4) shows the t_c curve for $Y_2 = 0.66$. There appears to be some deviation at the highest conversion, but the agreement for all other x values is good. The effective diffusivity is then readily obtainable from the definition of Y_2 and Eq. (C),

$$D_e = \frac{k r_s}{Y_2} = \frac{3.8 \times 10^5 \, e^{-30,000/R_g(273+477)} (0.0035)}{0.66}$$

$$= 3.6 \times 10^{-6} \text{ cm}^2/\text{sec}$$

This result can be checked by calculating a corrected curve at 495°C. If the diffusivity is assumed to be constant, Y_2 for 495°C will be

$$Y_2 = \frac{3.8 \times 10^5 \, e^{-30,000/R_g(273+495)} (0.0035)}{3.6 \times 10^{-6}} = 1.0$$

The solid line for 495°C in Fig. 14-4 represents the values of t_c calculated from Eq. (D), with $Y_2 = 1.0$.

The comparison between solid and dashed lines at any temperature shows the effect of introducing the resistance to diffusion through the product layer. The deviation between the lines increases as the temperature increases, since the resistance to reaction decreases. It should be emphasized that the calculations for this example involve several approximations and thus do not represent a thorough check of the shrinking-core model. However, the experimental data do show that it would be unsatisfactory to use a model which did not allow for either a change in reaction area or diffusion through the product layer. Such a model would give a rate independent of time, so that the predicted conversion-time relationship would be a straight line.

REACTOR DESIGN

The most important factors affecting the design of a fluid-solid noncatalytic reactor are the flow patterns of solid and fluid in the vessel. As noted in Sec. 14-1, the simplest case is where the composition of the fluid phase is uniform. Then the conversion-vs-time relationship for single particles, such as Eq. (14-19), can be employed, along with the residence-time and particle-size distributions of the solid phase, to evaluate the average conversion. This problem is considered in the next section. When the fluid phase does not have a uniform composition the design is more complex. However, quantitative treatment is possible when the flow patterns of both solid and fluid phases are well defined. These kinds of reactors are discussed in Sec. 14-6.

14-5 Uniform Composition in the Fluid Phase

Uniform C_b is satisfied either for a well-mixed fluid phase or for any flow pattern if the conversion in the fluid is small. Consider the situation in which the residence time of all the solid particles is the same, as in batch operation or plug flow of solids. The conversion-time relationship is given directly by Eqs. (14-19) and (14-20) and contains particle size r_s as a parameter. The regeneration of carbon-fouled catalyst particles in a batch-operated fluidized bed with a large air flow is an example. Another would be downflow of particles in a vertical-line kiln, where the gas flow is such that its composition is nearly uniform in the kiln. Equations (14-19) and (14-20) together give a relationship $x(r_s)$ between conversion and r_s for a residence time t. If the mass fraction of particles with a radius between r_s and $r_s + \Delta r_s$ is n, the average conversion of the particles for time t will be

$$\bar{x} = \sum_{i=1}^{n} x_i(r_{s_i})\, n_i \qquad (14\text{-}26)$$

In a continuous reactor the solid particles may not be in plug flow but may have a distribution of residence times. In general, the conversion for a given particle size will be a function of r_s and t, so that $x = x(r_s, t)$. However, the residence-time distribution is likely to be caused by the distribution of particle sizes; that is, there may be a unique relation between residence time and particle size. For this situation, Eq. (14-26) is applicable, but the conversion for a given particle size will be evaluated from Eqs. (14-19) and (14-20) by using the residence time applicable for that size. Example 14-2 illustrates these concepts.

Example 14-2 An upflow transport reactor (Fig. 14-5) is to be designed for producing HCl and Na_2SO_4 according to the reaction

$$2NaCl(s) + H_2O(g) + SO_3(g) \rightarrow Na_2SO_4(s) + 2HCl(g)$$

Suppose that the reaction is first order in SO_3 and that the shrinking-core model is applicable. The temperature is 900°F and uniform. Also, an excess of SO_3 and H_2O are used, so that the gas composition is uniform throughout the reactor. The solids flow rate is to be 1,000 lb/min into a cylindrical reactor 2 ft in diameter and 30 ft long. The density of the salt particles is 2.1 g/cm³. The particle-size distribution of the feed (assume it does not change in the reactor) is given in Table 14-1. In tests with a batch laboratory reactor with the same gas composition, it was found that the conversion of a sample of 88- to 105-micron particles was 85% in 10 sec. Measurements with other sizes of particles indicated that the chemical reaction controlled the rate. RTD studies were carried out in a flow-type laboratory reactor with the same flow conditions as expected in the large reactor. The results are given in the last column of Table 14-1. The *mean residence time* is defined as the volume of solids in the reactor divided by the volumetric feed rate of solids. The fraction of the reactor volume occupied by solids, as determined by measuring the solids holdup in the laboratory flow reactor, was 0.01. It is expected that the porosity will be the same in the large reactor.

Table 14-1

Sieve no.	Particle diameter, microns	Weight fraction, n_i	Residence time, $\theta/\bar{\theta}$
250–270	53–63	0.10	0.60
200–250	63–74	0.20	0.85
170–200	74–88	0.35	0.99
150–170	88–105	0.25	1.15
115–150	105–125	0.10	1.30

Calculate the average conversion of NaCl to Na_2SO_4 in the solid phase in the exit stream.

Solution For chemical reaction controlling the rate, Eq. (14-21) is applicable. From the laboratory measurements with 88- to 105-micron particles (average $r_s = 48.2 \times 10^{-4}$ cm),

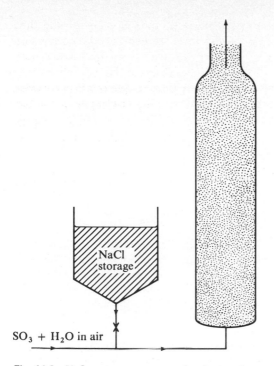

Fig. 14-5 *Upflow transport reactor for the reaction*
$SO_3(g) + H_2O(g) + 2NaCl(s) \rightarrow Na_2SO_4(s) + 2HCl(g)$

$$\frac{t}{r_s} = \frac{\rho_B}{b M_B k(C_A)_b}[1 - (1 - 0.85)^{\frac{1}{3}}] = 0.997 \frac{\rho_B}{b M_B k(C_A)_b}$$

or

$$\frac{\rho_B}{b M_B k(C_A)_b} = \frac{10}{0.997(48.2 \times 10^{-4})} = 2.08 \times 10^3 \text{ sec/cm}$$

The average conversion can be calculated from Eq. (14-26), but first it is necessary to evaluate x_i for each fraction of the particles. To do this we must obtain the residence time for each fraction from the given distribution. From the definition of $\bar{\theta}$,

$$\bar{\theta} = \frac{(\pi 2^2/4)(30)(0.01)}{1,000/[2.1(62.4)(60)]} = 7.1 \text{ sec}$$

The residence time of the 53- to 63-micron group of particles is $\theta = 7.1(0.6) = 4.3$ sec. The conversion of this group (average $r_s = 29 \times 10^{-4}$ cm) can be calculated from Eq. (14-21), which may be written

$$(1 - x)^{\frac{1}{3}} = 1 - \frac{t}{r_s} \frac{b M_B k(C_A)_b}{\rho_B}$$

$$1 - x = \left(1 - \frac{4.3}{29 \times 10^{-4}} \frac{1}{2.08 \times 10^3}\right)^3$$

$$x = 1 - (1 - 0.71)^3 = 1 - 0.0246 = 0.975$$

Although it does not occur in this instance, a negative value for the term in parentheses would mean that complete conversion of the particles of the smallest size group was obtained in less than their residence time of 4.3 sec. Actually, the conversion in this group will be 97.5%.

For the largest group, $\theta = 7.1(1.30) = 9.3$ sec. Hence the conversion for these particles will be

$$1 - x = \left(1 - \frac{9.3}{57.5 \times 10^{-4}} \frac{1}{2.08 \times 10^3}\right)^3 = (1 - 0.78)^3$$

$$x = 1 - 0.01 = 0.99$$

The conversions for the intermediate-sized groups, calculated in the same way, are given in Table 14-2 (for column height of 30 ft). The interesting feature of these results is that the effects of residence time and particle size tend to balance each other. The large particles, which might be expected to have a lower conversion, have a longer time to react, so that approximately the same conversion is obtained for them as for the smaller particles. This counterbalancing effect is due to the upflow in the reactor. If the particles were in downflow (due to gravity) and the gas passed upward, the smaller sizes would have a longer residence time than the large ones. The effects of residence time and particle size would then supplement each other, leading to a relatively lower conversion for the larger particles and a relatively higher conversion for the smaller particles.

Table 14-2

Particle diameter, microns	Weight fraction, n_i	Conversion	
		30-ft reactor	*15-ft reactor*
53–63	0.10	0.975	0.73
63–74	0.20	0.996	0.81
74–88	0.35	0.995	0.81
88–105	0.25	0.993	0.80
105–125	0.10	0.99	0.77

The average conversion, from Eq. (14-26), is

$$\bar{x} = \sum x_i(r_{s_i})n_i = 0.10(0.975) + 0.20(0.996) + 0.35(0.995) + 0.25(0.993)$$

$$+ 0.10(0.99)$$

$$= 0.991 \qquad \text{or } 99.1\%$$

If nearly complete conversion of NaCl to HCl is desired, a 30-ft reactor is needed.

If, however, the conversion need not be more than, say, 75%, a shorter column could be used. If it is assumed that the RTD would be the same in a shorter vessel, the average conversion in it could be evaluated by first obtaining a new $\bar{\theta}$. On this basis let us estimate the conversion obtainable in a 15-ft reactor. The average residence time would be

$$\bar{\theta} = 7.1 \left(\frac{15}{30}\right) = 3.6 \text{ sec}$$

For the smallest group of particles,

$$(1 - x)^{\frac{1}{3}} = 1 - \frac{t}{r_s} \frac{1}{2.08 \times 10^3} = 1 - \frac{3.6(0.6)}{29 \times 10^{-4}} \frac{1}{2.08 \times 10^3}$$

$$1 - x = (1 - 0.355)^3$$

$$x = 0.732$$

The x_i values for the other size groups are shown in the last column of Table 14-2. The average conversion, obtained using these results in Eq. (14-26), is 0.80. A 15-ft reactor would be ample for a 75% conversion.

14-6 Variable Composition in the Fluid Phase

Consider a fixed-bed reactor in which the solid particles constitute or contain one of the reactants, while a second reactant is in the fluid phase. In the general case the reactant concentration in the bulk fluid will decrease along the reactor length z. Since the rate at any location is a function of time, the variation of concentration with length will be a function of time; that is, $C_b = f(t,z)$. Such behavior is characteristic of several practical processes, such as ion-exchange reactors, regeneration of fouled catalysts by combustion with air, and adsorption from gas or liquid streams. Figure 14-6 shows concentration profiles for the catalyst-regeneration case. Hot air is passed over the bed of particles which have been uniformly deactivated by the deposition of carbon. The oxygen cleans the catalyst by reacting with the deposited carbon to give CO and CO_2, which go out with the air. The upper part of Fig. 14-6 shows how C_b varies with reactor length at various times after first introducing air to the bed. The O_2 concentration in the feed is $(C_b)_0$. The vertical dashed line at $z = L$ represents the exit of the reactor. If the O_2 concentrations at this point for various times are plotted, the breakthrough curve shown in Fig. 14-6b is obtained. The lower graph represents the fraction of the original carbon on the catalyst that has been removed. Similar curves describe the behavior of ion-exchange reactors. For example, for a water-softening process, C_b would represent the concentration of calcium or magnesium ions. The breakthrough curve (Fig. 14-6b) indicates the length of time that the bed could be operated before a

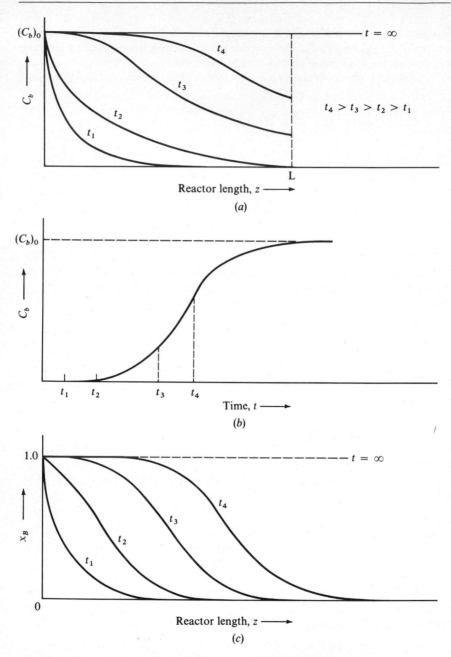

Fig. 14-6 *Concentration profiles in fixed-bed reactor for catalyst regeneration: (a) reactant (O_2) concentration vs bed depth, (b) breakthrough curve (O_2 concentration vs time), (c) conversion of solid reactant (carbon)*

given concentration of calcium or magnesium would appear in the water leaving the exchanger. Figure 14-6c would correspond to the fraction of the sodium form of the resin that has been converted.

In Chap. 13 the effects of mass and energy transfer on the design of catalytic *fixed-bed* reactors were considered, and several design procedures of varying degrees of complexity were discussed. The same problems, all related to the flow pattern of the fluid in the bed, exist for noncatalytic reactions, except that the global rate is a function of time. Methods[1] have been developed for predicting the profiles illustrated in Figure 14-6. Some methods include the effects of axial dispersion in the bed, external mass transfer between fluid and particle, intraparticle diffusion, and chemical reaction, but all are for isothermal conditions. Results are usually obtainable only by numerical methods, since simultaneous solution of several partial-differential equations is required. It is not appropriate to go into detail about these solutions. Rather the equations will be presented for a plug-flow reaction to illustrate the nature of reactor design for such cases. Also, the shrinking-core model will be used to represent the global rate, so that Eqs. (14-10) and (14-14) are applicable.

The reaction is described by

$$A(g) + bB(s) \rightarrow E(g) + F(s)$$

For plug-flow conditions a mass balance of reactant A in the fluid phase is similar to Eq. (13-16), except that it must now be written in partial-differential form,

$$-u \frac{\partial (C_A)_b}{dz} = \mathbf{r}_P \rho_B + \varepsilon_B \frac{\partial (C_A)_b}{\partial t} \tag{14-27}$$

where ρ_B = density of bed of particles

$\mathbf{r}_P \rho_B$ = global rate per unit volume of reactor

u = superficial velocity in the direction of flow

Equation (14-10) gives the rate per particle. For spherical particles packed with a void fraction ϵ_B, the number of particles per unit volume is $3\epsilon_B/4\pi r_s^3$, so that

$$\mathbf{r}_P \rho_B = \frac{3\epsilon_B k(r_c/r_s)^2 (C_A)_b}{r_s[1 + (r_c^2/r_s^2)(k/k_m) + (kr_c/D_e)(1 - r_c/r_s)]} \tag{14-28}$$

Then Eq. (14-27) becomes

[1] J. B. Rosen, *J. Chem. Phys.*, **20**, 387 (1952); *Ind. Eng. Chem.*, **46**, 1590 (1954); H. C. Thomas, *J. Am. Chem. Soc.*, **66**, 1664 (1944); N. K. Heister and T. Vermeulen, *Chem. Eng. Progr.*, **48**, 505 (1952); S. Masamune and J. M. Smith, *AIChE J.*, **10**, 246 (1964); *AIChE J.*, **11**, 34 (1965); *Ind. Eng. Chem., Fund. Quart.*, **3**, 179 (1964).

$$-u\frac{\partial(C_A)_b}{\partial z} = \frac{3\epsilon_B k(r_c/r_s)^2(C_A)_b}{r_s[1 + (r_c^2/r_s^2)(k/k_m) + (kr_c/D_e)(1 - r_c/r_s)]} +$$

$$\epsilon_B\frac{\partial(C_A)_b}{\partial t} \qquad (14\text{-}29)$$

This equation expresses $(C_A)_b$ in terms of the radius r_c of the unreacted core, which is the function of time described by Eq. (14-14). Equations (14-29) and (14-14) establish $C(t,z)_{A_b}$ and $r_c(t,z)$. The boundary conditions are

$$(C_A)_b = (C_A)_0 \qquad \text{at } z = 0 \qquad \text{for } t \geq 0 \qquad\qquad (14\text{-}30)$$

$$r_c = r_s \qquad \text{at } t = 0 \text{ for } z \geq 0 \qquad\qquad\qquad (14\text{-}31)$$

These equations may be solved numerically to give curves of the form shown in Fig. 14-6. The resultant $r_c(t,z)$ can be converted to $x(t,z)$ by means of Eq. (14-20).

Note that if $(C_A)_b$ is essentially constant, corresponding to a large excess of reactant A, Eq. (14-29) disappears, and the solution of Eq. (14-14) is the same as that given in Sec. 14-4.

A second kind of reactor in which C_b may vary with position is the *moving-bed* type. The inclined or vertical kilns and transport reactors with counterflow of fluid and solid phases (Fig. 14-7) are examples. In such cases steady state is achieved in that the properties at any location in the bed do not change with time. The behavior of this type of reactor can be described in a relatively simple way, provided both solid and fluid phases are in plug flow. The relations describing the conversion of solid and concentration of fluid reactant will be developed using the shrinking-core model. Suppose the reaction is $A(g) + bB(s) \rightarrow E(g) + F(s)$. Equation (14-14) expresses how r_c varies with the time that a solid particle is in the reactor. For plug flow this residence time is given by

$$t = \theta = \frac{z\epsilon_S}{G_S/\rho_B} \qquad\qquad (14\text{-}32)$$

or

$$dt = \frac{\epsilon_S\rho_B}{G_S} dz \qquad\qquad (14\text{-}33)$$

where z = reactor length (from entrance of solids)

ϵ_S = fractional holdup of solids (fraction of reactor volume occupied by solids)

G_S = mass velocity of solids

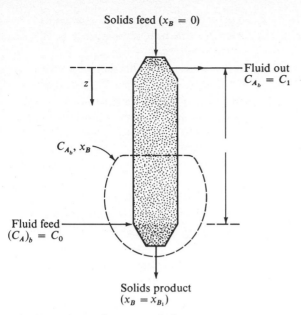

Fig. 14-7   Counterflow moving-bed reactor

Substitution of this expression for dt in Eq. (14-14) gives the change in r_c with length of reactor z,

$$\frac{dr_c}{dz} = -\frac{\epsilon_s b M_B k (C_A)_b}{G_S[1 + (r_c^2/r_s^2)(k/k_m) + (kr_c/D_e)(1 - r_c/r_s)]} \tag{14-34}$$

The radius r_c can be written in terms of conversion through Eq. (14-20). In this form Eq. (14-34) is

$$\frac{dx_B}{dz} = \frac{3\epsilon_s b M_B k (C_A)_b (1 - x_B)^{\frac{2}{3}}}{G_S r_s \{1 + (k/k_m)(1 - x_B)^{\frac{2}{3}} + (kr_s/D_e)(1 - x_B)^{\frac{1}{3}}[1 - (1 - x_B)^{\frac{1}{3}}]\}} \tag{14-35}$$

Before Eq. (14-35) can be integrated $(C_A)_b$ must be expressed in terms of x_B. A mass balance of reactant A around a section of the reactor including the bottom (Fig. 14-7) provides this relationship. Such a balance may be written

$$u[(C_A)_0 - (C_A)_b] = \frac{G_S}{b M_B}(x_{B_1} - x_B) \tag{14-36}$$

or

$$(C_A)_b = (C_A)_0 - \frac{G_S}{u b M_B}(x_{B_1} - x_B) \tag{14-37}$$

where x_{B_1} is the conversion of B in the solids leaving the reactor and $(C_A)_0$ is the concentration of A in the feed. We can now substitute Eq. (14-37) in Eq. (14-35) and integrate the resulting expression from $x_B = 0$ at $z = 0$ to obtain the conversion of B. A trial procedure is required if the conversion for a given length is needed, since the desired exit conversion occurs in Eq. (14-37). A value for x_{B_1} may be assumed and the x-vs-z profile evaluated. When the conversion at $z = L$ agrees with the assumed value, no further trials are necessary. If the length of reactor to achieve a certain conversion is required, the trial procedure is not necessary. Once the conversion is known, the concentration of reactant in the exit fluid phase can be obtained from Eq. (14-36) by setting $x_B = 0$.

When the flow patterns do not meet plug-flow requirements the same general procedure is applicable, but the RTD must be taken into account. Such problems have been considered for fluidized-bed reactors[1] and moving-bed reactors.[2]

Problems

14-1.   A gas-solid noncatalytic reaction of the type discussed in Sec. 14-3 is investigated by measuring the time required for complete conversion of solid B as a function of particle diameter. The results are as follows:

Particle diameter, mm	0.063	0.125	0.250
Time for complete conversion, min	5.0	10.0	20.0

   If the diffusion resistance in the gas phase around the particle is negligible, what mechanism controls the rate of reaction?

14-2.   In measurements for a reaction of the type discussed in Sec. 14-3, the time required for equal conversions is directly proportional to particle diameter at low conversion but becomes proportional to the square of the particle size as the conversion becomes larger. What can be said about the mechanism controlling the rate? Again, diffusion resistance in the gas surrounding the particles is negligible.

14-3.   The reaction described in Prob. 14-1 is to be carried out by passing the reactant gas crosswise over a moving grate carrying the solid particles. The velocity of the grate is such that the particles are exposed to the gas stream for 9 min. If the distribution of particle sizes is as given below, what will be the average conversion leaving the reactor? What will be the conversion for the 0.063-mm particles?

[1] S. Yagi and D. Kunii, *Chem. Eng. (Japan)*, **19**, 500 (1955); *Chem. Eng. Sci.*, **17**, 364, 372, 380 (1962).
[2] S. Yagi, K. Takagi, and S. Shimoyama, *J. Chem. Soc. (Japan)*, *Ind. Chem. Sec.*, **54**, 1 (1961).

Particle diameter, mm	0.063	0.125	0.250	0.500
Wt %	25	35	35	5

The gas composition does not change significantly during flow across the grate.

14-4. A solid feed of $\frac{1}{4}$-in. spherical particles of pure B is to be reacted in a rotary kiln (such as a lime or cement kiln). The gas A in contact with the solids is of uniform composition. The whole process is isothermal. The reaction is first order with respect to A, irreversible, and follows the stoichiometry

$$A(g) + B(s) \rightarrow C(s) + D(g)$$

The substance B is nonporous, but product C forms a porous layer around the unreacted core of B as the reaction proceeds. In the kiln the solids will move in plug flow from one end of the kiln to the other at a velocity of 0.1 in./sec. It is desired to design a kiln to obtain 90% conversion of B.

Small-scale studies of the reaction indicate that diffusion resistance between particle surface and the gas is negligible. In a batch agitated reactor operated at the same temperature and gas composition, the following data were obtained:

A conversion of 87.5% in 1 hr with $\frac{1}{8}$-in. particles
A conversion of 65.7% in 1 hr and 24 min with $\frac{1}{4}$-in. particles

(a) Calculate the length required for the kiln. (b) In the future it may be necessary to handle a feed of B which consists of 20 wt % $\frac{1}{8}$-in. particles, 50 wt % $\frac{1}{4}$-in. particles, and 30 wt % $\frac{3}{8}$-in. particles. Calculate the average conversion in the product from this mixed feed, using the reactor designed in part (a).

14-5. A plant produces HCl and Na_2SO_4 from salt and sulfuric acid in a transport reactor (Fig. 14-5). The reactor operates at about 900°F, so that the NaCl and Na_2SO_4 are solids, HCl is a gas and the H_2SO_4 exists as gaseous H_2O and SO_3 (see Example 14-2). The H_2O is present in great excess. At normal conditions, the residence time of the particles ($r = 0.05$ cm) in the reactor is 10 sec, and the conversion of NaCl is 100%. A new supply of salt is obtained which has particles twice the diameter of the normal material. If velocity of solids and gases and all other operating conditions are held constant, what would the residence time have to be to obtain complete conversion of the new salt particles?

There is a negligible concentration difference of SO_3 between bulk gas and the surface of the particles. The particles remain spherical and of constant diameter, regardless of extent of conversion to Na_2SO_4. Unlike the situation in Example 14-2, it is expected that the resistance to reaction at the salt interface, while significant, is not controlling for either size of salt particles. The chemical reaction is first order in SO_3. The diffusivity of SO_3 through the $NaSO_4$ layer is 0.01 cm^2/sec, and the first-order rate constant for the reaction is 0.5 cm/sec. The whole reactor is isothermal.

14-6. The reduction of FeS_2 to FeS is carried out in a tubular reactor with upflow of hydrogen and downflow of solids. The reactor will operate at 495°C and 1 atm with pure hydrogen. For these conditions gas-phase diffusion resistance is negligible. The diffusivity of hydrogen in the product layer and the rate constant for the reaction at the FeS_2 surface have the values established in Example 14-1. The sizes and RTDs for the particles in the reactor are as follows:

Particle radius, mm	0.05	0.10	0.15	0.20
Weight fraction	0.10	0.30	0.40	0.20
Residence time, $\theta/\bar{\theta}$	1.40	1.10	0.95	0.75

If the mole fraction of hydrogen in the gas may be safely assumed to be constant and unity, what will be the average conversion of FeS_2 to FeS in a reactor with a mean residence time of 60 min?

AUTHOR INDEX

SUBJECT INDEX